About IFPRI

The International Food Policy Research Institute (IFPRI), a CGIAR Research Center established in 1975, provides research-based policy solutions to sustainably reduce poverty and end hunger and malnutrition. IFPRI's strategic research aims to foster a climate-resilient and sustainable food supply; promote healthy diets and nutrition for all; build inclusive and efficient markets, trade systems, and food industries; transform agricultural and rural economies; and strengthen institutions and governance. Gender is integrated in all the Institute's work. Partnerships, communications, capacity strengthening, and data and knowledge management are essential components to translate IFPRI's research from action to impact. The Institute's regional and country programs play a critical role in responding to demand for food policy research and in delivering holistic support for country-led development. IFPRI collaborates with partners around the world.

About IFPRI's Peer Review Process

IFPRI books are policy-relevant publications based on original and innovative research conducted at IFPRI. All manuscripts submitted for publication as IFPRI books undergo an extensive review procedure that is managed by IFPRI's Publications Review Committee (PRC). Upon submission to the PRC, the manuscript is reviewed by a PRC member. Once the manuscript is considered ready for external review, the PRC submits it to at least two external reviewers who are chosen for their familiarity with the subject matter and the country setting. Upon receipt of these blind external peer reviews, the PRC provides the author with an editorial decision and, when necessary, instructions for revision based on the external reviews. The PRC reassesses the revised manuscript and makes a recommendation regarding publication to the director general of IFPRI. With the director general's approval, the manuscript enters the editorial and production phase to become an IFPRI book.

Library of Congress Cataloging-in-Publication Data

Names: Dorosh, Paul Anthony, editor. | Minten, Bart, editor.
Title: Ethiopia's agri-food system : past trends, present challenges, and future scenarios / edited by Paul Dorosh and Bart Minten.
Description: Washington, D.C. : International Food Policy Research Institute, 2020. | Includes bibliographical references and index. | Identifiers: LCCN 2020007433 (print) | LCCN 2020007434 (ebook) | ISBN 9780896296916 (paperback) | ISBN 9780896296923 (epub)
Subjects: LCSH: Agriculture—Ethiopia. | Agriculture—Economic aspects—Ethiopia. | Agriculture and state—Ethiopia. | Food supply—Ethiopia.
Classification: LCC HD2124.5 .E85 2020 (print) | LCC HD2124.5 (ebook) | DDC 338.10963—dc23
LC record available at https://lccn.loc.gov/2020007433
LC ebook record available at https://lccn.loc.gov/2020007434

Cover Design: Jason Chow/IFPRI
Project Manager: John Whitehead
Book Layout: BookMatters

Ethiopia's Agrifood System

Past Trends, Present Challenges, and Future Scenarios

Edited by Paul Dorosh and Bart Minten

A Peer-Reviewed Publication

International Food Policy Research Institute
Washington, DC

Recommended citation: Dorosh, P., and B. Minten. 2020. *Ethiopia's Agrifood System: Past Trends, Present Challenges, and Future Scenarios.* Washington, DC: International Food Policy Research Institute. https://doi.org/10.2499/9780896296916.

International Food Policy Research Institute, 1201 Eye Street, NW, 12th floor, Washington, DC 20005 USA, Telephone: +1-202-862-5600, www.ifpri.org

ISBN: 978-0-89629-691-6
DOI: https://doi.org/10.2499/9780896296916

Library of Congress Cataloging-in-Publication Data may be found on page ii.

CONTENTS

LIST OF TABLES

LIST OF FIGURES

ACRONYMS AND ABBREVIATIONS

AAAE	Addis Ababa Abattoirs Enterprise
AACCSA	Addis Ababa Chamber of Commerce and Sectoral Associations
AARES	Australasian Agricultural and Resource Economics Society
ADLI	agricultural development–led industrialization
AFS	agriculture and food system
AGP	Agricultural Growth Program
AGRA	Alliance for a Green Revolution in Africa
AgSS	Agricultural Sample Survey
AKLDP	Agriculture Knowledge, Learning, Documentation and Policy Project
ASF	animal-sourced foods
ASTI	agricultural science and technology indicators
ATA	Agricultural Transformation Agency of Ethiopia
BCC	behavioral change communication
CAADP	Comprehensive Africa Agriculture Development Programme
CAE	Center for Analytical Economics
CCAFS	climate change, agriculture, and food security
CET	constant elasticity of transformation
CFSVA	Comprehensive Food Security and Vulnerability Analysis
CGE	computable general equilibrium
CIAT	International Centre for Tropical Agriculture
CPI	Consumer Price Index
CRGE	Climate Resilient Green Economy

CSA	Central Statistical Agency
CV	coefficient of variation
DAP	di-ammonium phosphate
DFID	Department for International Development
DNA	deoxyribonucleic acid
DSGD	Development Strategy and Governance Division
DSSAT	Decision Support System for Agrotechnology Transfer
EAAE	European Association of Agricultural Economists
EAP	East Asia and the Pacific
ECX	Ethiopian Commodity Exchange
EDRI	Ethiopian Development Research Institute
EEA	Ethiopian Economics Association
EFSRA	Emergency Food Security Reserve Agency
EGTE	Ethiopian Grain Trading Enterprise
EHNRI	Ethiopian Health and Nutrition Research Institute
EPAU	Ethiopian Policy Analysis Unit
EPHI	Ethiopian Public Health Institute
EPTD	Environment and Production Technology Division
ERCA	Ethiopian Revenues and Customs Authority
ERSS	Ethiopia Rural Socioeconomic Survey
ESS	Ethiopia Socioeconomic Survey
ESSP	Ethiopia Strategy Support Program
FAI	Fertilizer Association of India
FANTA	Food and Nutrition Technical Assistance
FAO	Food and Agriculture Organization of the United Nations
FAOSTAT	Food and Agriculture Organization Statistics in the United Nations Corporate Statistical Database
FDI	foreign direct investment
FDRE	Federal Democratic Republic of Ethiopia
FRA	Food Reserve Agency
FtF	Feed the Future
FVI-Idele	France Vétérinaire Internationale–Institute de l'Elevage
GCPI	General CPI
GDP	gross domestic product
GRUMP	Global Rural and Urban Mapping Project
GTAP	global trade, assistance, and production
GTP	Growth and Transformation Plan
HCES	Household Consumption Expenditures Survey
HCI	Head Count Index

HDDS	Household Dietary Diversity Score
HDI	Herfindahl Diversification Index
HH	household head
HICES	Household Income, Consumption, and Expenditure Surveys
IAIP	integrated agro-industrial parks
ICT	information and communication technology
IDSP	Industrial Development Strategic Plan
IDR	industrial development roadmap
IFPRI	International Food Policy Research Institute
IHSN	International Household Survey Network
IISD	International Institute for Sustainable Development
ILO	International Labor Organization
ILRI	International Livestock Research Institute
IMF	International Monetary Fund
IPCC	Intergovernmental Panel on Climate Change
IPDC	Industrial Parks Development Corporation
IPMS	improving productivity and market success
IWMI	International Water Management Institute
LAC	Latin America and the Caribbean
LFS	Labor Force Survey
LICOS	Center for Institutions and Economic Performance, University of Leuven, Belgium
LSIL	Innovation Lab for Livestock Systems
LSMS-ISA	Living Standards Measurement Study–Integrated Surveys on Agriculture
MBS	marginal budget share
MODIS	Moderate Resolution Imaging Spectroradiometer
MoFED	Ministry of Finance and Economic Development
NABC	Netherlands Africa Business Council
NBE	National Bank of Ethiopia
NCOT	net-commercial off-take
NDRMC	National Disaster Risk Management Commission
NER	nominal exchange rate
NLFS	National Labor Force Survey
NOAA	National Oceanographic and Atmospheric Administration
NPC	National Planning Commission
OECD	Organisation for Economic Co-operation and Development
OLS	ordinary least squares

PASDEP	Plan for Accelerated and Sustained Development to End Poverty
PIM	Program Implementation Manual
PP-GCPI	Poor Persons' General CPI
PSI	Policy Studies Institute
PSNP	Productive Safety Net Program
RED&FS	Rural Economic Development and Food Security
RNF	rural nonfarm
RTCs	rural transformation centers
SA	South Asia
SAM	social accounting matrix
SEDAC	Socioeconomic Data and Applications Center
SLM	sustainable land management
SNNP	Southern Nations, Nationalities, and Peoples
SPAM	Spatial Production Allocation Model
SSA	Africa south of the Sahara
TFP	total factor productivity
TLU	tropical livestock units
TVE	township and village enterprises
ULGDP	Urban Local Government Development Programme
UNDESA	United Nations Department of Economic and Social Affairs
UNIDO	United Nations Industrial Development Organization
USAID	United States Agency for International Development
WFP	World Food Programme

FOREWORD

Ethiopia has achieved remarkable success in increasing agricultural production and enhancing food security over the past three decades. Major investments in agricultural technology and rural infrastructure, together with liberalization of agricultural markets and creation of a well-targeted safety net (the Productive Safety Net Program), have not only helped prevent a reoccurrence of the famines of the 1970s and 1980s, but have contributed to sustained economic growth and rising real incomes of the rural poor.

IFPRI has a long history of policy-relevant empirical research and analysis in Ethiopia, including work on famines by Joachim von Braun and Patrick Webb in the 1990s and analysis of cereal markets by Eleni Gabre-Madhin that led to the creation of the Ethiopian Commodity Exchange (ECX) in the early 2000s. More recently, researchers affiliated with the Ethiopia Strategy Support Program (ESSP) have conducted numerous surveys and analyses of safety nets, nutrition, poverty, and development strategy.

This book synthesizes a large volume of research on the transformation of agricultural production and markets in Ethiopia that has taken place over the past decade. It also is deliberately forward looking, examining the major drivers of change (including population growth, urbanization, per capita income growth, resource constraints, climate change, and technology) that have transformed (and will continue to transform) Ethiopia's agri-food system.

It is my hope that this book and IFPRI's ongoing research, policy analysis, and capacity strengthening efforts, together with our national and international partners, will contribute to a steady and sustainable increase in the welfare of Ethiopia's people.

Dr. Johan Swinnen
Director General, IFPRI

ACKNOWLEDGMENTS

At the end of 2016, El Niño–induced droughts plagued much of Ethiopia, reducing the country's agricultural production and necessitating assistance for almost 20 million people and 3 million tons of cereal imports for safety net programs and food aid. These events convinced many agricultural policymakers, members of civil society, and donor representatives that an objective assessment of the performance of Ethiopia's agrifood system and of possible scenarios for the future was needed. Thereafter, the Rural Economic Development and Food Security (RED&FS) Working Group, composed of representatives of the Government of Ethiopia and development partners, commissioned the Ethiopia Strategy Support Program (ESSP), a joint program of the International Food Policy Research Institute (IFPRI) and the Ethiopian Development Research Institute (EDRI), which has since become part of the Policy Studies Institute (PSI), to conduct this analysis.

Several meetings were held with stakeholders to discuss the concept note of the study, interim results, and final versions of chapters. Presentations were made to the RED&FS Working Group, the Ministry of Agriculture, the annual conference of the Ethiopian Economics Association (EEA), the World Bank, the Development and Planning Commission, and others. The major findings from this research effort were also presented and discussed at a daylong conference titled "The Future of Ethiopia's Agriculture: Towards a Resilient System to End Hunger and Undernutrition," jointly organized by IFPRI, EDRI, the IFPRI-led Compact 2025 initiative, and the RED&FS Sector Working Group on December 15, 2017.

Based on the feedback received at these different events, analyses were revised and updated. This book brings together these findings. We examine future scenarios through modeling, but we also look backward to draw lessons for future policy and assess the evidence on changes in Ethiopia's agricultural and food system and the rural economy since 2000. We sincerely hope that this report meets the objectives of the needs of the Government of Ethiopia and the development community and that it will serve as a resource for better informed decision-making as Ethiopia moves forward in its economic and agricultural transformation process.

Finally, we would like to acknowledge the donors and colleagues who played major roles in producing this book. In particular, we would like to thank Jenny Smart, Todd Benson, Ramela Carrion, Mahlet Mekuria, Pamela Stedman-Edwards, John Whitehead, and Nahume Yadene for their valuable contributions to the technical editing and the publication process. We also thank the donors who support the ESSP, including the United States Agency for International Development (USAID), the United Kingdom, Department for International Development (DFID), and the European Union. The research presented here was conducted as part of the CGIAR Research Program on Policies, Institutions, and Markets (PIM), which also provided support. We would further like to acknowledge the Policy Studies Institute, our partner in ESSP, for all their support and collaboration. We also acknowledge the support of the American people provided to the Feed the Future Innovation Lab for Livestock Systems (LSIL), Institute of Food and Agricultural Sciences, University of Florida, in partnership with the International Livestock Research Institute (ILRI) (Chapters 5 and 8). LSIL is funded by the United States Agency for International Development (USAID) through a five-year Leader with Associates Cooperative Agreement Award No. AID-OAA-L-15-00003. Any opinions expressed in this book belong to the authors and do not necessarily reflect those of IFPRI, the PSI, USAID, DFID, the European Union, PIM, or CGIAR.

Chapter 1

STRUCTURAL TRANSFORMATION AND THE AGRICULTURAL FOOD SYSTEM: AN INTRODUCTION

Paul Dorosh and Bart Minten

Ethiopia's agricultural sector has grown rapidly since 2000 due to a doubling in the use of modern inputs (chemical fertilizers and improved seeds), significant land expansion, increased labor use, and a 2.3 percent per year growth in total factor productivity (TFP) (Bachewe et al. 2018). At the same time, there has been a substantial spatial and structural transformation of the economy. Ethiopia's urban population has more than doubled in the past 20 years (from 7.3 million in 1994 to 17.5 million in 2015), and nonagricultural output has grown rapidly so that the share of nonagriculture in GDP has risen from less than half in 2000 to about two-thirds today. Moreover, household welfare indicators have improved dramatically as rural poverty fell from 45 percent in 1999–2000 to 23.5 percent in 2015–2016 (Figure 1.1). Measures of malnutrition of children also improved significantly. For example, from 2000 to 2016 the child stunting rate fell from 58 percent to 38 percent while the share of children that are underweight declined from 41 percent to 24 percent over the same period (Ethiopia, CSA and ICF 2016; NPC 2017) (Figure 1.2). However, wasting only declined by 2 percent.

A Story Too Good to Be True?

Some skeptical observers have challenged this rapid progress. Did agricultural production really increase that fast? Has poverty actually fallen? Are the gains sustainable? Will further urbanization bring an end to this agricultural growth?

The evidence for substantial agricultural growth is indeed strong, although it is not possible to definitively state that growth has been as high as reported. Ethiopia's fertilizer use per hectare of arable land has increased rapidly and was 2.8 times higher in 2014–2016 than in 1991–1992. The fertilizer use rate in neighboring Kenya was almost unchanged over the same period

FIGURE 1.1 Poverty estimates, 1996–2016

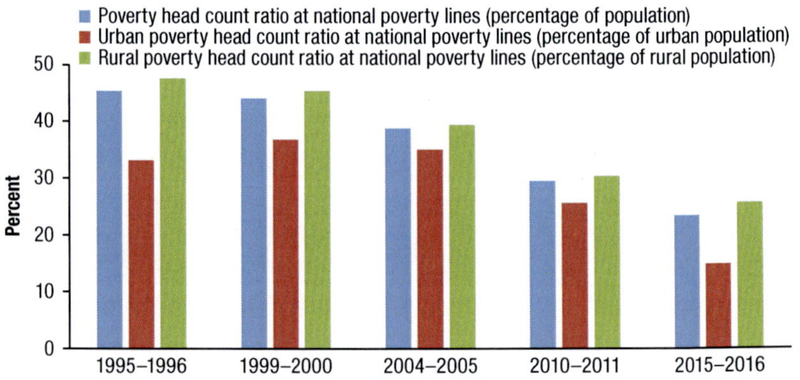

Source: NPC (2017), World Bank (2018), and authors' calculations.

(Figure 1.3).[1] Average maize yields in Ethiopia in 2015–2017 were 2.63 times those in 1999–2001. In the same period, average maize yields in Malawi rose 53 percent; yields in Kenya fell 11 percent (Figure 1.4).[2] Area cultivated also increased, as did labor input and use of improved seeds, so that estimated total factor productivity growth (the part of cereal production growth not accounted for by increased inputs) was only 2 percent per year—a high but not implausible figure (Chapter 3; Bachewe et al. 2018).[3]

Trends in real market prices of major cereals provide additional evidence of major gains in production, but likewise they do not suggest exact magnitudes of the size of the production increase. Earlier analysis of the implications of potential production gains for real prices, farmer incomes, and poverty income suggested that high agricultural growth (6 percent annual growth in real agricultural GDP) would lead to only moderate declines in real prices of major crops, as long as there was substantial growth in the nonagricultural economy as well (Dorosh and Thurlow 2012).[4] As shown in Figure 1.5, trend growth of

1 Note that the FAO data presented here on fertilizer nutrients per hectare of arable land up to 2015/2016 corresponds to the definitions used in Figure 3.12A. Figure 3.10, which shows fertilizer use (gross weight) per hectare of fertilized area, indicates a further sharp rise in fertilizer use in Ethiopia in 2016/2017 and 2017/2018.

2 Maize is used as a comparator for cereal yields across countries since it is Ethiopia's main cereal crop by tons of production and is widely grown across much of Africa south of the Sahara.

3 From 2004/2005 through 2015/2016, labor accounted for 34 percent of the 8.3 percent per year output growth; fertilizer 11 percent, improved seeds 10 percent, land 8 percent, and TFP 25 percent (Bachewe et al. 2018).

4 A 6 percent growth rate was the target set as part of the Comprehensive Africa Agriculture Development Programme (CAADP) compact agreed to by Ethiopia and 42 other countries of

FIGURE 1.2 Trends in nutritional status of children under five years old

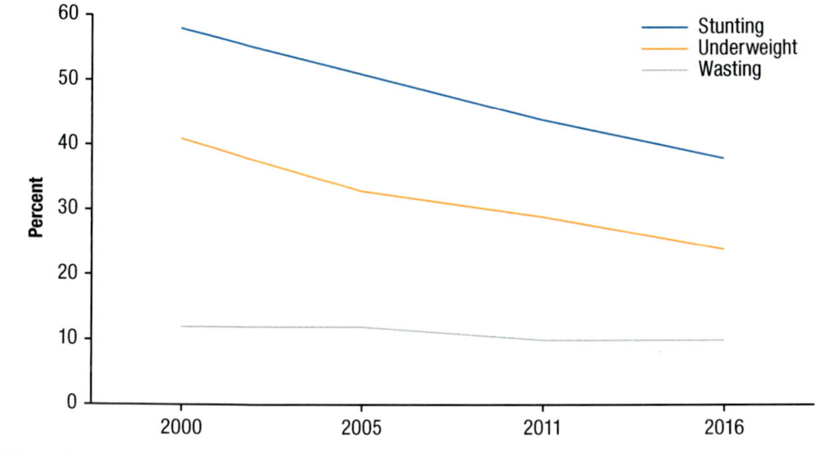

Source: Ethiopia, CSA and ICF (2016).

real prices of teff, wheat, maize, and sorghum has been negative (–0.2 percent, –1.5 percent, –2.9 percent, and –3.2 percent per year, respectively), an indication that growth in demand (from population growth, higher per capita incomes, and other factors) has not kept up with growth in supply.[5]

Finally, data on household consumption and nutritional outcomes strongly suggest that there have been significant increases in quantities of food consumed (Chapter 7, Table 7.3) and dietary diversity (Chapter 10, Tables 10.4 and 10.5) that are consistent with both gains in production and improvements in nutritional outcomes cited earlier.

Substantial public investments in agriculture have played a major role in this success. Beginning in the 1990s, the Meles government placed a major emphasis on agricultural growth as a mechanism for growth and poverty reduction. The strategy of agricultural development led industrialization (ADLI) initiated in the late 1990s saw agriculture as a major engine of overall economic growth through positive growth linkages with food processing and trade as well as through generation of fiscal and foreign exchange resources of increased investment.[6] Successive development plans in the first decade of

the African Union (Scheuermann and Mészáros 2017).

5 These growth rates are the annualized monthly growth rates for the period January 2005–April 2019, estimated from a regression of the form: $lnX = a + b * t + et$, where X is production, t is time, and a and b are estimated coefficients.

6 As discussed in Chapters 12–14, these growth linkages are indeed substantial and will likely have a larger effect on overall GDP growth as the share of agricultural crops that is marketed increases.

FIGURE 1.3 Fertilizer use in selected African countries

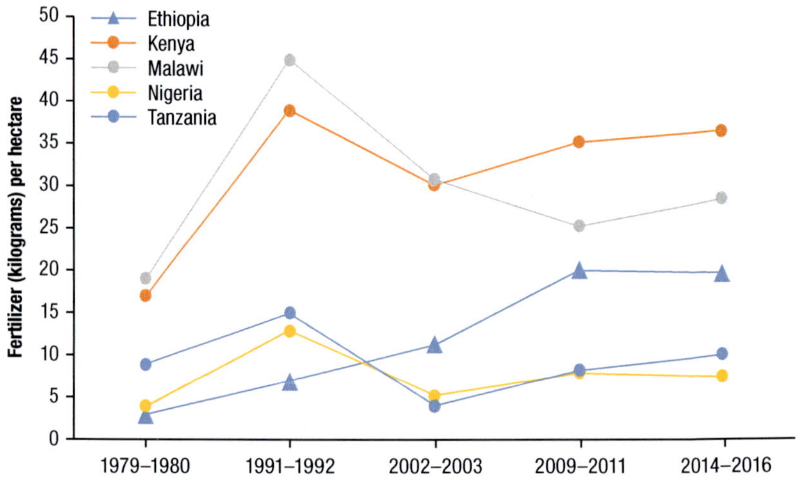

Source: Authors' calculations from FAO (2019).

FIGURE 1.4 Maize yields in selected African countries

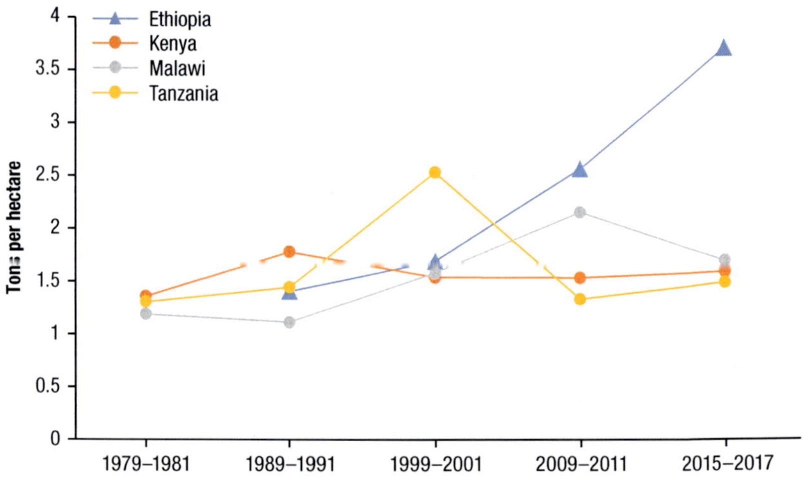

Source: Authors' calculations from FAO (2019).

FIGURE 1.5 Real price trends of major cereals in Ethiopia, 2005–2017

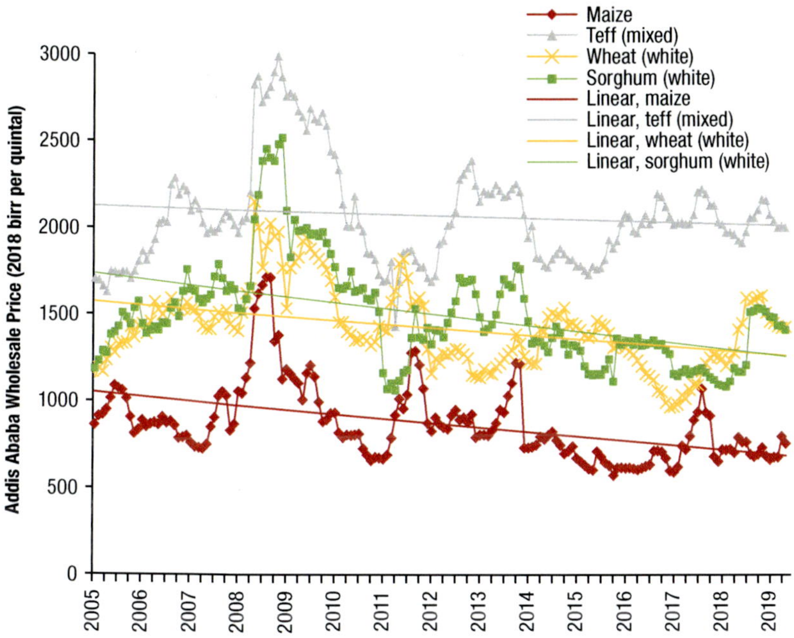

Source: Authors' calculations from EGTE (2019), IMF (2019), and Ethiopia, CSA (2019) data

the 2000s, the Sustainable Development and Poverty Reduction Program (Ethiopia, MoFED 2002) and the Plan for Accelerated and Sustained Development to End Poverty (PASDEP) (Ethiopia, MoFED 2006), likewise placed a major emphasis on public investment in agriculture. Ethiopia also participated in the Comprehensive Africa Agriculture Development Programme (CAADP), pledging to invest at least 10 percent of its government budget resources in agriculture, a level the government exceeded both before and after the signing of the CAADP Compact in 2009 (Benin 2014).[7] This investment included major expenditures on rural roads that reduced transport costs and improved market connectivity (Chapter 13). The government also financed a massive increase in fertilizer imports and an expansion

7 Moreover, to address the needs of its most vulnerable population, since 2005, Ethiopia has successfully implemented a large-scale safety net program—the Productive Safety Net Program (PSNP), the biggest safety net in Africa except for South Africa. In 2018 the PSNP covered almost eight million direct beneficiaries. This safety net is generally viewed as a model in Africa as it has helped in significantly improving food security in the country (Berhane et al. 2014).

in agricultural extension services, including delivery of improved seeds (Chapter 3).

The most recent five-year plans, Growth and Transformation Plans (GTP) I and II, place far less emphasis on agriculture (Ethiopia, MoFED 2010; NPC 2016). Investment in agriculture nonetheless remained substantial (see Figure 13.1). Moreover, as discussed in detail throughout this book, the agricultural sector has continued to perform well, with agricultural GDP growth of 5.2 percent from 2011 to 2016 under GTP I and more than 4 percent from 2016 to 2018 in the first two years of GTP II.[8] Nonetheless, growth of agricultural GDP has slowed down from its average of 7.8 percent from 2006 to 2011. Meanwhile, total GDP growth remained nearly the same: an average of 10.7 percent per year from 2006 to 2011 and 9.9 percent per year from 2010 to 2015, before slowing to only 5.4 percent per year from 2015 to 2018.

The agricultural sector faces daunting challenges, however, including increasingly binding land and water constraints in large parts of the country (Chapter 2), possible limits to economically achievable crop yields, and rising rural wage rates (Chapter 11) as workers shift from agricultural to nonagricultural employment and from rural to urban locations.[9] There is also a major concern in the country on increasing food import dependency. Food imports include most importantly wheat, rice, palm oil, and sugar. Wheat is often provided in the form of food aid for emergency aid (recently linked to a large number of internally displaced persons) or for food safety net programs (the Productive Safety Net Program). Figure 1.6 (left side) shows the evolution of the value of those imports in the country, illustrating a significant increase since 2006. In per capita terms wheat imports have not increased that much (right side); the high peak in 2016 reflects the increased food aid for the major drought during that year. Moreover, export earnings from agriculture have also increased during that period and are in value terms higher than agricultural imports.

The focus on rapid nonagricultural growth and employment in Ethiopia's development strategy under the GTP II is consistent with the normal pattern of economic development. Nonetheless, the experience of other developing countries suggests that substantial agricultural growth will also be required, especially to promote equity and accelerate poverty reduction. Investments

8 Throughout this book we adopt the convention of referring to an Ethiopian fiscal year (approximately from September 1 to the following September 1) by the second calendar—that is, 2017–2018 is denoted as 2018.

9 This rise in real wages is a measure of success for the rural sector overall but may slow future agricultural growth.

FIGURE 1.6 Value of major agricultural imports (left) and wheat imports per capita (right)

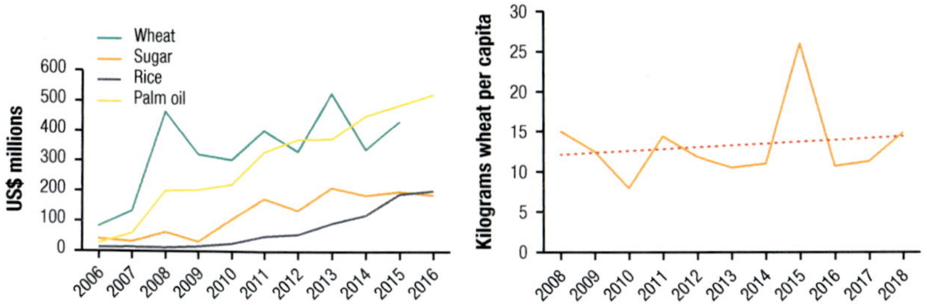

Source: UN Statistics Division (2019), USDA (2019), and World Bank (2017).

will be needed throughout Ethiopia's agrifood system—that is, all the sectors, actors, and activities involved in agriculture and providing food for the country's population, including input supply, food processing, transport, marketing, and retail sales of food by restaurants, shops, and street vendors.

Objectives of This Book

This book presents a forward-looking analysis of Ethiopia's agrifood system in the context of a rapidly changing economy. The overarching focus is to understand past trends based on solid and large-scale national representative data and to analyze the pattern of transformation unfolding in Ethiopia's agrifood system and its economy more broadly. We touch on a number of challenges for each of the issues studied. In some of the chapters we look at future scenarios by doing explicit modeling of the future or by putting Ethiopia in an international context to benchmark where Ethiopia might be in comparison with some of these other countries that are in a more advanced stage.

The analysis is based on substantial empirical research using satellite data on land use, official national surveys of agricultural production and household expenditures, and other household, firm, and trader data collected by IFPRI and national researchers. Alternative future scenarios are analyzed using an economywide model to simulate the impacts of changes in technology, urbanization, and economic structure to 2040. Overall, the book builds on several earlier volumes chronicling the evolution of Ethiopia's agricultural sector and food policy, including *People of the Plow* (McCann 1995), *Famine and Food Security in Ethiopia* (Webb and von Braun 1994), and *Food and Agriculture in Ethiopia: Progress and Policy Challenges* (Dorosh and Rashid 2012).

Contributions to the Literature

This book contributes to the literature in four main ways. First, it further documents and extends the story of successful agricultural development in Ethiopia, one of the largest countries in Africa. Since the aforementioned volumes were written, the microlevel data on agriculture and food security has been greatly enriched by additional rounds of the Agricultural Sample Surveys, surveys fielded to analyze the impact of the Agricultural Growth Program and the Feed the Future program of USAID, satellite data, the Household Income and Expenditure Survey as well as price surveys done by Ethiopia's Central Statistical Agency (CSA), enabling a richer analysis of key issues.

Second, the book looks at the agrifood system as a whole by going beyond crop agriculture alone and by including analysis of off-farm activities linked to the agricultural sector, such as input marketing and food trading, logistics, processing, and retailing. We also document the transformation that is happening in the rural nonfarm economy and elaborate on the important livestock sector and the functioning of its markets.

Third, this book explicitly covers the impacts of changes in key aspects of Ethiopia's economy outside of the agrifood system, such as rapid urbanization and the broader structural transformation of the economy. It is also broader in scope than recent work on individual crops, such as *The Economics of Teff* (Minten, Taffesse, and Brown 2018).

Fourth, the focus on future scenarios is on the medium run as well as the short run. Earlier analyses, such as the analysis done for CAADP showing that continued investments would spur agricultural output growth, and assuming moderate overall income growth in the economy, led to a sharp reduction in poverty, focused on shorter periods (for example, from 2005 to 2015 in both Diao et al. [2012] and Dorosh and Thurlow [2012]). The analysis in this book focuses on the near term but also includes analysis out to 2040, highlighting how major changes in Ethiopia's economy over time will affect the relative poverty-reduction effects of agriculture, the agrifood system, and nonagricultural investments.

Agricultural Growth and Structural Transformation

Structural transformation, an increase in the shares of nonagricultural sectors in GDP and in employment, is a central objective of Ethiopia's development

FIGURE 1.7 Share of different sectors in Ethiopia's economy, 2006/2007–2017/2018 (%)

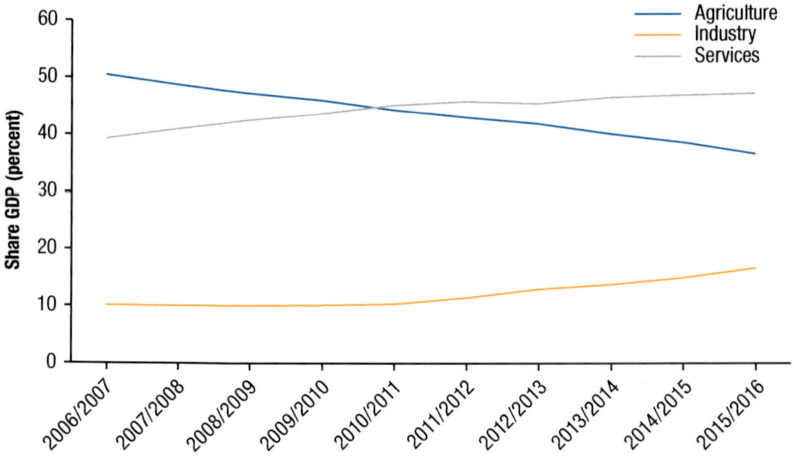

Source: NBE (2019), adjusted by authors for the rebasing of 2015/2016.

strategy.[10] Much progress has in fact been achieved. As shown in Figure 1.7, the share of nonagriculture in total GDP rose from 53.9 percent in 2009/2010 (the year before the start of GTP I) to 62 percent in 2014/2015.[11] In the first year of GTP II (2015/2016), this share rose by a further 2 percentage points. Likewise, data from the Labor Force Surveys of 2005 and 2013 indicate that the share of nonagriculture in the labor force increased from 19.8 percent to 23.6 percent, a rate of 0.45 percentage points per year. This rate of structural change is significantly slower than the rate observed over the 1993–2013 period (about 0.8 percentage points per year), or in China during the period in

10 These two basic measures of structural transformation are used in Ethiopia's Growth and Transformation Plans. There is a substantial economics literature on structural change. Early work by Johnston and Mellor (1961) argued that agricultural productivity growth is a primary driver of structural transformation through both backward and forward growth linkages (see also Timmer 1988; Mellor 1995; and Gollin, Jedwab, and Vollrath 2016). An alternative view is that economic growth is driven by investments and productivity growth in the modern or industrial sector (Lewis 1954) and the falling shares of agriculture in the labor force and in value-added result to a large extent from more rapid exogenous productivity growth in these nonagricultural sectors. Movement of labor from (lower productivity) agriculture to (higher productivity) nonagricultural sectors results in overall income growth, but a shift of labor into some low productivity service sectors may actually result in lower overall incomes. See Rodrik et al. (2016) and the discussion in Chapter 12, Rodrik, McMillan, and Sepúlveda (2013).

11 Note that the nonagricultural sector includes upstream (for example, production of fertilizer and pesticides) as well as downstream (for example, processing, transport and sales of agricultural commodities) activities.

FIGURE 1.8 Rates of structural change in selected countries and time periods (%)

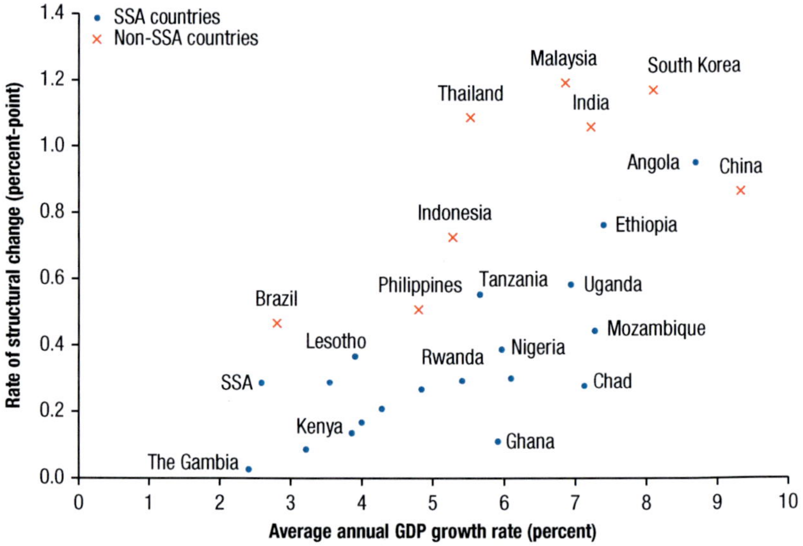

Source: Annex Figure A.1.Thurlow, Dorosh, and Davies (2018) using employment data from ILO (2017) and GDP data from World Bank (2017).

Note: Rate of structural change is the annual percentage point increase in the share of nonagricultural employment in total employment. All SSA countries (Africa south of the Sahara) are for 1993 through 2013. Non-SSA countries analysis is for different time periods: China (1994–2005), Indonesia (1985–2012), South Korea (1980–2000), Malaysia (1988–2001), Philippines (1999–2012), Thailand (1980–2010), India (1980–2010), and Brazil (1981–2011).

which its real GDP was at a similar level (Figure 1.8). Nonetheless, Ethiopia's real GDP growth rate of about 7.5 percent in the 1993–2013 period, and about 9 percent in 2016 has been substantial (about 7 percent per capita). Only China has achieved similar sustained high rates of structural change and real GDP growth.

There are indications that future growth may not be as rapid, however. Much of the growth in the Ethiopian economy in recent years has been due to increased investments financed by unsustainable foreign capital inflows that grew at 36.2 percent per year from 2010/2011 through 2015/2016 (Figure 1.9). Combined foreign and domestic borrowing in this period contributed to a surge in Ethiopia's total foreign and domestic debt, which reached 60.3 percent of GDP in 2016/2017 (Figure 1.10). Thus there is a strong likelihood that the pace of growth and structural change will decelerate in the coming decade. Even with a slowdown in the macroeconomy, though, increased population, urbanization, electrification, improved transport, and growth in per capita incomes are likely to lead to a rapid expansion of the agrifood system.

FIGURE 1.9 Balance of payments, 2004/2005–2015/2016 (US$ billions)

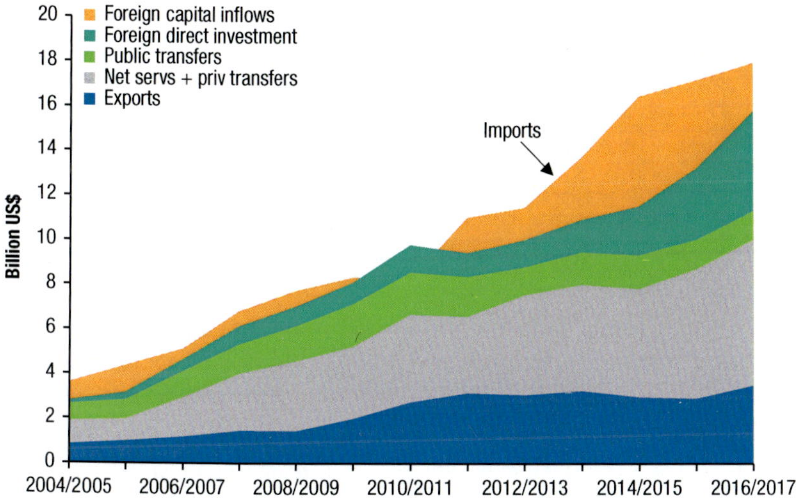

Source: IMF (various years) and authors' calculations.

FIGURE 1.10 Domestic and foreign debt, 2010/2011–2015/2016 (US$ billions)

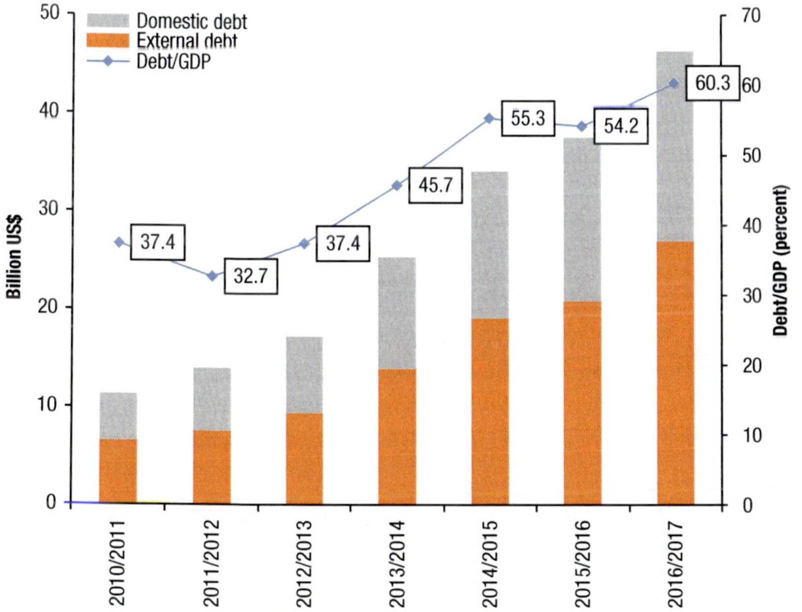

Source: IMF (various years) and authors' calculations.

Plan of the Book

This book is structured according to the major components of Ethiopia's agrifood system and their links to the overall economy. The first part, "Natural Resources and Production," presents an analysis of the drivers of recent increases in crop and livestock production. These chapters explore the question of whether such rapid growth can continue in the face of natural resource constraints and climate change. In the second part, "Evolving Markets and Household Consumption," we bring evidence related to how the rapidly changing structure of markets and overall demand will affect Ethiopia's agrifood system. The third part, "Economywide Perspectives," includes an analysis of urbanization and public investments, through simulations of alternative growth and investment strategies, and brings evidence on whether and for how long agricultural and rural development should remain a high priority for public investment in Ethiopia.

These three broad questions (whether rapid growth will continue, how changes in demand will affect the agrifood system, and whether agricultural development should remain a priority) are closely related. If resource constraints become binding, they could potentially slow growth in production and the market transformation process. The result would likely be rising agricultural prices, which would slow the expansion of urban markets, or increased food imports, which would come at the cost of increased producer incomes, thereby limiting positive growth linkages with the nonagricultural economy. Moreover, constraints on agricultural production, whether resulting from agronomic factors, limited market demand, or other considerations, ultimately affect the private and social profitability of investments in agriculture and their potential implications for poverty reduction.

Part 1: Natural Resources and Production

Chapter 2, "Cropland Expansion," uses spatially disaggregated satellite data to examine changes in land use over time at the national level. This is the first time that such dynamic analysis has been examined at this level. The data indicate a slowing in rate of cropland expansion in much of northern Ethiopia. The authors use results of econometric analysis of land use changes over time to estimate maximum potential economic cropland across space and potential land expansion. The analysis suggests little scope for further crop expansion in the drier parts of the highland but ample opportunities for increased crop cultivation on underused land with sufficient rainfall in parts of western regions (especially Benishangul-Gumuz and Gambella) and in the Oromia region.

Chapter 3, "Crop Productivity and Potential," documents and explains the causal factors for the rapid expansion in crop production in Ethiopia (official estimates show an increase in production of grain crops from 16.1 million metric tons to 30.6 million metric tons from 2007/2008 to 2017/2018). Using data on input use (such as land, improved seeds, fertilizer, and agricultural extension services), the authors estimate the combined effect of increased inputs. The residual, then, is an estimate of the gains in total factor productivity. Further rapid gains in yields seem less likely, though, given the high levels of input use already achieved. They illustrate some of the challenges in crop production, linked to the research and extension system, the limited access to irrigation, and the seed sector.

Chapter 4, "Climate Change Impacts on Crop Yields," presents results of model simulations of crop yields in Ethiopia through 2085. The analysis draws on climate outcomes from 32 global climate models and an agronomic crop model to estimate effects of expected higher temperatures and, for most of Ethiopia, increased rainfall. The simulation results suggest that climate change will likely have only relatively small effects on average yields of maize, wheat, and sorghum in Ethiopia up to 2055, as agronomic conditions for cultivation of these crops may actually improve in large parts of the country. Nonetheless, crop yields will need to increase over time to enable cereal production to keep pace with expected demand growth due to increases in population and per capita incomes. Moreover, even if future changes in climate have only moderate impacts on *average* crop yields in Ethiopia, weather outcomes and consequent crop yields are likely to become more variable in the future.

Chapter 5, "Evolving Livestock Sector," shows that livestock is important in Ethiopia's agricultural economy as almost all farmers own some livestock. Overall livestock output is shown to have grown rapidly over the past decade, estimated at almost 6 percent per year, but about 80 percent of that growth came from increases in the number of livestock. The stock of different livestock species was about 50 percent higher in 2015 than a decade earlier, while modern input use and improvements in production methods contributed little to growth in the livestock sector. Linked to improved access to extension and markets, adoption of improved breeds and improved feeding practices increased, but such adoption patterns started from a very low base. Within the livestock sector, cattle are dominant, making up an estimated three-quarters of the value of livestock stock. However, the share of cattle in total livestock output is declining, and small ruminants are on the rise, especially in pastoralist areas. Given the rapid growth in livestock numbers and the increasing livestock density per unit of land, the authors find that feeding practices are changing. Grazing land is declining in availability, so reliance on commercial feed markets is increasing. Access to vaccinations and veterinary service provision have improved, and livestock death rates declined slightly over the past decade. However, the number of livestock lost to deaths is still more than twice the number sold for meat production, indicating important challenges remaining for the development of the livestock sector in Ethiopia.

Chapter 6, "Farm Size, Food Security, and Welfare," looks at the association of farm sizes with food security and welfare. This is an important issue in Ethiopian agriculture given that a growing rural population combined with limited scope for further land expansion and slow movement out of agriculture means that average (mean) farm sizes are decreasing in Ethiopia. Using cross-sectional data for the Ethiopian highlands, the authors find—surprisingly—small differences between owners of small and large farms in key welfare and food security outcomes. For example, a 20 percent increase in owned land area is associated with only a 0.7 percent increase in food consumption (measured in calories). Five adjustments are made by households residing on small farms to assure similar calorie intake as those residing on larger farms.[12] First, they participate actively in land rental markets and, as a result, are able to double their cultivated land area on average. Second, they compensate their small landholdings with other income sources, mainly livestock and nonfarm businesses, permitting additional food purchases. Third, they cultivate their land more intensively, obtaining higher yields. Fourth, they favor more calorie-dense crops that are mostly used for their own consumption. Fifth, they produce as well as consume cheaper food items.

Part 2: Evolving Markets and Household Consumption

The second part of the book, "Evolving Markets and Household Consumption," begins with Chapter 7 on "Evolving Food Value Chains." Ethiopia's food systems are changing rapidly as high population growth, rapid urbanization, major infrastructure investments, and substantial income growth are leading to dietary, agricultural, and supply chain transformations. Over the past two decades, calorie consumption has risen, the share of starchy staples in the diet has fallen, and the shares of high-value products (such as animal-sourced foods, fruits and vegetables, processed convenience foods, and

12 The analysis in Chapter 6 examines differences in farm and household characteristics across land ownership quintiles, where the mean size of the first and fifth quintiles are 0.25 hectares and 2.93 hectares per household, respectively (see Table 6.1).

out-of-home food consumption) have risen. At the same time, use of fertilizer, improved seeds, and mechanization is increasing. Nonetheless, smallholders, who have an average agricultural area of 1.0 hectare per holder and who cultivate 93 percent of the agricultural land in Ethiopia, are facing increasing land constraints: farm sizes are declining over time and, in comparison to older farmers, younger farmers are more reliant on land rental markets to access the land that they farm. The transformation of agricultural supply chains is especially rapid, providing for increases in employment. Agricultural commercial surpluses are rapidly growing, and modern food marketing methods and technologies have emerged, including mobile phones, a commodity exchange, and an incipient modern retail sector and food service sector. Domestic markets are better integrated as spatial and seasonal price margins have narrowed. External trade has also expanded since 2000 despite high transport costs and foreign exchange restrictions.

Chapter 8, "Evolving Animal Sourced Foods and Livestock Markets," covers the changing situation in livestock marketing systems. The authors find that in many developing countries in which staple foods dominate the composition of diets, higher consumption of animal-sourced foods (ASF) is associated with significant nutritional benefits. Given the importance of prices for consumption decisions in these settings, they analyze ASF price patterns, relying on a large-scale price dataset collected in 116 urban retail markets in Ethiopia. The authors document important seasonal and spatial patterns, and they find, worryingly, that real prices of ASF have been increasing over the past decade by between 32 percent and 36 percent for three major ASFs—milk, eggs, and meat. Similar price increases are noted in rural and urban areas as well as for tradable and nontradable ASFs. This price trend is in contrast with staple cereals for which real prices stayed at similar levels over the past decade. As they estimate that a price increase of this magnitude would reduce consumption of ASF by approximately 25 percent, holding other things constant, it seems that more investments and attention to the production of ASF and the livestock sector are needed to reduce ASF prices and increase their consumption in Ethiopia.

Chapter 9, "Droughts, Cereal Prices, and Price Stabilization Options," looks at price volatility, causes, and policy options. Increases in cereal prices in Ethiopia often raise concerns about adverse effects for poor net consumers. In particular, the frequent natural calamities—especially droughts—in Ethiopia can lead to important price hikes. But domestic prices of some cereals (especially maize) fluctuate every year with prices at harvest times substantially dropping, to the detriment of producers. Price stabilization efforts are therefore an important consideration for Ethiopian policymakers. This chapter sheds light on options for cereal price stabilization in Ethiopia drawing on experiences of other developing countries. The international experience in food price stabilization shows that while some countries have achieved success, the efforts of many others have actually destabilized market prices at great fiscal costs. When assessing the extent to which price stabilization efforts in Ethiopia were effective during the major El Niño–induced droughts of 2015 and 2016 (including food aid distributed through the PSNP), the authors find that an opportunity was missed to enhance food security and consumer welfare by allowing private-sector imports to minimize the rise in cereal prices as well as to reduce the fiscal costs to the government and donors.

Chapter 10, "Food Security," describes the evolution of poverty and food security between 2010/2011 and 2015/2016 and examines the seasonality of food insecurity. Household survey data from 2016 show that Ethiopia continues to face high levels of food insecurity: daily energy consumption is low (3,055 kilocalories per adult on average), and diet quality is poor (starchy staples account for 71.6 percent of calories). There has been relatively little change in the composition of diets between 2011 and 2016, although the diets of the poorest half of the population (especially in rural areas) improved slightly with less reliance on starchy staples and greater dietary diversity. The authors find substantial seasonal patterns, however, with marked variations in energy (calorie) consumption, which is highest in postharvest periods. Regression analysis using the 2015/2016 data reveals that that average energy intake is positively associated with total household expenditure and productive farm assets. The authors find little evidence of diets being affected by rainfall shocks, a notable finding since Ethiopia experienced its worst drought in decades in 2015.

Part 3: Economywide Perspectives

The third part of the book presents "Economywide Perspectives" on the future of Ethiopian agriculture. Chapter 11, "Nonfarm Income and Rural Labor Markets," examines how the rapid transformation in Ethiopia's economy is affecting off-farm income and labor markets in rural areas.[13] Data from a large-scale household survey in high potential agricultural areas show that total off-farm income (defined as wage and enterprise income) makes up 18 percent of total rural household income. The shares of off-farm income and

---

13 The PSNP also contributed to this. Filipski et al. (2016) illustrate the extent to which the PSNP has led to expanded economic activities by illustrating the large spillovers of the program.

wage income are even larger for the poor and for female- and youth-headed households. Moreover, given strong growth in agriculture and the overall Ethiopian economy, real rural wages have increased by 54 percent since 2006. While this wage increase is good news for the poor, it also induces adjustments in agricultural production practices, including increased adoption of labor-substituting technologies such as herbicides and mechanization. Higher wages and incomes relax liquidity constraints in the off-season for some households, also contributing to higher productivity.

Chapter 12, "Urbanization and Structural Transformation," describes patterns of urbanization in Ethiopia and government policy to promote development of secondary cities. Official population data indicate rapid urban growth, 4.2 percent per year between 1994 and 2015, far outpacing the overall population growth rate of 2.5 percent. By 2050 urbanization is expected to reach 38 percent with major implications for relative wage rates in rural versus urban areas, infrastructure needs, and public service delivery. Improved road infrastructure is improving connectivity across the country and promoting secondary city development. In addition, recent public investments to promote industrialization and increase manufacturing labor opportunities via newly constructed and planned industrial parks, though small relative to the overall economy, are designed to be a catalyst for future growth.

Chapter 13, "Public Investments and Poverty Reduction," presents results of general equilibrium simulations of the impacts of alternative investment strategies, focusing on national income (GDP) and incomes of the poorest 40 percent of the population in both rural and urban areas. These simulations, which incorporate the major factors discussed in this book—including land constraints, productivity growth, migration, expanding markets, rising incomes and consumption, and greater urbanization—show that although urban investments generate faster economic growth and structural transformation, investments in the rural economy are likely to continue to be more pro-poor than urban public investments through the mid-2020s. After the mid-2020s, investments in cities become more pro-poor. The authors show that though rapid economic growth and structural transformation have diminished the relative importance of the agricultural sector in Ethiopia's economy, continued public investments in agriculture and the broader agri-food system remain crucial for equity and poverty alleviation (and for a reduction of food import dependency) in Ethiopia.

Finally, Chapter 14, "Toward a Medium-Term Agricultural and Rural Development Strategy," highlights the book's major themes and discusses implications for policy. In particular, the authors emphasize the importance of continued investments to increase agricultural productivity, market efficiency, and infrastructure (roads, small-scale irrigation, and electricity generation) to support both agriculture and the rural and small-town nonfarm economy. Without these investments substantial poverty increases and growing inequality are likely to result, even if there is rapid overall economic growth.

References

Bachewe, F. N., G. Berhane, B. Minten, and A. S. Taffesse. 2018. "Agricultural Transformation in Africa? Assessing the Evidence in Ethiopia." *World Development* 105 (c): 286–298.

Benin, S. 2014. "Implementation Performance and Progress towards Core CAADP Targets." Paper presented at the ReSAKSS annual conference, Addis Ababa, Ethiopia, October 8–10.

Berhane, G., D. O. Gilligan, J. Hoddinott, N. Kumar, and A. S. Taffesse. 2014. "Can Social Protection Work in Africa? The Impact of Ethiopia's Productive Safety Net Programme." *Economic Development and Cultural Change* 63 (1): 1–26.

Diao, X., J. Thurlow, S. Benin, and S. Fan, eds. 2012. *Strategies and Priorities for African Agriculture: Economywide Perspectives from Country Studies*. Washington, DC: International Food Policy Research Institute (IFPRI). http://dx.doi.org/10.2499/9780896291959.

Dorosh, P. A., and S. Rashid. 2012. *Food and Agriculture in Ethiopia: Progress and Policy Challenges*. Philadelphia: University of Pennsylvania Press. http://ebrary.ifpri.org/utils/getfile/collection/p15738coll2/id/127347/filename/127558.pdf.

Dorosh, P., and J. Thurlow. 2012. "Implications of Accelerated Agricultural Growth for Household Incomes and Poverty in Ethiopia: A General Equilibrium Analysis." In *Food and Agriculture in Ethiopia: Progress and Policy Challenges*, edited by P. A. Dorosh and S. Rashid, 219–255. Philadelphia: University of Pennsylvania Press.

EGTE (Ethiopian Grain Trade Enterprise). 2019. Commodity Statistics. Accessed September 4, 2019. www.egte-ethiopia.com/en.

Ethiopia, Central Statistical Agency (CSA). 2019. Unpublished market price data.

Ethiopia, Central Statistical Agency (CSA), and ICF International (ICF). 2016. *Ethiopia Demographic and Health Survey 2016*. Addis Ababa: CSA; Rockville, MD, US: ICF.

Ethiopia, MoFED (Ministry of Finance and Economic Development). 2002. *Ethiopia: Sustainable Development and Poverty Reduction Program, Ministry of Finance and Economic Development (MoFED)*. Addis Ababa.

———. 2006. *Ethiopia: Building on Progress—A Plan for Accelerated and Sustained Development to End Poverty (PASDEP) (2005/06–2009/10), Volume 1: Main Text*. Addis Ababa.

———. 2010. *Growth and Transformation Plan II (GTP II) (2010/11–2014/15), Volume 1: Main Text.* Addis Ababa.

FAO (Food and Agriculture Organization of the United Nations). 2019. "Value of Agricultural Production." Accessed September 9, 2019. www.fao.org/faostat/en/#data/QV.

Filipski, M., J. E. Taylor, G. A. Abegaz, T. Ferede, A. S. Taffesse, and X. Diao. 2016. *Synopsis: Economy-wide Impacts of the Productive Safety Net Programme (PSNP).* Ethiopia Strategy Support Program (ESSP) Research Note 57. Addis Ababa: IFPRI.

Gollin, D., R. Jedwab, and D. Vollrath. 2016. "Urbanization with and without Industrialization." *Journal of Economic Growth* 21 (1): 35–70.

ILO (International Labor Organization). 2017. "Employment Statistics." Accessed July 5, 2019. www.ilo.org/ilostat.

IMF (International Monetary Fund). 2011. *The Federal Democratic Republic of Ethiopia: Poverty Reduction Strategy Paper: Growth and Transformation Plan 2010/11–2014/15—Volume 1.* IMF Country Report 11/304. Washington, DC. www.imf.org/external/pubs/ft/scr/2011/cr11304.pdf.

———. 2012. *The Federal Democratic Republic of Ethiopia 2013 Article IV Consultation.* IMF Country Report 12/287. Washington, DC.

———. 2014. *The Federal Democratic Republic of Ethiopia 2014 Article IV Consultation.* IMF Country Report 14/303. Washington, DC.

———. 2016. *The Federal Democratic Republic of Ethiopia 2016 Article IV Consultation.* IMF Country Report 16/322. Washington, DC. www.imf.org/external/pubs/ft/scr/2016/cr16322.pdf.

———. 2019. *International Financial Statistics.* Accessed September 4, 2019. http://data.imf.org/regular.aspx?key=61545850.

Johnston, B. F., and J. W. Mellor. 1961. "The Role of Agriculture in Economic Development." *American Economic Review* (51): 566–593.

Lewis, W. A. 1954. "Economic Development with Unlimited Supplies of Labour." *The Manchester School* 22 (2): 139–191.

McCann, J. 1995. *People of the Plow: An Agricultural History of Ethiopia, 1800–1990.* Madison, US: University of Wisconsin Press.

Mellor, J. W. 1995. *Agriculture on the Road to Industrialization.* IFPRI. Baltimore: Johns Hopkins University Press.

Minten, B., A. S. Taffesse, and P. Brown. 2018. *The Economics of Teff: Exploring Ethiopia's Biggest Cash Crop.* Washington, DC: IFPRI.

NBE (National Bank of Ethiopia). 2019. *Statistical Data Series.* Accessed September 4, 2019. www.nbe.gov.et/.

NPC (National Planning Commission). 2016. *Growth and Transformation Plan II (GTP II) (2015/16–2019/20), Volume 1: Main Text.* Addis Ababa, Ethiopia. http://dagethiopia.org/new/images/DAG_DOCS/GTP2_English_Translation_Final_June_21_2016.pdf.

———. 2017. *Ethiopia's Progress Towards Eradicating Poverty—An Interim Report on 2015/16 Poverty Analysis Study.* Addis Ababa, Ethiopia.

Rodrik, D., M. McMillan, and C. Sepúlveda. 2013. "Structural Change, Fundamentals, and Growth: An Overview." In *Structural Change, Fundamentals, and Growth: A Framework and Case Studies,* edited by M. McMillan, D. Rodrik, and C. Sepúlveda, 1–38. Washington, DC: IFPRI.

Scheuermann, H., and T. Mészáros. 2017. *African Union: Continental Support for Agricultural Transformation in Africa.* Eschbom, Germany: Deutsche Gesellschaft für Internationale Zusammenarbeit (GIZ) GmbH.

Thurlow, J., P. Dorosh, and B. Davis. 2018. "Demographic Change, Agriculture, and Rural Poverty." In *Sustainable Food and Agriculture,* edited by C. Campanhola and S. Pandey, 31–53. London: Academic Press.

Timmer, C. P. 1988. "The Agricultural Transformation." *Handbook of Development Economics* (1): 275–331.

UN Statistics Division. 2019. United Nations Commodity Trade Statistics database (COMTRADE). Accessed November 4, 2019. Comtrade.un.org.

USDA (United States Development Agency). 2019. "Agricultural Commodity Prices." Accessed November 4, 2019. www.indexmundi.com/agriculture/?country=et&commodity=palm-oil&graph=imports.

Webb, P., and J. von Braun. 1994. *Famine and Food Security in Ethiopia: Lessons for Africa.* Washington, DC: IFPRI.

World Bank. 2017. World Development Indicators. Accessed August 1, 2019. http://wdi.org.

———. 2018. World Bank Poverty and Equity Data Base. Accessed August 1, 2019. http://databank.worldbank.org/data/source/poverty-and-equity#.

PART 1

Natural Resources and Production

Chapter 2

CROPLAND EXPANSION

Emily Schmidt and Timothy S. Thomas

Introduction

The agricultural sector is the cornerstone of Ethiopia's economy with approximately three-quarters of the economically active population engaged in agricultural production activities (Schmidt and Bekele 2016). Under the strategy of agriculture development–led industrialization (ADLI), agricultural production in Ethiopia increased substantially, with increases in agricultural GDP averaging 7 percent per year between 2004/2005 and 2013/2014.[1] During the earlier part of this period, land area expansion was the primary contributor to increases in agricultural GDP. However, in more recent years, rising crop yields coupled with continuing agriculture area expansion contributed to agricultural GDP growth (Bachewe et al. 2018). Given Ethiopia's reliance on agriculture as a mainstay of livelihoods as well as the country's rapidly declining area of unexploited cultivable land in the agricultural highlands, a question arises of whether this type of agricultural growth is sustainable for the foreseeable future.

Increases in agricultural production via agricultural expansion is not unique to Ethiopia. A variety of analyses have evaluated land use expansion and expansion potential within Africa south of the Sahara (SSA). Brink and Eva (2009) used historical satellite images to evaluate landcover change in SSA and concluded that agricultural area expanded by 57 percent, while natural vegetation decreased by 21 percent in SSA between 1975 and 2000. Deininger et al. (2011) use geospatial data to estimate potential crop area expansion in SSA, and find large potential for expansion of agricultural cropland overall in Africa, but with significant differences between African regions. Chamberlin, Jayne, and Headey (2014) find that SSA land resources are concentrated in a

1 GDP growth occurred throughout the economy over the past decade. Large increases were also seen in the industry and services sectors, which grew at approximately 14 percent and 13 percent per year, respectively, between 2004/2005 and 2013/2014. Given the growth in industry and services, agriculture's share of GDP has fallen over the past decade. However, agriculture remains an important driver of economic growth and employment (Schmidt and Bekele 2017).

few countries and currently are comprised of forested areas that would require significant development investments. Van Ittersum et al. (2016) pose the question of whether SSA will be able to achieve cereal self-sufficiency (approximately 80 percent) by 2050 considering current population growth rates. They find sobering evidence suggesting that a major effort to increase crop production via intensification will be necessary to avoid massive cropland expansion (with significant environmental costs) and a heavy dependence on imports of cereals.

While the Ethiopian highlands have benefited from impressive growth in agricultural production during the past several decades, recent studies have identified some of the costs of agricultural extensification. Farmers are cultivating on steeper slopes in the highlands without using proper sustainable land management techniques—that is, terracing and fallowing (Schmidt and Zemadim 2015; Taddesse 2001; Hamza and Anderson 2005). Production losses approaching 1.1 percent per year have been linked to increasing erosion and topsoil loss in the highlands (Holden and Shiferaw 2002). In addition to unsustainable cultivation practices, increased levels of deforestation are attributed to cropland conversion (Cleaver and Schreiber 1994). Pasturelands are also increasingly being put under crops (Tschopp et al. 2010).

Agricultural intensification, in particular increased input use, has contributed to important increases in overall production volumes (see Bachewe et al. 2015 and 2018; Chapter 3 of this book); however, microlevel analysis suggests intensification may not be enough to continue the agricultural GDP growth witnessed in previous decades. Although Headey, Dereje, and Taffesse (2014) find evidence to support Boserup's hypothesis of greater agricultural intensification strategies within land-constrained villages, they do not find increases in household income due to greater adoption of agricultural inputs in Ethiopia. Similarly, Josephson, Ricker-Gilbert, and Florax (2014) reported that smaller farm sizes had a positive influence on input demand; however, increased input use did not correspond to an increase in staple crop yields in Ethiopia, suggesting that farmers may be addressing declines in soil fertility with increasing inputs but only managing to maintain a base level of productivity.

Options for relieving rural agricultural land pressures in Ethiopia are few. Agriculture is the primary source of income for most Ethiopians and nonfarm employment remains limited throughout the rural highlands.[2] Moreover, the

2 Schmidt and Bekele (2017) find that a large share of the rural working population (78 percent) is engaged solely in own-farm activities. Only 12 percent report having a secondary job outside of their own farm.

potential for rural–rural migration to seek out less densely populated agricultural areas is constrained by current land tenure restrictions that permit usufruct land rights; however, these do not guarantee ownership if an individual is not seen to be working on their own land (Diao et al. 2013). Gebregziabher and Holden (2011) reported a need to facilitate land rental markets due to evidence of land rental being used as a last-resort coping mechanism after depleting other household resources (livestock, firewood, reducing household consumption). Recent studies have shown an increase in rural–urban migration of youth (particularly in the highlands) due to agricultural land scarcity; however, tenure security in urban areas remains a challenge for recent migrants as well (Moller 2012; Bezu and Holden 2014).

Understanding the potential for further agricultural expansion within Ethiopia is necessary to inform investment priorities aimed at maintaining agricultural performance. We build upon studies such as Chamberlin, Jayne, and Headey (2014) and Deininger and Byerlee (2011) that employ remote sensing data on land cover and other biophysical and economic data to estimate unused potentially arable cropland across Africa south of the Sahara (SSA). As Chamberlin, Jayne, and Headey (2014) establish, calculation of available cropland is very sensitive to model assumptions of potential availability. These results are illustrative when comparing across the region as a whole; however given the global nature of their studies, they are unable to evaluate more localized, potentially arable cropland within-country.

We address this knowledge gap by using a combination of data from country-level databases and from remote sensing satellites to evaluate change in agricultural area over time. We define a spatial regression to identify correlates of crop area expansion at the kebele (subdistrict) level considering current estimates of cropped area within the kebele. Finally, we calculate the maximum potential for cropland expansion, controlling for a collection of biological and economic factors.

When assessing future potential for agricultural area expansion, our analysis suggests that agricultural area expansion in the highlands is reaching its maximum economic potential, especially in the drought-prone highland agroecological zone. Select areas in the lowlands have greater potential to expand the share of their land areas devoted to agriculture. However, this will require investments in transportation and social infrastructure to attract investment as well as linking the newly expanded agricultural areas to input and output markets.

The remainder of the chapter provides a background to Ethiopia's agricultural production, including a description of the country's diverse agroecological zones; evaluates Ethiopia's agricultural landscape at the disaggregated kebele level, using satellite landcover data to characterize agricultural area expansion over the past decade; details the satellite and other complementary data used to evaluate the determinants of agricultural land expansion, followed by a description of our empirical strategy for evaluating market access correlation with agricultural land expansion; and provides results and discussion. The final part of the chapter concludes.

Agricultural Production in Ethiopia

Since 2004/2005, the agricultural sector in Ethiopia has performed well. The increase in agricultural GDP (approximately 7 percent per year during the period from 2004/2005 to 2013/2014) was primarily due to crop production, contributing almost 80 percent to agricultural GDP growth (Table 2.1). Rising crop yields were the primary driver of increased agricultural GDP followed by land area expansion. These factors contributed 60 percent and 28 percent, respectively, to agricultural growth (Table 2.2). Cereals generated more than half of crop GDP growth between 2004/2005 and 2015/2016. The five major cereals (teff, barley, wheat, maize, and sorghum) constitute 73 percent of total cultivated area (Ethiopia, CSA 2015/2016).

In a country where rainfed agriculture is the predominant production system, agroecological conditions determine production patterns and, in the case of Ethiopia, dictate the location of major economic centers and transportation corridors (Figure 2.1). Ethiopia's topography has influenced demographic and agricultural patterns throughout its history. The highlands of Ethiopia, defined as locations with a minimum elevation of 1,500 meters above sea level, are more densely populated and reflect the physical and climatic advantages that led to the country's agricultural development (Figure 2.1).[3] The highlands are endowed with relatively more predictable rainfall and do not house vectors that carry diseases, such as malaria or tsetse fly (Pankhurst and Piguet 2009). In contrast, the lowlands experience more erratic and limited rainfall and have greater risk for disease. These factors have constrained expansive development in lowland agriculture (Josephson, Ricker-Gilbert, and Florax 2014; Headey, Dereje, and Taffesse 2014).

Most agricultural production is in the country's highlands, which constitutes the breadbasket of the country, where 90 percent of the area planted

3 For references on highland and lowland definitions in Ethiopia, see FAO (1986); Constable (1985); and Hurni (1998).

TABLE 2.1 Subsectoral contributions to real agricultural GDP, 2004/2005–2015/2016 (%)

Indicator	Initial agricultural GDP share 2004/2005	Final agricultural GDP share 2015/2016	Contribution to increase in agricultural GDP 2004/2005–2015/2016
Agriculture	100.0	100.0	100.0
Crop	63.8	72.0	79.5
Livestock	23.6	19.5	15.8
Forestry	12.5	8.4	4.7

Source: Authors' calculations using data from national accounts (Ethiopia, CSA 2016) and Agricultural Sample Survey reports (Ethiopia, CSA various years).

Note: Shares calculated using constant factor cost GDP.

TABLE 2.2 Contribution of cereals and noncereals to agricultural GDP change, 2004/2005–2013/2014 (%)

Indicator	All crops	Cereals	Noncereals
Share of agricultural GDP in 2004/2005	100.0	56.2	43.8
Share of agricultural GDP in 2013/2014	100.0	54.9	45.1
Contribution to total agricultural GDP	100.0	53.8	46.2
Increase in crop yields	60.3	40.7	19.7
Cultivated land expansion	27.5	15.4	12.0
Reallocating land to higher value crops	12.2	–2.3	14.5

Source: Authors' calculations using data from national accounts (Ethiopia, CSA 2016) and Agricultural Sample Survey reports (Ethiopia, CSA various years).

to cereals is found and 89 percent of total cereal production is obtained (Table 2.3). The majority of teff, the local cereal used to make injera, a principal staple food, is grown in the highlands, accounting for 94 percent of teff production and 94 percent of area dedicated to teff cultivation. In addition, highland cultivation accounts for 98 percent of total wheat production and 89 percent of total maize production.

Urbanization and population density is significantly greater in the highlands compared with the lowlands; however, a large degree of variation exists within the highland and lowland regions as well (Table 2A.1). People living in the highland areas have greater market access to urban centers (Table 2.4). While the average travel time to a city of at least 20,000 people in the highlands is approximately three hours, the average travel time to a city of at least 20,000 people in the lowlands is approximately six hours. This is for several reasons. First, there are fewer cities in the lowland areas. While there are 96 cities of at least 20,000 people in the highlands, there are only 20 urban

FIGURE 2.1 Highland and lowland areas of Ethiopia

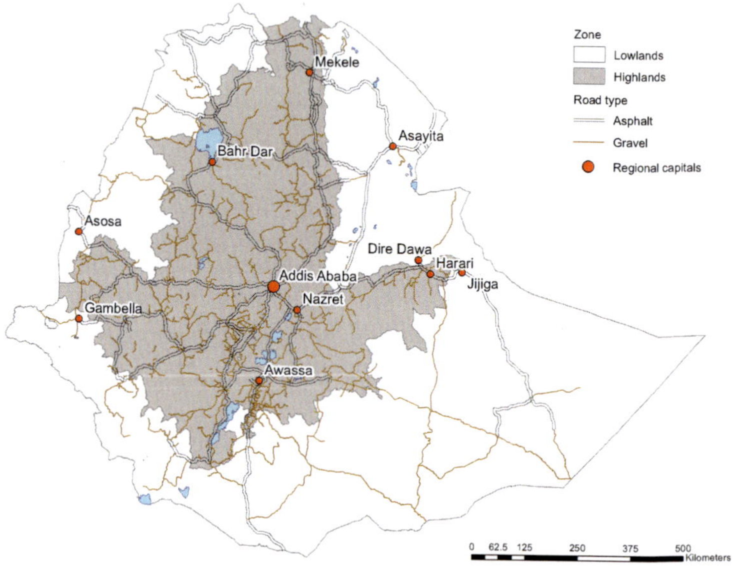

Source: Authors' calculation.

centers of this size in the lowlands.[4] The lowland areas of Ethiopia also have a sparser transportation infrastructure compared with the highlands. In the highlands road density is approximately 0.17 kilometers of road per square kilometer, while in the western areas of the lowlands (including parts of western Oromia, Gambella, and Benishangul-Gumuz regions) the road density is 0.05 kilometers per square kilometer (Figure 2.1). There are variations in climatic, demographic, and physical infrastructure within the highland and lowland regions as well (see Table 2A.1).

Rainfed agricultural production systems are vulnerable to variations in rainfall and climate. Over time, cultivated land in Ethiopia has expanded within the geographic area that permits relatively less risky agrarian livelihoods. Although the highlands make up only 37 percent of the total landmass of Ethiopia, three-fourths of the total population live in the highlands (Table 2.4). Looking forward, understanding the future potential for agricultural area expansion will be critical for future investment and establishing policy priorities to ensure continued economic growth. The following sections

4 When evaluating larger cities with a population of 50,000, the highlands have 36 cities of at least 50,000 people, while the lowlands have 16 cities of at least 50,000.

TABLE 2.3 Crop production and area in Ethiopia by highland and lowland areas, 2014/2015

Production (thousand metric tons)	Highlands	Lowlands	Total
All	**106,255**	**7,675**	**113,930**
Cereals	21,065	2,526	23,590
Barley	1,898	56	1,953
Maize	6,430	795	7,225
Sorghum	3,315	1,017	4,332
Teff	4,481	270	4,751
Wheat	4,132	100	4,232
Other cereals	809	289	1,098
Pulses	2,477	181	2,658
Oil Seeds	498	262	760
Other crops	54,324	2,589	56,912
Permanent crops	27,892	2,118	30,010
Area (thousand hectares)			
All	**12,537**	**1,618**	**14,155**
Cereals	9,102	1,034	10,136
Barley	961	33	994
Maize	1,840	266	2,106
Sorghum	1,447	381	1,829
Teff	2,835	181	3,016
Wheat	1,622	42	1,664
Other cereals	398	131	529
Pulses	1,441	103	1,544
Oil Seeds	519	336	856
Other crops	301	20	321
Permanent crops	1,173	124	1,297

Source: Authors' calculations using Agricultural Sample Survey 2014/2015 (Ethiopia, CSA 2016).
Note: Other cereals = rice, millet, and oats. Other crops = vegetables, root crops, and other temporary crops.

evaluate the expansion in agricultural land throughout Ethiopia during the past decade and estimate the potential for further expansion using a spatial regression approach.

Agricultural Area Expansion

To analyze changes in agricultural crop area across the Ethiopian landscape, we use the MODIS Land Cover Type product to classify agricultural areas of

TABLE 2.4 Characteristics of highland and lowland areas of Ethiopia

Indicator	Highlands	Lowlands
Rainfall, mean annual (millimeters)	1,221	1,152
Elevation, mean (meters)	2,065	1,385
Travel time to a city with a population of more than 20,000 persons, mean (hours)	2.9	6.1
Population density (persons per square kilometer)	321	61
Population 2016 (millions)		
Total	75.4	16.8
Urban	15.4	2.9
Rural	60.0	13.9
Total land area (square kilometers)	421,594	709,890

Source: Authors' calculations using a variety of remote sensing datasets including Jarvis et al. (2008), Bright, Rose, and Urban (2013), Funk et al. (2015), and Ethiopia, CSA (2013).

Ethiopia. A total of 17 landcover classes have been identified, including croplands and a cropland/natural vegetation mosaic classification, at a 500-meter resolution for the 2001–2013 period (see Friedl et al. [2010] for more details). The MODIS Land Cover Type product is derived from a year (during the years of 2001 to 2013) of observations collected from the Terra- and Aqua-MODIS sensors, satellites that view the entire Earth's surface every one to two days.

Although the MODIS Land Cover Type product (and similar remotely sensed data products) provides the ability to analyze year-to-year changes in landcover, there are several limitations to this dataset that are important to recognize, which we attempt to overcome with our analysis. Comparing landcover change over time presents a challenge because of the variation in the reflectance data collected by the satellite sensors due to differing environmental conditions. Given the multispectral mode of the MODIS sensors, satellite imagery of the Earth cannot be observed through clouds, dust, or other atmospheric factors. This lack of data in certain areas given differences in weekly/monthly/yearly weather conditions introduces noise in the data that may result in perceived landcover changes that do not exist in reality. Another source of data error in landcover analysis is the relatively low spatial resolution of satellite input data (in this case at a 500-meter resolution for the 2001–2013 period). Lower-resolution imagery presents several challenges: (1) in areas of low agricultural intensification, it is difficult to distinguish between croplands and grasslands; (2) in areas with complex landscapes, lower-resolution imagery has difficulty capturing intercropped agricultural systems. Finally, highly divergent results have been reported in published landcover datasets due to differences in classification algorithms, different satellite sensors used to collect primary imagery, and different datasets used to train the landcover identification algorithms.[5] Taking into account these challenges, remote sensing data provides the opportunity for an objective, frequent, and consistent measure of landcover over time; however, ground-truthing is also important to verify satellite interpretations and analysis. Patil and Gumma (2018) provide a comprehensive review of different landcover data products and their advantages and disadvantages.

We take into account the inadequacies in remote sensing data in this analysis in several ways. First, rather than evaluate each year of landcover output produced by the MODIS Land Cover Type product, we average total landcover area values by kebele (subdistrict) and average values over three periods: early (2001–2004), middle (2005–2009), and late (2010–2013). In doing so, we attempt to reduce noise in the landcover data due to differences in weather patterns, cloud cover, time of day of the satellite pass, and data collection accuracy. Second, it is important to note that production practices in Ethiopia are spatially and temporally heterogeneous with production occurring on smallholder farms characterized by diverse farming practices, intercropping, and mixed livestock-crop production systems. Given these complex agricultural production characteristics, we calculate total agricultural area as the cropland area extent plus half of the cropland/natural vegetation mosaic area extent to account for the difficulty of distinguishing between croplands and other similar landcover types (scrub, grasslands, and so on). Finally, we report descriptive statistics at an aggregated highlands and lowlands level considering recent analysis that finds improved accuracy of cropland acreage at higher levels of aggregation (Leroux et al. 2014).

Landcover analysis of Ethiopia during the 2010–2013 period suggests that most cropped area is in the highlands. Kebeles in the Amhara region have the greatest share of land under agriculture, with more than 60 percent of total land area dedicated to agricultural uses (Figure 2.2). Kebeles in central Oromia and northeast Southern Nations, Nationalities, and Peoples (SNNP) region have between 50 percent and 70 percent of total land area dedicated to

5 For example, GlobCover cropland distribution estimates are 20 percent higher than MODIS-derived global cropland area estimates (Patil and Gumma 2018).

FIGURE 2.2 Map of average share of area cropped, 2010–2013

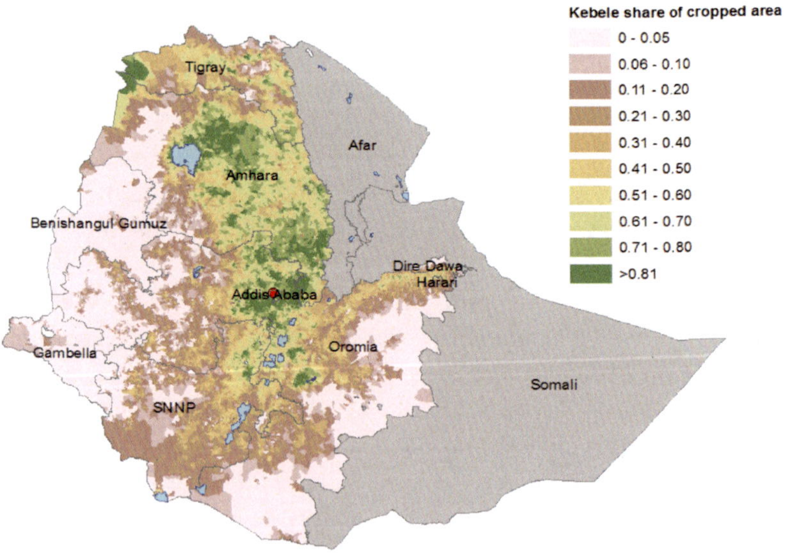

Source: Authors' calculations using MODIS Land Cover type product.
Note: SNNP = Southern Nations, Nationalities, and Peoples.

crop cultivation. In contrast, kebeles in the lowlands have very little area dedicated to cultivation.[6]

Changes in cropped area in Ethiopia show varying spatial patterns over time. Looking at changes between the first (2001–2004) and second (2005–2009) periods, most of the agricultural area expansion occurred in the highland areas of Amhara region east of Lake Tana. However, changes between the second and third (2010–2013) period were greater in the lowlands of northwest Amhara region and in the far western kebeles of Tigray region (Table 2A.2). Over the entire period the greatest growth in cropped area as a share of total area occurred in the lowlands, averaging 2.6 percent growth per year from 2001–2004 to 2010–2013 (Table 2.5).[7]

6 We do not include Afar and Somali regions in this analysis given their limited agricultural cultivation activities.

7 Evaluating by region, the greatest growth in crop area occurred in the Amhara and Oromia regions, which increased by 2.1 percent and 2.4 percent, respectively, over the 2001–2013 period (Table 2A.2).

TABLE 2.5 Changes in crop area in Ethiopia, 2001/2004–2010/2013

Indicator	Highlands	Lowlands	Total
Percentage of kebeles with declines in crop area of 5 percentage points or more			
2001–2004 to 2005–2009	21.3	6.9	19.3
2005–2009 to 2010–2013	19.2	23.5	19.8
2001–2004 to 2010–2013	17.3	14.0	16.8
Average annual growth in crop area (%)			
2001–2004 to 2005–2009	2.2	5.6	2.7
2005–2009 to 2010–2013	1.1	0.2	1.0
2001–2004 to 2010–2013	1.6	2.6	1.8
Total percentage point change in cropland area, 2001–2013	4.8	2.4	3.8

Source: Authors' calculations using MODIS Land Cover type product.

Although expansion occurred in several areas throughout the analysis period, these data suggest that cropland expansion is slowing down. Whereas the agricultural area in the highlands expanded at 2.2 percent per year from 2001–2004 to 2005–2009, this annual growth decreased to 1.1 percent from 2005–2009 to 2010–2013 (Table 2.5). Similarly, 21 percent and 19 percent of kebeles in the highlands experienced a contraction in their share of total land under crops during the first period and second period, respectively. In comparison, the lowlands experienced an average annual growth in crop area of 5.6 percent between 2001–2004 and 2005–2009, although average annual growth in crop area there dropped considerably to 0.20 percent between 2005–2009 and 2010–2013. This pattern may be due to droughts in 2009/2010 and 2010/2011 that affected areas of southern Ethiopia and lowland Oromia, Tigray, and Somali regions (Viste, Korecha, and Sorteberg 2012).

Overall, between 2001–2004 and 2010–2013, average annual growth in crop area was 2.6 percent in the lowlands and 1.6 percent in the highlands. This suggests that, although highland agriculture is the preferred location for production, there may be limited area in which to expand farther in the highlands. This descriptive analysis of agricultural land area expansion and contraction in Ethiopia suggests that significant spatial and temporal patterns exist, which may be associated with inherent biophysical endowments as well as climate and infrastructure characteristics. Given these trends and Ethiopia's dependence on highland agriculture, we evaluate the potential for future agricultural land expansion, taking into account bioeconomic factors and the current share of land area that is cropped in each kebele.

Understanding Ethiopia's Potential to Expand Cultivated Area: Data and Empirical Specification

Data Description

The objective of this analysis is to evaluate the factors associated with changes in agricultural area over time. Utilizing the MODIS landcover data, which identifies cultivated area as well as a variety of other landcover types, including grasslands, savanna, forest, and the like (see Friedl [2010] for details), from 2001 through 2013 at a 500-meter resolution, we evaluate the change observed in cultivated area between the period of 2001–2004 and 2010–2013 at the kebele (subdistrict) level. In doing so, we calculate by year the share of 500-meter grid cells within a kebele that are classified as cropland or cropland/natural vegetation mosaic, assuming 50 percent of the mosaic cells are dedicated to agricultural production. We then average this share over the analysis periods, 2001–2004 and 2010–2013, to reduce noise in the satellite reflectance values and to aid data interpretation. Finally, we subtract the share of cropland area calculated for the 2001–2004 period from the share of cropland area calculated for the 2010–2013 period by kebele to determine the change in the share of total area under crops in each kebele between the two time periods. This change represents our dependent variable in the following analysis.

A variety of factors are related to the attractiveness of expanding the area under agricultural production. An important factor that needs careful consideration is the share of the total area that is already under cultivation. As cultivated land area expands across a kebele, the propensity to continue expanding becomes less attractive as the more productive land areas are adopted into agricultural use. We hypothesize that the change in cropland area will be smaller for kebeles that have a greater share of their total land area already cultivated. To test (and to control for this in regard to other variables), we include the share of area under cultivation in the initial period (2001–2004) as well as its squared term as explanatory variables (median and 10th and 90th percentile values of regression covariates are reported in Table 2A.3).

As mentioned, Ethiopia's topographic and climatic variability influence the location of productive activities. Climate conditions such as the level and variations in rainfall, temperature, and elevation affect agricultural patterns within the country. We account for variations in climate by controlling for average precipitation, variation in precipitation, and average annual maximum temperature for the main growing season between June and September over the past 30 years. We include a square term for precipitation and elevation based on the assumption that, while area expansion initially increases with greater rainfall and higher elevation, eventually flood and frost would affect cultivable area in some of Ethiopia's more extreme climates. In addition, Ethiopia's topography varies dramatically from flat lowlands at 500 meters above sea level to rugged highlands reaching elevations above 3,500 meters. We control for topographic variation by including a terrain roughness indicator. This indicator is computed first for every 1-kilometer square and then averaged for all such squares within each kebele.

Related with climate factors, we assume that the cropping system available to smallholder farmers would also affect cropland expansion. The majority of kebeles in Ethiopia are dependent on one primary *meher* harvest, derived from the main *meher* rainy season that occurs from June through September. However, in some areas of the country a second *belg* season harvest is obtained. We include a dummy variable calculated using the Agricultural Sample Survey (AgSS) data collected by Ethiopia's Central Statistical Agency (CSA) to account for areas that benefit from two harvests.

Finally, the potential for profitable cropland expansion is highly dependent on the ability to access input and output markets. We measure market access potential by taking the difference in a measure of travel time between 2001 and 2013, from each location in Ethiopia to the nearest secondary market of at least 20,000 people. A more in-depth discussion of the market access variable construction used in this chapter is provided in Chapter 12 of this volume.[8] Given that we are evaluating change in agricultural area between 2001–2004 and 2010–2013, we include the base travel time in 2001 to account for the initial market access of the kebele. As travel time to a market decreases, we assume the cost of transporting goods to markets and the cost of purchased inputs decreases, making agricultural production and hence agricultural land expansion more attractive.

Empirical Specification

While we are interested in seeing how each of the variables influences the change in the proportion of the total land area of a kebele that is under crops,

8 We construct three separate market access measures. The 2001 market access grid was built from a roads database constructed for the 1997 census and updated to reflect city size targets from 2001. The second market access measure is constructed from the 2007 roads database and uses the same methodology as that of 2001; however, newly constructed and improved roads are included to reflect the road network of 2007 and city population of 2007. Finally, a 2013 roads database is constructed from the 2007 roads data by updating road infrastructure to the 2013 level using Google Earth satellite imagery and 2013 city population.

ΔC, we also intend to find the steady state maximum proportion of cropland, C_{max}, for each kebele. To make this clear in mathematical terms, we will make explicit both C (the proportion of cropland in each kebele) and ΔC in our specification:

$$\Delta C = \alpha_1 C + \alpha_2 C^2 + X\beta + \epsilon$$

where X is a matrix of biological and economic variables as described above, β is the vector of parameters to be estimated, and ϵ is the error vector.[9] The steady state maximum is found when $\Delta C = 0$.

We control for potential spatial dependence that may be present in ordinary least squares model estimates by using a spatial regression framework that controls for both spatial error and spatial lag. Allowing for spatial error recognizes that data observations associated with spatial units may reflect measurement error (that is, administrative boundary misalignment or road placement inconsistencies within the GIS road database) or that there are unmeasured variables that by definition are not part of the regression but are spatially correlated. Allowing for the possibility of a spatial lag takes into account that cropland expansion in one kebele might also influence expansion in neighboring kebeles.

For the full spatial case, which controls for spatial errors as well as spatial lags, we have

$$\Delta C = \rho W_1 \Delta C + \alpha_1 C + \alpha_2 C^2 + X\beta + \epsilon$$

where W_1 is a spatial weights matrix indicating the strength of influence that each neighbor has, and where

$$\epsilon = \lambda W_2 \epsilon + u$$

with u being independent and identically distributed (i.i.d.) normal with a mean of 0 and a variance of σ^2. W_2 is also a spatial weights matrix that may possibly be, and in our case is, equal to W_1.

Rearranging and substituting in the equation for the full spatial case gives us

$$\Delta C = (I - \rho W_1)^{-1} \alpha_1 C + (I - \rho W_1)^{-1} \alpha_2 C^2 + (I - \rho W_1)^{-1} X\beta + (I - \rho W_1)^{-1} (I - \lambda W_2)^{-1} u$$

9 This particular specification derives from a dynamic model of a representative farm in which there is an increasing cost in clearing land and which has a production function with decreasing marginal returns in land, except that by allowing for a quadratic specification, we acknowledge the possibility that for very low levels of cropland, the function might have increasing returns.

where I is the identity matrix. From this we note that

$$E(\Delta C) = (I - \rho W_1)^{-1}\alpha_1 C + (I - \rho W_1)^{-1}\alpha_2 C^2 + (I - \rho W_1)^{-1}X\beta$$

since the expected value of the residual is 0.

To solve for $E(\Delta C) = 0$, we can multiply the right of the previous equation by $I - \rho W_1$ to give us

$$0 = \alpha_1 C + \alpha_2 C^2 + X\beta$$

The terms involving the weights matrix drop out because all spatial units are assumed to have ΔC approaching 0. This is the key difference in comparison with the full spatial case where, if ΔC for the neighboring spatial units are not approaching zero, the computation is much more complicated.

We can use the quadratic formula to solve for the C at which each woreda converges. Note that the value of $X\beta$ will be different for each woreda, and therefore the value of C that it converges to will also be different for each woreda.

$$C_{max} = \frac{-\alpha_1 \pm \sqrt{\alpha_1{}^2 - 4\alpha_2 X\beta}}{2\alpha_2}$$

The following section describes the results and discusses the potential consequences of Ethiopia's cultivable land scarcity in the highlands.

Results

Table 2.6 presents the results of running the regression with spatial parameters restricted to non-negative values.[10] Table 2A.3 gives the median, 10th percentile, and 90th percentile values of the regression covariates. The parameters for the linear and quadratic terms for the proportion of cropland are highly significant and quantitatively large. The maximum change in cropland occurred in kebeles that in the initial 2001–2004 period had only approximately 5 percent of land under crops. A kebele with 5 percent of its land dedicated to cropland is likely to expand by 6 percentage points more than a

10 We ran this in Stata with the *spregress* command, using the generalized spatial two-stage least-squares estimator, treating errors as heteroskedastic. Before running the full spatial model, we ran it with spatial error only and spatial lag only. Both times the spatial parameters were positive and significantly different from 0 and also significantly less than 1. Running it as a full unrestricted model, we found the spatial lag parameter to be positive, but the spatial error parameter to be negative but not significantly different from 0. Therefore we opted to report the results of the model with only a spatial lag.

TABLE 2.6 Factors associated with change in proportion of kebele area in cropland in Ethiopia

Variable	Parameter	Standard error	
Initial cropland proportion	0.0173	0.0108	
(squared)	−0.1594	0.0140	***
Population density in 2000, persons per square kilometer	−5.05E-06	1.01E-06	***
(squared)	6.64E-11	1.82E-11	***
Precipitation (June–September) (meters)	2.08E-04	1.32E-05	***
(squared)	−1.27E-07	7.99E-09	***
Coefficient of variation for precipitation	0.2171	0.0215	***
Maximum temperature (June–September), °C	1.16E-03	2.85E-04	***
Belg cropping (two growing seasons), 0/1	−0.0096	0.0016	***
Elevation (kilometers)	2.14E-05	7.10E-06	***
(squared)	5.64E-11	1.70E-09	
Terrain roughness measure	−1.13E-04	1.20E-05	***
Travel time to town with population of more than 20,000 (hours)	−8.79E-04	2.90E-04	***
Reduction in travel time to town	−1.61E-04	3.67E-04	
National park, 0/1	−0.0323	0.0077	***
Intercept	−0.1210	0.0140	***
Rho (spatial lag parameter)	0.7417	0.0292	***

Source: Authors' calculations.

Note: Standard errors are in parentheses. *** indicates $p < 0.01$; ** indicates $p < 0.05$; and * indicates $p < 0.10$. Missing values for standard errors were due to negative elements on the diagonal of the inverted Hessian.

kebele with approximately two-thirds of its land under crops. Compared with kebeles with 30 percent of initial land under crops, a kebele with 5 percent of its land dedicated to cropland increases the share of area cropped by 1 percent more over the study period. Although both population density parameters are highly statistically significant in the regression results, there is little quantitative difference in cropped area expansion between low and moderate population densities, with slightly higher expansion rates at low population densities.

Precipitation is also an important influencer of the likelihood of increasing cropped area. Cropped area expansion increases with greater rainfall until reaching approximately 820 millimeters of rain during the main growing season from June to September, then falls thereafter. The difference between the driest kebele and a kebele with the optimal rainfall (820 millimeters) between the two periods is 8.5 percent greater cropland conversion. Similarly, the difference between the optimal and the wettest kebele is 7.8 percent of cropland expansion within a kebele. The optimal rainfall for conversion is very close to the median rainfall of kebeles in our study area of 713 millimeters over the four-month period.

It is not clear why higher coefficients of variation in rainfall lead to higher rates of conversion. This variable is highly negatively correlated with total rainfall, so perhaps it is in part reacting to how rainfall influences conversion of land to agriculture. Overall, however, the difference in cropland conversion levels between kebeles at the 10th percentile and the 90th percentile levels of rainfall variance is 3 percent of total kebele area. The quantitative significance of this variable is modest for explaining differences in conversion rates, except in kebeles with extremely high measures for this variable.

High temperature, although significant within the regression output, does not constrain cultivated areas in a large manner. Many crops are sensitive to high temperatures and generally hotter areas are less likely to be cultivated. However, in Ethiopia the hottest month of the year does not occur during the growing season. Therefore, it is not surprising that the temperature parameter had little quantitative effect on cropland conversion rates.

Although our regression results suggest cropland expansion in *belg* areas with two potential growing seasons is significant, this parameter has a very small effect on overall cropland expansion, amounting to a land conversion rate of 0.9 percent of total kebele area slower than non-*belg* areas. One possible reason might be that converting new land takes labor, and *belg* areas are likely to have less labor availability than other areas due to the need for agricultural labor in two seasons instead of just one.

As expected, elevation and terrain are important factors in cropland area conversion. In general, higher elevations are associated with greater area converted to cropland. The difference from a kebele with an elevation at the 10th percentile to a kebele at the 90th percentile (a difference of 1,300 meters in elevation) is associated with a change of 3 percentage points in cropped area within the kebele. This variable is an important indicator of differences in cropland conversion rates between the highlands and lowlands (see Table 2A.3 for 10th percentile, 90th percentile, and median values of the covariates). As average terrain roughness increases (measured first as the range of elevation within a 1 kilometer grid cell, then averaged over all grid cells in the kebele), crop area expansion decreases. This makes intuitive sense, because it is more difficult to cultivate on hilly terrain, making the land less desirable for conversion. Comparing the flattest to the hilliest kebele leads to a 6 percent difference of total kebele area converted to cropland during the study period; that is, rougher terrain areas, such as hilly areas, have 6 percent less land converted to cropland. Comparing the 10th percentile to the 90th in terrain roughness shows a 2.5 percent difference in the share of total land in a kebele converted to cropland.

Access to markets is important for the profitability of engaging in agricultural work. The farther a kebele is from town, the lower the probability that land will be converted into cropland. For each hour increase in travel time to a city of at least 20,000 people, the probability of expanding cropland decreases by 0.09 percent. The median kebele is 4.4 hours away. The median kebele will convert at a rate of 0.4 percent of total kebele area more slowly than a kebele just outside of an urban center. This is a relatively small effect and generally tells us that, contrary to expectations, change in cropland is not highly influenced by proximity to markets.

The parameter for the change in travel time to the nearest urban center due to improved roads between the two periods was not statistically significant. This was surprising, because reducing travel time should increase farmgate prices and reduce farmgate costs, giving farmers incentives to increase production. Even if we treated the parameter as being statistically significant, the difference in places with no change in travel time and those with 34 hours improvement in travel time appears to be only a half of a percent difference in conversion rates. It may be that the two different roads datasets that we used to produce this value had too much noise in them (that is, inaccuracies), that the change in travel time did not reflect the true change in travel over time.

Finally, all other things being equal, national parks appear to lower the percentage of conversion to cropped land by 3 percentage points. The national parks variable is a binary indicator created using spatial data of protected park areas within the country. As expected, under effective national park management, protected areas would represent "islands" of limited land conversion as seen in the regression output.

We recognize that conversion of land to cropland is a dynamic process, but it is a process that can reach a steady state value that is dependent on the characteristics included in the regression. A steady state would be reached in each kebele when the dependent variable—the change in cropland—would become zero. Thus we consider at what level of cropland percentage in each kebele—given the relatively fixed characteristics such as elevation, hilliness, and climate—would the change in cropland be zero. We use the regression parameters in Table 2.6 with the quadratic formula presented earlier. The only changes we make in the variables is that we allow C, the percent under cropland, to reach its steady state level—that is, we solve for C_{max}. We also assume that at a steady state the change in travel time to the nearest urban center between periods is zero.[11]

Figure 2.3 demonstrates the solution from the application of the quadratic formula to determine C_{max}. Notably, the highlands, particularly near Addis Ababa and along roads leading to Addis Ababa, have the highest projected steady state values—the highest maximum cropland percentages in the country. These levels extend to the northern border of Ethiopia but follow a relatively narrow band centered on the primary north-south highway. Several areas have little or no potential for conversion of land into cropland. Most of these are on the western border of Gambella region and selected kebeles of Benishangul-Gumuz as well as kebeles in eastern Oromia bordering Somali region.[12]

We compare the current reality with the calculated steady state solution. In Figure 2.4 we subtract the data in Figure 2.2 from the data in Figure 2.3 to do so. In our landcover dataset we observed many kebeles reducing the amount of cropland over time (Table 2.5). Therefore it is not surprising that Figure 2.4 shows a number of kebeles that have exceeded what we calculated as their maximum cropped area potential. Such kebeles are in both the highlands and along the borders with the Somali region. Kebeles might exceed the maximum cropped area because of the errors inherent in the statistical estimates but could also be because of several viable explanations on the ground. First, the profitability of the land could be lower than what farmers assessed it to be due to prices or soil fertility changing over time—or even a miscalculation by the farmers themselves when clearing the land. Second, the labor force size could have decreased in the kebele, reducing the optimal quantity of land required. Third, there could be shifting of optimal crops, which may cause a reduction in land needed. Fourth, forests or agroforests could have become more attractive due to prices for their products or value of their services to the households.

Despite the number of kebeles that are at or past their calculated maximum cropland area, there are also a lot of areas that are substantially below this cropland saturation point. These areas, without any intervention, could easily expand cropland in coming years. Appropriate policies could be

---

11 Some woredas had negative values for the term inside the square root of the quadratic formula. For those, we adjusted the estimate of $X\beta$ so that the square root equaled zero and therefore were assigned a maximum cropland area of 5.4 percent.

12 Although the maximum cropland area appears to be near to zero in kebeles in the border area between Oromia and Somali regions, it is important to note that this is a reflection of a lower bound constraint on the regression parameter.

FIGURE 2.3 Map of calculated maximum cropland area in Ethiopia

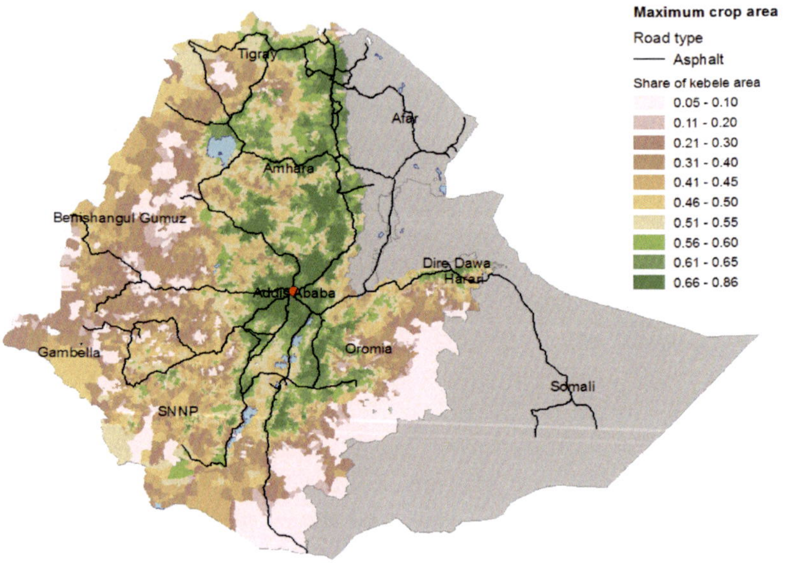

Source: Authors' calculations using MODIS Land Cover type product.
Note: SNNP = Southern Nations, Nationalities, and Peoples.

implemented to facilitate cropland expansion in these kebeles, such as improving connectivity and investment in rural towns.

Table 2.7 summarizes the results presented in Figure 2.3 and Figure 2.4 by highlands and lowlands. The highly cultivated highlands could convert an extra 9.8 percent of their total kebele areas to cropland. Currently, just more than 40 percent of the total land area is in cropland, so this would involve increasing cropland by approximately one-fourth. However, the lowlands could expand by almost 20 percent of their total area. Since in the lowlands currently only 16 percent of the land is under crops, this would represent an increase of cropland area in the lowlands by 117 percent.

Seen from another perspective, the percentage of land area in cropland is more than 150 percent higher in the highlands than in the lowlands. Moreover, the potential for putting land under crops is also higher in the highlands than in the lowlands but by a lesser factor of 60 percent. However, given the current state of cropland area in Ethiopia, the percentage of potential expansion is near 60 percent higher in the lowlands than in the highlands.

We wanted to see how sensitive the results were to assuming that maximum cropland area was 5.4 percent for those with very low values for *XB*

FIGURE 2.4 Map of cropped area expansion potential in Ethiopia

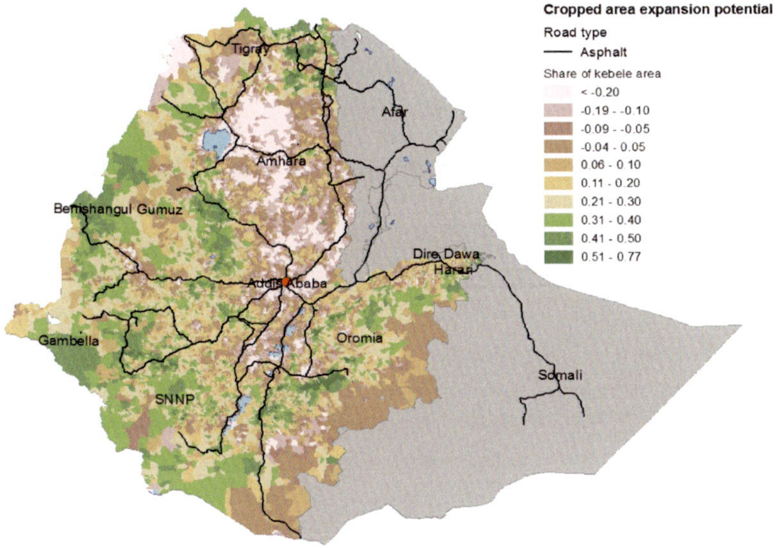

Source: Authors' calculations using MODIS Land Cover type product.
Note: SNNP = Southern Nations, Nationalities, and Peoples.

TABLE 2.7 Cropland, current area and potential area, by highlands and lowlands in Ethiopia

	Percentage of total area		
Region	Current cropland	Potential cropland	Potential expansion
Lowlands	11.6	29.3	17.7
Highlands	35.9	47.1	11.1
Ethiopia	26.4	40.1	13.7

Source: Authors' calculations using MODIS Land Cover type product.

(a measure of profitability). As an alternative, we assumed the maximum cropland area was zero (a lower bound to the upper bound assumption of 5.4 percent). We found small reductions in the aggregate, with potential cropland at 27.7 percent for the lowlands, 46.8 percent for the highlands, and 39.3 percent for the whole country.

We also experimented with different functional forms. First, we endeavored to keep the contribution of the profitability measure in the positive realm by exponentiating *XB* and linearizing the effect of cropland and rate of change in cropland. We did this with the *nl* command in Stata, so we were no longer able to control for any spatial autocorrelation. While the absolute level of predicted conversion was greater with this specification than with the original quadratic, the general results held that there was greater expansion potential in the lowlands than in the highlands and that there are kebeles that have exceeded the predicted maximum cropland.[13]

Potential to Increase the Steady State Cropland Maximum

It is not possible for policymakers to change the climate or terrain sufficiently to change the maximum steady state value of cropland potential for each kebele. However, it is possible to implement policies that better connect rural areas to markets. In the regression results presented in Table 2.6, this idea is proxied in the travel time to cities of 20,000 or more people. The idea is not to force the establishment of towns of this size per se, but rather to increase the access of farmers to places where they can buy inputs and household items and sell their farm produce and other items. Access to information and other amenities that urban areas bring to nearby rural areas also influences productive farmland potential.

To assess the potential for agricultural expansion given improvements in accessibility, we simulate the effect of improved access to markets by equating any kebele that had a travel time to the nearest urban area of more than three hours to be equivalent to three hours. We use three hours' travel time as a threshold for peri-urban areas assuming return travel to a market from this time distance would require a day of travel and a maximum for frequent commuters (Kedir, Schmidt, and Waqas 2016). We then re-evaluated *C*, the proportion of total land in a kebele under crops, using the equation for steady state. We subtract the cropland expansion potential under a simulated environment of all kebeles being within three hours travel time to a city of at least 20,000 from the current steady state of cropped area presented in Figure 2.3. The result of this simulation is presented in Figure 2.5.

Our simulation results suggest a potential for huge increases in cropped area in the western border areas of Ethiopia that previously had no potential. This simply points to the importance of connectivity to establishing the potential for putting land under crops, in addition to important biophysical features that play a key role in defining the upper bound of cropland potential.

---

13 Two other variations were also tried. They involved logitizing *XB* and keeping linear slopes, and logitizing both *XB* and slope. The qualitative results were the same even though the predicted quantities of maximum cropland varied.

FIGURE 2.5 Map of potential increase in steady state of cropped area by improving connectivity to markets

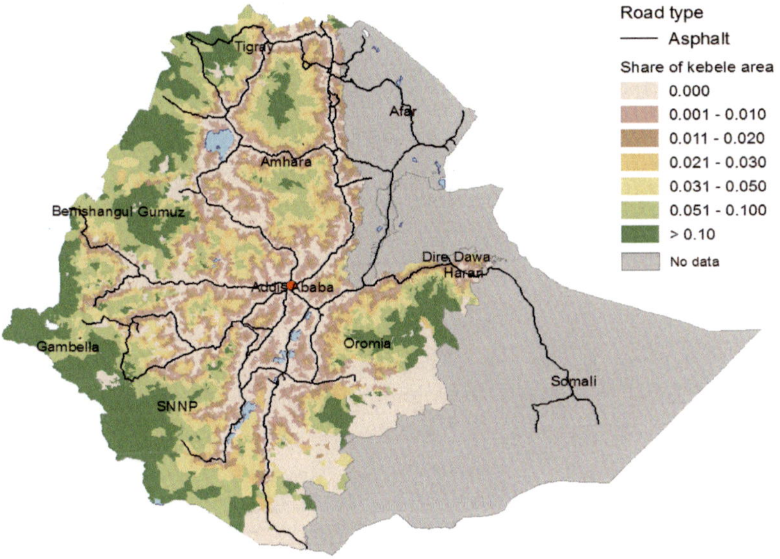

Source: Authors' calculations.
Note: SNNP = Southern Nations, Nationalities, and Peoples.

We also note in Figure 2.5 that there would be very little change in the area of land that potentially could be put under crops in much of the highlands and particularly along major highways, as these areas already are reasonably well connected.

Table 2.8 shows the differences such a scenario would make between the highlands and lowlands. Column A provides the maximum share of total area that could be expanded (considering current biological and economic conditions) under the analysis presented in the regression results explained in Table 2.6 and Table 2.7. Assuming improved connectivity, the lowland areas could expand an extra 6.9 percent of the total (maximum) potential for cropland reported in Table 2.7 by improving travel time for all rural inhabitants to reach a market within three hours (Table 2.8).[14] The highlands would only

14 Taking into account that the lowlands could expand from the current 11.6 percent of land area under crops (Table 2.7) to the maximum potential area at current state (current road

TABLE 2.8 Change in cropland potential with improved connectivity in Ethiopia

	Percentage of total area		
Region	A Potential cropland	B Potential with better connectivity	C Potential expansion
Lowlands	29.3	37.9	8.6
Highlands	47.1	50.3	3.2
Ethiopia	40.1	45.4	5.4

Source: Authors' calculations.

Note: Potential with better connectivity is maximum cropland potential if distance to market is no longer a major constraint. Potential expansion percent is the difference of columns B and A.

increase total cropland area by 3.2 percent based on our model of improved road connectivity. Given that the highlands cropland area is already used and connected to a market of at least 20,000 people (within three hours), improvements in road infrastructure do not have as great an effect on highland areas compared to lowland areas.

Conclusion

Agricultural production in Ethiopia has increased at an impressive rate over the past several decades due to yield increases through improved technology adoption and agricultural area expansion. Most growth occurred in the highlands of Ethiopia, which accounts for 92 percent of total cultivated cereal area. According to satellite data analysis, agricultural area expansion in the highlands is reaching its maximum potential, especially in the drought-prone highland areas. To maintain current agricultural growth rates, it will be important for Ethiopia to think strategically and spatially about future agricultural productivity.

Previous research has highlighted the risk of ongoing population pressure in the agricultural highlands, causing a shrinking of farm sizes with smallholder farmers responding by decreasing fallow periods and using unsustainable cultivation practices. An extensive literature has evaluated soil degradation outcomes due to unsustainable cultivation practices in Africa that

infrastructure) of 29.3, we find the highlands could expand cropland a further 8.6 percent assuming improved road infrastructure. Total expansion possibility under this scenario compared to current state would amount to 17.7 + 8.6 = 26.3 percent greater area under crop agriculture.

leads to decreases in overall productivity (Schmidt et al. 2017; Drechsel et al. 2001; Tittonell and Giller 2013).

Although Ethiopia's crop cultivation potential is largely constrained to the highland plateaus due to reliance on rainfed agriculture processes, specific areas in Ethiopia's lowlands have potential for agricultural area expansion, along with some underused highland areas. In coming years, connecting lowland areas to vital infrastructure will be necessary to create the conditions necessary to hasten agricultural development in such areas. The potential for smallholder expansion in the more remote locations of Ethiopia could be more attractive given a concerted effort in developing transportation or irrigation infrastructure. Meanwhile, in the densely populated highlands, a focus toward agricultural intensification coupled with sustainable land management campaigns will be necessary to maintain the recent agricultural performance experienced by smallholders in the moisture reliable agroecological zones of the highlands.

Appendix 2A: Agroecological Zones, Crop Area Changes, and Regression Covariates

TABLE 2A.1 Characteristics of agroecological zones of Ethiopia

	Highlands			Lowlands	
Characteristic	Drought prone	Rainfall sufficient, cereal	Rainfall sufficient, enset	Drought prone, pastoralist	Humid rainfall sufficient
Rainfall, mean annual (millimeters)	861	1,326	1,320	632	1,144
Elevation, mean (meters)	2,014	2,051	1,919	1,013	997
Travel time to a city with population of more than 20,000 persons, mean, (hours)	3.2	3.3	2.5	8.6	7.1
Population density (persons per square kilometer)	423	324	448	82	224
Population 2016 (millions)					
Total	19.3	34.3	16.7	12.6	3.7
Urban	3.1	5.1	2.5	1.8	0.8
Rural	16.2	29.3	14.3	10.8	2.9
Total land area (square kilometers)	128,115	226,929	65,192	574,305	133,980
Woredas (number)	146	264	126	139	52

Source: Authors' calculation using a variety of remote sensing datasets including Jarvis et al. (2008), Bright, Rose, and Urban (2013), Funk et al. (2015), and Ethiopia, CSA (2013).

TABLE 2A.2 Changes in crop area in Ethiopia, 2001–2004 to 2009–2013 (%)

	Percentage of kebeles with declines more than 6%					
Time Period	Tigray	SNNP	Amhara	Oromia	Other	Total
2001–2004 to 2005–2009	37.8	20.3	13.3	20.8	10.1	19.3
2005–2009 to 2009–2013	19.6	28.4	26.5	13.5	4.2	19.8
2001–2004 to 2009–2013	26.7	25.1	12.6	14.1	9.4	16.8
Average annual growth in crop area (%)						
2001–2004 to 2005–2009	2.7	1.2	3.9	2.2	–1.7	2.7
2005–2009 to 2009–2013	0.1	–0.9	0.6	2.5	3.6	1.0
2001–2004 to 2009–2013	1.2	0.0	2.1	2.4	1.2	1.8
Total 2001–2013	4.2	0.1	8.2	3.9	0.2	3.8

Source: Authors' calculations.
Note: SNNP = Southern Nations, Nationalities, and Peoples.

TABLE 2A.3 Regression covariates, 10th percentile, median, and 90th percentile values

Indicator	10th percentile	median	90th percentile
Initial cropland proportion	0.020	0.303	0.669
Population density in 2000 (persons per square kilometer)	17.875	94.581	348.71
Precipitation (June–September) (meters)	395.71	712.607	1,101.2
Coefficient of variation for precipitation	0.086	0.150	0.225
Belg cropping (two growing seasons), 0/1	0	0	1
Maximum temperature (June–September), °C	21.963	25.419	30.988
Terrain roughness measure	34.839	115.514	257.100
Elevation (kilometers)	1,344	1,936	2,654
Reduction in travel time to town	0.017	1.25	7.1
Hours to town of 20,000+	1.183	4.4	13.083
National park, 0/1	0	0	0

Source: Authors' calculations.

References

Bachewe, F. N., G. Berhane, B. Minten, and A. S. Taffesse. 2015. "Agricultural Growth in Ethiopia (2004–2014): Evidence and Drivers (No. 81)." Washington, DC: International Food Policy Research Institute (IFPRI).

———. 2018. "Agricultural Transformation in Africa? Assessing the Evidence in Ethiopia." *World Development* 105: 286–298.

Bezu, S., and S. T. Holden. 2014. *Rural-Urban Youth Migration and Informal Self-Employment in Ethiopia.* Report 01/2014. Bergen, Norway: Norwegian University of Life Sciences, Center for Land Tenure.

Bright, E. A., A. N. Rose, and M. L. Urban. 2013. *Landscan 2012 Data.* Oak Ridge National Laboratory. Accessed February 4, 2019. https://landscan.ornl.gov/downloads/2012.

Brink, A. B., and H. D. Eva. 2009. "Monitoring 25 Years of Land Cover Change Dynamics in Africa: A Sample-Based Remote Sensing Approach." *Applied Geography* 29: 501–512.

Chamberlin, J., T. S. Jayne, and D. D. Headey. 2014. "Scarcity amidst Abundance? Reassessing the Potential for Cropland Expansion in Africa." *Food Policy* 48 (October): 51–65. Special Issue on "Boserup and Beyond: Mounting Land Pressures and Development Strategies in Africa." http://dx.doi.org/10.1016/j.foodpol.2014.05.002.

Cleaver, K. U., and G. A. Schreiber. 1994. *Reversing the Spiral: The Population, Agriculture, and Environment Nexus in Sub-Saharan Africa.* Washington, DC: World Bank.

Constable, M. 1985. *Ethiopian Highlands Reclamation Study: Summary (draft).* UTF/ETH/037/ ETH Working Paper 24. Rome.

Deininger, K., B. Derek, L. Jonathan, N. Andrew, S. Harris, and S. Mercedes. 2011. *Rising Global Interest in Farmland: Can It Yield Sustainable and Equitable Benefits?* Washington, DC: World Bank.

Diao, X., A. Kennedy, O. Badiane, F. Cossar, P. Dorosh, O. Ecker, H. Ghebru, D. Headey, A. Mabiso, T. Makombe, M. Malek, and E. Schmidt. 2013. *Evidence on Key Policies for African Agricultural Growth.* IFPRI Discussion Paper 1242. Washington, DC: IFPRI.

Drechsel, P., L. Gyiele, D. Kunze, and O. Cofie. 2001. "Population Density, Soil Nutrient Depletion, and Economic Growth in Sub-Saharan Africa." *Ecological Economics* 38 (2): 251–258.

Ethiopia, CSA (Central Statistical Agency). 2002. *Ethiopian Agricultural Sample Enumeration.* Addis Ababa.

———. 2013. *Population Projection of Ethiopia for All Regions at Wereda Level from 2014–2017.* Addis Ababa.

———. 2016. Various years. *Agricultural Sample Survey Volume 1: Report on Area and Production of Major Crops (Private Peasant Holdings,* Meher *Season); and Volume 5: Report on Area and Production of* Belg *Season Crops for Private Peasant Holdings.* Addis Ababa.

FAO (Food and Agriculture Organization of the United Nations). 1986. *Ethiopian Highlands Reclamation Study.* Final Report for Government of Ethiopia. Rome.

Friedl, M. A., D. Sulla-Menashe, B. Tan, A. Schneider, N. Ramankutty, A. Sibley, and X. Huang. 2010. "MODIS Collection 5 Global Land Cover: Algorithm Refinements and Characterization of New Datasets." *Remote Sensing of Environment* 114: 168–182.

Funk, C., P. Peterson, M. Landsfeld, D. Pedreros, J. Verdin, S. Shukla, G. Husak, et al. 2015. "The Climate Hazards Infrared Precipitation with Stations—A New Environmental Record for Monitoring Extremes." *Scientific Data* 2: 150066. http://dx.doi.org/10.1038/sdata.2015.66.

Gebregziabher, G., and S. T. Holden. 2011. "Distress Rentals and the Land Rental Market as a Safety Net: Evidence from Tigray, Ethiopia." *Agricultural Economics* (42): 45–60.

Hamza, M. A., and W. K. Anderson. 2005. "Soil Compaction in Cropping Systems: A Review of the Nature, Causes, and Possible Solutions." *Soil and Tillage Research* 82 (2): 121–145.

Headey, D., M. Dereje, and A. S. Taffesse. 2014. "Land Constraints and Agricultural Intensification in Ethiopia: A Village-Level Analysis of High-Potential Areas." *Food Policy* 48: 129–141. http://dx.doi.org/10.1016/j.foodpol.2014.01.008.

Holden, S. T., and B. Shiferaw. 2002. "Poverty and Land Degradation: Peasants' Willingness to Pay to Sustain Land Productivity." In *Natural Resource Management in African Agriculture: Understanding and Improving Current Practices*, edited by C. B. Barrett, F. M. Place, and A. A. Aboud, 91–102. Oxon and New York: CABI Publishing in Association with International Centre for Research in Agroforestry.

Hurni, H. 1998. *Agroecological Belts of Ethiopia: Explanatory Notes on Three Maps at a Scale of 1:1,000,000.* Addis Ababa: Ethiopia, Ministry of Agriculture; Berne, Switzerland: Centre for Development and Environment.

Jarvis, A., H. I. Reuter, A. Nelson, and E. Guevara. 2008. "Hole-Filled Seamless SRTM Data V4. International Centre for Tropical Agriculture (CIAT)." Accessed July 8, 2019. http://srtm.csi.cgiar.org.

Josephson, A. L., J. Ricker-Gilbert, and R. J. Florax. 2014. "How Does Population Density Influence Agricultural Intensification and Productivity? Evidence from Ethiopia." *Food Policy* 48: 142–152.

Kedir, M., E. Schmidt, and A. Waqas. 2016. "Pakistan's Changing Demography: Urbanization and Peri-urban Transformation over Time." Pakistan Strategy Support Program Working Paper 39. Washington, DC: IFPRI.

Leroux, L., A. Jolivot, A. Bégué, D. L. Seen, and B. Zoungrana. 2014. "How Reliable Is the MODIS Land Cover Product for Crop Mapping Sub-Saharan Agricultural Landscapes?" *Remote Sensing* 6: 8541–8564.

Moller, L. C. 2012. "The Ethiopian Urban Migration Study 2008: The Characteristics, Motives and Outcomes to Immigrants to Addis Ababa." Washington, DC: World Bank.

Pankhurst, A., and F. Piguet. 2009. "Displacement, Migration and Relocation: Challenges for Policy, Research and Coexistence." In *Moving People: Displacement, Development and the State in Ethiopia*, edited by F. Piguet and A. Pankhurst, 246–264. Oxford, UK: James Currey.

Patil, P., and M. K. Gumma. 2018. "A Review of the Available Land Cover and Cropland Maps for South Asia." *Agriculture* 8 (111).

Schmidt, E., and F. Bekele. 2016. *Rural Youth and Employment in Ethiopia.* ESSP II Working Paper 98. Washington, DC: IFPRI; Addis Ababa: Ethiopian Development Research Institute (EDRI).

Schmidt, E., P. Chinowsky, S. Robinson, and K. Strzepek. 2017. "Determinants and Impact of Sustainable Land Management (SLM) Investments: A Systems Evaluation in the Blue Nile Basin, Ethiopia." *Agricultural Economics* 48 (5): 613–627.

Schmidt, E., and B. Zemadim. 2015. "Expanding Sustainable Land Management in Ethiopia: Scenarios for Improved Agricultural Water Management in the Blue Nile." *Agricultural Water Management* 158: 166–178.

Taddesse, G. 2001. "Land Degradation: A Challenge to Ethiopia." *Environmental Management* 27 (6): 815–824.

Tittonell, P., and K. E. Giller. 2013. "When Yield Gaps Are Poverty Traps: The Paradigm of Ecological Intensification in African Smallholder Agriculture." *Field Crops Research* 143: 76–90.

Tschopp, R., A. Aseffa, E. Schelling, and J. Zinsstag. 2010. "Farmers' Perceptions of Livestock, Agriculture, and Natural Resources in the Rural Ethiopian Highlands." *Mountain Research and Development* 30 (4): 381–390.

Van Ittersum, M. K., L. G. Van Bussel, J. Wolf, P. Grassini, J. Van Wart, N. Guilpart, L. Claessens, H. de Groot, K. Wiebe, D. Mason-D'Croz, and H. Yang. 2016. "Can Sub-Saharan Africa Feed Itself?" *Proceedings of the National Academy of Sciences* 113 (52): 14964–14969.

Viste, E., D. Korecha, and A. Sorteberg. 2013. "Recent Drought and Precipitation Tendencies in Ethiopia." *Theoretical and Applied Climatology* 112 (3–4): 535–551.

Chapter 3

CROP PRODUCTIVITY AND POTENTIAL

Guush Berhane, Bart Minten, Fantu Bachewe, and Bethelhem Koru

Introduction

Ethiopia has registered an average annual GDP growth rate of 10.3 percent over the 2005/2006–2015/2016 period (Bezawagaw et al. 2018), a remarkable achievement for a country with little dependence on natural resources such as oil or minerals. Critical to these changes were strong political commitments that put agriculture at the center of Ethiopia's development agenda since the 1990s (Berhane, Bachewe, and Minten 2018). Spearheaded by public investments, GDP growth has been largely driven by growth in the agricultural sector (Bachewe et al. 2018; Hill and Tsehaye 2018). For more than two decades, Ethiopia's growth strategy has remained agriculture-focused, as shown by a budget exceeding the CAADP agriculture investment target of 10 percent of the national budget (Mellor 2014; AGRA 2018). Within agriculture, crop productivity has received substantial attention as significant investments were made in its extensive extension system and in ensuring access to modern inputs (Berhane et al. 2018). Parallel investments in roads, safety nets, education, and health have also contributed to subsequent recovery and turnaround of the sector.

Largely driven by public investments and favorable economic conditions, Ethiopia's total value of smallholder crop output more than doubled—from 14 million metric tons in 2004/2005 to 32 million metric tons in 2015/2016. Average crop output grew between 8 percent to 13 percent a year and cereals accounted for a lion's share of the total crop output growth. In the same period, land under cultivation has expanded by about 27 percent, 90 percent of which was used for cereals (but later declined and leveled off), and average cereal yield has increased by about 5 percent a year. Output growth was attributed to land expansion as well as yield growth (Bachewe et al. 2018). However, despite high growth rates, Ethiopia's yield levels remain rather low and they are showing signs of slowing down recently. A key question is therefore whether Ethiopia can sustain productivity increases to achieve further agricultural transformation in the face of declining additional cultivable land and the recent slowing down of yields.

This chapter tries to answer this question by assessing the patterns, sources, and prospects of crop productivity in Ethiopia. We discuss the sources of crop output growth and assess the role of land, labor, technological inputs (for example, fertilizer, improved seeds, extension, and so on) as well as total factor productivity (TFP), a key measure of sustained long-term growth. We provide an overview of key policy achievements, focusing on fertilizers, improved seeds, credit, the extension system, irrigation, and mechanization. We outline existing structural bottlenecks for future growth. We contextualize the discussion by comparing Ethiopia's recent achievements with selected recent experiences of African (Burundi, Kenya, Malawi, Niger, Nigeria, Rwanda, Sudan, Tanzania, and Uganda) and Asian (China, India, Indonesia, Malaysia, and Thailand) countries—the latter when they were at about the same level of development as Ethiopia is now.

We find that expansion in cultivated land and labor have, respectively, accounted for about 13 percent and 30 percent of Ethiopia's observed crop output growth in the past decade. Expansion in use of chemical fertilizers, improved seeds, and extension were also important contributors. Moreover, TFP growth has contributed to about 18 percent of output growth in the same decade, suggesting efficiency gains of investments in the production process. However, TFP growth has declined in most recent years (from 38 percent in the beginning of the period to 16 percent at the end of the period), contributing to the recent slowdown in yield growth. The decline has continued despite continued investments in TFP increasing nonconventional inputs such as extension and chemical fertilizers. The contributions of land and labor inputs have also remained roughly the same, suggesting additional gains in and keeping up with sustained productivity is likely to be increasingly difficult. We discuss the most important bottlenecks behind this slowdown and highlight potential ways going forward.

An Assessment of Crop Production Patterns

Ethiopia's investment in agriculture since 2004/2005 has mainly focused on crop production. As a result, the lion's share of agriculture's contribution to GDP growth comes from growth in crop production (Bachewe et al. 2018). In the decade between 2004/2005 and 2013/2014, Ethiopia's agriculture sector on average grew by 7.6 percent annually, which accounted for 3.6 percent of the 10.7 percent real GDP growth in this period. The agricultural sector has on average contributed 47 percent to real GDP, 32 percent of which was accounted for by the crop subsector.[1]

This section focuses on this important subsector in Ethiopian agriculture (the livestock subsector is dealt with in a separate chapter). We begin with assessing trends in the real value of crop output (valued using the monthly producer prices of Ethiopia's Central Statistical Agency [CSA] and adjusted for inflation by regional price indexes) and report on growth rates in production, area cultivated, and yield. We discuss trends for major food grains in the main production season (called *meher*), while also highlighting the types of crops grown in Ethiopia. The data used are from CSA's annual crop output reports (Ethiopia, CSA Volume 1, 2005–2016) computed based on annual agricultural sample surveys.[2]

We update the evidence provided in Bachewe et al. (2018) to the most recent years for which data are available and discuss the potential and constraints for further growth in this sector in the years ahead. To help frame the discussion on further potential, we put the Ethiopian experience into an international context.

Trends in Crop Output, Area Cultivated, and Yield

Ethiopia's total value of crop output grew from 14 million metric tons in 2004/2005 to 36 million metric tons in 2017/2018—that is, an average growth rate of 9 percent per year (Table 3.1). Average crop output grew between 8 percent to 10 percent a year in the early years of the decade but had slightly fallen to 5 percent in 2011/2012 and then rose to about 10 percent to 13 percent annually in the last three years of the decade. Cereals accounted for about 72 percent of the total crop output and grew by 8 percent a year. Maize is among the top drivers of this growth in cereals (Abate et al. 2015). Fruit crops also grew rapidly—that is, by an average of 12 percent a year—although growth was sporadic over the 11 years and fruit crops are only grown by a few holders in relatively small pockets of the country. Pulses and oilseeds—some of them used for exports—have also performed well, particularly toward the end of the decade, suggesting the increasingly important contributions of commercial crops. However, compared to cereals, their share

1 Agriculture remains a major employer in the Ethiopian economy, accounting for 80 percent of all workers in 2005 and 77 percent in 2013 (Bachewe et al. 2018).

2 The analysis in this chapter, unless indicated otherwise, uses data from the annual smallholder agricultural production survey collected by Ethiopia's Central Statistical Agency (CSA). Recent investments by large commercial farmers mainly in the lowlands has limited spillovers to smallholders (Ali, Deininger, and Harris 2019) and is separately treated in Chapter 7.

TABLE 3.1 Crop output in Ethiopia (million quintals)

Crop	2004/ 2005	2005/ 2006	2006/ 2007	2007/ 2008	2008/ 2009	2009/ 2010	2010/ 2011	2011/ 2012	2012/ 2013	2013/ 2014	2014/ 2015	2015/ 2016	2016/ 2017	2017/ 2018	2018/ 2019
All crops	142.4	156.2	171.9	186.1	198.7	215.5	246.4	258.9	291.6	319.7	338	321.1	352.7	366.9	378.2
Grains	119.1	133.8	149.6	161.2	171.2	180.8	203.5	218.6	231.3	251.5	270.4	266.8	290.4	306.1	315.6
Cereals	100.3	116.2	128.8	137.2	145	155.3	177.6	188.1	196.5	215.8	236.1	231.3	253.8	267.8	277.6
Pulses	13.5	12.7	15.8	17.8	19.6	19	19.5	23.2	27.5	28.6	26.7	27.7	28.1	29.8	30.1
Oilseeds	5.3	4.9	5	6.2	6.6	6.4	6.3	7.3	7.3	7.1	7.6	7.8	8.4	8.6	7.9
Vegetables	4.3	4.5	3.5	4.7	6	5.6	6.8	7.6	8.5	7.2	6	7.4	8.1	7.4	8.9
Root crops	16.2	13.4	14.1	15.3	12.1	18.1	19.2	16.7	36.3	41.6	54.6	40	46.3	45.6	45.4
Fruit crops	2.6	4.3	4.6	4.6	3.5	4.1	4.9	5.4	4.8	5	7.1	6.8	7.9	7.8	8.3

Source: Authors' computation using CSA annual reports (Ethiopia, CSA volume 1, 2004–2019).

in total grain output remains low. Roots crops, a common staple in southern parts of Ethiopia, have become a dynamic subsector in recent years, doubling in production from 1.8 million metric tons in 2009/2010 to 3.6 million metric tons in 2012/2013, and then increasing to 4.2 million metric tons in 2013/2014. Vegetables have also achieved notable growth—with a growth rate of 6 percent a year—albeit at fluctuating rates over the decade considered. The latter could be due to changes in access to irrigation, which varies across years depending on availability of moisture as well as the increasing incidences of diseases.

Growth in output has been accompanied by growth in crop yields (defined as the ratio of value of output to area cultivated). Table 3.2 reports crop yields for the period 2004/2005 to 2018/2019 (see Table 3.3 for area cultivated). In general, yield has substantially increased (about 6 percent annually) for all crops over the 12 years considered, albeit from a low base. Among grains, cereals have seen the highest yield growth rate—almost doubling from 11.8 quintals per hectare in 2004/2005 to 26.8 quintals per hectare in 2018/2019—followed by pulses and oilseeds.[3] Among the nongrains, oilseeds register the lowest yield levels while root crops saw the highest yield levels—for example, 195.9 quintals per hectare in 2018/2019—followed by fruits at 60 quintals per hectare to 70 quintals per hectare. However, these latter crops constitute only a tiny share of the total cultivated land in the country, and the area covered by these crops has stagnated over the period considered.

Given the importance of cereals in Ethiopia's crop agriculture, we present statistics on the dynamics of individual crops separately. Bachewe et al. (2018) show that during the 2004/2005–2014/2015 period, cereals, composed of five important crops—teff, maize, wheat, sorghum, and barley—accounted for 63 percent of real value of crop output (and contributed the lion's share of the smallholder agriculture economy in Ethiopia) and that the extent of crop diversification remained stable over the decade considered. The four major cereal crops have seen an average annual yield growth rate of 18 quintals per hectare for 15 consecutive years (Figure 3.1). However, yield growth varies by crop. Maize stands out with the highest average annual yield growth (25.5 quintals per hectare), followed by wheat (18.9 quintals per hectare), barley (15 quintals per hectare), and teff (12.4 quintals per hectare). Clearly maize has seen an expanding share of production within staples in the same period and is now the second-most widely cultivated crop in Ethiopia (Abate et al. 2015).

3 A quintal is equal to 100 kilograms.

TABLE 3.2 Crop yields in Ethiopia (quintals per hectare)

Crop	2004/ 2005	2005/ 2006	2006/ 2007	2007/ 2008	2008/ 2009	2009/ 2010	2010/ 2011	2011/ 2012	2012/ 2013	2013/ 2014	2014/ 2015	2015/ 2016	2016/ 2017	2017/ 2018	2018/ 2019
Cereals	11.8	12.7	14.4	15.3	16.7	16.9	17.8	20	21.6	21.4	23.4	23.4	24.8	26.2	26.8
Barley	10.4	10.2	11.2	12	12.9	13.1	13.9	13.6	14.6	15.8	19.7	19.7	21.1	21.6	21.8
Maize	14.2	17.9	19.4	18.3	20.1	19.4	22.1	25.9	27.5	27.6	34.3	33.9	36.8	39.4	39.9
Teff	8.5	8.1	8.6	10.8	11.4	11.5	11.7	11.6	12.4	13	15.8	15.6	16.6	17.5	17.6
Wheat	13.1	11.8	13.5	15	15.4	17.3	16.1	17.2	18.9	21.3	25.4	25.4	26.8	27.4	27.6
Pulses	8.7	8.5	9.5	10.5	12	11.8	13.2	13.1	13.7	14.5	17.8	17.3	18.2	18.6	18.6
Oilseeds	5.4	4.8	5.5	8.2	8.2	7.8	8.6	8.1	8.4	8.7	9.5	9.8	10.4	10.1	10.5
Vegetables	57.2	46.6	43.3	50.4	50.6	53.6	58.4	67.6	64.3	63	59.5	37.2	33.9	35.4	36.9
Root crops	98	78	75.3	86.2	87	88.4	96.7	92.3	163.2	151.2	273.1	199.9	202.1	195.5	195.9
Fruits	46.9	86.2	83.3	67.2	67.3	73.3	68.1	79.1	69.5	59.5	70.7	68	73.4	74.5	69.6

Source: Authors' computation using CSA annual reports (Ethiopia, CSA volume 1, 2004–2019).

FIGURE 3.1 Yield growth of maize, wheat, barley, and teff, 2004/2005–2017/2018

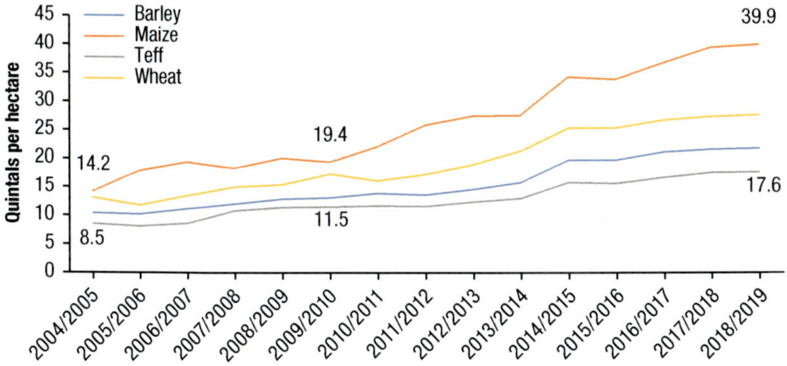

Source: Authors' computation using Ethiopia, CSA annual reports (Ethiopia, CSA 2005–2018).

While the aggregate crop output has more than doubled over the decade under consideration, the number of holders also grew from 11 million in 2004/2005 to 15.7 million in 2017/2018.[4] An increase in the number of holders given limited available land resources means that additional output gains from expanding cultivable land has declined (Headey, Dereje, and Taffesse 2014). Table 3.3 reports trends in cultivated area covered by grains, vegetables, root crops, and fruits (see also Figure 3.2). There are three important points to note from Table 3.3. First, considering all crops, an additional 3.1 million hectares of land has been cultivated between 2004/2005 and 2018/2019, a 31 percent increase from land cultivated in 2004/2005. Second, about 90 percent of the cultivable land added is used for cereals, and land under cash crops except for pulses (which increased by about 30 percent) has only slightly increased, or the increase is insignificant given the low base (as in the case of vegetables). Third, expansion in cultivable land was rapid at the beginning of the decade—about 3.7 percent—but that expansion has quickly dwindled toward the end of the decade. While yield growth is positive but variable and with a declining trend (see trend line in Figure 3.2), expansion in cultivated acreage has eventually leveled off. The widening declining trend lines for yield and area expansion in Figure 3.2 further suggest that improvements in use of modern inputs and the extension system were not sufficiently compensating for

4 A holder is a person who exercises management control over the operation of his/her plots (Ethiopia, CSA 2014/2015).

TABLE 3.3 Cultivated area in Ethiopia (millions of hectares)

Crop	2004/ 2005	2005/ 2006	2006/ 2007	2007/ 2008	2008/ 2009	2009/ 2010	2010/ 2011	2011/ 2012	2012/ 2013	2013/ 2014	2014/ 2015	2015/ 2016	2016/ 2017	2017/ 2018	2018/ 2019
All crops	10.1	10.5	11	11.3	11.6	11.9	12.3	12.6	12.8	12.9	12.8	12.8	13.2	13.2	13.3
Grains	9.8	10.2	10.6	11	11.2	11.5	11.8	12.1	12.3	12.4	12.6	12.4	12.6	12.7	12.7
Cereals	7.6	8.1	8.5	8.7	8.8	9.2	9.7	9.6	9.6	9.8	10.1	10	10.2	10.2	10.4
Pulses	1.3	1.3	1.4	1.5	1.6	1.5	1.4	1.6	1.9	1.7	1.6	1.7	1.5	1.6	1.6
Oilseeds	0.8	0.8	0.7	0.7	0.9	0.8	0.8	0.9	0.8	0.8	0.9	0.9	0.8	0.8	0.7
Vegetables	0.1	0.1	0.1	0.1	0.2	0.1	0.1	0.2	0.2	0.2	0.1	0.2	0.2	0.2	0.2
Root crops	0.2	0.2	0.2	0.2	0.1	0.2	0.2	0.2	0.2	0.2	0.2	0.2	0.2	0.2	0.2
Fruit crops	0.05	0.05	0.0[illegible]	0.01	0.05	0.01	0.01	0.01	0.01	0.01	0.01	0.01	0.1	0.1	0.1

Source: Authors' computation using Ethiopia, CSA annual reports (Ethiopia, CSA volume 1, 2005–2018).

FIGURE 3.2 Growth in area cultivated and yield of grains (%)

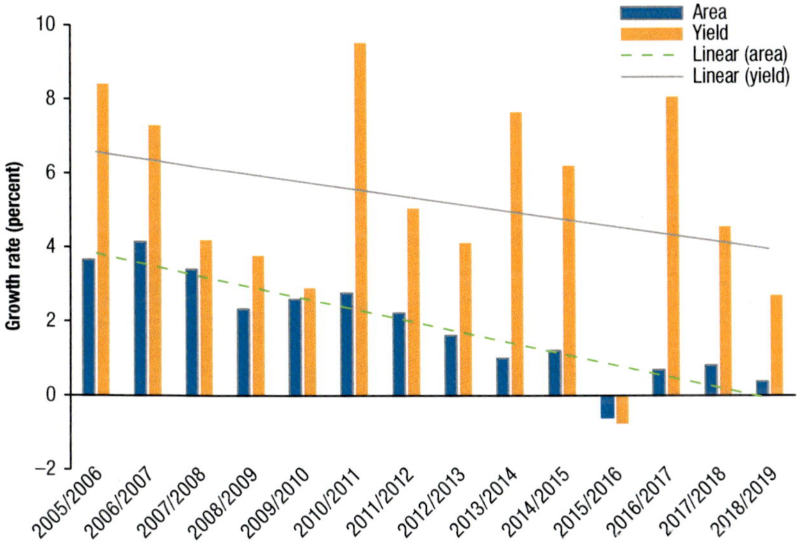

Source: Authors' computation using CSA annual reports (Ethiopia, CSA volume 1, 2004–2016).

declines in yield growth due to declines in area expansion (that is, given that the downward-sloping trend line for yield growth follows that of area expansion).

Yield Growth in Ethiopia Compared to Other Countries

For more than two decades, Ethiopia's overall agricultural growth surpassed the CAADP target of 6 percent growth (AGRA 2018).[5] Cereal yield growth has been significant in the most recent 12 years of the two decades (that is, it grew by 4.9 percent a year), making Ethiopia among the best recent performers in Africa south of the Sahara (Mellor 2014). An important feature of Ethiopia's agriculture growth is yield growth, unlike that of many other Africa south of the Sahara countries whose growth heavily relied on expansion of area under cultivation and not yield (see Nin-Pratt 2016; Fuglie and Rada 2013; Bachewe et al. 2018; Mellor 2014).

To put this in perspective, we compare Ethiopia's yield growth in the period 2001–2012 to selected African countries for which comparable data from the other countries is available. Figure 3.3 reports the annual average

5 CAADP is the Africa-wide strategy developed by the African Union, where all heads of state of African countries have agreed to allocate at least 10 percent of their spending on agriculture (African Union 2010).

FIGURE 3.3 Land productivity in selected countries in Africa, 2001–2012 (annual average growth rate, %)

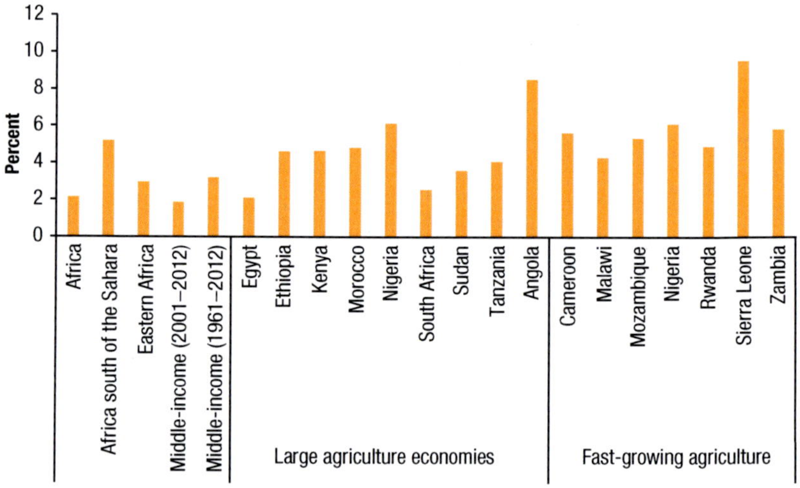

Source: Compiled from Benin and Nin-Pratt (2016).

Note: Large agriculture economies has at least 3 percent of Africa's total agricultural output; fast-growing agricultural economies surpass the CAADP agricultural growth rate target of 6 percent per year.

growth rate by group of African countries, mainly countries with middle-income status, and those that are deemed fast-growing as well as those with a large agricultural population share. With an average annual growth rate of 4.6 percent (only for the years for which we have data for other countries), Ethiopia clearly stands below the SSA average but surpasses the middle-income average of 3.2 percent per year for the period 2001–2012. It also performed well compared with many of the large agriculture as well as resource-rich and fast-growing agriculture economies except Angola, Nigeria, and Sierra Leone (which Ethiopia outperformed in most recent years). It is to be noted that these data are a bit outdated for Ethiopia, and its growth rate has gone up significantly since the years considered.

However, despite these recent high growth rates, yield levels are still rather low compared with what has been observed in many other contexts, particularly for crops in which Ethiopia has significant potential. Maize and wheat are among crops with high potential in Ethiopia given the promising fertilizer-improved seed responses observed for both crops in recent years and availability of relatively larger land suitable for both crops in the wheat and maize growing areas of the country (Abate et al. 2015; Mellor 2014). A cursory look at yield levels of these two important crops in Ethiopia against other countries

FIGURE 3.4 Maize and wheat yield levels and growth rates for selected countries, 2004–2013

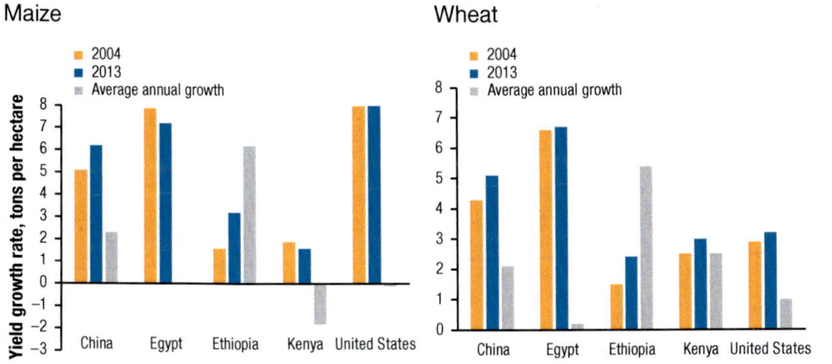

Source: Adapted from Bachewe et al. (2018).

can help shed light on this issue. Figure 3.4 shows yield levels and yield growth rates for maize and wheat for selected countries for the period 2004–2013. In 2004 maize yields in Ethiopia were respectively 25 percent and 20 percent lower than that of Egypt and the United Sates. This gap had declined by 2013 when Ethiopia's maize yield levels increased to 44 percent and 33 percent of that of Egypt and the United States, respectively. Clearly, although Ethiopia's maize and wheat yield growth rates have exceeded any of the countries presented in this discussion, including China (which has registered remarkable growth rates in both crops), Ethiopia's yield levels still remain low, suggesting a substantial yield gap that Ethiopia can catch up with.

Trends in Productivity and Implications for Agricultural Transformation

Sustained increase in agricultural productivity is the cornerstone for a successful agricultural transformation (Hazell 2009; Tomich, Kilby, and Johnson 1995). In agrarian economies like Ethiopia, decades of continued investments and government strategic leadership are needed for sustained productivity increases to achieve successful transformation (Hazell 2013). The descriptive evidence presented earlier suggests that Ethiopia's crop agriculture is on the move in the past decade. A fundamental question is what the key sources are of the recent output growth and whether such growth trends can be sustained into the future to achieve agricultural transformation in Ethiopia.

A key challenge in the agricultural transformation of Africa south of the Sahara is that growth has relied on expansion of the area under cultivation and not yield (Jayne, Chamberlin, and Benfica 2018; Nin-Pratt 2016; Fuglie and Rada 2013). In contrast, the yield (partial productivity) evidence presented so far shows Ethiopia's agricultural growth is associated with both area expansion and yield growth. However, comparisons with selected countries indicate that Ethiopia's yield growth occurs from a rather low base and as such yield levels lag far behind those required to achieve agricultural transformation as observed in Asia or elsewhere. Thus questions remain regarding the extent to which the recent takeoff can be sustained and whether Ethiopia is on the right track to transform its agricultural sector as envisioned by Ethiopia's Growth and Transformation Plans.

TOTAL FACTOR PRODUCTIVITY

To get better insights into these questions, we look at how Ethiopia's crop-agriculture has fared in terms of total factor productivity (TFP), a more comprehensive measure of overall performance than, for example, yield growth. TFP is a favored measure of long-term productivity growth, as sustained increase in TFP is considered the core driver of sustainable agricultural transformation (Laborde et al. 2018). TFP is estimated residually as the rate of change in output not explained by the rate of change in conventional input use such as labor, land, and capital and TFP growth suggests that the same or fewer inputs can produce more outputs through more efficient ways of production, including through adoption of new technologies or shifts to better production methods (Sheng, Ball, and Nossal 2015). In short, "changes in TFP can be interpreted as a measure of the collective contributions of nonconventional inputs in agriculture, such as improvements in input quality, introduction of better ways of doing things, leading to technical change and economies of scale" (Fuglie 2004: 1; Ruttan 2002). TFP is often further decomposed into finer measures of *technical change*—referring to the movement of the technological frontier; *efficiency change*—referring to movement of individual farmers toward or away from the technological frontier; and *scale-efficiency change*—referring to movement of farmers above the technological frontier to capture economies of scale (Benin and Nin-Pratt 2016). We come back to these further refinements later in the chapter, and we focus on the decomposition of productivity into changes in TFP and other factors involved.

We use 12 years of CSA data on crop output, land, labor, fertilizers, improved seeds, irrigation, extension services, pesticides, and other factors. Bachewe et al. (2018) used a modified growth accounting model (Solow 1957) to decompose growth in real value of crop output into changes in input use, changes in TFP, and other exogenous factors that affect output growth. This study follows the same method and extends the analysis using two additional data points that have become available after that study. However, the following important caveats are in order when interpreting TFP results using this method. First, it is assumed that markets are competitive, and production is represented by an aggregate crop production function that can be expressed as a relationship between inputs and outputs where output levels increase with increases in these inputs. Second, due to lack of data, not all factors that can affect this production relationship are included in the analyses. These include, farmers' use of organic fertilizers, mechanization, and improved land and water management practices, as well as other exogenous factors, such as weather changes. Insofar as these factors are not included in the analyses, changes in TFP obtained from the analyses also include the effects of those factors (Sumner 2014).

We provide results of the growth accounting analyses (described above) in Figure 3.5. It presents the average contributions of inputs considered and TFP to crop output growth over the period 2004/2005–2015/2016. On average, change in labor, land, fertilizer, improved seeds, and access to extension has, respectively, accounted for 30 percent, 13 percent, 11 percent, 12 percent, and 7 percent of real output growth over the 11 years considered.

All other factors—namely irrigation, pesticides, and rural roads—had relatively smaller contributions, which is expected given the limited roles these still have in Ethiopian agriculture. Note, however, that in this method roads measure direct effects only but they indirectly affect modern input use, which were important drivers of growth. The remaining 18 percent growth in real output is accounted for by a change in TFP. These results are consistent with those reported in Bachewe et al. (2018), except that the TFP contributions have declined from 22 percent—in 2004/2005–2013/2014 to 18 percent in 2004/2005–2015/2016.[6] As depicted in Figure 3.6, TFP growth has continued to decline substantially, slowing output growth from 15 percent a year in 2004/2005 to 7.2 percent in 2014/2015 and 3.4 percent in 2015/2016. When further disaggregated by a period of three to four years (see Figure 3.7), it is clear that average TFP contribution to output growth has decreased significantly from 38 percent in 2004/2005–2008/2009 to 16 percent in the most recent years of 2013/2014–2015/2016.[7] It is important to note that TFP growth has continued to decline despite the fact that traditionally

6 Details of the contributions of each factor are presented in Table 3A.1 and Table 3A.2.

7 Figure 3.7 shows relative contributions of inputs and TFP to output growth disaggregated by growth episodes of three to four years in the decade.

FIGURE 3.5 Average contributions of factors and total factor productivity to crop output growth, 2004/2005–2015/2016

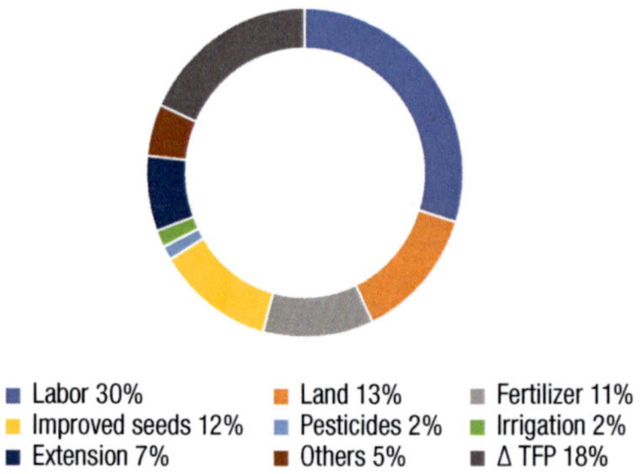

Source: Authors' computations using CSA data (Ethiopia, CSA volumes 1, 2, and 3, 2004/2005–2015/2016) and EDRI (2009).
Note: Others include rural roads and other capital investments.

FIGURE 3.6 Trends in total factor productivity and output growth, 2004/2005–2015/2016

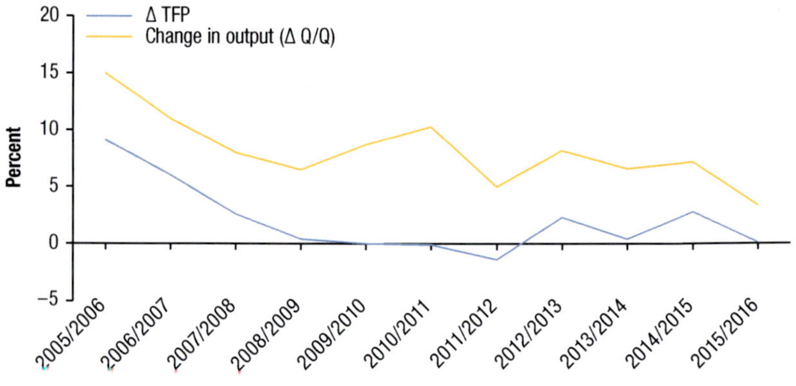

Source: Authors' computations using CSA data (Ethiopia, CSA volumes 1, 2, and 3, 2004/2005–2015/2016) and EDRI (2009).

TFP-increasing technological (nonconventional) inputs such as extension, use of fertilizer and pesticides, investments in roads, and irrigation have increased. The contributions of traditional inputs (land and labor) have also slightly increased or roughly remained the same over these periods.

A number of reasons can be given for this observed dynamic in TFP. First, Berhane et al. (2018) find that the extension system has increased productivity indirectly through its effects on increasing adoption of modern inputs but not directly through its (knowledge transfer) effects on improving

FIGURE 3.7 Relative contributions of inputs and total factor productivity to output growth (%)

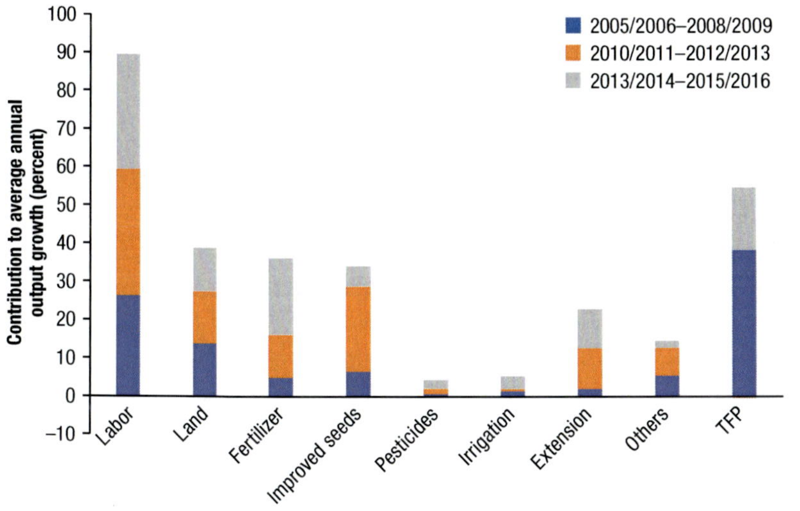

Source: Authors' computations using CSA data (Ethiopia, CSA volumes 1, 2, and 3, 2004/2005–2015/2016) and EDRI (2009).
Note: Others include rural roads and other capital investments.

better use of these technologies and modernizing the overall farming systems.[8] Once the low-hanging fruits in productivity increases (at a rather low base) are exhausted through this basic support, achieving additional productivity (TFP) gains becomes increasingly difficult. Second, Ethiopia's investment on agricultural R&D as a share of agriculture GDP is one of the lowest even among African peers (Laborde et al. 2018). As a result, the link between research and the extension system is weak (Berhane et al. 2018), limiting long-term productivity gains ultimately leading to declining TFP. Third, although significant capital investments have been made on roads and other physical infrastructures in recent years, poorly developed value chains and weak integration with central markets for a number of crops prevent farmers from realizing the full productivity potential—as in the form of changing production organization in farming—of such investments (Laborde et al. 2018). Fourth, despite the initial focus on agriculture-led development and the commitments

8 Ethiopia has put in place one of the largest extension systems in Africa (Davis et al. 2010). However, the system has remained largely under resourced and poorly linked with research, and as a result has rather focused more on channeling of modern inputs and less on providing knowledge-based support to farmers (Berhane et al. 2018). See also Abate et al. (2019) and Makhija et al. (2019) for discussions of video-mediated agricultural extension in Ethiopia.

to invest in agriculture in the early years, there has been a recent slowdown in Ethiopia's efforts to support agriculture (as evidenced by the relative budget decline as share of GDP in recent years and the shift in focus to manufacturing and infrastructure sectors) (Laborde et al. 2018).

Sources of Productivity Growth and Implications for the Future

The sources of recent productivity growth in Ethiopia's crop agriculture cannot be explained without understanding the initial conditions. Prior to the 1990s, Ethiopia's crop agriculture was characterized by a prolonged period of misguided policies that led to shrinking production, both in terms of cropped area and input intensity (Benin and Nin-Pratt 2016). The sector transited through a period of recovery and turnaround after major reforms following the change of government in the 1990s. Subsequent policies have had significant implications to growth in this sector. As such, the growth episodes and sources of growth following these reforms can be characterized in three periods (see also Mellor 2014): the period of "intensification of traditional practices" immediately after the reform and early policy efforts to reduce poverty; the period of "modern input expansion" following the agriculture-led development strategy of the 1990s and subsequent major investment plans, namely the Plan for Accelerated and Sustained Development to End Poverty (PASDEP); and the most recent years of the Growth and Transformation Plan (GTP) (Ethiopia, MoFED 2006, 2010). A centerpiece in the PASDEP was intensification of smallholder agriculture through tailored interventions to promoting technology packages of seed and fertilizers, supported by credit and field-level extension system and a network of demonstration centers. The GTP also deepened the focus on smallholder agriculture with, among others, additional emphasis on commercialization of high-value crops, support to smallholder irrigation, and better use of groundwater (more on this below).

The policy change in the early 1990s contributed to an impetus for recovery and turnaround, which led to a realization of early high growth. In other words, the peace dividend gave rise to deployment of more and better-quality crop husbandry practices, as in the form of increased labor intensity in land preparations, weeding, and harvesting. This was low-hanging fruit that led to an early growth momentum in the late 1990s to early 2000s, attained through intensification by using traditional labor-intensive practices using family labor on small farms (Mellor 2014). This sustained later growth, which was accompanied by strong public sector support of more than 10 percent of public spending toward agriculture, among the highest rates of spending on the continent.

In 1995 an ambitious frontline extension system was launched focusing on increasing cereal productivity through participatory demonstration and training (Berhane et al. 2018). In the early years the extension system, among others, was tasked with the coordination and delivery of packages of modern technologies for smallholders, mainly improved seeds and fertilizers along with financial services tied with these inputs.[9] The system also served as the nearest government agency to smallholders, providing series of meetings, motivational trainings, and on-farm demonstrations.

These efforts led to increased use of fertilizers, improved seeds, and pesticides, and to husbandry intensification, which ultimately resulted in increases in land use efficiency. Increased gains from land efficiency also meant that farmers were incentivized to expand land, including to marginal grazing areas that were not attractive before. A combination of these two factors—mainly efficiency change due to intensification of traditional practices and expansion of land under cultivation—contributed to productivity increases in the early years (Bachewe et al. 2018). For example, Benin and Nin-Pratt (2016) decompose TFP growth in Ethiopia for the period 1985–2012 (the closest period to this discussion) and find that more than 90 percent of TFP in this period was explained by *efficiency change*, suggesting that *technical change* contributed little to agriculture growth during this period.

In the remainder of the chapter, we look in detail at a number of input factors deemed important for crop productivity growth. We discuss agricultural extension and R&D spending, chemical fertilizers, improved seeds, agrochemicals, access to credit, irrigation, mechanization, and implementation capacity.

Agricultural Extension and R&D Spending

The productivity gains in the 1990s and the potential productivity gap motivated further investment in agriculture in the early to mid-2000s. Two consecutive ambitious five-year plans—namely PASDEP I (2006–2010) and PASDEP II (2011–2014)—were launched to, among other goals, sustain and accelerate productivity growth in agriculture. Technology packages were designed and introduced under the Participatory Demonstration and Training Extension System (PADETES) and implemented through a national agricultural intervention program. This has set forth the beginning of a large

9 Berhane et al. (2018) provide a detailed description of Ethiopia's extension system and its operations.

extension system where frontline extension agents are deployed to every kebele specializing in agronomic management practices.[10] Extension agents follow a "training and visit" type of extension approach, providing basic agronomic support and information dissemination. This is backed by a team of "subject matter specialists" stationed at the district level, tasked with providing technical support and supervision of the extension agents (for details, see Berhane et al. 2018). The PASDEPs have focused on an ambitious increase of the extension agents, deploying 63,000 of them to reach millions of farmers throughout the country. Figure 3.8 shows that both the number of smallholders reached and the cultivated area that benefited from the extension system tripled between 2004/2005 and 2015/2016. This system is hailed as one of the largest and most extensive public extension systems in Africa, at least in terms of extension agent-farmer ratio. In 2010 the extension agent-farmer ratio was estimated at one agent per 476 farmers—equivalent to 21 extension agents per 10,000 farmers (Davis et al. 2010).

Comparable figures for Tanzania stood at 2,500 farmers per one agent—that is, four agents per 10,000 farmers and 16 agents per 10,000 farmers in China (Davis et al. 2010). By 2016/2017 this estimate for Ethiopia went up to 46 agents per 10,000 farmers. Official records indicate that in 2017 there were around 72,000 extension agents deployed in the country, three to four agents per kebele specializing in crop production, livestock, natural resource management, irrigation agriculture, and veterinary services. Moreover, the country had planned to establish about 15,000 farmer training centers (FTCs), one per kebele, but in 2010 only about 30 percent were fully functional (Davis et al. 2010).

While these are impressive achievements of the extension system, recent studies (see Berhane et al. 2018) suggest that the system is weak in terms of the technical and knowledge support it offers to farmers, partly due to the sheer size it covers, making it difficult for frontline agents to reach farmers all over the country, characterized by complex and diverse problems. Key issues in the extension system are the lack of resources for extension agents, that they are not sufficiently trained to provide the required technical support, and that they are overburdened by nonagricultural activities limiting the amount of time left for technical support (Berhane et al. 2018). An important factor is also that the system is poorly linked with national agricultural research centers whose incentive structures remain misaligned to provide the required knowledge solutions to farmers.

10 Kebele is the lowest administration unit, composed of a few villages in Ethiopia.

FIGURE 3.8 Number of smallholders and cultivated area covered with extension package and number of beneficiary farmers

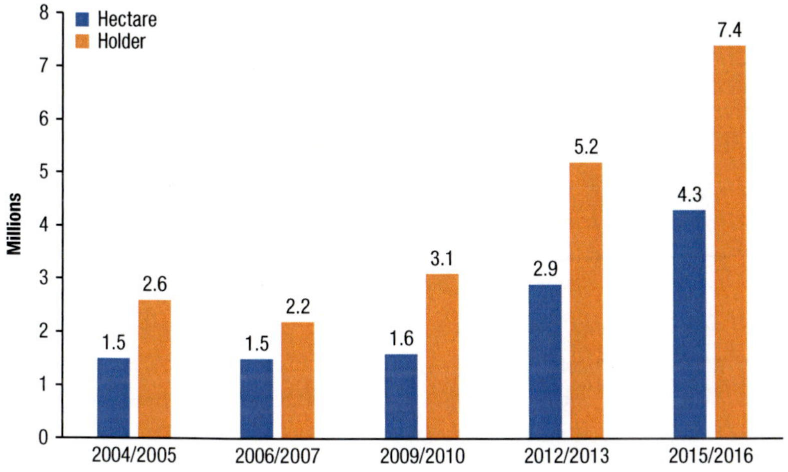

Source: CSA annual Agricultural Sample Survey household level data (Ethiopia, CSA 2004–2016).

A recent study finds that access to the extension system significantly contributed to the increased adoption of modern inputs such as chemical fertilizers and improved seeds through disseminating basic agronomic information to farmers but even more so through persuading farmers to adopt technologies largely channeled through government parastatals (Berhane et al. 2018). In other words, the impact on productivity has not been as expected mainly because the extension system has largely focused on facilitating the distribution of modern inputs and less on disseminating new knowledge due to its poor links with research centers. Thus extension has played an important role in expanding fertilizers and improved seeds and basic agronomic knowledge—hence it brought about some *efficiency change*, but not to the extent of bringing about *technical change* needed to move the frontier outward. Owing to this, despite improvements, production systems remain predominantly traditional. A key reason is seemingly the poor feedback loop between extension agents, farmers, and knowledge centers (Berhane et al. 2018).[11]

11 In other words, an extension system poorly linked with research centers means that the real value of extension messages conveyed by agents to farmers got less nuanced and the "effect has wore off" over time (Krishnan and Patnam 2014; Berhane et al. 2018).

These findings appear consistent with the national productivity trends presented earlier, suggesting that productivity growth rates are likely to slow down if the system is not revamped to provide additional technological and scientific support to help farmers advance their farming systems out of current traditional practices.

Ethiopia has also invested a significant amount of resources on agricultural R&D. With inflation-adjusted spending increase of 60 percent between 2011 and 2016, Ethiopia is among the top six SSAs that invested in R&D (Beintema and Haregewoin 2018). The number of agricultural researchers per 100,000 farmers has more than tripled between 2002 and 2014 (from 2.5 researchers per 100,000 farmers in 2002 to about 8 in 2014) and has seen one of the fastest increases (13.5 percent per year) in SSA, although still from a rather low-base as compared to other couturiers below the Sahara. Excluding university research institutes, there are about 62 federal and regional agricultural research centers, well spread across the various agroecologies in the country (Berhane et al. 2018).[12]

An important constraint often cited is that national research centers are underfunded, leaving them at a disadvantage internationally and, as a result, vulnerable to high staff turnovers. Ethiopia's investment on R&D as a share of agricultural GDP has declined in recent years (mainly because the research system has expanded without proportional change in R&D spending), making it one of the bottom nine countries in SSA (Figure 3.9). In 2014, Ethiopia's spending on R&D was only 0.24 percent of agriculture GDP.

In addition, despite recent rapid growth rate, Ethiopia still has one of the lowest number of researchers with PhDs in Africa south of the Sahara relative to its total number of researchers. Moreover, poor incentive systems, lack of proper coordination between centers, and misalignments in the research-extension-farmer feedback loops have rendered such investments ineffective (Deneke and Gulti 2016; Kassa et al. 2016; Davis et al. 2010). These declines in public spending in Ethiopia's R&D are consistent with the general contraction of the share of agriculture expenditures to GDP in recent years but the decline is more prevalent for agriculture (and water), which declined by about 6 percent between 2007/2008 and 2012/2013 (World Bank 2016). Some see this decline as a slowdown of the focus on agriculture in recent years (for example, see Laborde et al. 2018).

12 Indeed, some of the local research has helped generate improved local varieties, including on crops that are unique to Ethiopia (teff) and that receive limited international funding resources (Minten et al. 2013; Minten, Taffesse, and Brown 2018).

FIGURE 3.9 Total agricultural R&D spending as a share of agriculture GDP (%)

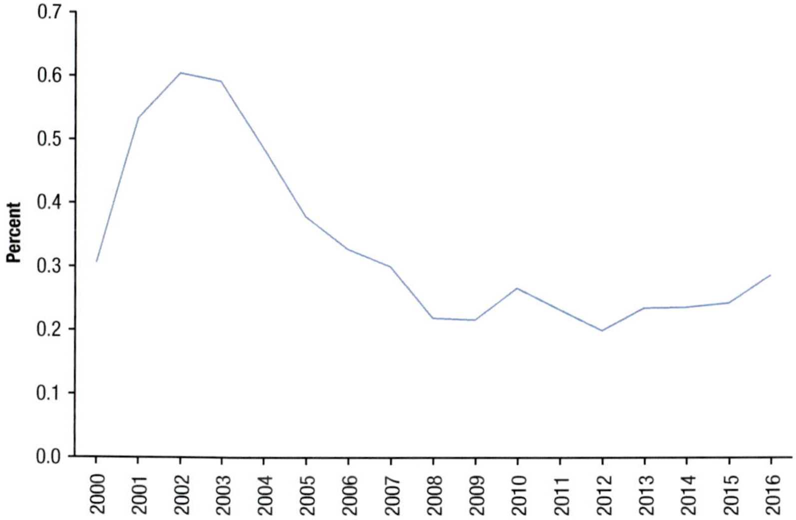

Source: ASTI (2019).

Chemical Fertilizers

More than 50 percent of agricultural productivity gains during the Green Revolution in Asia was due to an increase in adoption rate of inorganic fertilizer alone (Hopper 1993; Sheahan and Barrett 2014). In view of the same objective, Ethiopia's fertilizer imports and use have dramatically increased in the past decade. Average annual fertilizer imports have jumped from 346 thousand metric tons in 2004/2005 to 778 thousand metric tons in 2012/2013—an increase by 124 percent; and fertilizer use by smallholders has also increased by 144 percent over the same period (Bachewe et al. 2018). Fertilizer use has increased for all crops despite variations in application rates and proportion of area fertilized by crop types (for details, see Bachewe et al. 2018). However, consistent with the food security focus in production, most fertilizer use has been on cereals.

According to CSA reports, 4.7 million holders (46 percent) growing cereals used fertilizer in 2004/2005, which grew (by 76 percent) to 10.1 million holders in 2015/2016. Most of the chemical fertilizers are geared toward cereals. Cereal area applied with fertilizer—which nearly doubled during the same period from 2.7 million hectares in 2004/2005 to 5.2 million hectares in 2015/2016, or an increase from 36 percent to 53 percent of the total cereal area—accounted for at least 91 percent of total fertilized area in all years except in 2009/2010.

Figure 3.10 and Table 3A.1 present fertilizer intensity trends by fertilizer types (urea, DAP, urea and DAP, NPS, and NPS and DAP) in quintals per hectare of areas covered with these nutrient types for the period 2004/2005 through 2017/2018.[13] Intensity of areas covered by all fertilizers has increased from 64 kilograms per hectare in 2004/2005 to 148 kilograms per hectare in 2017/2018—an increase by about 131 percent. On those areas covered with urea and DAP, it has increased by 51 percent from 134 kilograms per hectare in 2004/2005 to 203 kilograms per hectare in 2017/2018. Intensity has also increased for those using only DAP (by 62 percent) and for those using only urea (by 34 percent) for the same period. Compared with the traditional urea and DAP blends, the recently introduced blends of NPS as well as NPS and DAP seem to enjoy higher use.

Despite the sharp increase in fertilizer imports and use, application rates remain significantly below recommended rates (of above 200 kilograms of fertilizer per hectare) with only 64 to 162 kilograms per hectare of fertilizer applied on average. Using the Agricultural Growth Program (AGP) data, Berhane and Assefa (2020) calculate that cereal fertilizer yield response in Ethiopia appears to be relatively profitable when more fertilizer is applied (on average, application of 300 kilograms of urea and DAP gives 2,100 kilograms of maize, 1,800 kilograms of wheat, or 1,300 kilograms of teff) (Figure 3.11). This is roughly above 4:1 ratio of additional grain per kilogram of fertilizer. In the same period, Bachewe et al. (2018) using the same data, report that the output-fertilizer (for the same five main cereals) ratios were twice as high in 2012 than in 2014. This implies that cereal productivity can be further enhanced by increasing fertilizer intensity to internationally comparable levels of 250 kilograms per hectare to 300 kilograms per hectare and beyond.

Figure 3.12A and Figure 3.12B further provide comparative fertilizer intensity figures per arable land available in each selected African and Asian

13 Note that figures reported based on CSA data are relatively higher than those from other sources (see, for example, Figure 3.11) because CSA fertilizer use is over fertilized area while those in Figure 3.11 are calculated for total arable land use in the country. The latter is only suggestive as arable land can significantly vary by size of country. Fertilizer application rates per hectares of applied land obtained from other household surveys (for example, AGP datasets) are comparable to the CSA figures. The latter are in the order of 100 to 133 kilograms per hectare (see Berhane, Bachewe, and Minten 2018).

FIGURE 3.10 Fertilizer use trend in Ethiopia, 2004/2005–2017/2018 (quintals per hectare of fertilized area)

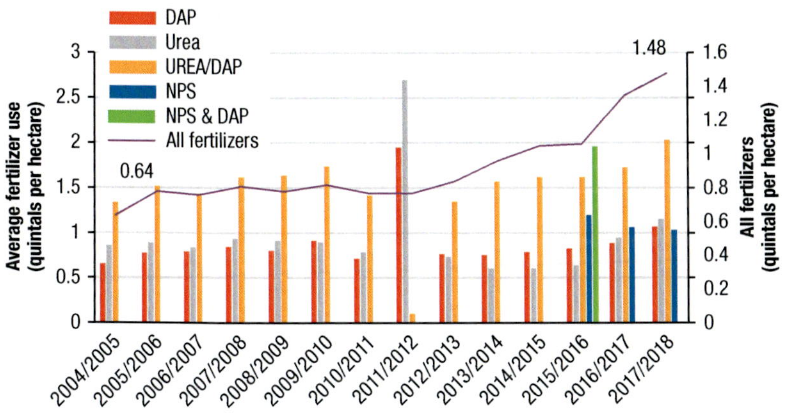

Source: Authors' computation using CSA annual reports (Ethiopia, CSA volume 3, 2004–2018).

FIGURE 3.11 Trends in yield responses to levels of fertilizer intensity for selected crops in Ethiopia

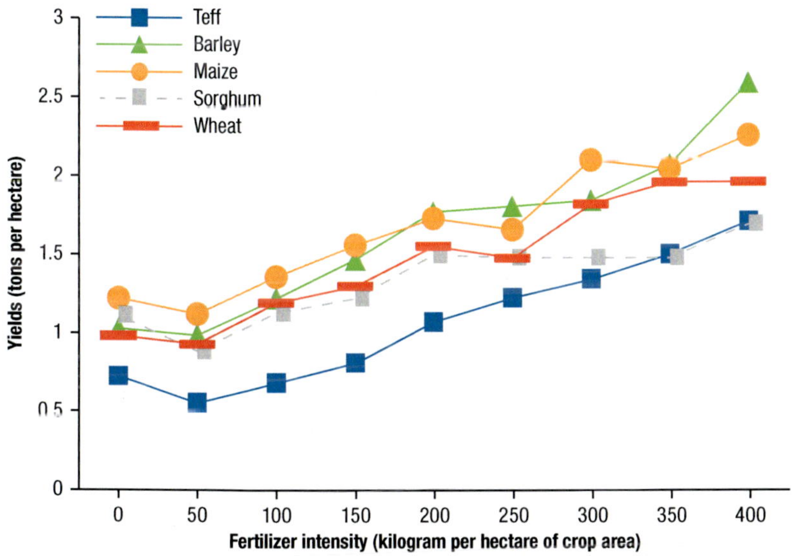

Source: Berhane and Assefa (2020), using the Agricultural Growth Program dataset (USAID 2013).

FIGURE 3.12A Average fertilizer applied to arable land for selected Africa south of the Sahara countries (kilograms per hectare)

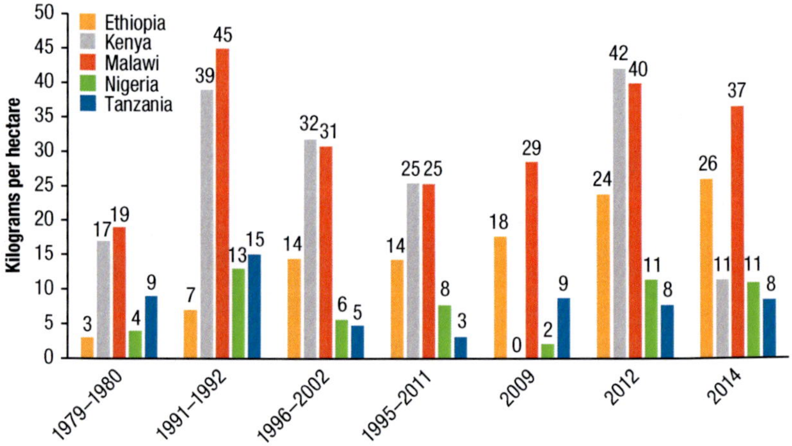

Source: Authors' calculations based on data from ourworldindata.org 2014.

FIGURE 3.12B Average fertilizer applied to arable land for selected Asian countries (kilograms per hectare)

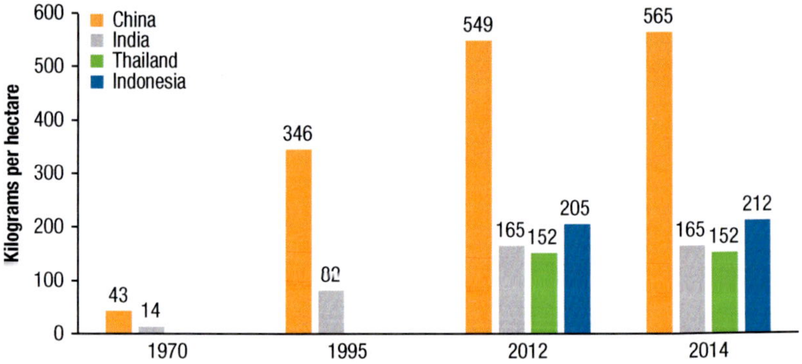

Source: Authors' calculations based on data from ourworldindata.org, 2014.

country respectively.[14] Compared with the selected SSA countries, Ethiopia's performance in applying fertilizer per arable land available is above the rest of SSA countries, except for Malawi, partly because of its large fertilizer subsidy program in recent years (Ricker-Gilbert, Jayne, and Chirwa 2011). This is consistent with findings by Sheahan and Barrett (2014), who indicated that Ethiopia's average kilograms of fertilizer per hectare application rate is slightly above SSA's average. Ethiopia's average application rate in 2014—at 26 kilograms per hectare—was slightly higher than the Asian average of 24 kilograms per hectare at the beginning of the Green Revolution in the 1970s but much lower than that of China—at 40 kilograms per hectare—in that period. Moreover, Ethiopia's average application rate in 2014 is also slightly below India's (82 kilograms per hectare) and China's (346 kilograms per hectare) performance in 1995. In 2014, Ethiopia's average fertilizer use per hectare of arable land is sixfold lower than that of India and Thailand, eightfold lower than Indonesia, and twenty-two-fold lower than that of China (Figure 3.12B). This shows that Ethiopia still has way to go in fertilizer intensification compared with Asia's recent performance in this regard.

While fertilizer adoption improved significantly over the past decade, there are seemingly a number of important constraints to adoption by Ethiopian farmers: (1) lack of availability of a variety of nutrient blends suitable for all soil types given the diversity of soil types in Ethiopia is one important constraint (for example, Croppenstedt, Demeke, and Meschi 2003; IFDC 2015); (2) rainfall variability, localized droughts, and lack of moisture remains another of the challenges facing fertilizer profitability and response rates influencing farmers' risk behavior in using fertilizers (for example, Alem et al. 2010); (3) inefficient logistics, bureaucratic structures, and unpredictable fertilizer-credit supply systems further constrain timely access to fertilizers (Rashid et al. 2013).[15]

Improved Seed

Access to improved seed varieties is a key challenge of agricultural transformation in Africa, largely due to the complexities involved to develop the sector.

---

14 Note that figures reported in Figures 3.12A and 3.12B are kilograms per total hectares cultivated (and not comparable to those in Figure 3.10, which reports kilograms per fertilized hectares). These figures need to be interpreted cautiously as fertilizer use per available cultivable land (not all fertilized) is not an accurate measure of intensity because cultivable land varies by size of country. Nevertheless, in the absence of other comparable accurate measures available, these figures provide some insights into cross-country comparisons of fertilizer intensity.

15 See Rashid et al. (2013).

A key challenge in Ethiopia's seed sector is that it has remained predominantly public sector–dominated, restricting potential involvement by the private sector (Alemu et al. 2007; Spielman et al. 2010; Alemu, Rashid, and Tripp 2010). Nevertheless, Ethiopia has shown progress in this sector as well. Over the decade discussed in this chapter, the number of improved seed varieties released to farmers has increased rapidly due to significant public investments in local capabilities of research and seed multiplication structures. Official data from the public seed enterprises and the Ministry of Agriculture shows that certified seed production has jumped more than five-fold over the past decade (ATA 2017a). Bachewe et al. (2018) indicate that release rates for improved seed varieties have been particularly dynamic for wheat but lower for other crops. In the past ten years about 50 varieties of wheat and at least 20 varieties of each for maize, barley, and teff have been released by Ethiopia's national agricultural research institutes (ATA 2017a).

Reliable national data on use of released varieties is scarce partly because of the challenges to account for reused varieties in national surveys. Reported adoption rates of improved seed varieties by farmers are low overall.[16] Using the CSA data, Table 3.4 reports total cultivated area and percentage of land under purchased improved seeds. The proportion of farmers using purchased improved seeds (and the area covered by improved seeds) has seen significant improvements, with more than a doubling noted for teff, barley, wheat, and maize over the past decade. Large increases in the proportion of farmers adopting purchased improved seeds are noted particularly for maize producers. This is consistent with the yield increases of all crops presented in Figure 3.1, suggesting that some of the yield improvements, mainly maize (followed by wheat) is also associated with the use of improved verities. Abate et al. (2015) show that increased use of improved maize varieties has greatly contributed to doubling of maize production in Ethiopia in the same period, and they have pointed out that despite these changes, opportunities for further maize yield growth still exist in Ethiopia.

The other increases are from a rather low base in 2004/2005 and remain very low both in terms of area and the proportion of land covered by improved seeds. Several factors contribute to this poor performance.

First, despite the large number of seed varieties released by the formal sector, only few of them are adopted and commercialized by seed enterprises and hence many did not reach farmers. A key reason is that variety development

16 Recent studies using DNA fingerprinting to identify reuse rates suggest that reported use of released varieties may underestimate actual improved seed use rates (Yirga et al. 2016).

TABLE 3.4 Total cultivated area in Ethiopia and percentage of land under purchased improved seed

Crop	Total cultivated area under improved seed (hectares)							Percent of land under improved seed						
	2004/ 2005	2006/ 2007	2009/ 2010	2012/ 2013	2015/ 2016	2016/ 2017	2017/ 2018	2004/ 2005	2006/ 2007	2009/ 2010	2012/ 2013	2015/ 2016	2016/ 2017	2017/ 2018
Teff	15,448	13,172	44,755	38,400	59,107	88,194	56,064	0.72	0.55	1.73	1.41	2.06	2.92	1.85
Barley	5,664	3,225	8,698	8,713	11,233	14,687	18,515	0.52	0.32	0.77	0.86	1.19	1.53	1.94
Wheat	53,361	47,953	37,888	70,743	108,126	165,067	111,388	3.82	3.25	2.25	4.35	6.5	9.73	6.56
Maize	221,682	267,981	209,850	673,747	883,576	1,089,899	1,169,817	15.91	15.87	11.84	33.47	41.85	51.04	54.95
Sorghum	6,312	2,541	16,390	711	2,906	5,282	5,805	0.5	0.17	1.01	0.04	0.16	0.28	0.31

Source: Authors' computation using Ethiopia, CSA annual reports (Ethiopia, CSA volume 3, 2004–2018).

is not demand-driven, making it unsuitable to many agroecological contexts, offering low value for money to farmers. Key structural bottlenecks are a low level of national breeding research quality and poor alignment of research incentives with the core transformation agenda. In other words, recent expansion in the quantity of seeds supply has come at the expense of quality deterioration. For instance, only two hybrid maize varieties released about 20 years ago account for about 95 percent of all hybrid maize varieties adopted by farmers (Ali, Deininger, and Harris 2019). Note, however, that recent evidence from DNA fingerprinting suggests that most farmers do not reliably identify the varieties they plant (see also footnote 16). Second, despite significant recent public investments in plant breeding research and multiplication, low level of coordination between research centers themselves and the extension system as well as poorly organized distribution mechanisms remain key bottlenecks. More specifically, a centralized seed need assessment and supply system executed by a lengthy bureaucratic system extending from the Ministry of Agriculture to public seed enterprises and cooperatives renders the seed sector value chain ineffective (ATA 2010, 2017b). Third, while many studies indicate improved seeds are profitable when adopted with other complementarity technologies including fertilizers and irrigation (for example, Abay et al. 2018), such synergistic values are low in Ethiopia due to limited access to irrigation infrastructures and optimized fertilizer blends.

LAND TENURE

In Ethiopia land is state owned, and holders are given only usufruct rights, including the right to rent and sharecrop it for a fixed number of years. Subsequent policies introduced to discourage distress leasing of land and associated transaction costs have further complicated land rental markets (Deininger et al. 2008; Ghebru, Holden, and Tilahun 2019). In addition, with increasing population pressure, the youth and vulnerable groups are faced with increased constraints to access land (Headey, Dereje, and Taffesse 2014; Gebrehiwot and Holden 2019). Relaxing land rental market restrictions and delinking land user rights from land use would encourage farmers with small plots, or those who do not have enough resources to operate the land, or those who have skills and resources that could more efficiently be applied in other sectors, to transfer their use rights to potential tenants (Ghebru 2010; Holden, Otsuka, and Deininger 2013; Bezu and Holden 2014; Haddis 2013).

Moreover, formalizing land rights are often considered as avenues to land tenure security and increases in land market participation (Holden,

Deininger, and Ghebru 2007). A large-scale land certification program was set up in the country in the 1990s to ensure more secure land property rights. This land certification program has been one of the largest, cheapest, and fastest in Africa (Deininger et al. 2008)—it is estimated that about half of farmers in the main four regions benefited from this certification program (Ghebru, Koru, and Taffesse 2016). While land remains the property of the state, the certificates have ensured more secure property rights, since they are found to encourage higher investments, more land rental market activity, higher productivity, and improved food security (Holden, Deininger, and Ghebru 2007; Deininger, Ayalew, and Alemu 2011; Ghebru and Holden 2013; Melesse and Bulte 2015). As this land certification program in most areas happened before the period under study here, it might not have directly contributed to increased productivity in the past decade, but it seems that these more secure property rights helped to provide the necessary required conditions for the increase in agricultural productivity seen afterward.

However, a number of issues are still problematic for land tenure. For example, while there are laws to help protect landowners, these laws are in practice sometimes not effective as some of the landlords see rental markets are the most efficient way to use their land (Holden and Ghebru 2016). Despite all these positive developments in the past two decades, going forward, restrictions in land markets and poor land transactions remain key challenges of increasing agricultural productivity in Ethiopia.

Irrigation and Mechanization

Ethiopia's crop agriculture remains predominantly rainfed. In the absence of irrigation, given climate change and erratic rainfall, poor moisture conditions together with degraded soils remain a binding constraint to increasing productivity in Ethiopia. With frequent droughts in recent years, agriculture has become increasingly risky, making increases in productivity difficult. Studies indicate that irrigation can double average land productivity and at times provide 90 percent higher yields than comparable rainfed land (Fuglie 2008; Fuglie and Rada 2013). Public investments in irrigation structures have played pivotal roles in Asia's Green Revolution, particularly in India (Evenson and McKinsey 1999). Proclaimed as one of the water towers of Africa, Ethiopia is known to have substantial irrigation water potentials from both river basin runoffs (12 river basins with an annual runoff volume of 122 billion cubic meters) and groundwater (an estimated 2.6 billion cubic meters to 6.5 billion cubic meters), with 3.7 million hectares of developable land potential for irrigation (Awulachew et al. 2017).

However, despite recent efforts on advancing small-scale irrigation development, irrigation use remains low even for Africa south of the Sahara (Sheahan and Barrett 2017). Table 3.5 provides total area and percentage of land under irrigation using CSA's annual agricultural household sample survey. It indicates that land under irrigation (only *meher* season) for some crops has slightly increased over the decade, but the total area under irrigation has remained the same and is very low compared to other countries, for example China (37 percent) and India (32 percent) in 1995 (Rosegrant and Hazell 2000). Estimates based on CSA's agricultural household survey during the *meher* season shows land area under irrigation at 181,281 hectares in 2017/2018 (see Table 3.5). Unfortunately, CSA lacks good data on use of irrigation in the off-season, and these estimates also exclude the large-scale commercial farms. The *meher* estimates are therefore underestimates of access and use of irrigation. The government estimates that 2.8 million hectares are irrigated (Ethiopia, NPC 2018). Recent irrigation area identification mapping exercises using satellite information systems with validations on ground inventorying estimate irrigated area at 1.35 million hectares (7 percent of total area)—lower but closer to the official figure but significantly higher than the CSA estimate for the *meher* season (Kiran et al. 2018).

Ethiopia's Growth and Transformation Plan II (GTP II) sets out ambitious irrigation development plans for 2016/2017 and beyond: undertake small-scale irrigation on 2.94 million hectares; undertake study and design works on 142.8 thousand hectares; and carry out irrigation structures on 162.2 thousand hectares of land. Official reviews in the first year indicate lower performance than expected due to capacity and budgetary constraints (Ethiopia, NPC 2018). Anecdotal evidences also suggest that poor capabilities in irrigation and water resource management remain hurdles to sustainability of existing structures.

While use of mechanized farm tools can play critical transformative roles through increasing land and labor productivity (Binswanger and Rosenzweig 1986), owing to these low levels of irrigation and overall traditional farming practices, use of mechanized farm equipment remains low and Ethiopia's agriculture relies still heavily on animal-traction and labor (Berhane et al. 2017). Berhane et al. (2017) indicate despite recent increases in imports of agricultural machines, farm-level uptake of agricultural mechanization in Ethiopia remains low: fewer than 1 percent of plots are plowed with a tractor. They document, on the one hand, that the use of machines in agriculture is likely to increase as costs of labor and animal traction are recently on

TABLE 3.5 Total area and percentage of land in Ethiopia under irrigation

Crop	Total area under irrigation (hectares)							Percentage of land under irrigation (%)						
	2004/ 2005	2006/ 2007	2009/ 2010	2012/ 2013	2015/ 2016	2016/ 2017	2017/ 2018	2004/ 2005	2006/ 2007	2009/ 2010	2012/ 2013	2015/ 2016	2016/ 2017	2017/ 2018
All crops	121,493	134,545	154,424	151,645	181,172	195,374	181,281	1.12	1.14	1.19	1.09	1.25	1.37	1.27
Cereals	59,052	71,084	96,109	61,517	66,003	61,928	69,100	0.77	0.84	1.04	0.64	0.66	0.6	0.67
Pulses	6,832	6,496	7,891	4,537	6,610	4,448	4,232	0.51	0.47	0.53	0.24	0.4	0.28	0.26
Oilseeds	1,915	939	971	2,244	1,087	919	699	0.23	0.13	0.12	0.27	0.13	0.11	0.08
Vegetables	5,573	7,321	5,090	10,013	12,740	8,441	6,357	5.91	7.69	3.68	5.2	6.33	3.52	3.04
Root crops	10,570	9,501	11,408	18,522	24,446	23,954	21,227	6.77	5.03	5.38	9.08	11.44	10.45	9.09

Source: Authors' computation using CSA annual reports (Ethiopia, CSA volume 3, 2004–2018).

the rise and that, on the other hand, increasingly fragmented plots and small farm sizes, diversity of crops, difficult terrain, credit, and foreign exchange constraints prevent modernization via agricultural mechanization. As such, it seems that important synergistic complementarities between use of modern farm tools and irrigation observed in Asia's Green Revolution and elsewhere are yet to be exploited to sustain recent increases in agricultural productivity in Ethiopia.

Credit

Lack of adequate access to credit and other financial services is often considered as one of the limiting factors to increase productivity in agriculture. Following the worldwide movement on rural financial markets and the desire to increase agriculture productivity in the past decade, Ethiopia's rural credit market development has received significant attention over the past two decades.

Several government-affiliated and semiprivate microfinance institutions as well as member-based financial cooperatives have flourished throughout the rural areas of Ethiopia in recent years, mobilizing an unprecedented amount of savings and loans, serving (but not limited to) millions of Ethiopian rural households (Berhane and Abay 2019). While rigorous studies on impacts of credit on productivity in Ethiopia are limited, the evidence suggests that eliminating access to credit constraints could increase productivity by about 11 percent (Ali and Deininger 2014) and reduce poverty (Berhane and Gardebroek 2011). However, Berhane and Abay (2019) indicate that despite the enormous expansion in rural financial services, mainly credit, a vast majority of rural households remain underserved, with only 6–10 percent of households participating in credit uptake in 2013 and 2016 (Berhane and Abay 2019).

A critical question for policymakers is therefore given the rapid expansion of financial service providers—mainly loans—to rural areas, why the majority of rural households restrain themselves from taking up these loans. An important point to note here is despite the impressive progress in credit supply, access to credit in Ethiopia is often seen as an intermediate input to other input supply. As a result, rural credit services in Ethiopia are tied up with input supply services, restricting the flexible use of credit to mitigate household liquidity constraints. Berhane and Abay (2019) also point to several demand and supply side bottlenecks preventing households from accessing these loans, including inflexible and the one-size-fits-all nature of credit product designs (for example, the group-lending approach commonly used in

Ethiopia is not favored by many households given agriculture risk is covariant among group members), high transaction costs of loan processing, inflexible loan size and terms, as well as remoteness. A case in point is that most rural microfinance institutions do not have a mechanism to accommodate emerging "model" farmers that demand increasingly larger loans, which cannot be obtained from formal banks because they lack urban-based collateral formal banks require. This group of farmers simply falls into the "missing middle" in rural credit markets and remains underserved (Berhane and Abay 2019). Clearly, these bottlenecks contribute to the challenges of sustaining recent gains in productivity increases and need to be addressed structurally.

Conclusion

Ethiopia has made sustained investments in crop agriculture for more than a decade and consistently achieved impressive productivity gains, albeit from a low base. Guided by a strong political will, supported by development partners, to invest in agriculture, Ethiopia has surpassed the 6 percent agricultural growth target set by the CAADP for more than a decade. Between 2004/2005 and 2016/2017, aggregate smallholder crop output has more than doubled, largely attributable to cultivated area expansion and yield growth. In the same period, land under smallholder cultivation has expanded by about 27 percent, 90 percent of which was used for cereals (but later declined and leveled off), and cereal yield has increased significantly, by about 5 percent a year. It is important to note that the analysis focuses on smallholders and does not include recent commercial farm expansion to the lowlands, where productivity levels are not particularly higher than that of smallholders (see Ali, Deininger, and Harris 2019). Despite the high growth rates, compared with other contexts where agricultural transformation has taken root, Ethiopia's yield levels are still rather low and showing signs of slowing down recently.

A key question is whether Ethiopia can sustain growth rates in yield levels and overall productivity to achieve transformation in the face of declining additional cultivable land and recent slowing down yield levels. This chapter has explored the patterns of crop production and assessed the sources of productivity growth so far and what the implications are for future growth. In doing so, the important roles played by critical investments made in the sector and associated structural bottlenecks for future growth are discussed. We present the key findings from the analyses to understand what explains recent output growth and then discuss what needs to be done to sustain future growth.

Decades of sustained increase in total factor productivity (TFP)—a key measure of long-term agricultural productivity growth—is required to achieve agriculture transformation. How has Ethiopia's TFP fared in the past decade? A stylized growth accounting decomposition of crop output into factors that contribute to its growth in the past decade shows expansion in cultivated land and labor have, respectively, accounted for about 13 percent and 30 percent of Ethiopia's observed crop output growth. Expansion in use of chemical fertilizers, improved seeds, and extension were also important contributors. Moreover, TFP growth has contributed to about 18 percent of output growth in the same decade, suggesting efficiency gains of investments in the production process. However, TFP growth has declined in most recent years (from 38 percent in the beginning of the period to 16 percent in the end of the period), contributing to the recent slowdown in yield growth (Bachewe et al. 2018). Important to note is that the decline in TFP growth rate has continued despite the continued investments on traditionally TFP increasing nonconventional inputs such as extension and chemical fertilizers. The contributions of land and labor inputs have also remained roughly the same, suggesting additional gains in and keeping up with sustained productivity is likely to be increasingly difficult with the status quo. Accelerating TFP growth is a sustainable way of offsetting declining productivity growth due to decelerating input growth, mainly land in the case of Ethiopia (Fuglie 2008).

Although further studies are warranted to understand the sources of growth in view of the recent slowdown in the contributions of TFP and other inputs to agriculture growth, the evidence so far points to the following reasons. First, despite its remarkable success stories in promoting modern technologies and linking farmers to input markets, recent studies emphasize Ethiopia's extension system is overburdened and poorly equipped with the knowledge and skill sets needed to help bring about changes in the production system. Moreover, the system is weakly linked with knowledge centers with poor forward and backward feedback looping between extension agents, farmers, and researchers. The implication is that once the low-hanging fruits in productivity increases through the basic support provided by the extension system are exhausted, achieving additional productivity (TFP) gains would require revising the system to provide much needed technical skills by synching itself with knowledge centers and farmers. A declining yield growth following a similarly (but faster) declining area expansion (see Figure 3.2) also suggests investments in the existing extension system is unlikely to sustain yield growth. Improving this system may require additional investments on agricultural R&D and realignment of institutional structures to sustain growth.

Second, despite recent efforts (see, for example, ATA 2017a), critical bottlenecks to sustain productivity increases remain, including inefficient fertilizer markets and logistics and lack of appropriate fertilizer-nutrient blends suitable to each soil type (Zelleke et al. 2010; Rashid et al. 2013; Agbahey, Grethe, and Negatu 2015) on the one hand and poorly organized and inefficient seed markets, poor and supply-driven breeding research, and mismatches between seed multiplication and seed research and farmers on the other (for example, Spielman, Kelemework, and Alemu 2012; Alemu, Rashid, and Tripp 2019).[17]

Third, although significant capital investments have been made on roads and other physical infrastructures in recent years, poorly developed value chains and market infrastructures, weak integration of farmers with central markets, poor and inflexible credit markets prevent farmers from realizing the full productivity potential though changing their farm production organizations.

Fourth, irrigation is one of the weakest links in Ethiopia's agricultural production system. Given increasingly poorer moisture conditions in most parts of the country, water becomes a key hurdle in increasing productivity. Despite recent efforts to expand irrigation capabilities and structures, the progress in this front has remained unsatisfactory. Mechanization and use of labor-saving technologies can also add value through increasing land and labor productivity. However, the sector has several structural challenges (including lack of foreign exchange to import machines, topographical challenges, fragmented plots, lack of enough economies of scale, weak incentives to attract the private sector, and so on) that call for serious attention by different players, including the government.

These key factors have contributed to the general slowdown of TFP in agriculture, seemingly requiring a broad-based investment strategy, given Ethiopia's large and complex farm sector, to address that declining contribution to growth.

17 In 2012, Ethiopia's Agricultural Transformation Agency (ATA) launched a first of its kind initiative on Ethiopia Soil Information System (EthioSIS) that uses extensive soil sampling based on which location-specific fertilizers are recommended. Other similar efforts—including provision of input supply through agrodealers and video-based extension systems—are also being piloted. However, rigorous studies assessing the effectiveness of these experimentations are not yet available, and it is too early to include them in this analysis. See ATA 2019.

Appendix 3A: Total Factor Productivity and Output Growth

TABLE 3A.1 Trends in contributions of inputs, other factors, and TFP to crop output growth, 2005/2006–2015/2016 (%)

	2005/2006	2006/2007	2007/2008	2008/2009	2009/2010	2010/2011	2011/2012	2012/2013	2013/2014	2014/2015	2015/2016
Change in output (ΔQ/Q)	15.0	11.0	8.0	6.5	8.7	10.3	5.0	8.2	6.6	7.2	3.4
Labor	2.9	1.8	2.5	2.5	3.6	3.6	1.7	1.7	1.8	1.0	1.7
Land	1.2	1.5	1.2	1.2	1.2	1.2	1.0	0.8	0.6	0.5	0.6
Fertilizer	0.5	0.5	0.5	0.5	0.0	1.6	0.9	0.9	0.9	1.1	1.1
Improved seeds	0.4	0.4	0.4	1.0	2.8	2.6	1.0	1.0	1.0	1.0	−0.5
Pesticides	0.1	0.1	0.0	0.0	0.1	0.1	0.0	0.3	0.1	0.1	0.1
Irrigation	0.1	0.1	0.1	0.1	0.1	0.1	0.1	−0.3	0.3	0.3	0.0
Extension	0.0	0.0	0.3	0.3	0.3	0.3	1.1	1.1	1.1	0.3	0.3
Others	0.6	0.6	0.4	0.5	0.6	0.9	0.5	0.3	0.4	0.1	0.0
ΔTFP	9.1	6.0	2.6	0.4	0.0	−0.1	−1.4	2.3	0.4	2.8	0.1

Source: Authors' computations (see also Bachewe et al. 2018).

Note: TFP = total factor productivity. Decomposition uses factor shares in 2009/2010 Ethiopia SAM (EDRI 2009).

TABLE 3A.2 Trends in relative contributions of factors and TFP as percentage of crop output growth, 2005/2006–2015/2016 (millions of birr)

Season	Labor	Land	Fertilizer	Improved seeds	Pesticides	Irrigation	Extension	Others	Δ TFP
2005/2006	19.5	8.3	3.1	2.6	0.8	0.9	0.0	3.7	61.0
2006/2007	16.7	13.8	4.2	3.5	1.1	1.3	0.0	5.1	54.4
2007/2008	31.1	14.9	5.7	4.8	0.5	1.8	3.7	5.5	32.1
2008/2009	38.3	18.3	7.0	15.4	0.6	2.2	4.5	7.6	6.1
2009/2010	41.4	13.6	0.3	31.5	1.1	1.6	3.3	6.7	0.5
2010/2011	35.0	12.0	15.1	24.9	0.9	1.1	2.8	8.8	−0.6
2011/2012	34.9	20.1	18.6	20.4	0.3	2.4	22.6	9.8	−29.0
2012/2013	21.3	9.2	11.2	12.3	3.6	−3.2	13.6	3.9	28.1
2013/2014	27.2	9.3	13.8	15.2	1.7	5.0	16.4	5.5	5.9
2014/2015	14.1	7.6	14.9	13.9	1.5	3.8	4.6	0.8	38.9
2015/2016	48.5	16.9	30.7	−13.1	3.4	1.4	9.6	−1.2	3.8
Average (%)	**29.8**	**13.1**	**11.3**	**11.9**	**1.4**	**1.7**	**7.4**	**5.1**	**18.3**

Source: Authors' computations (see also Bachewe et al. 2018).
Note: Decomposition uses factor shares in 2009/2010 Ethiopia SAM (EDRI 2009). Real values of crop output reported.

References

Abate, G. T., T. Bernard, S. Makhija, and D. J. Spielman. 2019. *Accelerating Technical Change Through Video-Mediated Agricultural Extension: Evidence from Ethiopia.* IFPRI Discussion Paper 1851 (June). Washington, DC: International Food Policy Research Institute (IFPRI).

Abate, T., B. Shiferaw, A. Menkir, D. Wegary, Y. Kebede, K. Tesfaye, M. Kassie, G. Bogale, B. Tadesse, and T. Keno. 2015. "Factors That Transformed Maize Productivity in Ethiopia." *Food Security* 7 (5): 965–981.

Abay, K. A., G. Berhane, A. S. Taffesse, K. Abay, and B. Koru. 2018. "Estimating Input Complementarities with Unobserved Heterogeneity: Evidence from Ethiopia." *Journal of Agricultural Economics* 69 (2): 495–517.

African Union. 2010. *Comprehensive Africa Agriculture Development Programme (CAADP).* Addis Ababa.

Agbahey, J. U., H. Grethe, and W. Negatu. 2015. "Fertilizer Supply Chain in Ethiopia: Structure, Performance and Policy Analysis." *Afrika Focus* 28 (1): 81–101.

AGRA (Alliance for a Green Revolution in Africa). 2018. *Africa Agriculture Status Report: Catalyzing Government Capacity to Drive Agricultural Transformation (Issue 6).* Nairobi, Kenya.

Alem, Y., M. Bezabih, M. Kassie, and P. Zikhali. 2010. "Does Fertilizer Use Respond to Rainfall Variability? Panel Data Evidence from Ethiopia." *Agricultural Economics* 41 (2): 165–175.

Alemu, D., D. Byerlee, D. Spielman, and M. Gautam. 2007. *Policies to Promote Cereal Intensification in Ethiopia: A Review of Evidence and Experience.* IFPRI Discussion Paper 707. Washington, DC: IFPRI.

Alemu, D., S. Rashid, and R. Tripp. 2019. "Seed System Potential in Ethiopia: Constraints and Opportunities for Enhancing the Seed Sector." Gates Open Research 3. Accessed March 9, 2020. https://gatesopenresearch.org/documents/3-948.

Ali, D. A., and K. Deininger. 2014. "Causes and Implications of Credit Rationing in Rural Ethiopia: The Importance of Zonal Variation." *Journal of African Economies* 23 (4): 493–527.

Ali, D., K. Deininger, and A. Harris. 2019. "Does Large Farm Establishment Create Benefits for Neighboring Smallholders? Evidence from Ethiopia." *Land Economics* 95 (1): 71–90.

ASTI (Agricultural Science and Technology Indicators). 2019. ASTI database. Accessed June 21, 2019. https://www.asti.cgiar.org/data.

ATA (Agricultural Transformation Agency of Ethiopia). 2010. *2010/11 Annual Report.* Addis Ababa, Ethiopia.

———. 2017a. "Seed System Development Strategy: Vision, Systemic Challenges, and Prioritized Interventions." Working Strategy Document, Addis Ababa.

———. 2017b. *2016/17 Annual Report.* Addis Ababa.

———. 2019. "EthioSIS." Accessed October 15, 2019. www.ata.gov.et/programs/highlighted-deliverables/ethiosis/.

Awulachew, S. B., A. D. Yilma, M. Loulseged, W. Loiskandl, M. Ayana, and T. Alamirew. 2007. "Water Resources and Irrigation Development in Ethiopia." Colombo, Sri Lanka: International Water Management Institute (IWMI). IWMI Working Paper 123. http://dx.doi.org/10.3910/2009.305.

Bachewe, F. N., G. Berhane, B. Minten, and A. S. Taffesse. 2018. "Agricultural Transformation in Africa? Assessing the Evidence in Ethiopia." *World Development* 105: 286–298.

Beintema, N., and T. Hargewoin. 2018. *Ethiopia, Agricultural R&D Indicators Factsheet Update August 2018.* Washington, DC: Agricultural Science and Technology Indicators (ASTI), program of IFPRI.

Benin, S., and A. Nin-Pratt. 2016. "Intertemporal Trends in Agricultural Productivity." In *Agricultural Productivity in Africa: Trends, Patterns, and Determinants*, edited by S. Benin, 25–104. Washington, DC: IFPRI.

Berhane, G., and K. A. Abay. 2019. "Rural Finance and Smallholder Farming in Ethiopia." In *The Oxford Handbook of the Ethiopian Economy*, edited by F. Cheru, C. Cramer, and A. Oqubay, 487–504. Oxford, UK: Oxford University Press.

Berhane, G., and T. Assefa. 2020. "Fertilizer Profitability in Ethiopia." ESSP Working Paper. Addis Ababa: IFPRI, forthcoming.

Berhane, G., F. Bachewe, and B. Minten. 2018. "Agricultural Transformation in Ethiopia: What Do We Know?" In *Africa Agriculture Status Report 2018: Catalyzing Government Capacity to Drive Agricultural Transformation*. Nairobi, Kenya: Alliance for a Green Revolution in Africa (AGRA).

Berhane, G., M. Dereje, B. Minten, and S. Tamru. 2017. *The Rapid—But from a Low Base—Uptake of Agricultural Mechanization in Ethiopia: Patterns, Implications and Challenges*. IFPRI-ESSP Working Paper 105. Addis Ababa: IFPRI/Ethiopia Strategy Support Program (ESSP).

Berhane, G., and C. Gardebroek. 2011. "Does Microfinance Reduce Rural Poverty? Evidence Based on Household Panel Data from Northern Ethiopia." *American Journal of Agricultural Economics* 93 (1): 43–55.

Berhane, G., C. Ragasa, G. Abate, and T. W. Assefa. 2018. *The State of Agricultural Extension Services in Ethiopia and Their Contribution to Agricultural Productivity*. IFPRI-ESSP Working Paper 118. Addis Ababa: IFPRI/ESSP.

Bezawagaw, M. G., N. C. Dihel, M. T. Geiger, and G. K. Zerihun. 2018. *Ethiopia Economic Update: The Inescapable Manufacturing Services Nexus: Exploring the Potential of Distribution Services*. Washington, DC: World Bank.

Bezu, S., and S. Holden. 2014. "Are Rural Youth in Ethiopia Abandoning Agriculture?" *World Development* 64: 259–272.

Binswanger, H. P., and M. R. Rosenzweig. 1986. "Behavioural and Material Determinants of Production Relations in Agriculture." *Journal of Development Studies* 22 (3): 503–539.

Croppenstedt, A., M. Demeke, and M. M. Meschi. 2003. "Technology Adoption in the Presence of Constraints: The Case of Fertilizer Demand in Ethiopia." *Review of Development Economics* 7 (1): 58–70.

Davis, K., B. Swanson, D. Amudavi, D. A. Mekonnen, A. Flohrs, J. Riese, C. Lamb, and E. Zerfu. 2010. *In-depth Assessment of the Public Agricultural Extension System of Ethiopia and Recommendations for Improvement*. IFPRI Discussion Paper 1041. Washington, DC: IFPRI.

Deininger, K., D. Ayalew, and T. Alemu. 2011. "Impacts of Land Certification on Tenure Security, Investment, and Land Market Participation: Evidence from Ethiopia." *Land Economics* 87 (2): 312–334.

Deininger, K., D. Ayalew, S. Holden, and J. Zevenbergen. 2008. "Rural Land Certification in Ethiopia: Process, Initial Impact, and Implications for Other African Countries." *World Development* 36 (10): 1786–1812.

Deneke, T. T., and D. Gulti. 2016. "Agricultural Research and Extension Linkages in the Amhara Region, Ethiopia." In *Technological and Institutional Innovations for Marginalized Smallholders in Agricultural Development*, edited by F. Gatzweiler and J. von Braun, 113–124. Switzerland: Springer International Publishing.

EDRI (Ethiopian Development Research Institute). 2009. *Ethiopia: Input Output Table and Social Accounting Matrix*. Addis Ababa: EDRI.

Ethiopia, CSA (Central Statistical Agency). 2005–2020. *Agricultural Sample Survey: Report on Farm Management Practices (Private Peasant Holdings, Meher Season)*. Volume 3. Addis Ababa.

Ethiopia, MoFED (Ministry of Finance and Economic Development). 2006. *Ethiopia: Building on Progress—A Plan for Accelerated and Sustained Development to End Poverty (PASDEP) (2005/06–2009/10)*. Volume 1, *Main Text*. Addis Ababa.

———. 2010. *Growth and Transformation Plan II (GTP II) (2010/11–2014/15)*. Volume 1, *Main Text*. Addis Ababa.

Ethiopia, NPC (National Planning Commission). 2018. *The Second Growth and Transformation Plan (GTP II) Midterm Review Report*. Addis Ababa.

Evenson, R. E., and J. W. McKinsey. 1999. *Technology—Climate Interactions in the Green Revolution in India*. Center Discussion Paper 805. Addis Ababa.

Fuglie, K. O. 2004. "Productivity Growth in Indonesian Agriculture, 1961–2000." *Bulletin of Indonesian Economic Studies* 40 (2): 209–225.

———. 2008. "Is a Slowdown in Agricultural Productivity Growth Contributing to the Rise in Commodity Prices?" *Agricultural Economics* 39 (s1): 431–441.

Fuglie, K., and N. Rada. 2013. *Resources, Policies, and Agricultural Productivity in Sub-Saharan Africa*. Economic Research Service Report 145. Washington, DC: United States Department of Agriculture.

Gebrehiwot, B. D., and S. T. Holden. 2019. "Variation in Output Shares and Endogenous Matching in Land Rental Contracts: Evidence from Ethiopia." *Journal of Agricultural Economics* 71 (1): 260–282.

Gebru, M., S. T. Holden, and M. Tilahun. 2019. "Tenants' Land Access in the Rental Market: Evidence from Northern Ethiopia." *Agricultural Economics* 50 (3): 291–302.

Ghebru, H. H. 2010. "Land Policy Reform and Land Rental Markets in Ethiopia: Equity, Productivity and Welfare Implications." PhD dissertation, Norwegian University of Life Sciences, Ås.

Ghebru, H., and S. Holden. 2013. *Links between Tenure Security and Food Security: Evidence from Ethiopia.* IFPRI-ESSP Working Paper 59. Addis Ababa: IFPRI/EDRI.

Ghebru, H., B. Koru, and A. S. Taffesse. 2016. *Household Perception and Demand for Better Protection of Land Rights in Ethiopia.* IFPRI-ESSP Working Paper 83. Washington, DC: IFPRI; Addis Ababa: EDRI.

Haddis, Z. 2013. "Towards Improved Transactions of Land Use Rights in Ethiopia." Paper prepared for presentation at the Annual World Bank Conference on Land and Poverty, World Bank, Washington, DC, April 8–11.

Hazell, P. B. 2009. *The Asian Green Revolution.* Vol. 911. Washington, DC: IFPRI.

———. 2013. "Options for African Agriculture in an Era of High Food and Energy Prices." *Agricultural Economics* 44 (s1): 19–27.

Headey, D., M. Dereje, and A. S. Taffesse. 2014. "Land Constraints and Agricultural Intensification in Ethiopia: A Village-Level Analysis of High-Potential Areas." *Food Policy* 48: 129–141.

Hill, R. V., and E. Tsehaye. 2018. *Growth, Safety Nets and Poverty: Assessing Progress in Ethiopia from 1996 to 2011 (English).* Policy Research working paper series 8380. Washington, DC: World Bank.

Holden, S., K. Deininger, and H. Ghebru. 2007. "Impacts of Low-Cost Land Certification on Investment and Productivity." *American Journal of Agricultural Economics* 91 (2): 359–373.

Holden, S. T., and H. Ghebru. 2016. "Land Rental Market Legal Restrictions in Northern Ethiopia." *Land Use Policy* 55: 212–221.

Holden, S., K. Otsuka, and K. Deininger, eds. 2013. *Land Tenure Reform in Asia and Africa: Assessing Impacts on Poverty and Natural Resource Management.* London: Palgrave Macmillan.

Hopper, W. 1993. "Indian Agriculture and Fertilizer: An Outsider's Observations." Keynote address given at the Fertiliser Association of India (FAI) Seminar on Emerging Scenarios in Fertilizer and Agriculture: Global Dimensions. New Delhi, India.

IFDC (International Fertilizer Development Center). 2015. *Assessment of Fertilizer Consumption and Use by Crop in Ethiopia.* Report. Muscle Shoals, AL, US.

Jayne, T. S., J. Chamberlin, and R. Benfica. 2018. "Africa's Unfolding Economic Transformation." *Journal of Development Studies* 54 (5): 777–787.

Kassa, T. A., T. Luck, A. Bekele, and S. G. Riedel-Heller. 2016. "Sexual and Reproductive Health of Young People with Disability in Ethiopia: A Study on Knowledge, Attitude and Practice: A Cross-sectional Study." *Globalization and Health* 12 (1): 5.

Kiran, C., S. Siddiqui, J. Barron, C. Subsasinghe, and A. Haileselassie. 2018. *Ecological Footprint of Food Security: Mapping Irrigated Area in Ethiopia.* Unpublished, IWMI.

Krishnan, P., and M. Patnam. 2014. "Neighbors and Extension Agents in Ethiopia: Who Matters More for Technology Adoption?" *American Journal of Agricultural Economics* 96 (1): 308–327.

Laborde, D., T. Lallemant, K. McDougal, C. Smaller, and F. Traore. 2018. *Transforming Agriculture in Africa and Asia: What Are the Policy Priorities?* Winnipeg, Canada: International Institute for Sustainable Development (IISD).

Makhija, S., D. J. Spielman, G. T. Abate, and T. Bernard. 2019. "Opportunities and Challenges in Field Data Validation and Corroboration: Matching Household Survey Data with Project Monitoring Data in Ethiopia." IFPRI Project Note (May). Washington, DC: IFPRI. http://ebrary.ifpri.org/cdm/singleitem/collection/p15738coll2/id/133257/rec/3.

Melesse, M. B., and E. Bulte. 2015. "Does Land Registration and Certification Boost Farm Productivity? Evidence from Ethiopia." *Agricultural Economics* 46 (6): 757–768.

Mellor, J. W. 2014. "Rapid Cereals Production Growth, 1997–2012, Ethiopia." Accessed March 19, 2020. https://www.johnwmellor.com/s/AERgrowthratesethiopiamellor-z5h3.docx.

Minten, B., A. S. Taffesse, and P. Brown, eds. 2018. *The Economics of Teff: Exploring Ethiopia's Biggest Cash Crop.* Washington, DC: IFPRI.

Minten, B., S. Tamru, E. Engida, and T. Kuma. 2013. *Ethiopia's Value Chains on the Move: The Case of Teff.* IFPRI-ESSP Working Paper 52. Addis Ababa: IFPRI/Ethiopia Strategy Support Program II (ESSP II).

Nin-Pratt, A. 2016. "Inputs, Productivity and Agricultural Growth in Sub-Saharan Africa." In *Productivity and Efficiency Analysis,* edited by W. Greene, L. Khalaf, R. Sickles, M. Veall, and M. C. Voia. Cham, Switzerland: Springer. DOI: https://doi.org/10.1007/978-3-319-23228-7_11.

Rashid, S., N. Tefera, N. Minot, and G. Ayele. 2013. "Can modern input use be promoted without subsidies? An analysis of fertilizer in Ethiopia." *Agricultural Economics* 44 (6): 595–611. Special Issue on Input Subsidy Programs (ISPs) in Sub-Saharan Africa (SSA). http://onlinelibrary.wiley.com/doi/10.1111/agec.12076/abstract.

Ricker-Gilbert, J., T. S. Jayne, and E. Chirwa. 2011. "Subsidies and Crowding Out: A Double-hurdle Model of Fertilizer Demand in Malawi." *American Journal of Agricultural Economics* 93 (1): 26–42.

Rosegrant, M. W., and P. B. Hazell. 2000. *Transforming the Rural Asian Economy: The Unfinished Revolution.* Hong Kong: Oxford University Press.

Ruttan, V. W. 2002. "Productivity Growth in World Agriculture: Sources and Constraints." *Journal of Economic Perspectives* 16 (4): 161–184.

Sheahan, M., and C. B. Barrett. 2014. *Understanding the Agricultural Input Landscape in Sub-Saharan Africa: Recent Plot, Household, and Community-Level Evidence.* Policy Research Working Paper Series 7014. Africa Region Office of the Chief Economist, World Bank.

———. 2017. "Ten Striking Facts about Agricultural Input Use in Sub-Saharan Africa." *Food Policy* 67: 12–25.

Sheng, Y., E. Ball, and K. Nossal. 2015. "Comparing Agricultural Total Factor Productivity between Australia, Canada, and the United States, 1961–2006." *International Productivity Monitor* 29: 38–59.

Solow, R. M. 1957. "Technical Change and the Aggregate Production Function." *Review of Economics and Statistics* 39 (3): 312–320.

Spielman, D. J., D. Byerlee, D. Alemu, and D. Kelemework. 2010. "Policies to Promote Cereal Intensification in Ethiopia: The Search for Appropriate Public and Private Roles." *Food Policy* 35 (3): 185–194.

Spielman, D. J., D. Kelemework, and D. Alemu. 2012. "Seed, Fertilizer, and Agricultural Extension in Ethiopia." In *Food and Agriculture in Ethiopia: Progress and Policy Challenges*, edited by P. A. Dorosh and S. Rashid, 84–122. Philadelphia: University of Pennsylvania Press.

Sumner, D. A. 2014. "American Farms Keep Growing: Size, Productivity and Policy." *Journal of Economic Perspectives* 28 (1): 147–166.

Tomich, T., P. Kilby, and B. Johnson. 1995. *Transforming Agrarian Economies: Opportunities Seized, Opportunities Missed.* Ithaca, NY, US: Cornell University Press.

USAID. 2013. "Agricultural Growth Program–Livestock Market Development. End Market Analysis for Meat/Live Animals, Leather and Leather Products, Dairy Products Value Chains." Expanding Livestock Markets for the Small-Holder Producers. Prepared by AGP-Livestock Market Development Project.

Vandercasteelen, J., M. Dereje, B. Minten, and A. S. Taffesse. 2013. *Scaling-Up Adoption of Improved Technologies: The Impact of the Promotion of Row Planting on Farmers' Teff Yields in Ethiopia.* LICOS Discussion Paper 344. Leuven, Belgium: Katholieke Universiteit Leuven, LICOS Centre for Institutions and Economic Performance.

World Bank. 2016. *Ethiopia Public Expenditure Review.* Report ACS14541. Washington, DC: World Bank.

Yirga, C., K. Negisho, L. Oruko, D. Alemu, and G. Taxler. 2016. *Tracking the Diffusion of Crop Varieties Using DNA Fingerprinting.* Addis Ababa: Ethiopian Institute of Agricultural Research (EIAR).

Zelleke, G., G. Agegnehu, D. Abera, and S. Rashid. 2010. *Fertilizer and Soil Fertility Potential in Ethiopia—Constraints and Opportunities for Enhancing the System.* Working Paper July 2010. Washington, DC: IFRI.

Chapter 4

CLIMATE CHANGE IMPACTS ON CROP YIELDS

Timothy S. Thomas, Paul Dorosh, and Richard Robertson

Introduction

It is commonly thought that climate change is likely to have major effects on Ethiopian agriculture and the country's overall development over the next fifty years and beyond. Indeed, each of the 32 major climate change models used in our analysis indicate a high probability of significant increases in average world temperatures by 2055 as well as an increase in median temperatures in Ethiopia. The prognosis for rainfall in Ethiopia is much less certain, although on average the simulation models suggest a relatively small increase in rainfall for the country overall. Moreover, there is a growing consensus that climate variability will increase in all regions of the world (IPCC 2018).

Earlier studies of the likely effects of climate change in Ethiopia point to both positive and negative effects. Studies using hydrological models to assess the potential effects of climate change on water flow in various basins in Ethiopia (Dile, Berndtsson, and Setegn 2013; Wagena et al. 2016) generally suggest significant increases in water availability with potential benefits for irrigation as well as increases in variability and flooding (see also Taye, Willems, and Block 2015; Nawaz et al. 2010; Kim, Kaluarachchi, and Smakhtin 2008; Kim and Kaluarachchi 2009).[1] Admassu et al. (2013) used a crop model linked to a multimarket model of supply and demand of agricultural commodities to project climate effects on key crops in Ethiopia between 2000 and 2050. Based on output of four climate models, their analysis shows yield gains of over 25 percent in much of the eastern highlands and north-central highlands, but large yield reductions and loss of areas suitable

1 In their analysis of the upper Blue Nile Basin for 2010–2100, Dile, Berndtsson, and Setegn (2013) find that climate change may result in an annual increase in flow volume for the Gilgel Abay River, benefiting local small-scale irrigation activities. Likewise, Wagena et al. (2016) used watershed models to analyze water flows in the highlands of the Blue Nile Basin for two periods (2041–2065 and 2075–2099) based on four climate scenarios from six climate models. Their results indicate that the Tana and Beles Basins will have large (22–27 percent) increases in mean annual flow, as well as 16 percent to 19 percent increases in sediment concentrations, with potentially large net benefits for hydropower generation.

for growing maize in the eastern and southwestern parts of central Ethiopia.[2] Robinson, Willenbockel, and Strzepek (2012) included general equilibrium effects in their analysis of potential effects on water flows, crops, infrastructure, and economic output, concluding that, in the absence of adaptation investments, Ethiopia's GDP in 2050 would be up to 10 percent below the counterfactual baseline of no climate change.[3]

This chapter presents a further analysis of the potential effects of projected climate change scenarios for cereal crop yields in Ethiopia in 2035, 2055, and 2085. The methodology used in determining the direct effect of climate on crop yields (described in detail below) is similar to that of Admassu et al. (2013) but differs in using agroecological zones more closely corresponding to cropping patterns in Ethiopia as well as by drawing on the results of a wider array of climate models.[4] In particular, we use coefficients from regressions of simulated yields from a crop simulation model, the Decision Support System for Agrotechnology Transfer (DSSAT) crop systems model, with climate variables as explanatory variables. In the discussion of our simulation results in this chapter, we provide an explicit comparison with the results of Admassu et al. (2013).

First we summarize the results of the major climate models for rainfall and temperature across six main agroecological zones in Ethiopia. Thereafter, we present results of the crop yield models for three of the major cereal crops in Ethiopia: wheat, maize, and sorghum. (Unfortunately, no crop model is available for teff.) We then construct long-term yield growth trends consistent with historical yield growth and yield gaps between Ethiopia and comparator countries in Africa south of the Sahara. The last part of the chapter presents a brief summary and conclusions.

---

2 In these simulations maize yields rise through 2020 before leveling off between 2020 and 2050. Rainfed wheat yields generally decline in response to climate change, with one of the climate models Admassu et al. (2013) used showing large losses. The crop model results for sorghum are very similar to those for maize.

3 Robinson, Willenbockel, and Strzepek (2012) used a system of hydrology, crop, road, and other models to simulate the possible effects of changes in precipitation and temperature as reflected in projections from four of 22 global climate models available at the time of the study. Unlike the main analysis here, which is based on the average results from a set of 31 global climate models, the Robinson, Willenbockel, and Strzepek (2012) analysis was designed to provide an estimate of the range of outcomes, so was based on four extreme projections: the wettest and driest projections on a global scale, and the wettest and driest projections for Ethiopia. Agricultural GDP is highest under the driest global climate projection but falls in other projections, in part due to damage from extreme floods in some years.

4 Also, unlike Admassu et al. (2013), our modeling approach does not include effects of changes in incomes and prices.

Climate Change Scenarios: Rainfall and Temperature Projections

Our analysis draws on temperature and rainfall outcomes from simulations of 32 downscaled global climate models for 2035, 2055, and 2085 (Ramirez and Jarvis 2008).[5] We use the simulated monthly means of rainfall and temperature by 5 arc minute grid cells (pixels equivalent in size to about 9 kilometers × 9 kilometers at the equator) from each global climate model. We use the absolute deviations from the mean temperature and rainfall from each of the 31 years (1980 to 2010) from the AgMERRA database (Ruane, Goldberg, and Chryssanthacopoulos 2015) to generate a set of 992 (32 × 31) simulated temperature and rainfall outcomes for each of the grid cells that encompass Ethiopia for each of the three simulation years (2035, 2055, and 2085).[6] We use climate models to tell us how much the mean of the distribution should change under climate change, and we use historical weather to give plausible variation to those means, with the purpose of simulating a distribution of what future weather might look like.

Thus, for global climate model m, grid cell i, climate realization t and simulation year T (=2035, 2055, or 2085), mean rainfall (or temperature) R, is calculated as:

$$R(m,i,t,T) = r(m,i,T) + e(i,t),$$

where $r(m, i,T)$ = mean (or median) simulated value of rainfall (or temperature) from model m for grid cell i and year T, and
$e(i,t)$ = deviation from mean rainfall (or temperature) from historic climate realization from 1980 through 2010 ($t = 1{,}31$).

Table 4.1 presents the results for rainfall and temperature for the six agro-ecological zones in Ethiopia defined in Schmidt and Thomas (2018). As a base for comparison, we use the average climate conditions for the 1960–1990 period from WorldClim (v. 1.4, Hijmans et al. 2005). Taken together, the outputs from these 32 climate models suggest that Ethiopia will likely have

5 Throughout this chapter, "1975" may be used as a shorthand notation for averages for the 1960–1990 period; "2013" for averages for the 2012–2015 period; "2035" for averages for the 2020–2049 period; "2055" for averages for the 2040–2060 period; and "2085" for averages for the 2070–2099 period. The latter three periods are based on Ramirez-Villegas and Jarvis (2010).

6 Note that this procedure implicitly assumes that the distribution of percentage deviations from mean temperature and rainfall realizations from the 1980–2010 period remains the same for future periods.

TABLE 4.1 Projections of temperature and rainfall changes in Ethiopia between 1975 and 2055

	Precipitation (millimeters) May–October				Mean daily maximum temperature, warmest month (°C) May–October			
		Change between 1960–1990 and 2041–2070				Change between 1960–1990 and 2041–2070		
Agroecological zone	Mean for 1960–1990	2.5 percentile	Median	97.5 percentile	Mean for 1960–1990	2.5 percentile	Median	97.5 percentile
Drought prone, highland	646	–106	47	372	29.0	0.4	2.9	5.4
Drought prone, lowland	423	–119	18	373	29.3	0.7	2.5	4.5
Humid moisture reliable, lowland	1,018	–130	46	317	34.1	0.2	2.7	4.3
Moisture reliable, highland-cereal	1,091	–132	60	410	26.7	0.3	2.6	4.7
Moisture reliable, highland-enset	878	–176	29	540	26.4	0.4	2.4	4.3
Pastoralist	193	–58	31	259	34.3	1.1	2.6	4.5
Ethiopia	**595**	**–112**	**39**	**355**	**31.0**	**0.7**	**2.6**	**4.7**

Source: Authors' calculations based on Hijmans et al. (2005) and Ramirez-Villegas and Jarvis (2010).

Note: The tables summarize the results of simulations of 32 global climate models. The data shown are unweighted averages of precipitation and rainfall across grid cells for each agroecological zone. See a map of the agroecological zones (using a slightly different definition in their use within Chapter 13) in Figure 13.2.

a moderate increase in rainfall in all agroecologies in the main growing season (May–October), with a median increase in precipitation of 39 millimeters above the mean precipitation of 595 millimeters in the 1960–1990 period (Figure 4.1). The mean daily maximum temperatures of the warmest month in the main growing season (May–October), however, are likely to be considerably higher than in the base period, with a median increase of 2.6 degrees C. This increase is approximately equal to the median temperature increase value for eastern Africa, but less than the global median temperature increase.

The largest absolute increases in mean daily maximum temperatures are expected to be in the drought-prone highlands—a median increase of 2.9 degrees C, from 29.0 degrees C in the 1960–1990 period to 31.9 degrees C in the 2041–2070 period. Likewise, this agroecological zone has the largest extreme temperature gains, with an increase of 5.4 degrees C in the top 2.5 percent of the temperature distribution across the sample of model simulations (Figure 4.2).

Potential Effects of Climate Change on Maize, Wheat, and Sorghum Yields

To estimate the effects of climate change on crop yields, we employ results of simulations using the Decision Support System for Agrotechnology Transfer (DSSAT) suite of crop models.[7] We use yields calculated by DSSAT along with the corresponding weather input into those models to estimate parameters in a regression that describe how weather variables and other important agronomic variables influence yield. We then take these regression parameter estimates and use them to estimate the effects of simulated climate change outcomes on yields.

Yield effects in DSSAT are calculated for each 9 kilometers by 9 kilometers grid cell in a geographic grid under purely rainfed conditions. In our work with DSSAT we use monthly weather inputs that DSSAT converts to daily weather using an internal weather simulator. For this exercise we feed into DSSAT rainfall and temperature outcomes from the HadGEM2_ES global climate model (Collins et al. 2011; Martin et al. 2011) and under the RCP8.5 high emissions scenario for the 2041–2070 period and for the 2071–2099 period as well as for historical climate data from 1960 to 1990. Doing so gives

7 DSSAT is a suite of single crop models (Hoogenboom et al. 2012; Jones et al. 2003). Yields are estimated using site-specific weather data during the growing season, soil data, and management practices.

FIGURE 4.1 Simulated rainfall in 2055 by agroecological zone in Ethiopia (millimeters)

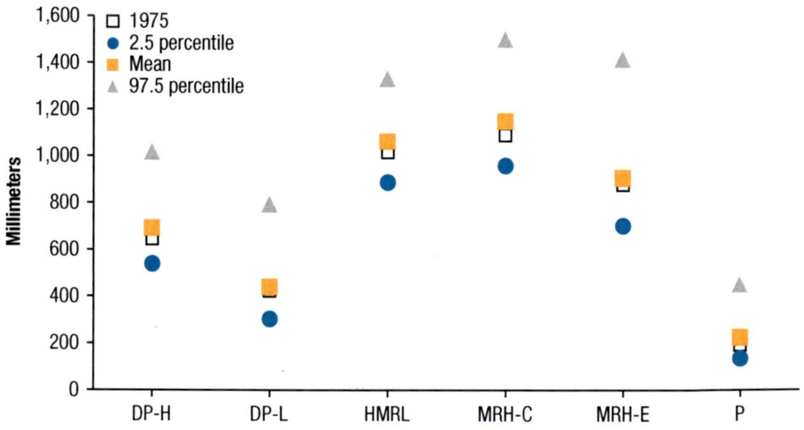

Source: Authors' calculations from climate model simulations.

Note: DP-H = drought prone, highland; DP-L = drought prone, lowland; HMRL = humid moisture reliable, lowland; MRH-C = moisture reliable, highland—cereal; MRH-E = moisture reliable, highland—enset; P = pastoralist. The year 2055 is the middle year of the 2041–2070 period, on which the climate data are based.

FIGURE 4.2 Simulated temperature in 2055 by agroecological zone in Ethiopia (degrees Celsius)

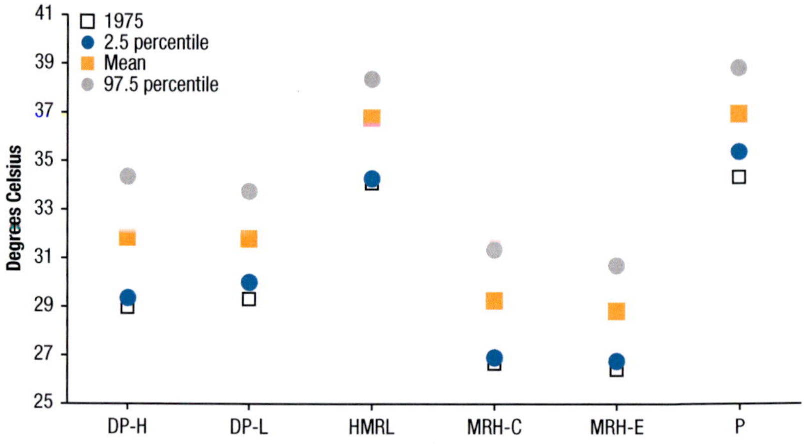

Source: Authors' calculations from climate model simulations.

Note: DP-H = drought prone, highland; DP-L = drought prone, lowland; HMRL = humid moisture reliable, lowland; MRH-C = moisture reliable, highland—cereal; MRH-E = moisture reliable, highland—enset; P = pastoralist. The year 2055 is the middle year of the 2041–2070 period, on which the climate data are based.

us a wide range of values for the regression (ranging from no climate change to a very large climate change represented by the HadGEM2_ES model at the end of the century).[8] A wide range of climate values helps us to ensure that the parameter estimates provide a good fit for feasible values spanning more than a century.

Using the output from the DSSAT model, the yield of crop *c* (wheat, sorghum, *meher* season maize, or *belg* season maize) planted on soil type *s* in grid cell *i* under model simulation *ms* is estimated as follows:

$$Yield\ (c,i) = a_0(c) + a_1(c)^*elev(i) + a_2(c)^*elev(i)^2 + a_3(c)^*N(i) + a_4(c)^*N(i)^2 + \textstyle\sum_{s=0}^{12} [a_{5s}(c)^*D(s,i)]$$

$$+ \textstyle\sum_{m=0}^{3} [\beta_{1m}(c)^*rain(i,m) + \beta_{2m}(c)^* rain(i,m)^2 + \beta_{3m}(c)^*rain(i,m)^3$$

$$+ \beta_{4m}(c)^*temp(i,m,ms) + \beta_{5m}(c)^*temp(i,m,ms)^2 + \beta_{6m}(c)^*temp(i,m)^3$$

$$+ \beta_{7m}(c)^*rain(i,m)^*temp(i,m) + \beta_{8m}(c)^*rain(i,m)^{2*}temp(i,m)$$

$$+ \beta_{9m}(c)^*temp(i,m)^{2*}rain(i,m)] + e(c,i),$$

where $rain(i,m,ms)$ and $temp(i,m,ms)$ are total rainfall and mean daily maximum temperature, respectively, for grid cell *i* and model simulation *ms*; *m* indexes the month during the growing season (with 0 being the planting month, 1 the month after the planting month, and so on); *elev* is the elevation of grid cell *i*; *N* is the starting nitrogen level for the soil in grid cell *i*; and $D(s,i)$ is a categorical value of 1 if the soil in grid cell *i* is of type *s* and 0 otherwise.[9] Parameters from this regression are then used to estimate yields for each crop for each 9 kilometer by 9 kilometer grid cell across Ethiopia using the rainfall and temperature outcomes from all 32 global climate models for each of 31 weather realizations.

For each pixel and crop, the DSSAT algorithm searches for the planting month that gives the highest yield based on the prediction from the regression, with the constraint that the planting month has to align with those recognized as feasible by agronomic professionals that work with Ethiopian agriculture.

Average yield changes for each of Ethiopia's six agroecological zones are calculated using weights from the spatially disaggregated estimates of area

8 Note that the yield calculations ignore possible carbon dioxide fertilization effects.

9 Because more spatially disaggregated data are not available, we assume that the soil type, rainfall, and temperature are uniform within each grid cell.

cultivated for each crop from IFPRI's Spatial Production Allocation Model (SPAM) (You, Wood, and Wood-Sichra 2006; You et al. 2014).

Implications of Climate Change on Crop Yields

Overall, the simulated net effects of increases in average rainfall and higher average temperatures are relatively small. Simulated maize yields are higher on average in the climate change simulations, with average yields 1.2 percent and 4.2 percent higher than 2013 yields in 2035 and 2085, respectively (Table 4.2; Figure 4.3; Figure 4.4). Average wheat yields are only 0.3 percent lower in 2035 than in 2013, and only 1.1 percent and 2.7 percent lower by 2055 and 2085, respectively, than in 2013 (Table 4.2; Figure 4.3; Figure 4.5). Sorghum

TABLE 4.2 Simulated crop yields in Ethiopia with climate change

Crop	2013	2035	2055	2085
Maize				
Yield: Climate change with no technical change (kilograms per hectare)[a]	3,372	3,414	3,460	3,515
Climate change effect, relative to 2013 (%)	n.a.	1.2	2.6	4.2
Yield: Technical change only (kilograms per hectare)[b]	3,372	4,644	6,255	9,777
Cumulative technical change, relative to 2013 (%)	n.a.	38.8	86.9	192.1
Yield: Technical change with climate change (kilograms per hectare)	3,372	4,702	6,418	10,191
Cumulative technical change with climate change (%)	n.a.	40.5	91.8	204.5
Wheat				
Yield: Climate change with no technical change (kilograms per hectare)[a]	2,475	2,467	2,447	2,409
Climate change effect, relative to 2013 (%)	n.a.	−0.3	−1.1	−2.7
Yield: Technical change only (kilograms per hectare)[b]	2,475	3,065	3,740	5,041
Cumulative technical change, relative to 2013 (%)	n.a.	23.9	51.1	103.7
Yield: Technical change with climate change (kilograms per hectare)	2,475	3,056	3,698	4,908
Cumulative technical change with climate change (%)	n.a.	23.5	49.4	90.0
Sorghum				
Yield: Climate change with no technical change (kilograms per hectare)[a]	2,336	2,350	2,349	2,314
Climate change effect, relative to 2013 (%)	n.a.	0.6	0.6	−0.9
Yield: Technical change only (kilograms per hectare)[b]	2,336	2,893	3,530	4,758
Cumulative technical change, relative to 2013 (%)	n.a.	23.9	51.1	103.7
Yield: Technical change with climate change (kilograms per hectare)	2,336	2,911	3,550	4,714
Cumulative technical change with climate change (%)	n.a.	24.6	52.0	101.8

Source: Authors' calculations from model simulations.

Note: n.a. = not applicable

a. Based on average of actual yields for 2012 to 2015 (FAO 2019) and percentage changes from climate model simulations.

b. Using assumed yield annual growth rates (maize 1.5 percent, wheat 1.0 percent, sorghum 1.0 percent).

yields in 2035 and 2055 are 0.6 percent higher than in 2013; by 2085, however, sorghum yields are 0.9 percent lower than in 2013 (Table 4.2; Figure 4.3; Figure 4.6).

Moreover, the simulated effects of climate change on yields vary somewhat across regions. For maize, median maize yields in the three highland

FIGURE 4.3 Simulated climate change impacts on crop yields in Ethiopia relative to 2013 yields (%)

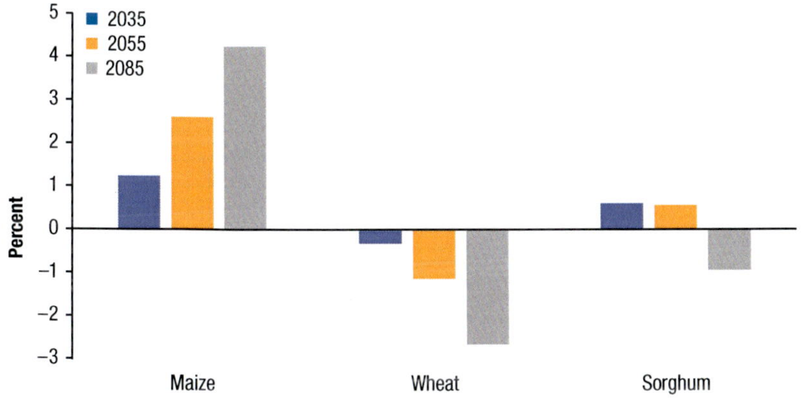

Source: Authors' calculations from model simulations.

FIGURE 4.4 Simulated climate change impacts on maize yields in Ethiopia, 2015–2085

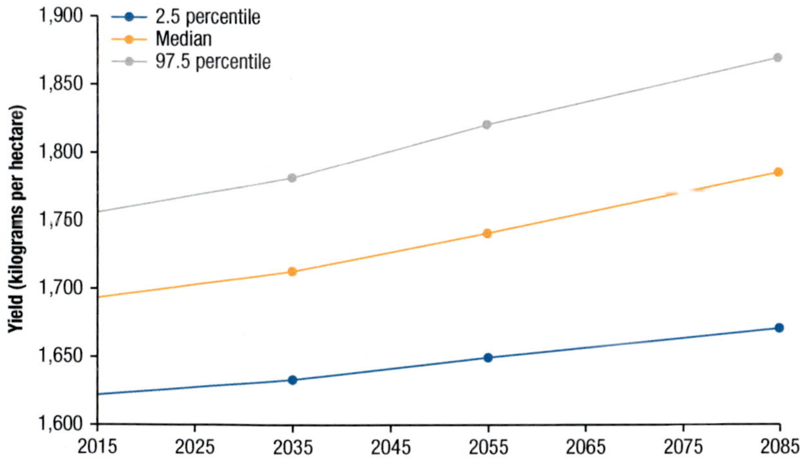

Source: Authors' calculations.

FIGURE 4.5 Simulated climate change impacts on wheat yields in Ethiopia, 2015–2085

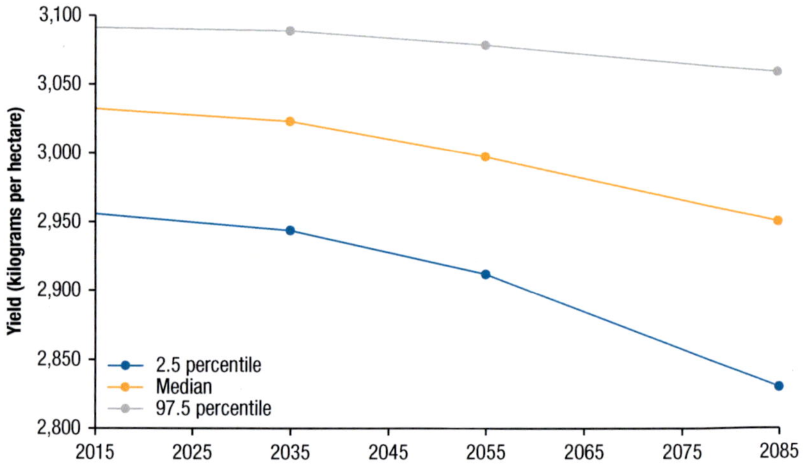

Source: Authors' calculations.

FIGURE 4.6 Simulated climate change impacts on sorghum yields in Ethiopia, 2015–2085

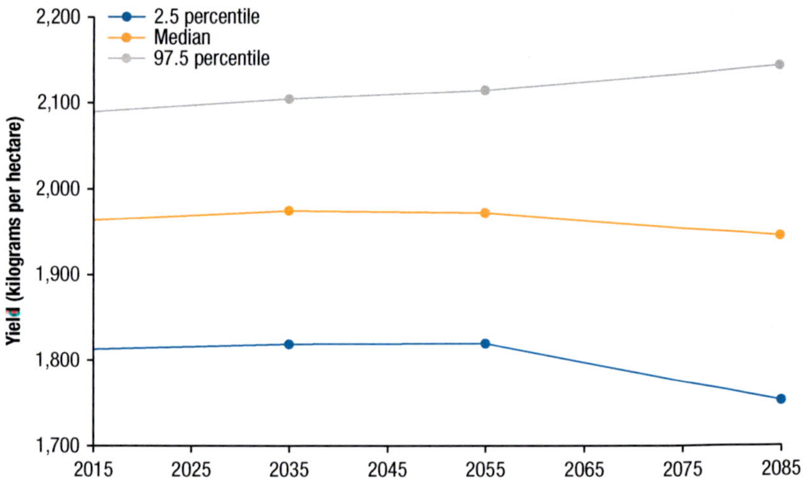

Source: Authors' calculations.

TABLE 4.3 Simulated climate change effects on maize yields by agroecological zone in Ethiopia, relative to 2013 yields (%)

	Drought prone		Humid moisture reliable				
Year	Highland	Lowland	Lowland	Highland-cereal	Highland-enset	Pastoralist	Ethiopia
2035							
P2.5	0.0	−1.5	0.3	1.1	−0.5	−0.4	0.8
P25	1.2	0.2	0.3	1.3	0.6	0.2	1.2
Median	*1.8*	*0.5*	*0.2*	*1.3*	*1.0*	*0.9*	*1.3*
P75	1.7	1.3	0.5	1.4	1.7	1.6	1.4
P97.5	2.4	1.6	0.4	1.7	2.3	2.8	1.6
2055							
P2.5	−0.8	−4.8	0.1	2.4	−0.8	−2.3	1.8
P25	2.3	−0.4	0.2	3.0	1.7	0.0	2.8
Median	*3.2*	*0.9*	*0.3*	*3.1*	*2.3*	*1.5*	*2.9*
P75	3.2	3.0	0.6	3.4	3.6	2.4	3.1
P97.5	5.1	3.9	0.7	4.1	4.9	5.8	3.8
2085							
P2.5	−1.2	−8.5	−2.7	4.6	−1.1	−5.0	3.1
P25	3.9	−0.9	−0.8	5.4	3.8	−0.7	5.2
Median	*5.5*	*2.8*	*−0.5*	*5.7*	*5.6*	*1.9*	*5.5*
P75	6.6	6.6	0.0	6.0	7.2	5.5	6.0
P97.5	9.0	8.7	0.9	7.3	9.1	10.5	6.6

Source: Authors' calculations from model simulations.

Note: Percentage changes from climate model simulations. Baseline yield values were computed from observed climate data for 1960–1990 and from model projections for 2020–2039, then interpolated to 2012–2015. P2.5 = 2.5 percentile level of climate model simulation yield results; P25 = 25th percentile; P75 = 75th percentile; and P97.5 = 97.5 percentile.

agroecological zones in 2035 are 2.7 percent to 5.0 percent higher than average yields in the 1960–1990 period, but only 0.6 percent to 1.5 percent higher in the two lowland agroecological zones (excluding the pastoralist zone) (Table 4.3). Median wheat yields are lower in each region, ranging from −0.3 percent in the drought-prone highlands to −1.9 percent in the drought-prone lowlands in 2035 (Table 4.4). Sorghum yields vary most across regions, with median yields in 2035 having fallen sharply by 6.2 percent due to climate change in the humid moisture-reliable lowlands but rising in all other regions by between 1.4 percent and 3.7 percent (Table 4.5).

TABLE 4.4 Simulated climate change effects on wheat yields by agroecological zone in Ethiopia, relative to 2013 yields (%)

	Drought prone		Humid moisture reliable				
Year	Highland	Lowland	Lowland	Highland-cereal	Highland-enset	Pastoralist	Ethiopia
2035							
P2.5	−0.1	−1.4	−0.8	−0.5	−1.1	−0.9	−0.4
P25	−0.2	−1.0	−0.6	−0.4	−0.6	−0.4	−0.3
Median	*−0.1*	*−0.7*	*−0.4*	*−0.3*	*−0.6*	*−0.6*	*−0.3*
P75	−0.1	−0.4	−0.4	−0.3	−0.5	−0.2	−0.3
P97.5	0.2	−0.2	0.3	−0.1	−0.3	−0.2	−0.1
2055							
P2.5	−0.3	−4.0	−2.5	−1.7	−2.8	−2.4	−1.5
P25	−0.9	−2.5	−1.8	−1.4	−2.2	−1.4	−1.2
Median	*−0.6*	*−1.7*	*−1.1*	*−1.2*	*−1.8*	*−1.4*	*−1.2*
P75	−0.6	−1.1	−0.7	−1.0	−1.4	−0.7	−1.0
P97.5	0.2	−0.5	−0.8	−0.5	−0.9	−0.6	−0.4
2085							
P2.5	−1.8	−5.8	−5.3	−4.5	−7.1	−3.4	−4.3
P25	−2.5	−4.1	−3.9	−3.5	−4.9	−2.5	−3.2
Median	*−2.0*	*−3.1*	*−2.6*	*−2.8*	*−3.9*	*−2.3*	*−2.7*
P75	−1.3	−2.2	−1.3	−2.2	−3.2	−0.9	−2.1
P97.5	0.0	−1.3	−0.1	−1.1	−1.7	−0.5	−1.1

Source: Authors' calculations from model simulations.

Note: Percentage changes from climate model simulations. Baseline yield values were computed from observed climate data for 1960–1990 and from model projections for 2020–2039, then interpolated to 2012–2015. P2.5 = 2.5 percentile level of climate model simulation yield results; P25 = 25th percentile; P75 = 75th percentile; and P97.5 = 97.5 percentile.

Implications of Technical Change Coupled with Climate Change on Crop Yields

Given that cereal yields in Ethiopia are considerably lower than those in most developed countries, there is substantial potential for technical change through increased use of improved seeds and fertilizer and better water management to lead to significant increases in cereal yields over time (van Ittersum et al. 2016). Maize, wheat, and sorghum yields increased by 4.6 percent, 7.2 percent, and 2.3 percent per year, respectively, over the 2010/2011–2015/2016 Ethiopian calendar years.[10] Further yield increases of

10 These calculations are based on Ethiopia, Central Statistical Agency (CSA) data. Note that the Ethiopian calendar year begins and ends in mid-September.

TABLE 4.5 Simulated climate change effects on sorghum yields by agroecological zone in Ethiopia, relative to 2013 yields (%)

	Drought prone		Humid moisture reliable				
Year	Highland	Lowland	Lowland	Highland-cereal	Highland-enset	Pastoralist	Ethiopia
2035							
P2.5	0.4	1.0	−3.3	1.3	0.1	0.6	0.4
P25	0.6	1.0	−2.9	1.4	0.8	0.8	0.5
Median	*0.8*	*1.1*	*−2.3*	*1.3*	*0.9*	*0.5*	*0.6*
P75	0.7	1.3	−1.2	1.4	1.4	1.1	0.6
P97.5	1.6	1.7	−0.7	1.7	2.3	1.7	0.8
2055							
P2.5	0.1	1.0	−10.1	2.8	1.3	−0.1	0.4
P25	0.3	1.7	−8.8	2.8	1.9	0.4	0.4
Median	*0.8*	*1.8*	*−6.9*	*2.6*	*2.0*	*0.6*	*0.5*
P75	0.5	2.0	−4.9	2.7	2.5	1.3	0.7
P97.5	2.5	4.1	−2.9	3.3	6.2	3.0	1.3
2085							
P2.5	−4.7	0.4	−25.2	2.8	1.6	−3.7	−3.3
P25	−2.2	1.8	−19.3	3.1	2.8	−1.2	−2.1
Median	*−0.6*	*2.3*	*−16.0*	*3.4*	*3.0*	*−0.3*	*−0.9*
P75	−0.3	2.8	−12.5	3.8	3.7	2.1	−0.5
P97.5	4.4	7.4	−1.9	5.2	11.5	5.6	2.6

Source: Authors' calculations from model simulations.

Notes: Percentage changes from climate model simulations. Baseline yield values were computed from observed climate data for 1960–1990 and from model projections for 2020–2039, then interpolated to 2012–2015. P2.5 = 2.5 percentile level of climate model simulation yield results; P25 = 25th percentile; P75 = 75th percentile; and P97.5 = 97.5 percentile.

this magnitude are not likely, however, so we model a much more moderate growth in yields due to technical change of 1.5 percent per year for maize and 1.0 percent per year for wheat and sorghum. Under these assumptions technical change (with no climate change) would result in a 38.8 percent increase in maize yields between 2013 and 2035; wheat and sorghum yields would each increase by 23.9 percent (see Table 4.2; Figure 4.7).

Even with these rather modest rates of yield gains from technical change, the effects of technical change are far greater than the effects of climate change, so that average yields of maize, wheat, and sorghum overall, taking into account the effects of climate change on yields, will increase by 40.5 percent, 23.5 percent, and 24.6 percent, respectively, between 2013 and

FIGURE 4.7 Simulated yields in 2035 in Ethiopia with technical change and climate change relative to 2013 yields (%)

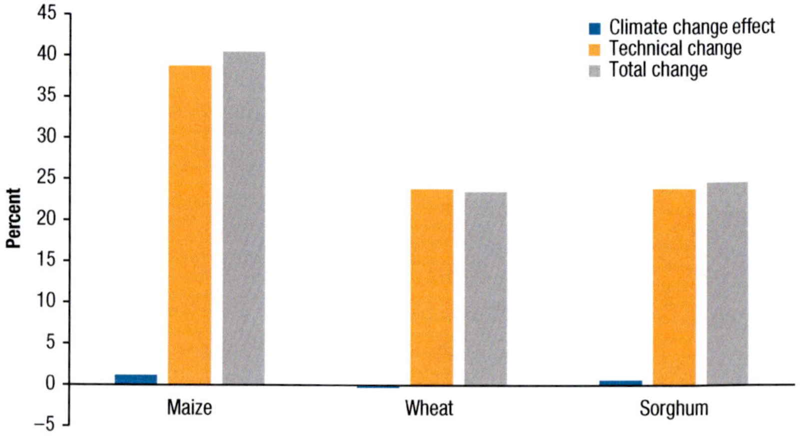

Source: Authors' calculations from model simulations.
Note: Yields estimated from percentage changes in yields from climate change model simulations.

2035 (Table 4.2; Figure 4.7). By 2055 the total yield gains would be substantially larger—91.8 percent for maize, 49.4 for wheat, and 52.0 percent for sorghum.

Conclusion

The model simulations suggest that climate change will likely have only relatively small effects on average yields of maize, wheat, and sorghum in Ethiopia in 2035 and even in 2055. Although temperatures are expected to increase, average rainfall is also expected to increase in most regions of the country. Thus agronomic conditions for cultivation of these crops may actually improve in large parts of the country, especially the highlands, that currently have moderate average temperatures.

Nonetheless, crop yields will need to increase to enable cereal production to keep pace with expected demand growth due to increases in population and per capita incomes. This in turn will require continued public and private investments in agriculture and rural infrastructure as well as policies that maintain incentives for modern input use and adoption of new technology. These investments and policies are particularly important because in many parts of the country land degradation is reducing yields, necessitating further investments in sustainable land management as well as increased fertilizer use to enable even moderate yield gains (Schmidt et al. 2017).

Moreover, even if future changes in climate have only moderate impacts on *average* crop yields in Ethiopia, there is growing evidence that weather outcomes, particularly rainfall, are likely to become more variable in the future. For example, Pendergrass et al. (2017) suggest that precipitation variability over land rises by 4 percent to 5 percent for every degree Celsius increase in temperature. Thus there could still be substantial effects on crop production and household welfare (as well as on livestock) due to extreme events—droughts, floods, or extremely high temperatures. And because the effects of rainfall on yield is more U-shaped than linear, a rise in variability, even without a shift in the mean, almost always leads to a reduction in average yields and not just yields under extreme weather. The findings presented in this chapter are no cause for complacency in Ethiopian agriculture in the face of climate change.

References

Admassu, H., G. Mezgebu, T. S. Thomas, M. Waithaka, and M. Kyotalimye. 2013. "Ethiopia." In *East African Agriculture and Climate Change: A Comprehensive Analysis*, edited by M. Waithaka, G. C. Nelson, T. S. Thomas, and M. Kyotalimye, 149–182. Washington, DC: International Food Policy Research Institute (IFPRI).

Collins, W., N. Bellouin, M. Doutriaux-Boucher, N. Gedney, P. Halloran, T. Hinton, J. Hughes, et al. 2011. "Development and Evaluation of an Earth System Model—HadGEM2." *Geoscience Model Development* 4 (4): 1051–1075.

Dile, Y. T., R. Berndtsson, and S. G. Setegn. 2013. "Hydrological Response to Climate Change for Gilgel Abay River, in the Lake Tana Basin—Upper Blue Nile Basin of Ethiopia." *PLoS ONE* 8 (10): e79296.

FAO (Food and Agriculture Organization of the United Nations). 2019. FAOSTAT database. Accessed January 24, 2019. www.fao.org/faostat/en/.

Hijmans, R. J., S. E. Cameron, J. L. Parra, P. G. Jones, and A. Jarvis. 2005. "Very High Resolution Interpolated Climate Surfaces for Global Land Areas." *International Journal of Climatology: A Journal of the Royal Meteorological Society* 25 (15): 1965–1978.

Hoogenboom, G., J. W. Jones, P. W. Wilkens, C. H. Porter, K. J. Boote, L. A. Hunt, U. Singh, et al. 2012. *Decision Support System for Agrotechnology Transfer (DSSAT)*. Version 4.5. Honolulu: University of Hawaii, Honolulu. CD-ROM.

IPCC (Intergovernmental Panel on Climate Change). 2018. *Global Warming of 1.5°C. An IPCC Special Report on the Impacts of Global Warming of 1.5°C above Pre-Industrial Levels and Related Global Greenhouse Gas Emission Pathways, in the Context of Strengthening the Global Response to the Threat of Climate Change, Sustainable Development, and Efforts to Eradicate Poverty.* Geneva, Switzerland: World Meteorological Organization.

Jones, J. W., G. Hoogenboom, C. H. Porter, K. J. Boote, W. D. Batchelor, L. A. Hunt, P. W. Wilkens, U. Singh, A. J. Gijsman, and J. T. Ritchie. 2003. "The DSSAT Cropping System Model." *European Journal of Agronomy* 18 (3–4): 235–265.

Kim, U., and J. J. Kaluarachchi. 2009. "Climate Change Impacts on Water Resources in the Upper Blue Nile River Basin, Ethiopia." *Journal of the American Water Resource Association* 45 (6): 1361–1378.

Kim, U., J. J. Kaluarachchi, and V. U. Smakhtin. 2008. *Climate Change Impacts on Hydrology and Water Resources of the Upper Blue Nile River Basin, Ethiopia.* International Water Management Institute (IWMI) Research Report 126. Colombo, Sri Lanka: IWMI.

Martin, G., N. Bellouin, W. Collins, I. Culverwell, P. Halloran, S. Hardiman, and T. Hinton. 2011. "The HadGEM2 Family of Met Office Unified Model Climate Configurations." *Geophysical Model Development* 4: 723–757.

Nawaz, R., T. Bellerby, M. Sayed, and M. Elshamy. 2010. "Blue Nile Runoff Sensitivity to Climate Change." *Open Hydrology Journal* 4: 137–151.

Pendergrass, A. G., R. Knutti, F. Lehner, C. Deser, and B. M. Sanderson. 2017. "Precipitation Variability Increases in a Warmer Climate." *Scientific Reports* 7: 17966.

Ramirez, J., and A. Jarvis. 2008. *High Resolution Statistically Downscaled Future Climate Surfaces.* Cali, Colombia: International Center for Tropical Agriculture: CGIAR Research Program on Climate Change, Agriculture and Food Security (CCAFS).

Ramirez-Villegas, J., and A. Jarvis. 2010. *Downscaling Global Circulation Model Outputs: The Delta Method.* Decision and Policy Analysis Working Paper 1. Cali, Colombia: International Center for Tropical Agriculture.

Robinson, S., D. Willenbockel, and K. Strzepek. 2012. "A Dynamic General Equilibrium Analysis of Adaptation to Climate Change in Ethiopia." *Review of Development Economics* 16 (3): 489–502.

Ruane, A. C., R. Goldberg, and J. Chryssanthacopoulos. 2015. "Climate Forcing Datasets for Agricultural Modeling: Merged Products for Gap-Filling and Historical Climate Series Estimation." *Agricultural and Forest Meteorology* 200: 233–248.

Schmidt, E., P. Chinowsky, S. Robinson, and K. Strzepek. 2017. "Determinants and Impact of Sustainable Land Management (SLM) Investments: A Systems Evaluation in the Blue Nile Basin, Ethiopia." *Agricultural Economics* 48 (5): 613–627.

Schmidt, E., and T. S. Thomas. 2018. *Cropland Expansion in Ethiopia: Economic and Climatic Considerations for Highland Agriculture.* Ethiopia Strategy Support Program (ESSP) Working Paper 127. Addis Ababa: IFPRI.

Taye, T. M., P. Willems, and P. Block. 2015. "Implications of Climate Change on Hydrological Extremes in the Blue Nile Basin: A Review." *Journal of Hydrology: Regional Studies* 4 (B): 280–293.

van Ittersum, M. K., L. G. J. van Bussel, J. Wolf, P. Grassini, J. van Wart, N. Guilpart, L. Claessens, et al. 2016. "Can Sub-Saharan Africa Feed Itself?" *Proceedings of the National Academy of Sciences of the United States of America* 113 (52): 14964–14969.

Wagena, M. B., A. Sommerlot, A. Z. Abiy, A. S. Collick, S. Langan, D. R. Fuka, and Z. M. Easton. 2016. "Climate Change in the Blue Nile Basin Ethiopia: Implications for Water Resources and Sediment Transport." *Climatic Change* 139: 229–243.

WorldClim Version 1.4. 2019. Climate data. Accessed January 24, 2019. www.worldclim.org/.

You, L., S. Wood, and U. Wood-Sichra. 2006. "Generating Global Crop Maps: From Census to Grid." Selected paper presentation at the International Association of Agricultural Economists annual conference, Gold Coast, Australia, August 12–18.

You, L., U. Wood-Sichra, S. Fritz, Z. Guo, L. See, and J. Koo. 2014. "Spatial Production Allocation Model (SPAM) 2005 v2.0." Accessed January 24, 2019. http://mapspam.info.

Chapter 5

EVOLVING LIVESTOCK SECTOR

Fantu Bachewe, Bart Minten, Fanaye Tadesse, and Alemayehu Seyoum Taffesse

Introduction

Ethiopia is home to some of the largest numbers of different livestock species in the world.[1] For example, the stock of cattle in Ethiopia is the fifth largest in the world, larger than the stocks in important beef producing countries, such as Argentina and Australia (FAO 2018). The value of livestock was evaluated at US$11 billion in 2015, four times the size of all agricultural exports from the country. Almost all Ethiopian farmers hold some livestock, and it provides for significant employment in rural areas. In short, the animal stocks and the number of people engaged in livestock production indicate that the livestock subsector is a central component of the economy of Ethiopia.

However, livestock production contributed little to the rapid economic growth recorded in the past decade in Ethiopia. While livestock output grew at 5.8 percent per year over the past decade, this was slower than overall GDP growth. Livestock output, therefore, declined in importance. Given this relatively slow growth, the subsector accounted for only 0.5 of the 10.5 percent growth in GDP recorded during the period studied. That is, out of every 100 birr added to the economy, less than 5 birr came from the livestock sector, while crop output growth contributed about five times as much, 24 birr (NBE 2017).

Yet, following the rapid growth in incomes in Ethiopia over the past decade, real expenditure on animal-sourced foods (ASF) and its share in total food spending grew between 2000 and 2011 (Tafere and Worku 2012), partly driven by increases in ASF prices (Chapter 8). Moreover, with further increase in income, ASF are expected to increase in importance, given the changing food preferences of consumers as their incomes increase but also because of the

1 Ethiopia had the largest stock of donkeys in the world in 2016. The stocks of goats and horses are both eighth largest, and those of mules, camels, and sheep ranked fourth, seventh, and ninth, respectively (FAO 2018).

nutritional benefits of ASF. Furthermore, exports of meat, live animals, and animal products increased in importance from 11 percent of the total value of exports in 2004/2005 to 13 percent in 2015/2016 (NBE 2017). Despite these growing market opportunities and largely untapped resource potentials, livestock production, productivity, and marketed supply remain stagnant.

A number of previous studies investigated different aspects of the livestock subsector in Ethiopia.[2] Those studies indicate weak institutions, markets, and policymaking, particularly noting the absence of a clear policy framework as an important bottleneck resulting in weak performance of the livestock subsector and market chain. This chapter complements these studies by providing a comprehensive overview of animal-sourced food and live animal production, productivity, and input use at the national level.

This chapter uses descriptive analyses to study the dynamics in the stock of live animals, average holdings, and composition of live animals over the 2004/2005 to 2015/2016 period. It also studies production and productivity of live animals and ASF as well as input use. The chapter relies on econometric analyses to study factors associated with the adoption of modern inputs in livestock production, while a growth accounting method is used to analyze the different sources contributing to growth in real livestock output. For this purpose we use data from the Central Statistical Agency of Ethiopia (CSA) that is representative of mixed crop-livestock farmers in Ethiopia, by far the most important segment of livestock production systems in the country.[3]

We find that recent growth in livestock output came largely from increases in the number of livestock and livestock owners, both of which were over 50 percent higher in 2015/2016 than in 2004/2005, while modern input use and improvements in production methods contributed little. This growth path contrasts with the crop sector where modern input adoption played a large role in recent growth (Bachewe et al. 2018). Given increasing livestock

---

2 They include studies on livestock production and marketing (Negassa and Jabbar 2008; Negassa, Rashid, and Gebremedhin 2011; Solomon et al. 2003), milk and meat production and marketing (Anteneh et al. 2010), live animal and meat export value chains (Legesse et al. 2008); determinants of herd stocking decisions (Tamirat 2013), livestock holding size and off-take rates (Negassa and Jabbar 2008), characteristics of pastoralist livestock production and marketing (Barrett et al. 2004), and the impacts of risk and wealth on cattle herd size and marketing behavior (Lybbert et al. 2010).

3 Unfortunately, there is a lack of updated representative data on pastoralist livestock systems in the country, so we focus on these mixed systems. In any case, we estimate that the stock of cattle in areas dominated by mixed crop livestock agriculture accounted for more than 90 percent of the nationwide stock in 2015. Similarly, the stock of sheep, goats, equines, and poultry accounted for 78 percent, 63 percent, 93 percent, and more than 97 percent of the nationwide stock of the animals, respectively.

density per unit of land, we find that feeding practices are changing as access of producers to grazing land is declining. Reliance on commercial feed markets is increasing. Adoption of improved breeds and improved feeds increased rapidly over the past decade, but it started from a low base. Although veterinary service provision also increased considerably, and livestock death rates slightly declined, the number lost to deaths is still more than twice the number sold for meat consumption. Results of the econometric analyses further indicate that adoption of modern inputs in livestock production is positively associated with, among other factors, education level of farmers, access to livestock extension, and proximity to markets and urban centers. Improvements in these factors might therefore help to stimulate modern input adoption in the future.[4]

The remainder of this chapter describes the main dataset used in the study, discusses trends in the numbers and growth rates of livestock, discusses livestock output and productivity, and deals with inputs used in livestock production. The last part of the chapter concludes.

Data Sources and Coverage

The Central Statistical Agency of Ethiopia (CSA) collects agricultural data in its annual Agricultural Sample Survey (AgSS). The AgSS is conducted through a survey of many households. In an average year during the period covered in this study (2004/2005–2015/2016), 63,000 households were surveyed. Data is collected, among others, on rural households' production of crops and livestock. CSA publishes in its annual statistical reports a summary of these data, which are representative at national, regional, and zonal levels. The data pertain to smallholder farmers, who dominate the Ethiopian agriculture sector.[5] Our analyses rely on CSA data taken from both household level (raw) data and the statistical reports, while we complement our analyses of these data in this chapter with data from other sources.[6] Unless indicated

4 With the recent focus of the Ethiopian government on livestock, resulting in a Livestock Sub-ministry being established within the Ministry of Agriculture, Livestock, and Natural Resources, if investments envisaged in the livestock sector roadmap are implemented, livestock production and productivity are expected to improve considerably and make significant contributions to reduction of rural poverty (Shapiro et al. 2015).

5 While several large-scale livestock ranches are being established (see, for example, www.verdebeef.com/), their importance in the national livestock sector is limited.

6 We compare CSA data on live animal stocks with that of FAO (2018) data. The numbers from the two sources are identical in most cases, particularly those since 2009. In only 11 of the 96 cases compared is the number from one source more than 1 percent lower or higher than the

otherwise, the descriptive analyses presented in this chapter cover the period from 2004/2005 to 2015/2016.[7]

Households in Afar and Somali regions as well as in pastoralist (southern) parts of Oromia rely heavily on livestock production for their livelihood. Consequently, these areas also are home to a considerable proportion of the country's livestock population. Therefore, a study that investigates livestock production in Ethiopia needs to give due attention to these areas. However, during the period covered in this study, the sampling frame of the AgSS excluded the nonsedentary areas of three (of five) zones of Afar and six (of nine) zones of Somali regions (Ethiopia, CSA 2005–2016). CSA collected data on livestock production in the two predominantly pastoralist regions in its pastoralist survey, which was conducted in 2003. However, we do not use that dataset in our analyses for two reasons. First, the data are too outdated to provide timely insights. Second, although the survey covered households in all five zones of Afar, it covered only one zone and left out all six Somali zones not included in the annual AgSS. Therefore, we rely on data from Awsi and Gabi zones in Afar; Siti, Fafan, and Liben zones in Somali; and Borena zone in Oromia, which are covered in CSA's annual AgSS, for our description of pastoralist areas. Using these data, we characterize the pastoralist areas in terms of average livestock ownership and importance in nationwide stock of live animals and animal-sourced food production.

Livestock Size, Composition, and Growth

We describe trends in the proportions of households that own livestock, and the number, composition, and growth rates of livestock during the 2004–2015 period.

Number, Structure, and Composition of Livestock

Data from CSA annual reports between 2005 and 2016 indicate enormous growth in the number of livestock in the country. The data, summarized in Table 5.1, show that the cattle stock in Ethiopia stood at 57.8 million at the end of 2015, 49 percent higher than the number at the end of 2004. The number of sheep in 2015, about 29 million, increased by 60 percent during the period. The stocks of goats, equines (horses, mules, and donkeys), camels, and poultry

other. We could not do similar comparisons for other variables, as equivalently defined data are unavailable.

7 In the remainder of this chapter we write 2004/2005 as 2004, 2015/2016 as 2015, and so forth.

TABLE 5.1 Livestock numbers in Ethiopia, by type (2004–2015)

Livestock type	Number (millions)			Growth rate 2004–2015 (%)
	2004	2010	2015	
Tropical livestock units	**34.5**	**48.7**	**54.0**	**56.4**
Cattle	38.7	53.4	57.8	49.4
Sheep	18.1	25.5	28.9	59.7
Goats	14.9	22.8	29.7	99.3
Equines	5.8	8.6	10.4	79.3
Camels	0.5	1.1	1.2	140.0
Poultry	30.9	49.3	60.5	95.8

Source: Authors' analyses using CSA data (Ethiopia, CSA 2005–2016).

were 99 percent, 79 percent, 140 percent, and 96 percent higher in 2015 than in 2004, at which time the stocks were 14.9, 5.8, 0.5, and 30.9 million, respectively.

Most Ethiopian farmers own livestock. About 92 percent of farm households surveyed owned at least one type of livestock in 2015 (Table 5.2). Cattle are the most common livestock owned, followed by poultry. The trend in the proportion that owned at least one type of livestock remained about the same during the period. However, the absolute number of livestock-owning households increased by about 51 percent, from 10.3 million in 2004 to 15.5 million in 2015. The trend was also similar for the proportion that owned each livestock species, except for a slight increase in the proportion owning camels and equines, mainly donkeys.

Considering households that own livestock, the average number of cattle and chickens owned was the highest at about four head per household in 2015 (Table 5.2). Yet the average number of equines and camels owned was less than one. Except for a slight increase in goats and poultry, there was no notable variation in average livestock ownership over the years. Most of the increase in livestock numbers seems to have come from an increase in the number of households owning livestock.

Table 5.2 also provides information on the composition of the livestock owned by households—the contribution of each livestock type to the total tropical livestock units (TLUs) owned by households.[8] An average household owned around 3.7 TLUs in 2015. Total TLU increased by 56.4 percent or an

8 TLU normalizes the number of livestock into camel units. It is computed using the formula: TLU = camels + (0.7*cows) + (0.8*horses) + (0.5*donkeys) + (0.5*mules) + (0.1*sheep) + (0.1*goats) + (0.01*chickens).

TABLE 5.2 Livestock ownership and composition in Ethiopia, by type (2004–2015)

Livestock type	Households that own livestock (%)			Average livestock owned, head count, among owners			Livestock composition, percent of tropical livestock units		
	2004	2010	2015	2004	2010	2015	2004	2010	2015
All livestock	92.6	92.9	92.3	—	—	—	—	—	—
Cattle	82.3	82.1	79.7	4.0	4.0	3.9	73.1	71.7	68.7
Sheep	35.7	36.0	35.2	1.9	1.9	2.0	7.2	7.4	7.6
Goats	27.4	27.5	27.7	1.5	1.7	2.0	5.6	5.7	6.1
Equines	33.9	37.3	41.3	0.6	0.7	0.7	8.3	9.2	10.7
Camels	1.0	1.4	1.7	0.1	0.1	0.1	0.4	0.5	0.6
Poultry	57.2	58.3	58.3	3.2	3.7	4.0	5.5	5.5	6.3

Source: Authors' analyses using CSA holder-level data (Ethiopia, CSA 2017).
Note: — = data not available.

average annual rate of 4.2 percent during the period. In 2015 cattle and small ruminants (sheep and goats) accounted for around 69 percent and 14 percent of the total TLUs, respectively. The composition of TLU changed little between 2004 and 2015; notable changes include a decline in the contribution of cattle to total TLU by 4.0 percent and an increase in equines to total TLU by 2.4 percent.

We further measure the real value of the stock of livestock in the country using data on producer prices of livestock collected by CSA (Ethiopia, CSA 2017) and data on the number of livestock. We also use zonal numbers of livestock to weight the importance of each zone in the real price of each type of livestock.[9] The real value of livestock type *x* in a given year is computed as a multiple of the number of livestock type *x* in that year and its weighted average price in December 2011. We use the US dollar to birr exchange rate in December 2011 to convert that value to US dollars.

We provide a summary of these computations in Table 5.3. In 2004 the stock of livestock in the zones covered by the CSA data were worth 121.2 billion birr (in December 2011 prices). This grew to 192 billion birr (US$11.2 billion) in 2015. That is, the real value of livestock grew by 58.4 percent or an average annual rate of 4.3 percent, which is close to the growth rate of livestock size measured in TLU. Growth in real value was

9 The weights ensure that zones with unduly high or low prices but with relatively few livestock do not overly influence prices. Each administrative zone's weight in livestock type *x* price is computed for each year as: weight = (number of *x* in zone/total number of *x* nationally). Then for each zone we take as the weight the average of the weights so computed.

TABLE 5.3 Real value of the stock of live animals in Ethiopia, December 2011 prices

	Value (billions)		Share out of real value of livestock (%)					
Year	birr	US$	Cattle	Sheep	Goats	Equines	Camels	Poultry
2004	121.24	7.1	77.8	6.5	5.8	7.1	1.6	1.2
2010	171.67	10.0	75.9	6.5	6.3	7.2	2.7	1.4
2015	192.03	11.2	74.4	6.6	7.4	7.5	2.6	1.5

Source: Authors' analyses using CSA data (Ethiopia, CSA 2005–2016, 2017).

lowest in cattle (52 percent) and sheep (60 percent), while it was highest for goats (100 percent) and camels (168 percent). The contribution of each livestock species remained about the same during the period. However, the lower growth rate in the real value of cattle is reflected in the decline in the share of cattle in the total real value of livestock. The rapid growth in the real value of goats and camels is reflected in the higher increase in the share of the animals in real value of livestock.

Livestock Stocks in the Pastoralist Areas of Ethiopia

We summarize CSA data on livestock ownership in the six pastoralist zones in Table 5.4. Almost all households in the six pastoralist zones owned livestock in 2015. The proportion of pastoralist households that owned cattle was 83 percent in 2004, but this declined to 69 percent in 2015. The proportion that owned sheep, goats, camels, and poultry in 2015 was 53 percent, 67 percent, 25 percent, and 22 percent, respectively. The data indicate that the proportion of pastoralist farmers that owned cattle and poultry in 2015 was 15 percent and 18 percent lower, respectively, relative to 2004, while the proportion that owned sheep, goats, and camels increased by 15 percent, 5 percent, and 6 percent, respectively. This indicates that pastoralists are shifting away from cattle and poultry, focusing more on camels and small ruminants (see, for example, Aklilu and Catley 2010).

Cattle ownership per household averaged 5.4 in 2015, which was 1.5 lower than that in 2004. In contrast, sheep, goat, and camel ownership in 2015 was 3.0, 4.5, and 0.4, respectively, higher than average ownership in 2004, which reinforces the observation that pastoralist farmers may be shifting toward camels and small ruminants. The six pastoralist zones accounted for 4.7 percent of the total number of cattle in Ethiopia in 2015, but accounted

TABLE 5.4 Livestock ownership by type, pastoralist areas in Ethiopia (2004–2015)

	Households that own livestock (%)			Average livestock owned, head count, among owners			Livestock composition, % of tropical livestock units		
Livestock type	2004	2010	2015	2004	2010	2015	2004	2010	2015
All livestock	**96.1**	**94.3**	**96.1**	—	—	—	—	—	—
Cattle	83.4	73.4	68.7	6.9	6.7	5.4	4.8	4.9	4.7
Sheep	38.3	46.0	53.2	3.1	5.9	6.1	4.7	9.2	10.5
Goats	61.6	61.9	66.8	6.8	10.0	11.3	12.3	17.4	19.4
Equines	36.3	33.4	34.8	0.6	0.6	0.6	2.9	2.9	2.9
Camels	18.9	22.5	24.9	1.2	2.0	1.6	71.3	70.8	67.8
Poultry	39.5	29.1	21.8	2.3	2.0	1.7	2.0	1.6	1.4

Source: Authors analyses using CSA holder level data (Ethiopia, CSA 2017).
Note: — = data not available.

for nearly 11 percent, 20 percent, and 68 percent of the total number of sheep, goats, and camels, respectively.

We further estimate the contribution of all pastoralist zones (including those not in the AgSS data) to the total stock of animals in Ethiopia under the simple assumption that the stock of each animal species in each of the zones in Afar and Somali regions not covered in CSA's AgSS is equal to the average of the stocks in the zones covered in the AgSS.[10] Under this assumption the cattle stock in pastoralist areas—that is, Afar and Somali regions and the Borana zone in Oromia region—accounted for 9.6 percent of the nationwide stock of cattle in the country in 2015. Similarly, the share of the national stock of sheep, goats, and camels in the pastoralist areas is estimated at 22 percent, 37 percent, and 84 percent, respectively.

Livestock Technical Performance

Two important livestock technical performance parameters are annual death rates and annual growth rates. These parameters are measured by looking at retrospective 12-month recall data for each of the years considered in the study. The annual death rate of each livestock type is calculated by dividing

10 Average livestock holdings of farmers in nonsedentary parts of Afar and Somali regions is likely to be higher than those in sedentary parts. However, since the nonsedentary parts are sparsely populated, we assume that the two effects cancel each other.

FIGURE 5.1 Annual death rates by livestock type in Ethiopia, 2005–2015

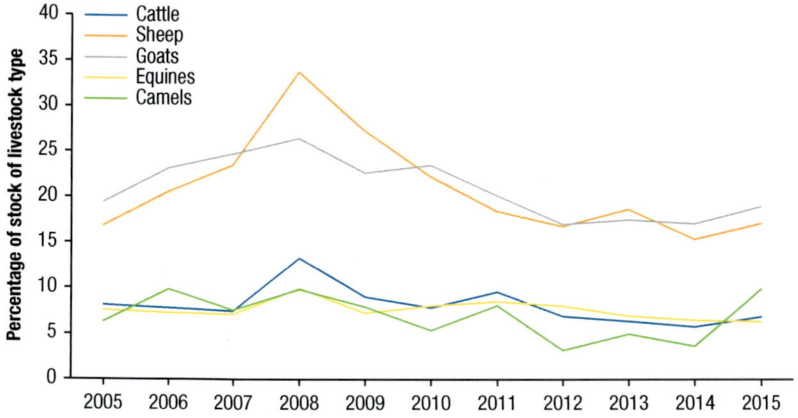

Source: Authors' analyses using CSA holder-level data (Ethiopia, CSA 2017).

the total number of deaths by the average stock of that livestock type.[11] CSA data indicate that livestock death rates generally increased between 2004 and 2008, when there was a spike in death rates of all species but has generally declined since (Figure 5.1). Average death rates during the 2005–2015 period were close for cattle, equines, and camels at around 7 percent, while they were higher at about 20 percent for sheep and goats. Relative to 2005, death rates in 2014 were about 1 percent lower for sheep and equines, 2 percent lower for cattle and goats, and 3 percent lower for camels.

TABLE 5.5 Annual growth rates by livestock type in Ethiopia, 2005, 2010, and 2015 (%)

Livestock type	2005	2010	2015	Average
Cattle	4.0	1.2	2.0	3.8
Sheep	14.6	1.3	1.5	4.6
Goats	9.4	2.0	2.0	6.7
Equines	7.4	6.8	5.2	5.6
Camels	–4.8	10.6	5.5	9.4
Poultry	3.8	12.8	6.4	6.6

Source: Authors' analyses using CSA holder-level data (Ethiopia, CSA 2017).

Annual growth rate in the aggregate stock of each livestock type is calculated by dividing the difference between the ending and beginning stocks by the beginning stock of that individual animal. Growth rates have been positive for all livestock species in most years (Table 5.5). This excludes sheep and

11 This number includes livestock deaths due to diseases or other reasons and does not include slaughters.

camels, in which negative growth rates were observed in four of the 11 years. However, average annual growth rates were positive for all livestock species. Average annual growth rate was the lowest for cattle at 3.8 percent, while it was the highest for camels at 9.4 percent. Growth averaged 4.6 percent, 5.6 percent, 6.6 percent, and 6.7 percent for sheep, equines, poultry, and goats, respectively.

Livestock and Animal-Sourced Food Production and Productivity

Livestock serve a multitude of purposes, including, among others, providing food for farm households, being sold when the household is in need of money, and being used to plow the land used in crop production. Consequently, livestock constitute an important source of income for farm households and ultimately for the national economy. For example, livestock income accounted for 11 percent of the total income of rural households in the Agricultural Growth Program baseline survey datset, which included 93 woredas in Tigray, Amhara, Oromia, and Southern Nations, Nationalities, and Peoples (SNNP) regions (Chapter 11). When this analysis is expanded to include households in Afar, Somali, Gambella, and Benishangul-Gumuz regions, other regions in which livestock production is important, the share of livestock income is significantly higher at 21 percent.

The livestock subsector has historically been an important input supplier for Ethiopia's manufacturing sector. Established in the mid-1920s, leather processing and leather articles producing firms were among the earliest manufacturing enterprises in Ethiopia (Oqubay 2015). However, over many years the performance of the leather processing and leather products manufacturing subsector remained poor and thus was unable to take advantage of the rich primary inputs Ethiopia's livestock subsector could provide (Oqubay 2015). However, the contribution of the livestock subsector to the manufacturing sector is expected to grow following recent improvements in the leather processing and manufacturing, meat processing, and dairy processing subsectors (AACCSA 2015, 2016).

Next we discuss trends in animal-sourced food production, specifically of milk and eggs, and productivity and trends in livestock used in meat production (net commercial off-take). Trends in real output and productivity are also discussed. Finally, results of growth accounting analyses conducted to analyze the sources of growth in real livestock output are discussed.

Milk and Egg Production and Productivity

Total milk production increased by about 41 percent between 2005 and 2015, while growth in egg production was 84 percent over the same period (Figure 5.2).[12] Despite the significant growth in outputs, milk and egg productivity (output per animal) stagnated during the period. Milk and egg productivity in 2015 were 5 percent and 8 percent lower than in 2005, respectively. Figure 5.2 indicates that there were no significant changes in productivity during the interim period. The stagnant productivity per animal implies that the rapid growth in output resulted from the increase in the number of livestock and livestock farmers, a claim corroborated by the growth accounting analyses discussed below.

Milk production of the six pastoralist zones accounted for about 12 percent of the national milk production in both 2005 and 2015. The contribution of these zones toward nationwide egg production has been less than 3 percent in all years studied and has generally declined. This is consistent with the summary in Table 5.4, which shows a decline in the share of farmers owning poultry and average ownership of poultry in pastoralist areas.

We use FAO (2018) data to compare milk productivity in Ethiopia with countries at similar and higher levels of development (Figure 5.3). Out of countries depicted, milk yields in Ethiopia are higher only than those of Nigeria. Other countries with lower milk yields than Ethiopia's include Bangladesh, Ghana, Senegal, and several other West African countries, while yields in all East African countries are higher than Ethiopia's. Milk yields in Uganda, Kenya, and Malawi in 2014 were 27 percent, 125 percent, and 239 percent higher, respectively, than yields in Ethiopia, while yields in the Asian countries were at least five times higher. Perhaps more important is the fact that milk productivity is stagnant in Ethiopia. While milk yields in 2014 were lower than in 2004 in Ethiopia and Uganda and grew only marginally in Nigeria, growth rates in the remaining countries and regions over this period were considerable. While these numbers indicate the opportunities in milk production of which Ethiopia can take advantage, they also indicate the scope of the challenges to increase productivity to levels observed in other regions.

---

12 Annual growth in milk and egg output averaged about 3.8 percent and 6.4 percent, respectively.

FIGURE 5.2 Milk and egg production and productivity, 2005–2015

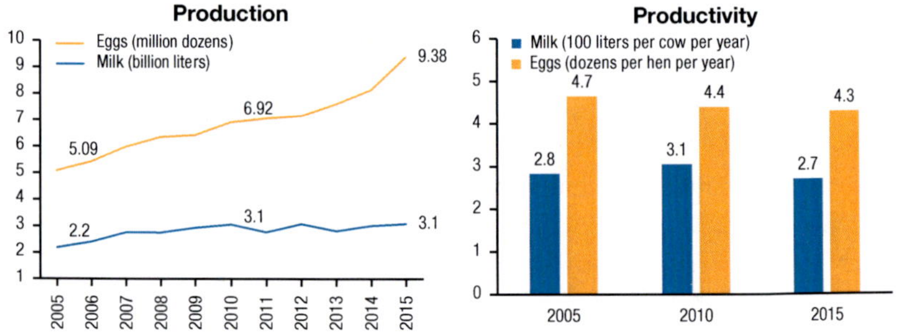

Source: Authors' analyses using CSA data (Ethiopia, CSA 2017).

FIGURE 5.3 International comparison of milk productivity, 2014 (kilograms per cow)

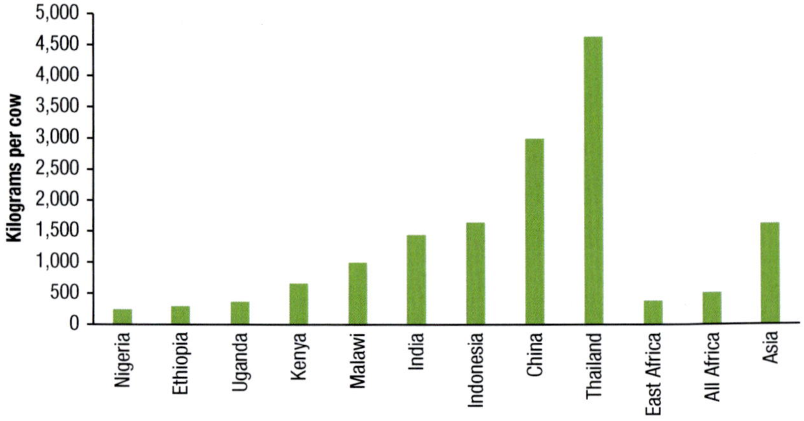

Source: Authors' analyses using FAO (2018) data.

Note: The numbers for East Africa, All Africa, and Asia pertain to 2012.

Meat Production and Exports: Net-Commercial Off-Take

The meat production capacity of the livestock subsector can be directly influenced by the number of live animals produced for such purposes as well as by the quantity of meat per animal produced. This section discusses the net-commercial off-take (NCOT) rates of livestock, while productivity in terms of meat per animal is discussed next. The NCOT rate is essentially the rate at which livestock leave the farm ultimately for use in meat production or for live exports. While the NCOT rate shows the net market position of households for each livestock type, we also discuss below three related (gross) market participation rates: sales off-take, slaughter off-take, and commercial intake (purchase) rates.

Results of computations of gross and net commercial market participation rates are presented in Table 5.6. Poultry has the highest sales off-take rate, followed by sheep, goats, and cattle, while camels have the lowest sales off-take rates. The proportion sold increased over the period for all animals, except equines, although the increase was generally low. The highest increase was observed for cattle and goats at 1.7 percent.

Chicken had the highest slaughter off-take rate of 28 percent in 2015 (Table 5.6). The second and third most slaughtered animals are sheep and goats, respectively. Cattle and camels are the least slaughtered animals. The proportion of slaughtered cattle changed little over the years. The proportion increased by more than 4 percent for chicken and by about 2 percent for small ruminants, but it declined for camels.

The livestock commercial purchase rate in 2015 was the highest for chickens (20 percent), followed by sheep (16 percent) and equines (10 percent), while it is the lowest for camels at 4 percent. The commercial intake rate has shown a slight decline over the period for all livestock types, except for cattle and poultry. In 2015 the net-commercial off-take (NCOT) rate was only 3.3 percent for cattle while it was higher at 7.7 percent, 9.7 percent, and 6.8 percent for sheep, goats, and poultry, respectively. The net market position in 2015 was 1 percent higher than in 2004 for cattle. It increased by nearly 2 percent for goats, was stagnant for sheep, and declined by about 2 percent for poultry.

Considering the six pastoralist zones, average NCOT rate of cattle was 5 percent in 2004 and 7 percent in 2015. These are higher than twice the corresponding national rates in Table 5.6. Similarly, the NCOT rates of sheep and goats were about the same at 17 percent for pastoralist areas in 2015, which is considerably higher than the nationwide rate. Looking into the numbers that constitute the NCOT rate reveals that, while sales rates in pastoralist areas are about the same as in other parts of the country, the higher NCOT rates in pastoralist areas are mainly due to relatively lower purchase rates in such areas.

The foregoing discussion indicates that growth in the NCOT rate or in the proportion of live animals sold for use in meat production has generally been low. The latter, together with the rapid increase in the real price of meat observed in the country during the period (see Chapter 8) imply, among

TABLE 5.6 **Proportion of livestock sold, slaughtered, and purchased in Ethiopia, by type (%)**

	Sales			Slaughters			Purchases			Net commercial off-take (NCOT)		
Livestock type	2005	2010	2015	2005	2010	2015	2005	2010	2015	2005	2010	2015
Cattle	9.4	10.6	11.1	0.7	0.8	0.9	7.3	7.9	8.1	2.4	3.0	3.3
Sheep	23.4	22.6	23.5	10.2	12.7	12.2	16.5	15.1	16.1	7.4	8.0	7.7
Goats	16.5	20.6	18.2	6.6	8.8	8.2	9.2	8.6	8.7	8.0	12.6	9.7
Equines	8.0	7.0	6.3	—	—	—	12.0	11.7	10.4	−4.0	−4.7	−4.1
Camels	4.3	6.5	4.9	1.5	0.9	0.3	4.4	4.3	3.7	1.4	3.0	1.5
Poultry	24.0	24.5	25.4	24.1	26.6	28.4	16.6	18.5	20.0	8.7	7.3	6.8

Source: Authors' analyses using CSA holder-level data (Ethiopia, CSA 2017).
Note: — = data not available.

others, that growth in the supply of livestock put to such uses was slower relative to growth in demand for meat. In other words, productivity in livestock produced for meat production grew slower than demand during the period studied.

Meat Productivity

We investigate trends in meat productivity per animal in Ethiopia using data from the Food and Agriculture Organization of the United Nations (FAO 2018). The data indicate that meat yield (carcass weight) per animal changed little between 2004 and 2014, the latest year for which data are available. Average meat yields of sheep, goats, poultry, and camels was the same during the entire period at 10 kilograms, 8.5 kilograms, 0.8 kilograms, and about 170 kilograms per animal, respectively (FAO 2018). Carcass weights for cattle were between 108.0 kilograms and 109.2 kilograms during the period.[13]

Data to triangulate the numbers on meat productivity per animal from FAO are unavailable. Nonetheless, we suspect meat productivity per animal may not have remained constant between 2004 and 2014 but may have increased, at least for poultry and cattle. This follows from CSA (Ethiopia, CSA 2017) data that shows increased adoption of hybrid and foreign poultry by smallholders, from the increasing number in Ethiopia of modern poultry farms (NABC 2011) that produce larger chicken varieties, and focus-group discussions conducted by the authors indicating increases in the number of

13 It is not very clear what data collection processes the FAO has used to come to these estimates.

institutions that fatten cattle for live exports and for domestic butcheries selling premium quality beef. However, whether and by how much these improvements have impacted meat yields needs to be empirically assessed.

Real Livestock Output and Productivity

In addition to producing eggs and milk and being used in meat production, livestock provide other services. Most notably, farmers use cattle to plow land used in crop production and as insurance against unexpected cash needs.[14] The production and productivity measures discussed so far ignore such services of livestock. We provide two additional measures of livestock productivity that account for most outputs or services of livestock. We use data from the National Bank of Ethiopia (NBE) on the real value of livestock output. We proxy labor productivity in the livestock subsector by dividing the real value of livestock output (NBE 2017) by the total number of livestock farmers (Ethiopia, CSA 2017). Similarly, the real productivity of livestock is estimated by dividing real livestock output by the total tropical livestock units (Ethiopia, CSA 2005–2016).

NBE (2017) data indicate that real livestock output in 2015 (53.4 billion birr, in December 2011 prices) was 73 percent higher than in 2004 (30.9 billion birr), and growth in real output averaged 5.1 percent per year. Real livestock output per TLU was 894 birr in 2004 and 990 birr in 2015, which indicates a total growth in livestock output per TLU of 10.7 percent and annual growth in livestock output per TLU of 1 percent per year (Figure 5.4). Similarly, real output per livestock farmer grew at an average annual rate of 1.3 percent for total growth of 15 percent from 2,996 birr in 2004 to 3,443 birr in 2015.

Livestock Output Growth Accounting

This section discusses the results of a growth accounting analyses of Ethiopia's livestock subsector. The analysis, which uses a modified growth accounting model (Solow 1957), decomposes growth in real value of livestock output into changes in input use, exogenous factors that affect output growth, and

14 CSA (Ethiopia, 2005–2016) data indicate that out of the three-year and older cattle households kept during 2004–2015, the highest proportion of 39.5 percent were kept to provide draft power, with the proportion changing little during the period. A (suspiciously) high proportion (33 percent) were kept for breeding. CSA survey instruments do not include options such as "saving" or "insurance against unexpected cash needs" as coded answers for the question that asks why livestock are kept. Given this, we suspect that some of the households that keep livestock as a means of saving state that they are kept for "breeding."

FIGURE 5.4 **Real value of output per farmer and per tropical livestock unit, 2004–2015**

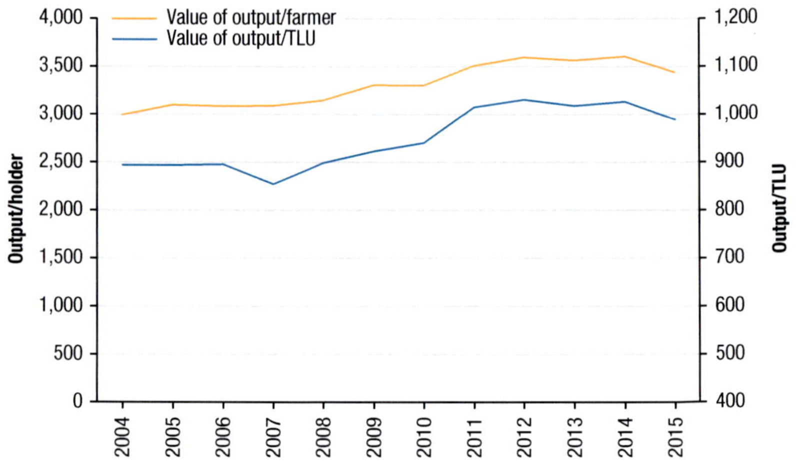

Source: Authors' analyses using real livestock output data from NBE (2017) and CSA data (Ethiopia, CSA 2017, 2005–2016).
Note: TLU = tropical livestock unit.

total factor productivity (TFP). The method starts by assuming that markets are competitive and that there exists a well-behaved aggregate livestock production function that expresses real livestock output as a function of inputs used in production and a variable that represents techniques of production. Starting from such a production function and under the assumptions given, the method expresses growth in real livestock output using the formula:

$$\frac{\Delta Output}{Output} = \Delta TFP + \sum_J w_J \frac{\Delta Input_J}{Input_J} + \alpha \Delta Exog$$

where J indexes types of inputs and w_J is the relative share or payment to input J.[15] For instance, if J is labor, w_J represents wages paid/imputed for labor used. The equation states that growth in output between two time periods ($\Delta Ouput/Output$) is equal to the sum of growths in total payments for all inputs ($\Sigma_J w_J \Delta Input_J/Input_J$); change in exogenous factors that affect livestock production ($\Delta Exog$) times the magnitude of the impact of exogenous factors (α); and change in TFP (ΔTFP), which captures changes in methods of production and the effects of other factors not included in the analyses.

15 Interested readers can see, among others, Solow (1957) and Carlaw and Lipsey (2003). Bachewe et al. (2018) describe the method in the Ethiopian context and apply it to the crop production subsector.

We conduct the analyses assuming that real livestock output is a function of labor, which we proxy using the total number of livestock farmers.[16] We proxy capital using total grazing land and the stock of live animals (TLU); real value of crop output used as livestock feed (used as a proxy for intermediate inputs from the crop subsector); manufacturing sector by-products used as animal feed (used as proxy for intermediate inputs from the manufacturing sector); total number of livestock vaccinated or treated for diseases (used as proxy for intermediate inputs from the services sector); and total number of livestock extension users. We provide a summary of the data used in Appendix Table 5A.1. Furthermore, we derive w_j, payments to factors used in production, from the 2005/2006 social accounting matrix for Ethiopia (EDRI 2009), while we follow Bachewe et al. (2018) to approximate the exogenous factors. Note that livestock production may have been affected by other inputs used by farmers, such as crop residues used as animal feed, data on which are unavailable, and by other exogenous factors, such as weather changes. This implies that changes in TFP obtained from the analyses also include the effects of those factors not included in the analyses (Sumner 2014).

We provide the results of the growth accounting analysis in Figure 5.5. This shows the average contribution of the factors considered, expressed as a percentage of real livestock output growth. During the period 2004 to 2014, growth in labor used in livestock production accounted for 2.9 percent of the 5.8 percent average annual growth in real livestock output. That is, growth in labor accounted for almost half—that is, 49 percent—of real livestock output growth. Similarly, growth in TLU contributed 1.5 percent to the total 5.8 percent average annual growth in livestock output—that is, about a quarter of output growth came from growth in the stock of livestock. Growth in improved feed and grazing land together accounted for 18 percent of growth in real livestock output, extension and veterinary services together accounted for 2 percent, and exogenous factors for 5 percent.

The finding that labor and TLU growth combined accounted for 78 percent of the growth in real output and that improved feed, veterinary services, and extension services, which can be considered as modern inputs, accounted for only 11 percent, indicates that most of the livestock output growth resulted from increases in the numbers of livestock and livestock owners (extensive margin) and not through increased use of modern inputs

16 While households constitute the ultimate sampling units in CSA surveys, there can be more than one livestock farmer/holder in a given household. However, the proportion of households with more than one farmer averaged 4 percent and stayed about the same during the period.

FIGURE 5.5 Average contribution of factors as percentage of livestock output growth

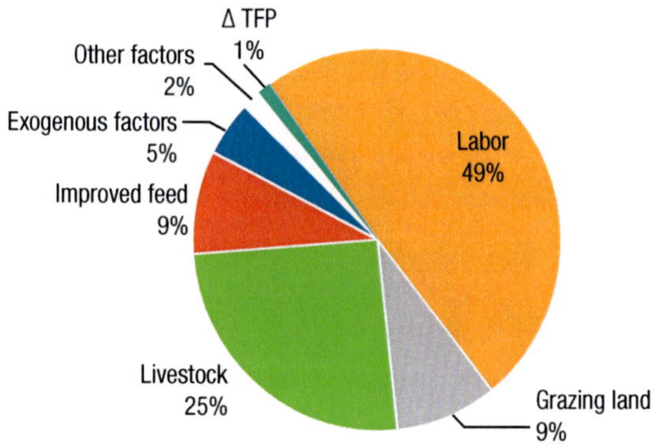

Source: Authors' analyses using real livestock output data from NBE (2017) and CSA data (Ethiopia, CSA 2005–2016, 2017).
Note: TFP = total factor productivity.

(intensive margin). Furthermore, changes in TFP accounted for only 1 percent of the growth in output, indicating that almost all the real livestock output growth is accounted for by the factors used in the analyses and that little improvements in methods of livestock production occurred during the period. The growth seen in the livestock sector stands in contrast to growth in the crop sector, in which modern input use and TFP growth were much more important factors in explaining higher output levels (Bachewe et al. 2018). This indicates a lesser role of modernization in explaining recent growth in Ethiopia's livestock sector.

Inputs Used in Livestock Production

Here we discuss the inputs used in livestock production, including improved breeds, livestock feed, veterinary and livestock extension services, and employment in livestock production.

Improved Livestock Breeds

We use CSA data to study the breed composition of cows and chickens, the only livestock types for which data are available on breed type. Table 5.7, which summarizes the data on breed type, indicates that indigenous or local breed cows and chickens are dominant. Foreign cow breeds are almost nonexistent among smallholder farmers' herds, while hybrid cows represented only 1 percent and 2 percent of the total number of cows in 2004 and 2015, respectively. A relatively higher percentage of chickens, around

TABLE 5.7 Breed composition of cows and poultry in Ethiopia, 2004, 2010, and 2015 (%)

Livestock type	Breed	2004	2010	2015
Cows	Local	99	99	98
	Foreign	0	0	0
	Hybrid	1	1	2
Broiler chickens	Local	98	97	89
	Foreign	2	1	6
	Hybrid	0	2	5
Egg-laying chickens	Local	98	97	89
	Foreign	2	1	6
	Hybrid	0	2	5

Source: Authors' analyses using CSA holder-level data (Ethiopia, CSA 2017).

11 percent, were foreign and hybrid breeds in 2015, compared to just 2 percent in 2004.

These estimates indicate that effort is needed in this area. However, it is to be noted that artificial insemination is increasingly being adopted in the livestock subsector of Ethiopia. Figure 5.6 shows the evolution in the production, distribution, insemination, pregnancy, and calves born using artificial insemination. The number of calves born using artificial insemination was 11 times higher in 2016 than in 2005. However, only 230,000 calves were born through artificial insemination in 2016, still a small number compared to the total cattle population.

Animal Feed

Table 5.8 summarizes CSA data on the proportion of households that used different livestock feed and sources of the feed. Most livestock-owning households use grazing as the principal source of feed for their livestock followed by crop residue. However, the percentage of farmers using green fodder (grazing) as animal feed is declining slowly over time, from 93 percent in 2005 to 90 percent in 2015. The percentage of households that use crop residues remained about the same at 79 percent, while those that used hay slightly increased to 30 percent in 2015 from 27 percent in 2005. But the largest increase, 6.4 percent, occurred in the proportion of households using improved feed—both improved pasture and by-products—from 7.2 percent in 2005 to 13.6 percent in 2015. Most of that increase happened between 2010 and 2015.

FIGURE 5.6 Artificial insemination of cows, annual doses and outcomes, three-year moving average

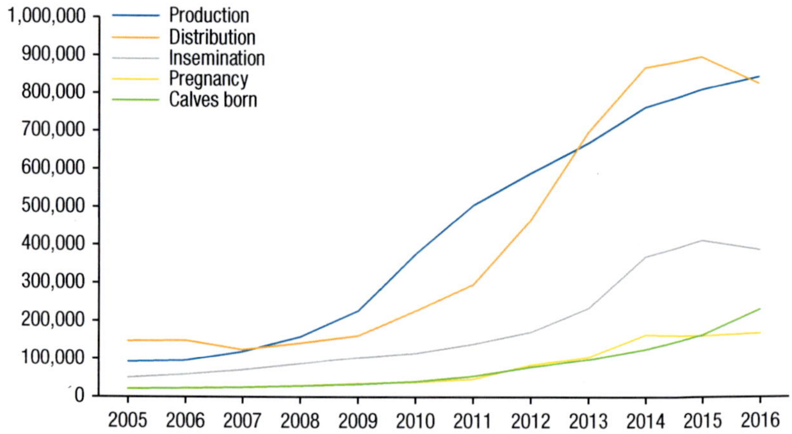

Source: Data from the Ethiopia Ministry of Agriculture and Livestock.

TABLE 5.8 Trends in animal feed type and source of feed, percentage of livestock-owning households, 2005, 2010, and 2015

Category	2005	2010	2015
Animal feed type			
Grazing land	93.3	94.1	90.5
Crop residue	78.9	80.2	78.8
Hay	27.5	32.4	30.6
Improved feed	7.2	7.3	13.6
Source of feed			
Own property	59.1	61.5	62.1
Purchased	6	6.6	7.8
Public property	17.6	15.9	11.1
Own property and purchased	4.7	5.2	6.2
Own property and public property	8.9	8.5	9.7

Source: Authors' computation using CSA data (Ethiopia, CSA 2005–2016).

Most farmers use their own property to acquire livestock feed. In 2015, 62 percent, 11 percent, and 8 percent of households used livestock feed obtained from own property, public property, and purchases, respectively. About 16 percent of the households used feeds acquired from multiple sources. The proportion using purchased feed in 2015 was more than 2 percent higher than in 2005, an increase of 30 percent over the period. In contrast, the proportion using public property in 2015 was 37 percent lower than that in 2005. The trends in feed sources observed, which shows a rapid decline in the use of common grazing grounds and an increase in importance of purchased feeds, is consistent with the decline in cultivated area per farmer that occurred during the same period in the country (Minten et al. 2018a). This likely is a result of more of the common grazing grounds of Ethiopia being brought into crop cultivation.

Despite the growth in nationwide agricultural area in the past decade by 2.9 percent per year, landholdings for the average farmer declined by 25 percent (Minten et al. 2018a). A similar pattern was also observed in grazing area. Total grazing area increased from about 1.37 million hectares in 2004 to 1.77 million hectares in 2015. Despite this increase in the aggregate grazing area, average grazing area per farmer in 2015 (0.10 hectares) was about 16 percent lower than the average in 2004 (0.12 hectares).

Veterinary Services

A considerable proportion of cattle, small ruminants, and poultry are afflicted by various diseases (Table 5.9). Chickens are the most affected with 57 percent of the stock diseased in 2015, while equines (12.5 percent) and camels (11 percent) are the least affected. Relative to 2004 the percentage of poultry afflicted in 2015 increased by 18 percent, while the increase was between 1 percent and 5 percent for other livestock types.

CSA data also indicate that, of the animals affected by diseases, the proportion that received veterinary services increased in all livestock types during the period (Figure 5.7). In 2015 the percentage of animals treated of those afflicted by disease was the highest for cattle (68 percent) and the lowest for poultry (13 percent). Relative to 2004, the proportion treated for diseases increased in 2015 by about 34 percent for cattle and equines, by 31 percent for sheep, and by 24 percent for goats. The increase was lower for poultry (10 percent) and camels (5 percent). The increase in the proportion of sick animals treated is despite the increase in the proportion afflicted by diseases. This implies that the number of

TABLE 5.9 Trends in livestock afflicted with disease, by livestock type in Ethiopia, 2005, 2010, and 2015 (%)

Livestock type	2005	2010	2015
Cattle	14.8	16.8	16.5
Sheep	17.3	26.1	20.9
Goats	18.6	23.7	20.2
Equines	10.9	13.1	12.5
Camels	12.6	6.9	11.1
Poultry	40.9	48.7	56.8

Source: Authors' analyses using CSA holder level data (Ethiopia, CSA 2017).

FIGURE 5.7 Trends in livestock afflicted by disease that were treated, by livestock type (%)

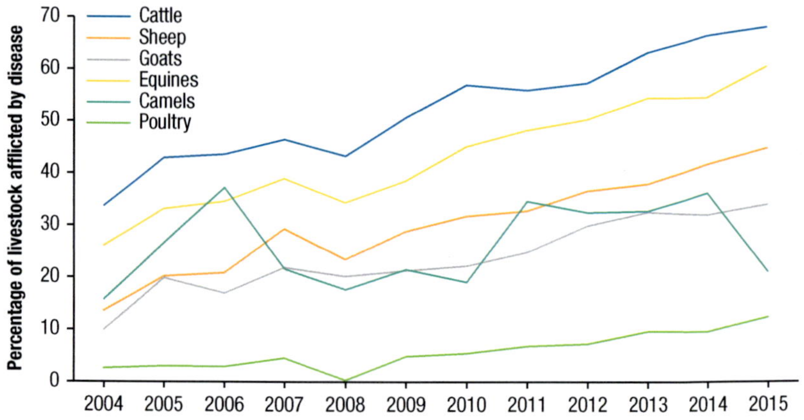

Source: Authors' analyses using CSA holder level data (Ethiopia, CSA 2017).

livestock treated increased faster than the number of sick animals. The data also indicate that the number of treated animals grew faster than the total stock of animals.[17] These numbers are indicative of the increasing spread of veterinarians and veterinary clinics in the country.

The proportion of livestock vaccinated of total stock also increased considerably for all animal types—58 percent, 25 percent, 23 percent, and 8 percent of cattle, sheep, goats, and camels, respectively, were vaccinated in 2015 (Figure 5.8). These proportions were 34 percent, 19 percent, 14 percent, and 4 percent higher than the proportions vaccinated in 2004, respectively. The number of cattle vaccinated in 2015, 33.6 million, was more than three times the number vaccinated in 2004, 9.2 million. Similarly, the number of sheep, goats, and camels vaccinated in 2004 was more than four times the number vaccinated in 2004.

Disease-related animal deaths declined during the period, which may imply a negative relationship between such deaths and the expansion in animal vaccination services (Figure 5.8).[18] The proportion of cattle afflicted by diseases that died was 28 percent in 2015. The proportion was higher for other

17 The proportion treated out of total stock increased by 5 percent to 7 percent in all livestock types except for camels (1 percent).

18 Disease-related deaths on average accounted for 76 percent to 82 percent of total deaths during 2005–2015 for all livestock types except camels (70 percent). The share of disease-related deaths remained about the same during the period.

FIGURE 5.8 Proportion of livestock vaccinated and disease-related deaths, 2004/2005–2015/2016 (%)

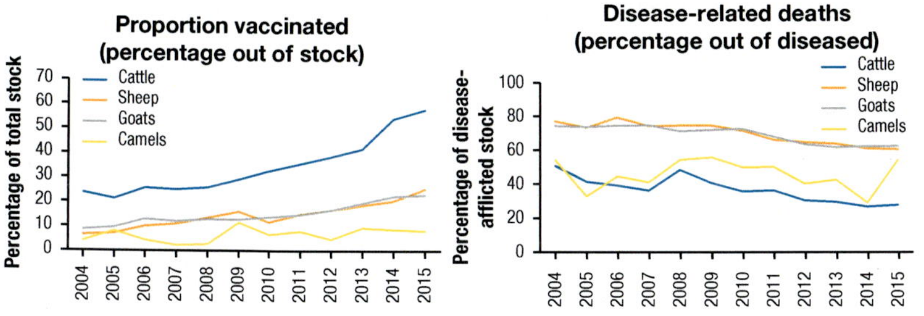

Source: Authors' computation using CSA data (Ethiopia, CSA 2005–2016).

livestock types. Despite the proportion of disease-related deaths being considerable in 2015, they were 23 percent, 16 percent, 12 percent, and 25 percent lower relative to 2005 for cattle, sheep, goats, and equines, respectively.

Livestock Extension

Extension service provision expanded rapidly in Ethiopia in the past two decades, particularly in the crop subsector. The promotion of modern inputs and production methods through the extension system is cited as one of the contributors to the rapid increase in modern input use and consequent crop production and productivity growth over the past decade (Bachewe et al. 2018; Berhane et al. 2018). However, the extension system has mostly focused on the crop subsector. Table 5.10 shows that only 3 percent and 2 percent of livestock farmers used extension services in 2005 and 2015, respectively. This contrasts to the higher proportion and steady growth in the proportion of farmers using crop extension. The proportion of livestock extension users is also sporadic, which may have resulted from campaign-type extension services provided by programs or projects.

Labor

The description earlier on the number of livestock owners indicates only the number of households that own livestock and does not represent the number of people actually engaged in such activities as feeding, watering, milking, and providing other care for the animals, which often involve more members of the household than just the head. We use CSA's most recent Labor Force Survey (LFS) data to gauge the contribution of the livestock subsector

TABLE 5.10 Trends in number of farmers using livestock extension services, 2005, 2010, and 2015

Extension service type	2005	2010	2015
All livestock extension	227,903	195,251	292,256
Dairy	37,783	48,021	57,801
Meat	29,646	56,700	45,023
Poultry	123,736	41,626	154,967
Honey/wax	23,172	25,336	13,329
Two or more	13,566	6,338	12,533
Livestock and crop extension users, percentage of farmers (%)			
Livestock	2.9	1.4	1.9
Crop	27.5	25.5	44.3

Source: Authors' computation using CSA data (Ethiopia, CSA 2005–2016).

for nationwide employment. The LFS was conducted in 2013. It included 240,660 persons sampled to represent 80.4 million people residing in all regions and zones of the country, except for six zones in Somali region.

Using the three-digit employment classification used by CSA (Ethiopia, CSA 2014), we identify nine employment types that involve the production and processing of livestock products. This includes livestock farming, mixed crop-livestock farming, processing of animal-sourced food (ASF), and leather processing and manufacturing of leather products. The number of people employed in these sectors accounted for 54.6 percent of the 47.4 million people employed in Ethiopia. The number engaged in mixed crop-livestock production accounted for the largest share (46.9 percent), followed by those engaged in only livestock production (7.6 percent). Those engaged in the remaining seven subsectors accounted for less than 0.2 percent of employment nationwide in 2013.

The LFS data do not provide information on the proportion of time/employment used in livestock production by those engaged in mixed crop-livestock production. Thus the contribution of the livestock subsector for employment is imprecise. We can only estimate the contribution of the livestock subsector to national employment patterns. If we assume mixed crop-livestock farmers used between 20 percent and 30 percent of their time for livestock production, then the share of employment in livestock out of total employment in 2013 would be between 17 percent and 22 percent, respectively. These shares are higher than the share of livestock output in real GDP, which was 12.3 percent in 2004 but steadily declined to 7.1 percent in 2015.

Correlates of Modern Input Adoption

Previous discussions indicate that the period 2005–2015 has been marked by rapid growth in the proportion of farmers that use improved feeds, improved cattle and poultry breeds, and livestock vaccinated, although growth for all factors started from a low base. Use levels of these modern inputs are generally still low, and modern inputs contributed little to livestock output growth. This implies that future growth in livestock output needs to be driven by increased use of these and other modern inputs, since the major sources of current output growth, the increases in numbers of farmers and livestock is unsustainable given land scarcity and increasing environmental problems.

Next we discuss the results of analyses conducted to investigate the factors associated with the adoption of the three modern inputs mentioned: improved feed, improved breeds, and vaccination. For this purpose we deploy probit and multinomial probit models on farmer-level CSA data. Average marginal effects obtained from the probit model analyses are provided in Table 5.11, while those obtained using the multinomial probit model are provided in Appendix Table 5A.2. Marginal effects of the analysis on factors associated with adoption of improved feed, data which are collected at the household level, are provided in the second column of Table 5.11. Columns 3 and 4 provide marginal effects of the analyses on factors associated with vaccination of cattle and sheep and goats, respectively. The last two columns of Table 5.11 provide marginal effects of factors associated with the adoption of improved (hybrid or foreign) cattle and poultry breeds. The results in both tables are mostly consistent with expected signs. Moreover, the results pertaining to cattle in Table 5.11 and those in Appendix Table 5A.2 have qualitatively the same implications. Therefore, our discussion below focuses on the results in Table 5.11.

The results indicate that, although all demographic characteristics of farmers appear to be associated with adoption of the three technologies, education and household size are the most important. More educated farmers are more likely to adopt all inputs, which is consistent with the finding for modern input adoption in crop production (Bachewe et al. 2018). Larger-sized households are more likely to adopt all inputs, probably because such households own more labor, which is needed to make the inputs available to or provide the increased care needed by improved breeds.

The likelihood of adopting any of the technologies considered is positively associated with receiving relevant extension advice. That is, the likelihood of having one's livestock vaccinated and the adoption of improved livestock

TABLE 5.11 Adoption of improved feed, improved livestock breeds, and vaccination in livestock production in Ethiopia—probit model estimates of variables associated with adoption

	Used improved feed? (yes = 1)	Dependent variable			
		Have livestock been vaccinated? (yes = 1)		Owned improved breed of livestock? (yes = 1)	
Variables	Farmer level	Cattle	Sheep and goats	Cattle	Poultry
Received extension advice, 0/1	0.040***	0.125***	0.071***	0.037***	0.241***
Average travel time to nearest city (hours)	−0.000**	−0.002***	0.001***	−0.002***	−0.001***
Age of farm holder (years)	0.000***	0.000***	0.000**	0.000***	−0.000
Farm holder is female, 0/1	−0.002**	−0.005***	0.005***	0.000	−0.002
Highest educational attainment in household (years)	0.003***	0.009***	0.004***	0.002***	0.005***
Household size	0.003***	0.007***	0.004***	0.001***	0.002***
Cattle (number)	0.001***	0.002***		0.001***	
Sheep and goats (number)	−0.000*		0.001***		
Poultry (number)	0.001***				0.002***
Improved feed users in woreda, lagged percentage	Yes				
Vaccinated livestock in woreda, lagged percentage		Yes	Yes		
Improved livestock breed in woreda, lagged percentage				Yes	Yes
Rainfall and other climate variables	Yes	Yes	Yes	Yes	Yes
Average population density	Yes	Yes	Yes	Yes	Yes
Year dummies	Yes	Yes	Yes	Yes	Yes
Zone dummies	Yes	Yes	Yes	Yes	Yes
Chi-squared	55,418	66,782	35,993	33,586	37,363
Observations	677,985	520,216	377,885	562,964	409,217

Source: Authors' analyses.

Note: Table shows (average) marginal effects. Coefficients with ***, **, and * are significant at 1 percent, 5 percent, and 10 percent levels, respectively.

breeds are positively associated with receiving extension advice pertaining to each livestock species. Adoption of improved livestock feed is positively associated with receiving one or more of the types of extension advice. These results do not imply that adoption of the inputs can be attributed to extension advice. However, the positive associations imply either that farmers first adopt the inputs and then seek advice or that they may adopt the inputs because of the advice. This, in turn, implies that expansion of livestock extension services is likely to be followed by increases in the number of farmers that adopt the inputs, whichever way the causation runs.

Distance to urban centers is negatively associated with adoption of all inputs. This also is consistent with the adoption of modern inputs in crop production, given significant differences in factor and output prices and therefore profitability over space (Bachewe et al. 2018; Vandercasteelen et al. 2018). The results corroborate the evidence that farmers in urban and peri-urban areas and in rural areas closer to major urban centers use improved inputs at higher rates than farmers in remote areas (Minten et al. 2018b).

The results also indicate that larger herd sizes are positively associated with the adoption of all inputs, excluding the negative association of number of sheep and goats and improved feed. These results may imply that farmers with larger herds look for alternative sources to feed all their animals; are likely to own at least one improved breed than those with smaller herds; and will have to vaccinate their herd, given the higher opportunity cost they suffer during disease epidemics.

Adoption of the inputs is also positively influenced by past levels of adoption of the inputs in the woreda (district). This is likely given that higher past adoption rates may imply easier access to the inputs. Moreover, farmers in areas with higher past adoption rates can easily get information about the inputs through neighborhood effects.

Conclusion

In this chapter we study the performance of the livestock sector in Ethiopia over the past decade. For this we mostly rely on CSA data, which are representative of households engaged in mixed crop-livestock production, by far the largest segment of livestock production systems in the country.[19] Livestock is important in Ethiopia. We find that almost all farmers in Ethiopia own livestock, and we evaluate the value of livestock stock in the country at US$11 billion, or at about US$720 per farmer. The largest share of the total value of livestock is from cattle, but this share is coming down, while small ruminants are on the increase, especially in pastoralist areas. Livestock activities are also important for employment, as 55 percent of the population reported at least partial employment in livestock or related sectors. However,

19 This study also covers nonsedentary areas of three zones of Afar and six zones of Somali regions only parenthetically due to data unavailability. This can be taken as a caveat of the study. However, we make estimates of the importance of pastoralist areas in total livestock production in Ethiopia.

while Ethiopia recorded rapid economic growth in the past decade, the livestock subsector contributed little to this growth relative to its size, whether measured in terms of employment or of stock of live animals.

Over the past decade, livestock output grew by almost 6 percent annually. Results of a growth accounting analysis indicate that 80 percent of that growth was explained by an increase in numbers of livestock and labor, and not by productivity increases. The number of livestock farmers and the stock of each livestock species was at least 50 percent higher in 2015 than it was in 2004. The growth in the use of modern inputs, such as improved breeds and feeds, and TFP growth was much less important in explaining output growth. This growth path for the livestock sector is in contrast with the crop sector in which modern input use, particularly fertilizer and improved seeds, as well as total factor productivity (TFP) growth played a much more important role in explaining recent growth (Bachewe et al. 2018). Part of the success in the modernization of the crop sector was explained by the increasing presence of extension agents providing advisory services on improved crop production. However, in the livestock subsector, in contrast, farmers' access to livestock extension use has been low and sporadic over the past decade.

While the share of livestock-owning households making use of improved feeds and breeds increased in importance, its level is still low. For example, the number of calves born from cows through artificial insemination increased eleven-fold over the past 10 years, but the technology still accounts for a small share of all calves born in the country. The proportion of all livestock types afflicted with diseases increased during the period studied. However, the proportion of livestock treated for diseases and the proportion vaccinated increased considerably, while a negative relationship between disease-related deaths and vaccination in animals was also observed. Consequently, livestock death rates slightly declined during the period. Nevertheless, a considerable proportion of livestock—higher than twice the proportion sold for meat production—is still lost to unplanned deaths, indicating an important development challenge facing the sector.

The growth by numbers but not in productivity has important implications for feed market use, as the proportion of farmers using purchased feed is increasing, while those grazing their animals on public property is declining, consistent with the decline in grazing area per farmer observed during the period. The growth path in the livestock sector is contrary to what was projected in the Climate Resilient Green Economy (CRGE) plan (Ethiopia 2011), in which it was envisioned that productivity per animal would grow and the number of animals would decline. To achieve that vision, it seems that more effort is needed to stimulate productivity growth through improved input use leading to higher productivity levels in the sector. As shown in our adoption analysis, this could possibly be achieved through better livestock extension and service delivery systems, better access to input and output markets for livestock producers, and improved education and awareness levels of farmers owning livestock. A promising avenue for future work would also be an ex-ante evaluation of tradeoffs between investments in livestock versus crop R&D. Such assessment would be useful given the rapid changes that are happening in demand for livestock products, the past overemphasis on the cereal sector, and the lack of work on R&D, and the potential of impact, in the livestock sector. Such analysis might therefore help to potentially inform the rebalancing of these priorities.

Appendix 5A: Variables Used in Growth Accounting, and Average Marginal Effects Regression Estimates

TABLE 5A.1 Variables used in livestock growth accounting analyses, 2004, 2010, 2014

Variable	2004	2010	2014
Number of livestock holders (million)	10.3	13.9	15.0
Grazing area (thousand hectares)	844.6	1,708.6	1,758.5
Tropical livestock units (million)	34.5	48.7	52.4
Value of crop used as animal feed (million birr)	481.3	740.2	806.5
Veterinary services (millions of livestock)	14.6	33.8	53.7
Number of extension users (thousands)	302	195	226
Rural roads (hundred kilometers)[a]	18.4	29.2	32.6
Real livestock (billion birr)[a]	30.9	45.8	54.3

Source: Authors' analyses using CSA data (Ethiopia, CSA 2017, 2005–2016) except those with superscript of (a), which are from NBE (2017).

TABLE 5A.2 Marginal effects of multinomial probit model estimates of modern input adoption in cattle production in Ethiopia

	Vaccination		Improved breed		Both	
Variables	Marginal effect	Standard error	Marginal effect	Standard error	Marginal effect	Standard error
Received extension advice, 0/1	0.087***	0.0066	0.018***	0.0008	0.018***	0.0006
Average travel time to nearest city (hours)	−0.001***	0.0003	−0.001***	0.0001	−0.001***	0.0001
Age of farm holder (years)	0.000***	0.0000	0.000***	0.0000	0.000***	0.0000
Farm holder is female, 0/1	−0.005***	0.0016	−0.000	0.0004	0.000	0.0003
Highest educational attainment (years)	0.007***	0.0002	0.001***	0.0000	0.001***	0.0000
Household size	0.005***	0.0003	0.001***	0.0001	0.001***	0.0000
Cattle (number)	0.002***	0.0001	0.000***	0.0000	0.000***	0.0000
Vaccinated cattle in woreda, lagged percent	0.003***	0.0000	−0.000***	0.0000	0.000***	0.0000
Improved cattle breed in woreda, lagged percent	−0.000	0.0001	0.001***	0.0000	0.000***	0.0000
Rainfall and other climate variables	Yes		Yes		Yes	
Average population density	Yes		Yes		Yes	
Year dummies	Yes		Yes		Yes	
Zone dummies	Yes		Yes		Yes	
Chi-squared			**84,491**			
Observations			**572,336**			

Source: Authors' analyses.

Note: Table shows average marginal effects. Coefficients with ***, **, and * are significant at 1 percent, 5 percent, and 10 percent levels, respectively.

References

AACCSA (Addis Ababa Chamber of Commerce and Sectoral Associations). 2015. *Value Chain Study on Meat Processing Industry in Ethiopia.* Accessed April 15, 2015. http://addischamber.com/wp-content/uploads/2017/01/Value-Chain-study-on-Meat-Processing.pdf.

———. 2016. *Value Chain Study on Dairy Industry in Ethiopia.* Accessed April 15, 2015. http://addischamber.com/wp-content/uploads/2017/01/Value-chain-study-on-Dairy-industry-in-Ethiopia.pdf.

Aklilu, Y., and A. Catley. 2010. *Mind the Gap: Commercialization, Livelihoods and Wealth Disparity in Pastoralist Areas of Ethiopia.* Project report. Somerville, MA, US: Feinstein International Center of the Friedman School of Nutrition Science and Policy, Tufts University. Accessed June 2, 2018. http://fic.tufts.edu/assets/mind-the-gap.pdf.

Anteneh, B., A. Tegegne, F. Beyene, and B. Gebremedhin. 2010. *Cattle Milk and Meat Production and Marketing Systems and Opportunities for Market-Orientation in Fogera Woreda, Amhara Region, Ethiopia.* Improving Productivity and Market Success of Ethiopian Farmers Project Working Paper 19. Nairobi: International Livestock Research Institute.

Bachewe, F., G. Berhane, B. Minten, and A. S. Taffesse. 2018. "Agricultural Transformation in Africa? Assessing the Evidence in Ethiopia." *World Development* 105: 286–298.

Barrett, C., S. Osterloh, P. D. Little, and J. G. McPeak. 2004. *Constraints Limiting Marketed Off-Take Rates among Pastoralists.* Research Brief 04-06-PARIMA. Global Livestock Collaborative Research Support Program. Davis, CA, US: University of California.

Berhane, G., C. Ragasa, G. Abate, and T. W. Assefa. 2018. *The State of Agricultural Extension Services in Ethiopia and Their Contribution to Agricultural Productivity.* IFPRI-ESSP Working Paper 118. Addis Ababa: IFPRI/ESSP.

Carlaw, K., and R. G. Lipsey. 2003. "Productivity, Technology, and Economic Growth: What Is the Relationship?" *Journal of Economic Surveys* 17 (3): 457–495.

EDRI (Ethiopian Development Research Institute). 2009. *Ethiopia: Input-Output Table and Social Accounting Matrix.* Addis Ababa.

Ethiopia. 2011. *Ethiopia's Climate-Resilient Green Economy: Green Economy Strategy.* Addis Ababa. Accessed September 11, 2019. www.undp.org/content/dam/ethiopia/docs/Ethiopia%20 CRGE.pdf.

Ethiopia, CSA (Central Statistical Agency). 2005–2016. *Agricultural Sample Survey Volume 2: Report on Livestock and Livestock Characteristics (Private Peasant Holdings,* Meher *Season).* Addis Ababa.

———. 2017. *Agricultural Sample Survey: Holder Level Data of Livestock and Livestock Characteristics (Private Peasant Holdings,* Meher *Season).* Addis Ababa.

FAO (Food and Agriculture Organization of the United Nations). 2018. FAOSTAT database. Accessed January 13, 2019. http://faostat.fao.org/.

Legesse, G., H. Teklewold, D. Alemu, and A. Negassa. 2008. *Live Animal and Meat Export Value Chains for Selected Areas in Ethiopia. Constraints and Opportunities for Enhancing Meat Exports.* Improving Market Opportunities. Discussion Paper 12. Nairobi: International Livestock Research Institute.

Lybbert, T. J., F. B. Galarza, J. McPeak, C. B. Barrett, S. R. Boucher, M. R. Carter, S. Chantarat, A. Fadlaoui, and A. Mude. 2010. "Dynamic Field Experiments in Development Economics: Risk Valuation in Morocco, Kenya and Peru." *Agricultural and Resource Economics Review* 39 (2): 1–17.

Minten, B., M. Dereje, F. Bachewe, and S. Tamru. 2018a. *Evolving Food Systems in Ethiopia.* IFPRI-ESSP Working Paper 117. Addis Ababa: IFPRI/ESSP.

Minten, B., Y. Habte, S. Tamru, and A. Tesfaye. 2018b. *Transforming Agri-food Systems in Ethiopia: Evidence from the Dairy Sector.* IFPRI-ESSP Working Paper 129. Addis Ababa: IFPRI/ESSP.

NABC (Netherlands Africa Business Council). 2011. *Poultry in Ethiopia: A Survey of Production, Value Chains and Marketing of Commercial Poultry in Ethiopia.* The Hague, Netherlands. Accessed June 2, 2018. https://nabc.nl/uploads/content/files/NABC_poultryrapport%20Ethiopia.pdf.

NBE (National Bank of Ethiopia). 2017. *Annual Report 2015/16.* Addis Ababa.

Negassa, A., and M. Jabbar. 2008. *Livestock Ownership, Commercial Off-take Rates and Their Determinants in Ethiopia.* ILRI Research Report 9. Nairobi: International Livestock Research Institute.

Negassa, A., S. Rashid, and B. Gebremedhin. 2011. *Livestock Production and Marketing.* Ethiopia Strategy Support Program II Working Paper 26. Addis Ababa: IFPRI/ESSP.

Oqubay, A. 2015. *Made in Africa: Industrial Policy in Ethiopia.* New York: Oxford University Press.

Shapiro, B. I., G. Gebru, S. Desta, A. Negassa, K. Nigussie, G. Aboset, and H. Mechal. 2015. *Ethiopia Livestock Master Plan.* ILRI Project Report. Nairobi, Kenya: International Livestock Research Institute (ILRI).

Solomon, A., A. Workalemahu, M. A. Jabbar, M. M. Ahmed, and H. Belachew. 2003. *Livestock Marketing in Ethiopia: A Review of Structure, Performance and Development Initiatives.* Socio-economics and Policy Research Working Paper 52. Addis Ababa: Livestock Marketing Authority.

Solow, R. M. 1957. "Technical Change and the Aggregate Production Function." *Review of Economics and Statistics* 39: 312–320.

Sumner, D. A. 2014. "American Farms Keep Growing: Size, Productivity and Policy." *Journal of Economic Perspectives* 28 (1): 147–166.

Tafere, K., and I. Worku. 2012. *Consumption Patterns of Livestock Products in Ethiopia: Elasticity Estimates Using HICES (2004/05) Data.* IFPRI-ESSP Working Paper 12. Addis Ababa: IFPRI/ESSP.

Tamirat, M. 2013. "Econometric Estimation of Herd Stocking Decisions in South Ethiopia." *Journal of Development and Agricultural Economics* 5 (8): 321–327.

Vandercasteelen, J., S. Tamru, B. Minten, and J. Swinnen. 2018. "Cities and Agricultural Transformation in Africa: Evidence from Ethiopia." *World Development* 105 (May): 383–399. https://doi.org/10.1016/j.worlddev.2017.10.032.

Chapter 6

FARM SIZE, FOOD SECURITY, AND WELFARE

Kibrewossen Abay, Kalle Hirvonen, and Bart Minten

Introduction

Understanding the consequences of decreasing land availability per capita because of population growth is of considerable interest to policymakers. Already in the 19th century, there were worries voiced about the effects of population growth on food security. Malthus (1826) famously predicted that the planet would run out of food because of increasing land constraints and that population growth was therefore to be controlled. However, technological changes in agriculture have since contributed to substantial increases in agricultural production per unit of land, and no global food shortages have occurred. Available evidence in different settings, including Africa, shows that population growth has induced agricultural change and led to a reduction in fallow periods, to adoption of agricultural innovations, and to an increase in input use—labor and other agricultural inputs—per unit of land, all leading to higher yields (Boserup 2005; Fresco 1986; Binswanger and Pingali 1988; Tsakok 2011; Jayne, Chamberlin, and Headey 2014; Headey and Jayne 2014; Ricker-Gilbert, Jumbe, and Chamberlin 2014; Muyanga and Jayne 2014; Headey, Dereje, and Taffesse 2014; Josephson, Ricker-Gilbert, and Florax 2014; Lipton 1980, 2006).

Still, there must be limits to increases in agricultural output per unit of land. If true, then an important question arises on how shrinking farms can provide a decent livelihood for households residing on these farms. Given rural population growth and the importance of agriculture in providing livelihoods, especially in most African countries, there is a considerable interest in understanding the linkages between declining farm sizes and welfare and food security outcomes.

Making progress on this topic, however, is difficult due to data constraints. Differences in welfare and food security outcomes across the land quintiles could be driven by differences in landownership or by other household characteristics that systematically differ between households that own different amounts of land. Establishing a causal relationship between land size and welfare outcomes requires a natural experiment—for example, a policy reform that re-allocates agricultural land across households. Such natural experiments are hard to come by, and when they do take place, they are often accompanied by other drastic changes in the society—for example, a civil war or a revolution—making it difficult to isolate the impact of the land reform from other factors shaping the society. Given this, most existing studies, such as the one presented in this chapter, make use of cross-sectional or longitudinal datasets to establish descriptive patterns rather than causal relationships.

In their seminal study, Jayne et al. (2003) show how rural incomes are strongly positively correlated with land size, especially among small-scale farmers in Ethiopia, Kenya, Mozambique, Rwanda, and Zambia. Another branch of related research examines the relationship between population density and agricultural productivity. Headey, Dereje, and Taffesse (2014) find that farmers in more land-constrained areas invest more in inputs and use more labor per hectare, and enjoy higher yields, on average, than farmers in less constrained areas. Despite this, total per capita incomes are considerably lower in land-constrained areas. Muyanga and Jayne (2014) find similar evidence in Kenya: higher population density is associated with more intensified agriculture but lower per capita total incomes.

We revisit this topic using a rich cross-sectional survey from rural Ethiopia and focusing on the question of how land size is associated with household welfare and food security outcomes. We update and extend the analysis carried out by Jayne et al. (2003) by using more recent data and by considering a larger set of welfare and food security outcomes. Research on this area is important for Ethiopia given rapid population growth and limited urbanization. Most rural residents depend on agriculture as their main source of livelihood (Schmidt and Bekele 2016; Bachewe et al. 2018), and access to land is therefore crucial for rural livelihoods. A large—and rapidly growing—population means that more and more people are going to reside in rural areas. For example, ISS (2017) projects that the rural population in Ethiopia will grow by 26 million people—a number higher than the current population of Australia—between 2016 and 2030, an increase of 32 percent (from 80 million to 106 million). The scope for further land expansion seems limited. Recent analysis by Bachewe et al. (2018) shows how the rate of agricultural land expansion has decreased over the past decade. Unless the rate at which people move out of agriculture increases considerably, the average land size owned by households is going to decrease (Masters et al. 2013). Absent of increase in the agricultural intensification, diversification of rural income sources to the nonfarm sector, or rural-to-urban migration, decreasing landholdings will evidently lead to lower rural incomes and worsen food security. Therefore, this increasing lack of available land, especially for younger generations (Bezu and Holden 2014; Kosec et al. 2018), is a major concern for all stakeholders concerned about livelihoods and food security in Ethiopia.

We find surprisingly minor differences in key welfare outcomes such as food consumption and food security between small and large farms. We document five adjustments made by small farms to assure similar caloric intake as for households living on large farms. First, small farms participate actively in land rental markets and in this way double their cultivated land compared to owned land—that is, for the smallest quintile from 0.25 hectare to 0.5 hectare. Second, they compensate their small landholdings with other income sources, mainly livestock and nonfarm businesses that allow for additional food purchases. Third, they cultivate their land more intensively. Fourth, small farms switch to bulkier and more calorie-dense crops that are mostly used for their own consumption. Fifth, they produce as well as consume cheaper food items.

Background

The past two decades have been characterized by high population growth in Ethiopia. The population grew by 4.1 percent per year from 1994 to 2007, but this rate has decreased to 2.5 percent from 2007 to 2015. While urban growth rates are higher than in rural areas (4.6 percent versus 2.1 percent, respectively, between 2007 and 2015), the growth in absolute population numbers in rural areas is still considerably higher than in urban areas. Figure 6.1 shows that the rural population grew by 40 million people and doubled to more than 80 million people in the past 30 years. Over the same period, the urban population increased from 1.5 million to 19 million. Even though the urban population is expected to grow faster in the future, projections put the rural population at 106 million (26 million more than in 2016), constituting 74 percent of the total population in 2030 (World Bank 2015).

As shown in Chapter 2, there seem to be limits to increasing agricultural land area. The Ethiopian Central Statistical Agency (CSA) estimated that land expanded by more than 4 percent in 2006/2007, but that the land expansion rate has come down gradually over the next decade. The rapid growth in rural population and the slower growth in cultivated land area implies that the already small farm sizes are further decreasing in Ethiopia. We rely on data from the Annual Agricultural Sample Surveys conducted by CSA to understand this dynamic in land allocation for smallholder farmers in particular, who cultivate about 95 percent of the agricultural land in Ethiopia (for a discussion on large farms [more than 25 hectares], see Chapter 7).

FIGURE 6.1 Population growth in Ethiopia, rural and urban, 1961–2015

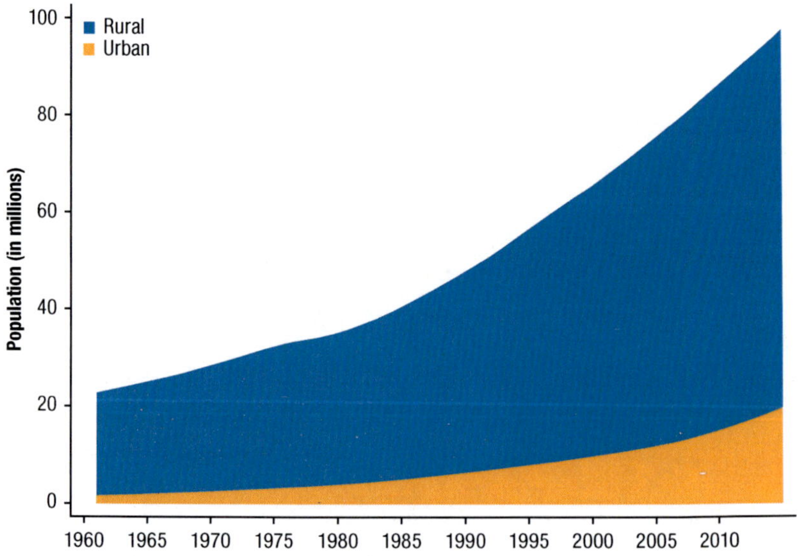

Source: Authors' calculation from World Bank (2017) data.

These data show that farm sizes of smallholders are, on average, declining. Figure 6.2 (left) shows that both total agricultural land area and the number of farmers increased considerably during the period 2004/2005–2016/2017. Total agricultural land in 2016/2017 was 40 percent higher than in 2004/2005—a growth of 2.9 percent annually—but the number of farmers increased even faster during the same period. It is especially in the past five years that we see significantly slower growth in the expansion of agricultural land.[1] Since population growth patterns change slowly and growth in agricultural land is constrained by the limited availability of arable land, the observed decline in the availability of land per farmer is likely to be an increasingly binding constraint for agriculture in the future (see Chapter 2 and Chapter 3). Further building on these numbers, Figure 6.2 (right) shows trends in the average farm size of smallholders in Ethiopia over the past decade. Average crop area declined from about 1.0 hectare per farmer in 2004/2005 to 0.85 hectare per farmer in 2015/2016. Although the size of agricultural

1 Indeed, the difference between average annual growth rates in the number of farmers and agricultural land (–1.4 percent) is about the same as the observed average annual decline in agricultural landholdings.

FIGURE 6.2 Agricultural area, number and farm size of smallholder farmers, 2004/2005–2016/2017 (three-year moving average)

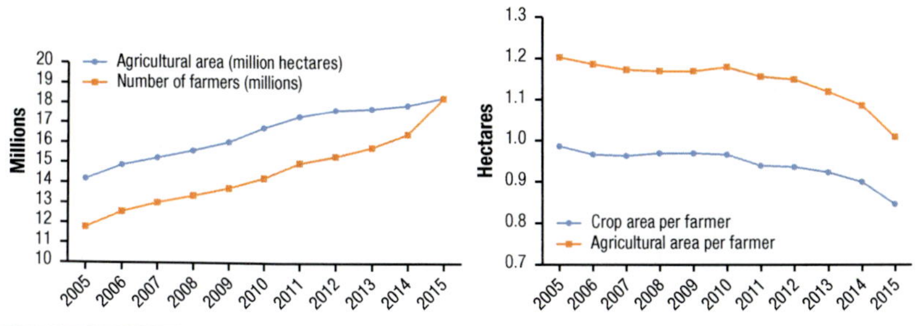

Source: Authors' calculations based on Ethiopia, CSA (2005–2017).

landholdings declined at an average annual rate of 2.3 percent, they stayed about the same in the first half of the decade (2004–2010), with almost all of the decline occurring more recently from 2010 to 2016.[2]

The decline in agricultural landholding per farmer is a widespread phenomenon rather than a localized occurrence. Figure 6.3 shows zonal average agricultural landholding sizes in the 64 administrative zones covered in CSA's Agriculture Sample Survey (AgSS).[3] Darker colors reflect relatively larger farms. The map shows that farms are relatively smaller in the south and in the central north of the country, illustrating the higher population density in those areas, and are larger in the west and northwest. The maps also illustrate that agricultural landholding per farmer declined in almost all zones of the country—that is, in 57 of the 64 administrative zones covered in CSA's AgSS.

All land in Ethiopia, in principle, is owned by the state. More specifically, individual farmers enjoy all the rights of the owner but cannot officially sell the land (Ambaye 2012; Deininger et al. 2008). Land rental is permitted, although with restrictions regarding the size of the land that can be rented out and with respect to rental duration (Deininger et al. 2008; Kosec et al. 2018). As a result, land in Ethiopia is mostly acquired through inheritance

2 The distance between the two graphs in Figure 6.2 (right panel) is about the same over the years, which indicates that noncrop area remained about the same during the period (at about 0.2 hectare). However, the data also show that the area of fallow land declined considerably and that the land area under grazing, woodland, and other uses increased, which reflects existing land constraints and that increasingly little land is left to bring into production.

3 Data for 2004/2005 and 2005/2006 are unavailable for ten zones (four, three, two, and one zone in Gambella, Oromia, Benishangul-Gumuz, and Amhara regions, respectively). We use the data in the earliest year (2006/2007) for these zones.

FIGURE 6.3 Average agricultural land size of smallholders, 2004/2005 and 2016/2017 (hectares)

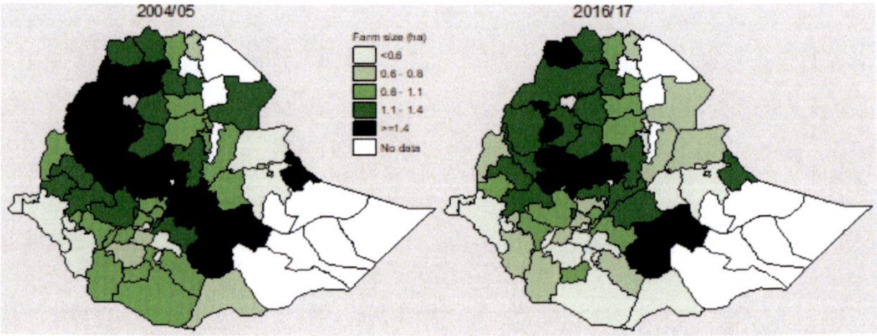

Source: Authors' calculations based on Ethiopia, CSA (2005–2017).

from parents or by community allocation (Ghebru, Koru, and Taffesse 2016).[4] In this chapter we define land holdings as the total farmland for which the household has legal rights through inheritance or community allocation. We then assess how the participation in land rental markets varies across this landholding distribution.

Data and Descriptive Statistics

The data used in this study are from nearly 7,000 rural households resident in five regions of the country: Amhara; Oromia; Tigray; Southern Nations, Nationalities, and Peoples (SNNP); and Somali. These data were collected by the CSA with support from the International Food Policy Research Institute to evaluate the impact of the Feed the Future (FtF) investments in Ethiopia. We use the midline data collected in June and July 2015 and focus on the highland regions of the country: Amhara, Oromia, Tigray, and SNNP. Of these households, 5,609 own farmland, and the sample used in this study is restricted to these households.[5] The geographical coverage of the sample is wide with the survey having been implemented in 252 kebeles (subdistricts, the lowest administrative unit) in 84 woredas (district). Still, despite this, the sample is not nationally representative. Bachewe et al. (2014) provide more information about these data.

We begin by dividing household landholdings into five quintiles according to size. Table 6.1 shows that the average household in the first quintile owns

4 For further discussion, see Chapter 3 and Chapter 11.

5 Throughout this chapter, by "owning" we refer to land to which the household has legal rights.

TABLE 6.1 Land ownership distributions in Ethiopia, by quintile (hectares)

Land quintile	1st	2nd	3rd	4th	5th
Minimum	0.01	0.48	0.75	1.13	1.88
Mean	0.25	0.60	0.94	1.48	2.93
Median	0.25	0.56	1.00	1.50	2.54
Maximum	0.48	0.75	1.13	1.88	12.97

Source: Authors' calculation from Feed the Future 2015 midline survey data (FtF 2015).

0.25 hectare of land. This increases to nearly 3 hectares in the fifth quintile. The smallest landholdings in our sample is 0.01 hectare and the largest nearly 13 hectares, although very few households have land in excess of 10 hectares.

By construction, each quintile has 20 percent of the households in the full sample. However, average farm sizes vary across regions raising a concern that a land quintile is dominated by households in one region. If so, the assessment of welfare differences across land quintiles is confounded by welfare differences across regions. In Table 6.2 we see that defining the quintiles using the full sample works reasonably well also within the regions. For example, in Tigray the distribution of households is close to 20 percent in each quintile. Households in Oromia are somewhat overrepresented in the fifth quintile, while households in SNNP are somewhat overrepresented at the bottom quintiles. We conclude that welfare differences across land quintiles are unlikely to only reflect regional differences. Mindful of this, we verify our core findings with simple regressions that net out regional differences.

We next study how basic household characteristics differ across the land distribution. Table 6.3 shows that households with smaller landholdings are, on average, younger and somewhat more likely to have dependents—that is, household members younger than 16 or older than 61 years old (see also Headey and Jayne 2014). Household size increases with land size. Education levels are extremely low, and there are no clear patterns across the land distribution. Interestingly, more than 12 percent of the households in the top quintile are model farmers, the farmers that extension agents use as role models for other farmers, while the corresponding figure in the bottom quintile is less than 3 percent.

In Table 6.4 we look at land characteristics across the land distribution. As expected, the average number and the size of plots increase with overall farm size. In our dataset more than 75 percent of the land owned by households in the first quintile is inherited. This decreases to about 40 percent in the top

TABLE 6.2 Percentage of households in each landownership quintile in Ethiopia, by region

Land quintile	1st	2nd	3rd	4th	5th	Total
Tigray	21	20	20	18	21	100
Amhara	17	18	23	24	18	100
Oromia	15	16	20	22	28	100
SNNP	30	28	19	13	10	100
Total	20	20	20	20	20	100

Source: Authors' calculation from Feed the Future 2015 midline survey data (FtF 2015).
Note: SNNP = Southern Nations, Nationalities, and Peoples.

TABLE 6.3 Household characteristics in Ethiopia, by landownership quintile

Land quintile	1st	2nd	3rd	4th	5th
Age of head (years)	40.3	43.0	44.4	46.6	49.4
Young-headed households (%)	40.3	30.9	28.5	21.1	17.9
Female-headed households (%)	31.8	30.2	28.5	23.8	20.8
Dependency ratio	0.66	0.65	0.64	0.61	0.60
Education of the head (years)	1.73	1.75	1.56	1.41	1.60
Household size (adult equivalent)	3.44	3.84	4.12	4.39	4.88
Household size (number)	4.37	4.79	5.10	5.33	5.82
Model farmers (%)	2.9	4.3	5.6	8.1	12.1

Source: Authors' calculation from Feed the Future 2015 midline survey data (FtF 2015).
Note: Young-headed = households headed by an individual 15–34 years old. Dependency ratio = (number of less than 16 years + above 61 years old) / (household size).

land quintile. Interestingly, a larger fraction of the land owned by households in the top quintile has been allocated by the community.

Looking at a number of land quality measures, we note that households at the bottom quintile are somewhat more likely to report having fertile or flat sloped land than other households, but the differences across the land distribution are marginal. Interestingly, the short rainy season (*belg*) is relatively more important for households who own less land. But again, the differences are quite marginal; the long rainy season (*meher*) is the dominant cropping season for most households. The use of irrigation is extremely low across all households, and there is no clear pattern across land quintiles. Overall, these descriptive results suggest that there are few differences in land quality between small and large farms. Therefore, we will not make adjustments for differences in land quality in the following.

TABLE 6.4 Land characteristics in Ethiopia, by landownership quintile

Land quintile	Statistic	1st	2nd	3rd	4th	5th
Number of plots	mean	3.6	4.0	4.5	5.3	6.2
	median	3	4	4	5	6
Average size of plots (hectare)	mean	0.17	0.21	0.26	0.31	0.44
	median	0.12	0.17	0.21	0.25	0.38
Fertile land (%)	mean	68.8	66.4	65.4	63.8	66.8
Flat sloped land (%)	mean	69.0	69.0	68.1	68.4	70.2
Number of plots cultivated in *belg* (%)	mean	16.0	16.3	13.2	13.2	12.2
Area of the total land cultivated in *belg* (%)	mean	15.5	15.3	12.1	12.2	11.5
Percentage of land irrigated	mean	1.2	2.7	1.9	1.7	1.4
Percentage of land inherited	mean	76.6	68.9	59.4	52.7	42.2
Percentage of land allocated by community	mean	17.0	26.1	35.4	41.4	52.4

Source: Authors' calculation from Feed the Future 2015 midline survey data (FtF 2015).

Farm Size, Crop Income, and Welfare

We start our analysis with looking at how crop income varies over land quintiles. Table 6.5 shows how total agricultural production—measured either in birr or in calories per adult equivalent—increases sharply with farm size. Across the board, mean values are considerably larger than medians, suggesting that these crop income distributions are skewed to the right—that is, a small number of very high values. The median household at the bottom land quintile produced 941 birr per adult equivalent worth of crops over the two cropping seasons in 2014/2015. The median household at the top quintile produced nearly six times as much (5,524 birr per adult equivalent). This result holds if one controls for regional differences in a regression framework (results available upon request). However, these averages mask considerable dispersion within the land quintiles. Figure 6.4 shows the full crop income distributions by land quintile. The dispersion of crop income is particularly high among households in the bottom quintile.

Another way of aggregating crop production is to measure them in terms of calories using calorie conversion factors.[6] The mean and median total kilocalories produced per day per adult equivalent are reported in the bottom rows of Table 6.5. As before, the number of calories produced increases with land size, but the gradient is not as steep as when the monetary measure is used.

6 We thank the Ethiopian Public Health Institute (EPHI), Mulugeta Tilahun, and Roseline Remans for sharing these conversion factors with us.

FIGURE 6.4 Distributions of total annual crop production in Ethiopia by landownership quintile, birr per adult equivalent

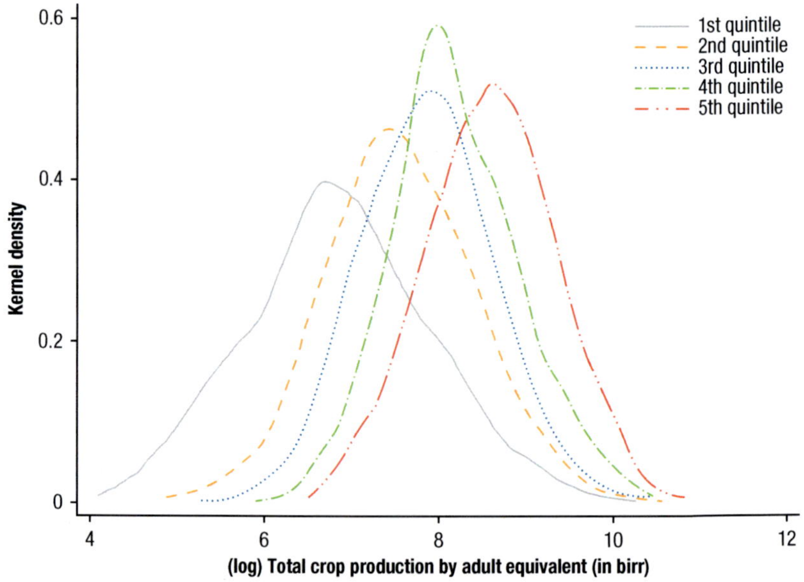

Source: Authors' calculations from Feed the Future 2015 midline survey data (FtF 2015).

Production in calories produced per adult equivalent is only 2.5 times as high for the highest quintile, compared to six times in value terms. This suggests that households with smaller landholdings focus on producing food items that are less valuable in monetary terms but contain relatively more calories.

Nevertheless, we note important differences in crop output for both measures, and we explore next how these large differences in crop output translate into differences in welfare, consumption, and food security. Table 6.6 presents the results of these calculations. We see surprisingly small differences across the land quintiles. In fact, the average (mean and median) food consumption levels are higher among households in the bottom quintiles compared to those in the top quintile.[7] This suggests a negative gradient across the land distribution. However, the opposite is true for nonfood consumption: the average (mean and median) nonfood consumption levels are highest in the top

7 Food and nonfood consumption data are based on a detailed consumption module, similar to those fielded as a part of the Living Standards Measurement Study (LSMS) surveys. The recall period for food consumption expenditure was seven days. For nonfood expenditure the recall period varied between 1 and 12 months, depending on the type of item.

TABLE 6.5 **Total annual crop production in Ethiopia, per adult equivalent, by landownership quintile**

Land quintile	Statistic	1st	2nd	3rd	4th	5th
Total crop production (birr)	mean	1,706	2,848	3,573	4,817	7,123
	median	941	1,844	2,620	3,411	5,524
Total calorie production (kilocalorie per day)	mean	2,423	2,728	3,055	3,757	4,879
	median	1,501	1,895	2,283	2,971	3,769

Source: Authors' calculation from Feed the Future 2015 midline survey data (FtF 2015).

TABLE 6.6 **Welfare indicators in Ethiopia, by landownership quintile**

Land quintile	Statistic	1st	2nd	3rd	4th	5th
Per adult equivalent consumption (birr)	mean	7,018	6,542	6,417	6,222	6,119
	median	5,749	5,493	5,425	5,369	5,242
Per adult equivalent food consumption (birr)	mean	6,248	5,818	5,656	5,416	5,168
	median	5,080	4,846	4,692	4,600	4,370
Per adult equivalent nonfood consumption (birr)	mean	770	724	761	806	951
	median	510	483	543	594	761
Nonfood consumption out of total consumption (%)	mean	11.5	11.6	12.4	13.4	16.0
	median	8.9	9.1	10.1	11.5	14.3
Calorie consumption per adult equivalent per day	mean	3,703	3,736	4,024	3,717	4,149
	median	2,904	2,910	2,993	3,037	3,154
Household size (adult equivalents)	mean	3.4	3.8	4.1	4.4	4.9
	median	3.3	3.7	4.0	4.3	4.7
Number of months of food shortage	mean	0.86	0.77	0.56	0.49	0.35
Food shortage during the rainy season (%)	mean	19.6	17.8	14.5	11.5	7.7
Stunting among children (%)	mean	50.1	47.7	49.0	49.0	45.6

Source: Authors' calculation from Feed the Future 2015 midline survey data (FtF 2015).
Note: All consumption values (except calorie consumption) are expressed in annual terms.

land quintals. Interestingly, the average household size (in adult equivalents) increases steadily as we move up the land quintiles.

In Table 6.7 we study these surprising gradients further in a regression framework. Specifically, we include a household size variable as well as binary indicator variables for each region to control for differences in household size and across regions. Once we add these controls, the coefficients on the land size variable turn positive, suggesting that the negative consumption gradients observed in Table 6.6 are driven by differences in household size or differences across regions. However, of equal note is that the slope of the gradient is quite

TABLE 6.7 Welfare indicators in Ethiopia, controlling for household size and regional differences, by landownership quintile

	(log) annual per adult equivalent consumption		(log) annual per adult equivalent food consumption		(log) annual per adult equivalent nonfood consumption		nonfood consumption out of total consumption	
Dependent variable	1	2	3	4	5	6	7	8
(log) land owned (hectares)	−0.052***	0.047***	−0.069***	0.037***	0.123***	0.178***	0.171***	0.127***
	(0.008)	(0.006)	(0.008)	(0.006)	(0.019)	(0.020)	(0.017)	(0.017)
household size in adult equivalent units		−0.205***		−0.215***		−0.123***		0.081***
		(0.003)		(0.003)		(0.010)		(0.009)
Region dummies?	No	Yes	No	Yes	No	Yes	No	Yes
R^2	0.008	0.434	0.013	0.446	0.008	0.040	0.019	0.039

Source: Authors' calculation from Feed the Future 2015 midline survey data (FtF 2015). Observations: 5,609.
Note: Robust standard errors in parentheses. *** p<0.01, ** p<0.05, * p<0.1

small: a 20 percent increase in land size is associated with only a 0.9 percent increase in household total consumption and an even lower 0.7 percent increase in food consumption.

The bottom part of Table 6.6 looks at food insecurity. Households with larger landholdings are less likely to report food insecurity. The average household in the lowest quintile had 0.86 months of food shortage in the 12 months preceding the interview. This reduces to 0.35 in the top land quintile. Similarly, nearly 20 percent of households in the bottom quintile report having experienced food shortage in the previous rainy season, while the corresponding figure in the top quintile is nearly 8 percent. Finally, about half of children residing in households in the bottom quintile are stunted (short for their age). This stunting prevalence falls only by 4 percentage points as we move to the top quintile.

These tabulations and simple regressions suggest that land size is associated with considerably higher crop income. However, these crop income gains are not translated into better consumption outcomes. While households with larger landholdings report less food insecurity, the differences are not as large as one would expect based on income differences. Moreover, chronic undernutrition (stunting) rates are similar across the land quintiles.[8] Together,

8 This finding is in line with recent work by Brown, Ravallion, and Van De Walle (2017), who document that a surprisingly large fraction of the undernourished children in Africa south of the Sahara reside in wealthier households.

these are quite surprising findings. To be sure that this is not just some artifact of our data, we verified this finding using nationally representative data collected by the CSA together with the World Bank through the Ethiopia Socioeconomic Survey 2013/2014 (Ethiopia, CSA and the World Bank 2013). Appendix 6A provides replicates of Tables 6.1, 6.5, 6.6, and 6.7 using these data. The results are remarkably similar: a large gradient in crop income with increasing landholding size but a negligible one in welfare (consumption) outcomes. The remainder of the chapter attempts to understand these puzzling findings.

What Are Households with Small Land Sizes Doing Differently?

LAND RENTAL

The analysis presented above shows that land sizes increase by more than tenfold when we move from the first land quintile to the fifth (Table 6.1). At the same time, calorie production per adult equivalent only doubles (Table 6.5). One possible explanation is that households have access to rental markets and compensate lack of owned land with rental land. Table 6.8 explores this. An average household in the bottom quintile rents in about 0.30 hectare of land. This rental land doubles the total operated land size in this quintile. However, even after this, an average household in the bottom quintile operates less land than an average household in the second quintile owns. Importantly, the ranking of the quintiles remains the same if we considered net operated land instead of owned land.

CROP CHOICE

Another hypothesis is that households with smaller landholdings favor more calorie-dense crops. Figure 6.5 shows the land area allocated to different crops by land quintile. As expected (Taffesse, Dorosh, and Gemessa 2012), cereals dominate crop portfolios across all land quintiles. Cereals account for nearly 60 percent of the cultivated land area in the first land quintile and about 70 percent in the fifth land quintile. Interestingly, root crops—about half of this is due to enset (false banana)—are relatively more important at the bottom quintiles. Given the importance of cereals in the overall crop production, we next zoom into their composition. Figure 6.6 shows two interesting patterns. First, the importance of maize decreases as we move from the bottom land quintiles to the top quintiles. Second, the opposite pattern emerges for teff. It is of note that of the main cereals cultivated and consumed in Ethiopia, maize is the most and teff is the least energy (kilocalorie) dense (Baye 2014).

TABLE 6.8 Owned land versus operated land size in Ethiopia, by landownership quintile (hectares)

Land quintile	1st	2nd	3rd	4th	5th
Owned land size	0.25	0.60	0.94	1.48	2.93
Rented in	0.30	0.18	0.17	0.20	0.24
Rented out	0.06	0.10	0.14	0.23	0.36
Net operated land size	0.49	0.68	0.97	1.44	2.81

Source: Authors' calculation from Feed the Future 2015 midline survey data (FtF 2015).

FIGURE 6.5 Area allocation of different food crop groups in Ethiopia, by landownership quintile

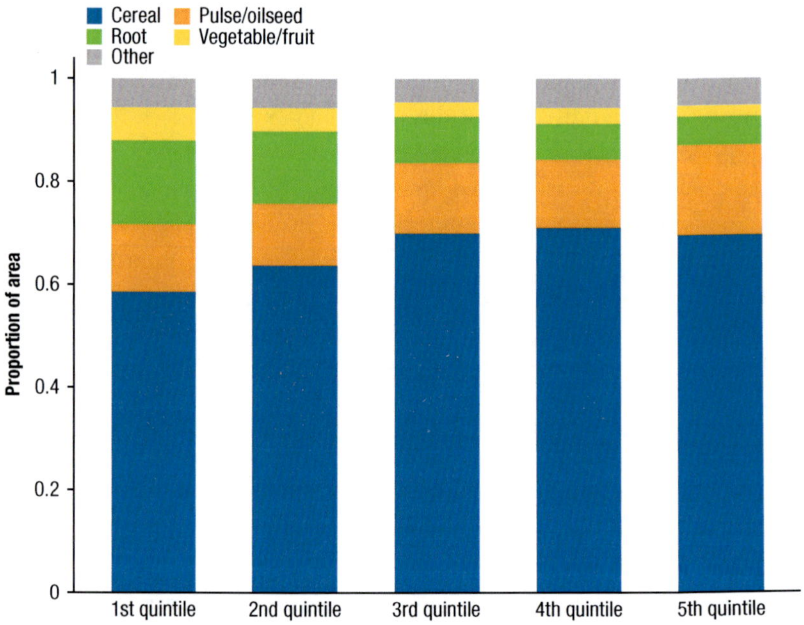

Source: Authors' calculations from Feed the Future 2015 midline survey data (FtF 2015).

INTENSIFICATION

Another reason for the observed disconnect between land size and food security outcomes could be that that yields (production per area) are considerably higher on smaller farms, as is often found in the literature (for example, Barrett, Bellemare, and Hou 2010).[9] Table 6.9 shows that for all cereals, except

9 There is an intense debate in the literature whether this inverse relationship between productivity and land size is an artifact of systematic measurement error, either in land size or crop output

FIGURE 6.6 Area allocation of different cereals in Ethiopia, by landownership quintile

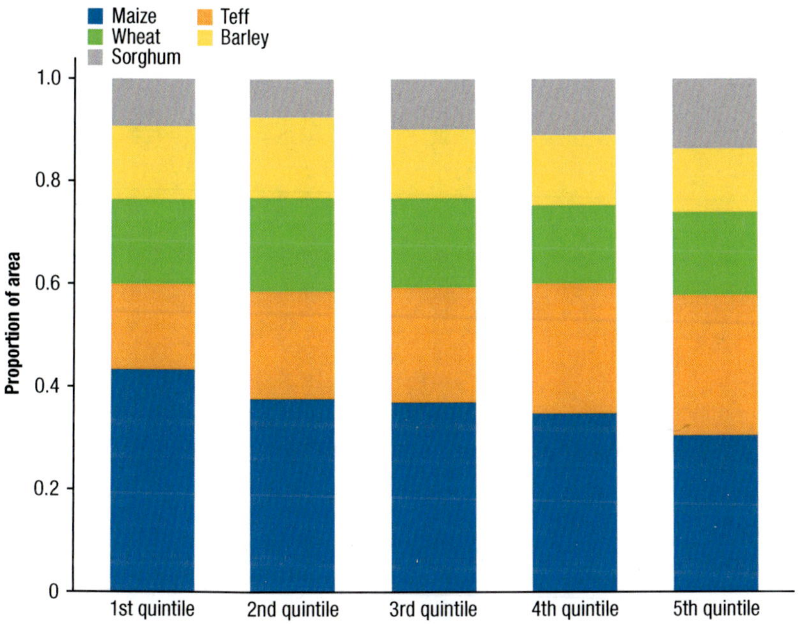

Source: Authors' calculations from Feed the Future 2015 midline survey data (FtF 2015).

wheat, households with smaller farms have higher yields. The story is mixed for other noncereal crops. Still, these yield differences are not large enough to explain the small difference in calorie production relative to farm size observed in Table 6.5.

We further look at investments in fertilizers and improved seeds—proxies for agricultural intensification—and explore if households with larger landholdings are more or less likely to invest in these. Table 6.10 shows that the likelihood that the household applies fertilizer increases with land size. Nearly 60 percent of households in the lowest land quintile use fertilizers on their plots. This increases to 71 percent among households in the fifth land quintile. However, the opposite is true when we look at fertilizer use per hectare. Here, households with smaller landholdings use nearly three times more fertilizer per hectare than households in the fifth quintile. When we look at improved seeds, we do not see a clear trend as for three of the five cereals,

(for example, Carletto, Savastano, and Zezza 2013; Desiere and Jolliffe 2016; Abay et al. 2019; Gourlay, Kilic, and Lobell 2019) or driven by other factors (for example, Bevis and Barrett 2020; Barrett, Bellemare, and Hou 2010).

TABLE 6.9 Yields by crop and landownership quintile in Ethiopia, quintals per hectare

Land quintile:	1st	2nd	3rd	4th	5th
Cereals					
Teff	7.8 (4.4)	7.6 (4.4)	7.6 (4.2)	7.0 (3.8)	6.5 (3.9)
Barley	13.3 (6.9)	13.2 (7.3)	12.7 (6.4)	11.9 (6.3)	12.1 (6.3)
Wheat	13.7 (7.5)	13.9 (8.3)	14.4 (7.9)	13.5 (7.7)	14.0 (8.1)
Maize	16.9 (10.8)	15.6 (9.8)	16.9 (10.6)	15.7 (9.7)	15.0 (9.5)
Sorghum	12.5 (8.8)	11.6 (7.8)	10.0 (6.8)	11.3 (7.7)	9.1 (6.6)
Other crops					
Horse bean	10.1 (6.4)	10.1 (6.5)	8.8 (5.0)	8.8 (5.6)	8.6 (5.6)
Coffee	13.6 (9.2)	12.4 (9.6)	10.7 (8.9)	9.4 (7.5)	10.8 (6.7)
Enset	54.2 (63.2)	40.6 (46.4)	38.4 (46.6)	43.3 (63.1)	40.3 (65.7)
Millet	9.7 (4.6)	10.9 (6.4)	10.7 (6.1)	9.3 (5.3)	9.9 (5.9)
Potato	40.8 (32.5)	39.9 (31.6)	43.3 (32.6)	42.2 (32.4)	42.2 (29.4)
Cowpea	8.4 (4.3)	7.6 (4.6)	6.8 (4.0)	7.6 (4.1)	7.6 (4.6)
Niger-seed	2.9 (2.0)	2.1 (1.3)	3.3 (2.2)	3.1 (1.8)	3.1 (1.8)
Cabbage	102.7 (141.6)	90.0 (108.9)	62.8 (95.9)	58.8 (111.0)	73.7 (110.7)
Rice	37.8 (12.8)	33.8 (10.3)	37.2 (24.0)	20.0 (9.2)	17.9 (9.6)

Source: Authors' calculation from Feed the Future 2015 midline survey data (FtF 2015). Standard errors in parentheses.

adoption rates of improved seeds are higher for larger farms, while the opposite is true for the other two. When we assess the adoption of agrochemicals, they are mostly more intensively used by larger farm sizes. This is especially the case for herbicides, possibly because larger farms lack relatively more access to labor, and herbicides are considered a substitute for weeding labor (Tamru et al. 2017).

TABLE 6.10 Adoption of fertilizers, improved seeds, and agrochemicals in Ethiopia, by cereal and landownership quintile

Land quintile	Crop	1st	2nd	3rd	4th	5th
Fertilizer use, percentage of households (%)	All	59.7	66.0	69.2	70.3	70.9
	Teff	51.1	46.7	45.6	40.9	42.7
	Barley	57.0	60.6	60.1	58.3	62.0
	Wheat	78.4	80.1	79.3	75.3	83.8
	Maize	49.8	53.6	53.8	50.9	49.7
	Sorghum	15.2	24.5	16.2	18.1	16.7
Fertilizer (kilograms per hectare)	All	155	92	73	62	52
	Teff	72.8	72.1	67.6	57.4	61.8
	Barley	96.6	94.5	91.5	83.4	78.0
	Wheat	148.9	152.6	143.1	124.1	139.1
	Maize	101.6	93.4	99.7	95.7	93.7
	Sorghum	17.0	31.7	18.8	20.2	16.6
Use of improved seed (%)	All	32.6	37.7	42.4	42.4	43.2
	Teff	15.2	12.7	15.0	18.0	12.9
	Barley	11.9	14.3	14.8	16.6	22.7
	Wheat	25.6	30.1	33.4	30.0	30.8
	Maize	38.6	42.0	44.1	44.1	42.6
	Sorghum	9.9	11.7	11.6	11.9	6.3
Use of agrochemicals (%)	All	24.1	28.0	30.8	32.0	41.2
	Teff	26.3	21.9	23.2	17.8	23.4
	Barley	21.9	20.6	23.3	25.8	30.0
	Wheat	38.1	38.7	39.7	38.4	51.0
	Maize	3.1	3.4	4.3	2.8	3.5
	Sorghum	8.8	5.5	6.6	7.7	8.5
Use of herbicides (%)	All	19.0	22.6	24.7	27.0	36.4
	Teff	25.1	20.6	21.7	17.0	22.9
	Barley	19.9	19.7	21.5	24.9	27.0
	Wheat	36.3	37.0	36.8	37.1	48.7
	Maize	2.1	1.1	2.7	1.3	2.8
	Sorghum	5.8	3.7	2.9	5.2	8.0

Source: Authors' calculation from Feed the Future 2015 midline survey data (FtF 2015).

Table 6.11 looks at the use of labor and mechanized agriculture across the land quintiles. Households with less land use more labor per hectare and are less likely to engage in mechanized agriculture than are households with larger landholdings. However, the differences are not that large. Overall labor use per hectare is only 27 percent higher on smaller farms compared to larger ones, despite cultivated areas being almost six times as large. Differences in labor use are noted in almost all activities of the agricultural production process, including clearing, weeding, and harvesting.

Noncrop and Nonfarm Income

We now shift focus to overall income portfolios and examine whether households with less land are able to compensate for differences in crop income through income from other sources: livestock, wage employment, own businesses, and remittances.[10] Table 6.12 shows that crop income accounts for 67 percent of the total income in the case of households that own the least land. This jumps to 81 percent in the second land quintile and is more than 90 percent for households with the largest landholdings. Moreover, we see that households in the first land quintile compensate for lower crop income with income from their own businesses (13 percent of total income), livestock rearing (12 percent), and remittances (4.1 percent).[11] Another key takeaway from Table 6.12 is that total incomes earned by the households are strongly associated with land size owned by the households.

Output Use and Food Consumption

Finally, we look at agricultural output use and food consumption. Table 6.13 gives an overview by land quintile of the use of calories expressed in calories per day produced. The results show that the majority of the calories produced on the farm are consumed by the household itself. In the case of the

---

10 As shown in Chapter 11, more connected farms are more likely to use nonfarm labor on their farms, creating more opportunities for land-poor households. It is also possible that farm sizes are systematically smaller in these more connected areas. This raises a concern that the observed differences in household income portfolios are more a reflection of market access than farm size. We explored this possibility by regressing (log) land size on (log) travel time to the nearest city. We get a coefficient of 0.090 ($p < 0.001$) on the (log) travel time variable indicating that increasing travel time by 10 percent is associated with a 1 percent increase in land size. In other words, the average farm sizes are somewhat larger in more remote areas but the difference is relatively small and thus unlikely to be driving the observed differences in income portfolios or other outcomes analyzed in this chapter.

11 This finding is in contrast to recent evidence from cross-country analyses (Headey and Jayne 2014).

TABLE 6.11 Use of labor and mechanization in Ethiopia, by landownership quintile (person-days per hectare)

Land quintile	1st	2nd	3rd	4th	5th
All activities	102	96	89	87	80
Clearing	21	20	18	17	15
Planting	13	12	11	11	10
Weeding	38	36	34	33	30
Harvesting	30	28	26	26	25
Any mechanization (%)	3.7	4.8	4.9	5.7	9.5

Source: Authors' calculation from Feed the Future 2015 midline survey data (FtF 2015).

TABLE 6.12 Income from different sources in Ethiopia, by landownership quintile (%)

Land quintile	1st	2nd	3rd	4th	5th
Total income (birr)	6,794	10,786	14,256	20,349	32,282
Crop production	67	81	85	87	91
From livestock	12.0	8.4	8.0	8.2	6.0
Remittance income	4.1	3.2	2.5	1.6	0.8
Nonfarm wage activities	1.9	1.2	0.8	0.6	0.3
Off-farm wage activities	1.6	0.6	0.3	0.2	0.1
Own business	13.0	6.0	3.6	2.5	1.7

Source: Authors' calculation from Feed the Future 2015 midline survey data (FtF 2015).

first quintile, 63 percent of production is consumed by the household itself. This compares to just below 50 percent for households in the quintile with the biggest farm sizes. These results indicate that in order to reach the same consumption levels, households from smaller farms need to rely more intensively on food purchases than do larger farms, at least for their staples.

We next look at food consumption patterns by land quintile. As shown in Chapter 10 and earlier work in Ethiopia (Worku et al. 2017), cereals account for the bulk of the overall sources of calories in all household types (Figure 6.7). Root crops are relatively more important for the bottom land quintiles than for the top quintiles. Within cereals (Figure 6.8), maize is the single most important crop in terms of calories for all households. However, its importance declines as we move across the land size distribution. Teff becomes more important in the food consumption basket in the top land quintiles.

TABLE 6.13 Agricultural output use in Ethiopia, by landownership quintile (kilocalorie per day)

Land quintile	1st	2nd	3rd	4th	5th
Total calorie production (kilocalorie per day)	2,423	2,728	3,055	3,757	4,879
Auto-consumed calorie (kilocalorie per day)	1,223	1,470	1,636	1,876	2,103
Sold calorie (kilocalorie per day)	429	591	679	916	1,336
Seed (kilocalorie per day)	88	155	182	221	358
Gift (kilocalorie per day)	40	85	33	67	62
Animal feed (kilocalorie per day)	5	8	8	12	36
Storage or loss (kilocalorie per day)	638	419	517	665	984
Auto-consumed calorie (%)	62.6	62.6	60.1	55.9	49.9
Sold calorie (%)	15.3	18.0	19.8	20.5	24.3
Seed (%)	3.6	5.2	5.8	6.2	7.3
Gift (%)	1.0	1.3	0.9	1.3	1.2
Animal feed (%)	0.3	0.2	0.3	0.4	0.7
Storage or loss (%)	20.8	17.9	18.9	21.9	23.9

Source: Authors' calculation from Feed the Future 2015 midline survey data (FtF 2015).

FIGURE 6.7 Sources of calories from different food groups, by landownership quintile

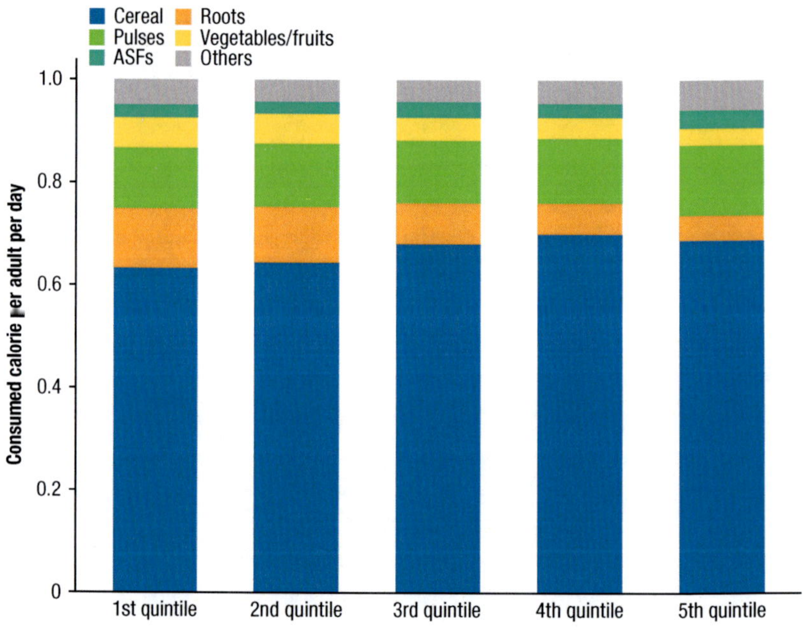

Source: Authors' calculations from Feed the Future 2015 midline survey data (FtF 2015).

Note: ASF = animal-sourced foods.

FIGURE 6.8 Sources of calories from cereals, by landownership quintile

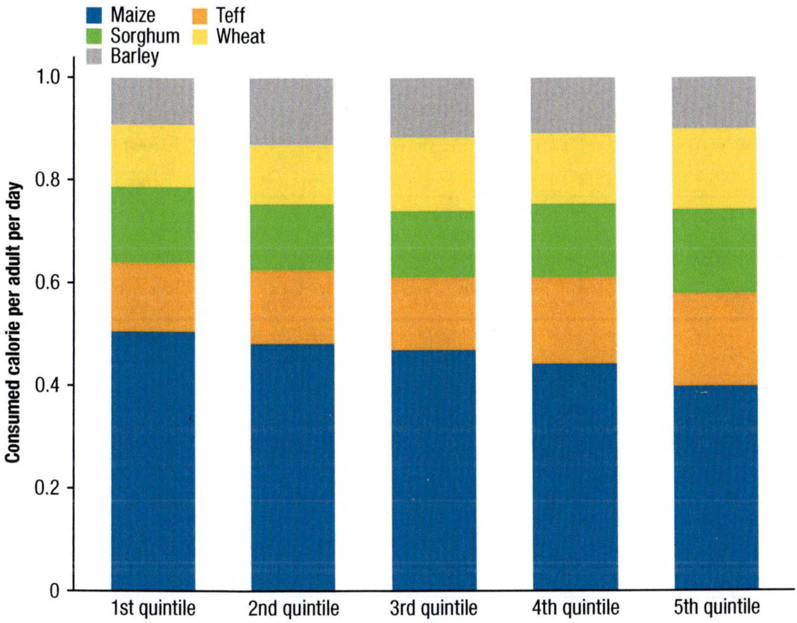

Source: Authors' calculations from Feed the Future 2015 midline survey data (FtF 2015).

Conclusion

This chapter studies associations between farm size, food security, and welfare. Given agricultural land constraints and the rapidly increasing rural population in Ethiopia, with 26 million more people being projected to be residing in rural Ethiopia in 2030 relative to 2016, this is a major concern for the country. Unsurprisingly, we find that larger landholdings are associated with considerably higher crop income. However, despite these large differences in crop income, there are small differences in welfare and food security. In particular, per capita food consumption levels are surprisingly similar across the land quintiles. Further analysis suggests that households with smaller landholdings compensate lower crop incomes by engaging in other income generating activities, such as livestock rearing and running nonfarm businesses. They also engage in more intensified agriculture by using more labor per unit of land and by applying considerably more fertilizer per hectare than do other households. They also grow different crops and focus on bulkier, calorie-dense, and lower-valued crops. Households on smaller farms consume cheaper foods. Moreover, the smallest landownership quintile engages actively in land rental markets and is therefore able to double its cultivated land compared to what it owns.

This study has limitations. The analysis is purely descriptive and static in nature. The associations documented here should not be interpreted as causal. In particular, we cannot be sure whether the differences across the land quintiles are driven by differences in landownership or by other household characteristics that systematically differ between households that own different amounts of land. Moreover, decreasing average land sizes may bring in dynamics that cannot be predicted using cross-sectional or short longitudinal surveys. For example, as a result of decreasing land sizes, an increasing share of the younger population may seek employment outside of agriculture. Recent evidence suggests that this might indeed be the case (Bezu and Holden 2014; Kosec et al. 2018).

Mindful of these limitations, this study suggests a number of implications to consider going forward. First, land rental markets should be stimulated as especially less land-endowed farms seem to rely substantially on these markets to obtain access to land that they are then able to cultivate more intensively. More active land rental markets could therefore contribute toward equity and higher productivity. Second, an active rural off-farm sector is important to allow farmers to further diversify their agricultural activities. Making sure that there is an enabling environment that allows farmers to diversify in such off-farm activities is therefore useful. Third, providing easier access to land-intensifying agricultural technologies—such as modern inputs as well as access to irrigation, allowing more intensive cultivation on the same piece of land—might be important.

Appendix 6A: Replicating the Section 4 Analysis Using the Ethiopia Socioeconomic Survey 2013/2014

TABLE 6A.1 Land distribution by landownership quintile, Ethiopia Socioeconomic Survey 2013/2014

Land quintile	1st	2nd	3rd	4th	5th
Minimum	0.001	0.29	0.61	0.97	1.56
Mean	0.15	0.45	0.78	1.22	2.33
Median	0.16	0.45	0.77	1.20	2.08
Maximum	0.29	0.61	0.97	1.56	7.82

Source: Authors' calculation from the Ethiopia Socioeconomic Survey 2013/2014 data. Observations: 2,173 households.

TABLE 6A.2 Total annual agricultural production by landownership quintile, Ethiopia Socioeconomic Survey 2013/2014

Land quintile	Statistic	1st	2nd	3rd	4th	5th
Total agricultural production (birr)	mean	2,559	3,572	4,144	4,476	5,183
	median	1,863	2,882	3,402	4,006	4,499

Source: Authors' calculation from the Ethiopia Socioeconomic Survey 2013/2014 data. Observations: 2,173 households.

TABLE 6A.3 Welfare indicators by landownership quintile, Ethiopia Socioeconomic Survey 2013/2014

Land quintile	Statistic	1st	2nd	3rd	4th	5th
Per adult equivalent consumption (birr)	mean	4,481	6,033	5,542	5,336	5,179
	median	3,854	4,192	4,022	4,340	4,090
Per adult equivalent food consumption (birr)	mean	3,616	5,151	4,637	4,412	4,108
	median	3,023	3,298	3,131	3,553	3,214
Per adult equivalent nonfood consumption (birr)	mean	865	882	906	924	1,071
	median	591	682	683	770	729
Nonfood consumption out of total consumption (%)	mean	19.3	14.6	16.3	17.3	20.7
	median	15.3	16.3	17.0	17.7	17.8
Household size (adult equivalents)	mean	3.3	4.0	4.2	4.5	4.9
	median	3.2	3.7	4.0	4.5	4.9

Source: Authors' calculation from the Ethiopia Socioeconomic Survey 2013/2014 data. Observations: 2,173 households.
Note: All consumption values are expressed in annual terms.

TABLE 6A.4 Welfare indicators by landownership quintile, controlling for regional differences, Ethiopia Socioeconomic Survey 2013/2014

	(log) annual per adult equivalent consumption		(log) annual per adult equivalent food consumption		(log) annual per adult equivalent nonfood consumption		nonfood consumption out of total consumption	
Dependent variable	1	2	3	4	5	6	7	8
(log) land owned (hectares)	0.034**	0.049***	0.078***	0.071***	0.022	0.041***	−0.008***	−0.005*
	(0.013)	(0.014)	(0.020)	(0.021)	(0.014)	(0.014)	(0.002)	(0.002)
household size in adult equivalent units		−0.088***		−0.046***		−0.097***		−0.006***
		(0.009)		(0.011)		(0.009)		(0.001)
Region dummies?	No	Yes	No	Yes	No	Yes	No	Yes
Observations	2,173	2,173	2,166	2,166	2,173	2,173	2,173	2,173
R^2	0.003	0.053	0.010	0.045	0.001	0.044	0.005	0.019

Source: Authors' calculation from the Ethiopia Socioeconomic Survey 2013/2014 data.

Note: Robust standard errors in parentheses. *** p<0.01, ** p<0.05, * p<0.1.

References

Abay, K., G. Abate, C. Barrett, and T. Bernard. 2019. "Correlated Non-Classical Measurement Errors, 'Second Best' Policy Inference, and the Inverse Size-Productivity Relationship in Agriculture." *Journal of Development Economics* 139: 171–184.

Ambaye, D. 2012. "Land Rights in Ethiopia: Ownership, Equity, and Liberty in Land Use Rights." International Federation of Surveyors Working Week 2012: Knowing to Manage the Territory, Protect the Environment, Evaluate the Cultural Heritage, Rome, Italy, May 6–10, 2012.

Bachewe, F., G. Berhane, K. Hirvonen, J. Hoddinott, J. Hoel, F. Tadesse, A. S. Taffesse, et al. 2014. *Feed the Future (FtF) of Ethiopia—Baseline Report 2013.* Addis Ababa: Ethiopia Strategy Support Program (ESSP) and International Food Policy Research Institute (IFPRI).

Bachewe, F., G. Berhane, B. Minten, and A. S. Taffesse. 2018. "Agricultural Transformation in Africa? Assessing the Evidence in Ethiopia." *World Development* 105 (2018): 286–298. https://doi.org/10.1016/j.worlddev.2017.05.041.

Barrett, C. B., M. F. Bellemare, and J. Y. Hou. 2010. "Reconsidering Conventional Explanations of the Inverse Productivity–Size Relationship." *World Development* 38 (1): 88–97.

Baye, K. 2014. *Teff: Nutrient Composition and Health Benefits.* IFPRI-ESSP Working Paper 67. Addis Ababa: IFPRI/ESSP.

Bevis, L. E., and C. B. Barrett. 2020. "Close to the Edge: High Productivity at Plot Peripheries and the Inverse Size–Productivity Relationship." *Journal of Development Economics* 143 (March): 102377. https://doi.org/10.1016/j.jdeveco.2019.102377.

Bezu, S., and S. Holden. 2014. "Are Rural Youth in Ethiopia Abandoning Agriculture?" *World Development* 64: 259–272.

Binswanger, H., and P. Pingali. 1988. "Technological Priorities for Farming in Sub-Saharan Africa." *World Bank Research Observer* 3 (1): 81–98.

Boserup, E. 2005. *The Conditions of Agricultural Growth: The Economics of Agrarian Change under Population Pressure.* London: George Allen & Unwin Ltd.

Brown, C., M. Ravallion, and D. Van De Walle. 2017. *Are Poor Individuals Mainly Found in Poor Households?* World Bank Policy Research Working Paper 8001. Washington, DC: World Bank.

Carletto, C., S. Savastano, and A. Zezza. 2013. "Fact or Artifact: The Impact of Measurement Errors on the Farm Size–Productivity Relationship." *Journal of Development Economics* 103: 254–261.

Deininger, K., D. A. Ali, S. Holden, and J. Zevenbergen. 2008. "Rural Land Certification in Ethiopia: Process, Initial Impact, and Implications for Other African Countries." *World Development* 36 (10): 1786–1812.

Desiere, S., and D. Jolliffe. 2016. "Land Productivity and Plot Size: Is Measurement Error Driving the Inverse Relationship?" *Journal of Development Economics* 130: 84–98.

Ethiopia, CSA (Central Statistical Agency of the Federal Democratic Republic of Ethiopia). 2005–2017. *Agricultural Sample Survey Volume I: Report on Area and Production of Major Crops (Private Peasant Holdings,* Meher *Season).* Addis Ababa: CSA.

Ethiopia, CSA, and World Bank. 2013. *Living Standards Measurement Study—Integrated Surveys on Agriculture (LSMS-ISA): Ethiopia Rural Socioeconomic Survey (ERSS)—Basic Information Document.* Addis Ababa: CSA; Washington, DC: World Bank.

Fresco, L. O. 1986. *Cassava Shifting Cultivation: A Systems Approach to Agricultural Technology Development in Africa.* Amsterdam: Royal Tropical Institute.

FtF (Feed the Future). 2015. Unpublished Impact Evaluation Survey data. Addis Ababa.

Ghebru, H., B. Koru, and A. S. Taffesse. 2016. *Household Perception and Demand for Better Protection of Land Rights in Ethiopia.* IFPRI-ESSP Working Paper 83. Addis Ababa: IFPRI/ESSP.

Gourlay, S., T. Kilic, and D. Lobell. 2019. "A New Spin on an Old Debate: Errors in Farmer-Reported Production and Their Implications for Inverse Scale—Productivity Relationship in Uganda." *Journal of Development Economics* 141 (November). https://doi.org/10.1016/j.jdeveco.2019.102376.

Headey, D., M. Dereje, and A. S. Taffesse. 2014. "Land Constraints and Agricultural Intensification in Ethiopia: A Village-Level Analysis of High-Potential Areas." *Food Policy* 48: 129–141.

Headey, D. D., and T. S. Jayne. 2014. "Adaptation to Land Constraints: Is Africa Different?" *Food Policy* 48: 18–33.

ISS (Institute for Security Studies). 2017. *Ethiopia Development Trends Assessment: Ethiopia Performance Monitoring and Evaluation Service.* Report prepared for USAID/Ethiopia. Pretoria, South Africa.

Jayne, T. S., J. Chamberlin, and D. D. Headey. 2014. "Land Pressures, the Evolution of Farming Systems, and Development Strategies in Africa: A Synthesis." *Food Policy* 48: 1–17.

Jayne, T. S., T. Yamano, M. T. Weber, D. Tschirley, R. Benfica, A. Chapoto, and B. Zulu. 2003. "Smallholder Income and Land Distribution in Africa: Implications for Poverty Reduction Strategies." *Food Policy* 28 (3): 253–275.

Josephson, A. L., J. Ricker-Gilbert, and R. J. Florax. 2014. "How Does Population Density Influence Agricultural Intensification and Productivity? Evidence from Ethiopia." *Food Policy* 48: 142–152.

Kosec, K., H. Ghebru, B. Holtemeyer, V. Mueller, and E. Schmidt. 2018. "The Effect of Land Inheritance on Youth Employment and Migration Decisions: Evidence from Rural Ethiopia." *American Journal of Agricultural Economics* 100 (3): 931–954.

Lipton, M. 1980. "Migration from Rural Areas of Poor Countries: The Impact on Rural Productivity and Income Distribution." *World Development* 8 (1): 1–24.

———. 2006. "Can Small Farmers Survive, Prosper, or Be the Key Channel to Cut Mass Poverty?" *Journal of Agricultural and Development Economics* 3 (1): 58–85.

Malthus, T. R. 1826. *An Essay on the Principle of Population: Or a View of Its Past and Present Effects on Human Happiness with an Inquiry into Our Prospects Respecting the Future Removal or Mitigation of the Evils Which It Occasions.* 6th edition. London: John Murray.

Masters, W. A., A. A. Djurfeldt, C. De Haan, P. Hazell, T. Jayne, M. Jirström, and T. Reardon. 2013. "Urbanization and Farm Size in Asia and Africa: Implications for Food Security and Agricultural Research." *Global Food Security* 2 (3): 156–165.

Muyanga, M., and T. S. Jayne. 2014. "Effects of Rising Rural Population Density on Smallholder Agriculture in Kenya." *Food Policy* 48: 98–113.

Ricker-Gilbert, J., C. Jumbe, and J. Chamberlin. 2014. "How Does Population Density Influence Agricultural Intensification and Productivity? Evidence from Malawi." *Food Policy* 48: 114–128.

Schmidt, E., and F. Bekele. 2016. *Rural Youth and Employment in Ethiopia.* IFPRI-ESSP Working Paper 98. Addis Ababa: IFPRI/ESSP.

Taffesse, A. S., P. Dorosh, and S. A. Gemessa. 2012. "Crop Production in Ethiopia." In *Food and Agriculture in Ethiopia*, edited by P. Dorosh and S. Rashid, 53–83. Philadelphia: University of Pennsylvania Press.

Tamru, S., B. Minten, D. Alemu, and F. N. Bachewe. 2017. "The Rapid Expansion of Herbicide Use in Smallholder Agriculture in Ethiopia: Patterns, Drivers, and Implications." *European Journal of Development Research* 29 (3): 628–647.

Tsakok, I. 2011. *Success in Agricultural Transformation: What It Means and What Makes It Happen*. Cambridge, UK: Cambridge University Press.

Worku, I., B. Minten, M. Dereje, and K. Hirvonen. 2017. "Diet Transformation in Africa: The Case of Ethiopia." *Agricultural Economics* 48 (S1): 73–86.

World Bank. 2015. *Ethiopia Urbanization Review: Urban Institutions for a Middle-Income Ethiopia*. Washington, DC: World Bank. Accessed June 21, 2019. https://openknowledge.worldbank.org/handle/10986/22979.

———. 2017. World Bank Open Data: Population Growth (Annual Percent). Accessed May 25, 2017. https://data.worldbank.org/indicator/SP.POP.GROW.

PART 2

Evolving Markets and Household Consumption

Chapter 7

EVOLVING FOOD VALUE CHAINS

Bart Minten, Mekdim Dereje, Fantu Bachewe, and Seneshaw Tamru

Introduction

When food value chains and agricultural markets transform, they tend to go through fairly typical stages (Reardon et al. 2019; Reardon and Timmer 2014; Reardon and Minten forthcoming). In the *traditional* system, farmers are mostly focused on producing for own consumption. The urban share of the population is low and supply chains are mainly short and local, serving own consumption, local villages, and nearby towns. In the *transitional* system, markets become more important. Value chains extend spatially because of growing demand from cities and an expansion in the catchment areas for merchants (Braudel 1982). Emergence of public standards of quality are observed, but spot market relations are still dominant. In the *modern* system, value chains are spatially extensive, but there is disintermediation and consolidation in various segments—such as in retail markets that are characterized by the dominance of supermarkets. Private quality and safety standards, contracts, and capital intensification are widespread in this stage. A synthesis of the features of the food economy and of the food economy's structure and conduct in these three different stages of food value chains (FVC) transformation is shown in Table 7.1.

This chapter looks at the evolving food value chains in Ethiopia. Driven by population growth, improved connectivity, higher incomes, urbanization, and consequent diet change, we see a food and agricultural system that is rapidly transforming. Guided by the framework of Timmer (2014), we assess recent evidence regarding dietary, agricultural, and supply chain transformation. For this assessment we rely on a literature review as well as on an analysis of secondary data. We speculate on expected future developments based on simple assumptions of income growth, urbanization, and population growth and by benchmarking Ethiopia with other countries with lower and higher GDP levels—economic growth has been shown to be an important associate of change in food value chains (Timmer 2014).

TABLE 7.1 Characteristics of the various stages of food value chains (FVC) in Ethiopia

Indicator	Traditional FVC	Transitional FVC	Modern FVC
Features of food economy			
Urban share in food market	Low	Medium	High
Share of grains and staples	High	Medium	Low
Seasonality	High	Medium	Low
Food service sector	Small	Modest	Large
Reach FVC	Local	National	Global
Dominant product cycle	Local niche	National commodity	Differentiated product
Structure			
Spatial length FVC	Short	Long	Long
Intermediation length FVC	Short	Long	Short
Value addition	Low	Medium	Large
Concentration	Moderate	Low (large number of Small and Medium Enterprises)	High (private large-scale food industry firms)
Conduct			
Quality differentiation	Low	Low	High
Quality and safety standards	Few	Public	Private
Technologies	Labor-intensive	Labor-intensive	Capital-intensive
Contracts	Spot markets only	Spot markets dominate; emergence contracts	Spot markets small; contracts dominate

Source: Reardon and Minten (Forthcoming).

Following the definition of Reardon et al. (2019) and acknowledging it is hard to put a food economy in one system based on all indicators, we are seeing in Ethiopia the start of a change from the traditional to a transitional system.[1] First, we see dietary transformation with higher consumption levels and better food security; the relative share of cereals in food baskets declining and those of high-value products, such as animal-sourced foods and fruits and vegetables, rising; an emergence of processed and convenience foods in markets; and greater out-of-home consumption. Second, we note changes in supply chains with more reliance on markets by consumers; better integrated markets; smaller spatial and seasonal price margins; and an increase in prices of noncereals. While food imports and the number of food aid beneficiaries in Ethiopia are not coming down, we find that Ethiopia was a net importer of agricultural products in value terms only in one of the past ten years.

At the agricultural production level, we note processes of intensification and modernization. We see changes in the characteristics of smallholder (with less than 25 hectares of land) farms and farmers—declining average farm sizes, increasing average age of farmers (head of households), and younger farmers being more reliant on rental markets to access the land they require.

1 Grain, pulses, and horticultural trade has existed for a while in Ethiopia as has international trade, especially of coffee.

We further see an increasing role of bigger commercial farmers (larger than 25 hectares) with at least 1.3 million hectares leased out to them. Although only 1 million hectares was effectively cultivated in 2013 (Ali, Deininger, and Harris 2017)—representing about 5 percent of all cultivated land in Ethiopia—it has been estimated that the area cultivated by them doubled over an eight-year period (Ethiopia, CSA 2015). However, Ali, Deininger, and Harris (2017) indicate their growth is slowing as the number of new licenses given out peaked in 2007/2008, and an increasing number of these farms are abandoning agriculture altogether (Fikade 2018). Furthermore, these commercial farms are mostly focused on export crops, and few spillovers have been noted for small farmers (Ali, Deininger, and Harris 2019).

While structural transformation of Ethiopia's economy is still at an early stage as seen by the continuing high share of employment in agriculture, food value chains are expected to continue to transform rapidly given similar dynamics going forward. This will have enormous implications for the required growth, functions, and efficiency of private-led agricultural input supply, logistics, and trading sectors.

This chapter describes the data used in the analysis, looks at contextual changes in Ethiopia, assesses dietary transformation, and examines agricultural and supply-chain transformations. We then discuss the current state of structural transformation of Ethiopia's economy and speculate on likely upcoming changes in Ethiopia's food value chains, focusing on agriculture, dietary, and supply chains. We finish with conclusions.

Data

Diverse datasets are used in this analysis. First, for the consumption analysis we use nationally representative Household Income, Consumption, and Expenditure Surveys (HICES) conducted by the Central Statistical Agency (CSA) of Ethiopia.[2] These surveys were administered in 1995/1996, 1999/2000, 2004/2005, and 2010/2011. The HICES are repeated cross-sectional surveys that serve as the official source for poverty statistics in Ethiopia (Ethiopia, MoFED 2012). The sampling was done by stratifying the country into rural and urban areas. After that, enumeration areas were selected using a probability proportional to size approach where more populated units had a higher

2 The World Bank changed the name of these surveys to "Household Consumption Expenditures Surveys," which would be the official name of the 2011 and 2016 surveys. For simplicity, however, we refer to "HICES" whenever referring to these surveys.

probability of being selected into the final sample. Sampling weights provided by the CSA, which are based on selection probabilities, are used to compute representative estimates for rural and urban areas of the country.

Second, we use data collected by the CSA in its annual Agricultural Sample Survey (AgSS) of agricultural households. The data covers the main agricultural season (*meher*) for the period 2004/2005–2015/2016 and pertain to smallholder farmers—those farming less than 25 hectares—that dominate agricultural land use in Ethiopia. During the period covered in this study, the sampling frame of the AgSS included the entire rural parts of the country except for the nonsedentary population of three zones in Afar and six zones in Somali regions (Ethiopia, CSA 2018). We use these data to conduct descriptive analyses of trends in smallholder landholdings in Ethiopia.

Third, the Ethiopian Grain Trading Enterprise (EGTE), a grain procurement arm of the government, gathers prices of cereals in a number of major wholesale markets in the country. Prices are collected during the early morning, late morning, and afternoon on major market days, with simple averages of these prices over the course of a month being reported as monthly prices. The prices are collected by noting prices from observed transactions. Wholesale prices at 12 selected markets are made available publicly. These price data are used in the analysis. We rely on the national Consumer Price Index (CPI) as constructed by the CSA to deflate prices.

Fourth, import data were obtained from the Ethiopian Revenues and Customs Authority or downloaded from the United Nations's Comtrade website. We also rely for some data on the National Bank of Ethiopia (NBE 2018), the Ethiopian Road Authority, the World Bank, FAOSTAT (the Food and Agriculture Organization Statistics in the United Nations Corporate Statistical Database), the Ethiopian Meat and Dairy Institute, and the Ministry of Trade.

Contextual Changes

A number of changes in Ethiopia's economy are resulting in significant dynamism in its agricultural and food markets. We discuss three major factors: infrastructure development, population growth and urbanization, and income growth.

First, Ethiopia has invested heavily in road network development over the past two decades. It is estimated that the length of asphalted roads expanded fourfold over the past 15 years from 3,900 kilometers in 2000–2001 to 15,900 kilometers in 2016–2017 (NBE 2018). However, investments in rural roads were important as well. Data from the Ethiopian Roads Authority indicates that the length of gravel roads in Ethiopia increased from 22,900 kilometers in 2007 to 62,100 kilometers in 2015, while the length of asphalted roads increased from 4,800 kilometers to 15,400 kilometers over the same period. As a result of these and earlier massive investments in roads, travel times to major cities throughout Ethiopia have been greatly reduced (see Chapter 12). Second, the population of Ethiopia is rapidly growing. Since 2000, a population equivalent to that of Canada has been added to the country. The rural population grew by approximately 27 million people between 2000 and 2017. Over the same period, the urban population increased by approximately 11.5 million. While the rural areas are rapidly growing, cities are growing even more rapidly. In 2017, 20 percent of the Ethiopian population was living in cities (World Bank 2017b; see Chapter 12).

Third, Ethiopia has been one of the fastest growing economies in the world over the past decade—an impressive feat for a low-income African country that exports relatively few natural resources. Average annual GDP growth over the decade 2007/2008 to 2016/2017 was estimated at 10 percent (NBE 2018). Agriculture was a main contributor to that growth. While national official data show that real agricultural output growth was lower than overall GDP growth, it still grew on average by 6.3 percent per year over the same decade. This growth in agriculture, and especially in cereal production, has been a major contributor to the important reductions in poverty observed in the past decade (World Bank 2014).

Dietary Transformation

We study consumption patterns and their changes over time using four rounds of the HICES dataset, covering the period from 1996 through 2011. To ensure comparability over time, expenditures are deflated using the national Consumer Price Index (CPI) and values are expressed in constant 1996 birr.[3] Quantities consumed per adult equivalent were calculated as well. We use as our preferred measure calories per adult equivalent, following the conversion factors suggested by CSA (Ethiopia, CSA 2012b). As the focus of our analysis in this chapter is on dynamics, the advantage of using adult equivalents is that changes over time reflect effective changes in consumption patterns instead of changes in household demographics. The results of this exercise, presented in

3 The exchange rate was 6.3 birr per US dollar in 1996, as per the National Bank of Ethiopia (www.nbebank.com/pdf/monthlymacroeconomic/January%202007.pdf).

TABLE 7.2 Food consumption and real per adult equivalent expenditure in Ethiopia, 1996–2011

	1996		2000		2005		2011	
Category	Birr	Share (%)	Birr	Share (%)	Birr	Share (%)	Birr	Share (%)
Real expenditures (birr per adult equivalent per year)								
Food								
Teff	85	11.6	96	12.6	72	8.9	69	7.5
Wheat	53	7.2	66	8.7	71	8.9	68	7.4
Barley	34	4.7	29	3.8	35	4.4	22	2.4
Maize	74	10.1	82	10.8	69	8.6	71	7.7
Sorghum	52	7.1	46	6.1	65	8.1	46	5.0
Other cereals	22	3.0	24	3.2	13	1.7	13	1.4
Processed cereals	15	2.1	15	2.0	26	3.2	42	4.6
All cereals	337	45.7	357	47.1	351	43.8	333	36.0
Pulses	56	7.6	75	9.9	62	7.7	88	9.5
Oilseeds	2	0.3	2	0.3	2	0.2	1	0.2
Animal products	56	7.5	60	7.9	70	8.7	100	10.8
Oil and fat	34	4.6	27	3.6	31	3.9	61	6.6
Vegetables and fruits	28	3.7	34	4.5	37	4.6	59	6.4
Pepper	36	4.9	30	4.0	21	2.6	61	6.6
Enset/kocho	38	5.1	57	7.5	36	4.4	39	4.2
Coffee/tea/chat	72	9.8	52	6.9	62	7.7	84	9.1
Root crops	19	2.6	26	3.4	25	3.1	16	1.8
Sugar and salt	18	2.5	15	2.0	15	1.9	25	2.7
Other foods	42	5.7	23	3.0	90	11.3	57	6.2
Total food	*739*	*100.0*	*759*	*100.0*	*802*	*100.0*	*925*	*100.0*
Food/nonfood								
Food	739	59.6	759	63.6	802	54.1	925	47.9
Nonfood	502	40.4	434	36.4	681	45.9	1,005	52.1
Total	*1,240*	*100.0*	*1,193*	*100.0*	*1,483*	*100.0*	*1,930*	*100.0*

Table 7.2, illustrate a number of interesting findings (see Worku et al. [2017] for more details).

The share of nonfood items in the total consumption basket increased substantially over time, especially since 2000. In 2000 the share of nonfood consumption expenditures accounted for 36 percent of the total. Over the following decade, this type of expenditure grew to 52 percent by 2011. Such rapid increases of nonfood expenditures, outstripping increases in food

Consumption (kilogram per adult equivalent per year)	1996		2000		2005		2011	
	Kilo-grams	Share (%)	Kilo-grams	Share (%)	Kilo-grams	Share (%)	Kilo-grams	Share (%)
Teff	31	10.6	38	10.0	32	8.0	33	7.3
Wheat	25	8.7	32	8.4	37	9.1	31	6.9
Barley	17	5.9	12	3.3	16	3.9	12	2.7
Maize	41	14.0	47	12.5	47	11.6	63	14.2
Sorghum	20	7.1	28	7.5	40	9.9	35	7.8
Other cereals	10	3.4	17	4.5	7	1.6	9	2.0
Processed cereals	5	1.9	6	1.6	8	2.0	9	2.0
All cereals	149	51.7	180	47.8	187	46.2	192	43.0
Pulses	23	8.1	21	5.6	21	5.1	22	5.0
Oilseeds	1	0.3	1	0.1	0	0.1	0	0.1
Animal products	17	6.0	16	4.2	18	4.6	21	4.6
Oil and fat	2	0.8	2	0.5	3	0.7	5	1.2
Vegetables and fruits	31	10.9	37	9.7	42	10.5	45	10.0
Pepper	3	1.2	2	0.6	2	0.6	5	1.2
Enset/kocho	13	4.6	71	18.8	52	12.8	58	13.0
Coffee/tea/chat	10	3.6	10	2.6	10	2.5	15	3.4
Root crops	15	5.3	28	7.4	34	8.3	30	6.7
Sugar and salt	8	2.7	6	1.7	7	1.7	10	2.1
Other foods	14	4.8	4	1.0	28	7.0	43	9.7
Total food	*288*	*100.0*	*376*	*100.0*	*404*	*100.0*	*447*	*100.0*

Source: Ethiopia, CSA 1996, 2000, 2005, and 2011; also cited in Worku et al. (2017).
Note: Exchange rate in 1996 was approximately 6.4 birr per US$.

expenditures, are typical of transforming economies and indicate significant improvements in welfare in the country (World Bank 2014; Ethiopia, MoFED 2012).

An important increase in the total quantity of food consumed per capita equivalent is seen at the bottom of Table 7.2. HICES data show that consumption increased from 288 kilograms per capita equivalent per year in 1996 to 447 kilograms in 2011, an increase of 55 percent. The quantities of cereals consumed have shown much less growth, especially in the past 10 years. Consumption of cereals grew from 180 kilograms per capita equivalent in 2000 to 192 kilograms per capita equivalent in 2011, an increase of 7 percent. Moreover, expenditures on food have grown in real terms in the last two surveys conducted compared to 2000. Per capita equivalent food expenditures in 2011 were 22 percent higher than in 2000.

Some notable shifts are seen within the food basket. Overall, the share of cereals in total food expenditures is declining. While the share made up 47 percent of expenditures in 2000, this declined to 36 percent 10 years later. Most growth in the noncereal food categories was recorded in the "other food" category that grew from 3.0 percent in 2000 to 6.2 percent in 2011. There also is seen a growing importance of animal-sourced foods in the food basket over time. Although the share is still relatively low (see Humphries et al. [2014] for a comparison with Peru), it has grown from about 7.5 percent in 1996 to 10.8 percent in 2010. These patterns are a reflection of Bennett's law that describes a relative decline in starchy staples and an increase in animal proteins with income (Bennett 1941). While rural—and to a lesser extent urban—Ethiopia is generally characterized by a lack of diverse diets, which are associated with poor nutritional outcomes (Headey 2014), this seems to be changing over time, albeit slowly.

Cereal expenditures make up 36 percent of the total national consumption basket, but they make up 43 percent of the quantity consumed in 2011. This indicates that the relative cost of cereals is declining. In contrast, animal products constitute 10.8 percent of expenditures and 4.6 percent of the quantities consumed. These animal products are the most expensive items in the consumption basket. On the opposite side of the price spectrum, root crops are a relatively cheap food category in the consumption basket.

Table 7.3 illustrates differences in consumption patterns between rural and urban areas using the 2011 HICES dataset. Average per capita equivalent expenditures are significantly higher in urban areas than in rural ones, and the share of nonfood expenditures is also significantly higher in urban areas (61.8 percent) than in rural areas (48.1 percent). Compared to rural areas, urban food expenditures are relatively higher: rural food consumption expenditures are only two-thirds of the urban food expenditures at 863 birr versus 1,219 birr, respectively. Although food expenditures are significantly higher in urban areas, the actual quantities consumed are lower—462 kilograms in rural areas versus 376 kilograms in urban areas. This is due to the higher prices paid in urban areas for food, but also possibly because of lower calorie requirements in these urban settings (Deaton and Drèze 2009).

Interestingly, there are almost no differences in the share of cereals in total food expenditures between urban and rural areas, and the quantities consumed of cereals are also at similar levels. However, consumption of animal products is substantially higher in urban areas. Within the cereal category, however, consumption patterns differ significantly. Rural households consume significantly more sorghum (40 kilograms versus 14 kilograms in urban

TABLE 7.3 **Per adult equivalent expenditures and food consumption in Ethiopia in 2011, urban versus rural**

	Real expenditures by food group (birr/adult equivalent/year)				Consumption by food group (kg/adult equivalent/year)			
	Rural		Urban		Rural		Urban	
Category	Birr	Share (%)	Birr	Share (%)	Kilograms	Share (%)	Kilograms	Share (%)
Food								
Teff	51	6.0	153	12.6	25	5.4	69	18.4
Wheat	59	6.8	113	9.3	29	6.2	41	10.9
Barley	25	2.9	10	0.8	14	3.0	4	1.1
Maize	80	9.3	30	2.4	72	15.7	22	5.8
Sorghum	52	6.0	18	1.5	40	8.6	14	3.6
Other cereals	14	1.6	11	0.9	10	2.2	4	1.0
Processed cereals	28	3.3	109	8.9	5	1.1	26	7.0
All cereals	309	35.8	445	36.5	195	42.2	180	47.9
Pulses	84	9.8	105	8.6	23	4.9	21	5.6
Oilseeds	2	0.2	1	0.1	0	0.1	0	0.1
Animal products	84	9.8	175	14.3	20	4.4	23	6.1
Oil and fat	51	5.9	110	9.0	4	0.8	11	2.9
Vegetables and fruits	51	5.9	99	8.1	42	9.2	56	14.8
Pepper	59	6.8	73	6.0	5	1.1	6	1.6
Enset/kocho	45	5.3	7	0.6	69	14.9	8	2.0
Coffee/tea/chat	85	9.9	78	6.4	16	3.6	10	2.5
Root crops	16	1.9	17	1.4	32	7.0	19	5.1
Sugar and salt	21	2.4	44	3.6	9	1.9	13	3.5
Other foods	55	6.4	67	5.5	46	10.0	30	7.9
Total food	*863*	*100.0*	*1,219*	*100.0*	*462*	*100.0*	*376*	*100.0*
Food/nonfood								
Food	863	51.9	1,219	38.2	n.a.	n.a.	n.a.	n.a.
Nonfood	799	48.1	1,975	61.8	n.a.	n.a.	n.a.	n.a.
Total	*1,662*	*100.0*	*3,194*	*100.0*	n.a.	n.a.	n.a.	n.a.

Source: Worku et al. (2017).

Note: n.a. = not applicable.

households) and maize (72 kilograms versus 22 kilograms). In contrast, the share of teff in the urban food consumption basket is significantly higher than in rural areas—more than three times as high. Urban residents consume relatively more expensive cereals, such as teff, than do rural ones (Worku et al. 2017). Overall, we see a substantially different diet in urban households than in rural ones. This implies that national food value chains will likely change in important ways in the future with an increase in the share of Ethiopia's population residing in urban areas and the changing food preferences accompanying that demographic shift.

Agricultural Transformation

CSA official statistics show that grain production doubled from 13.4 million metric tons in 2005/2006 to 27.0 million metric tons in 2014/2015. Bachewe et al. (2018) triangulate these numbers and assess some of the drivers for that change. They find that there has indeed been substantial growth in agricultural production in Ethiopia, driven by increasing land expansion but even more so by increasing yields. These increasing yields have been achieved through improved total factor productivity but also by increasing use of modern inputs. They argue that this modernization and intensification process in agriculture was driven by expanded availability of agricultural extension agents, improved market access, better price incentives, and improvements in education levels of farmers.

This process of intensification and modernization of agricultural production has continued in more recent years. Figure 7.1 illustrates to what extent imports of chemical fertilizers and agrochemicals have changed over the past decade. The figure shows that imports of such modern inputs more than quadrupled since 2004. While some of these agrochemicals are used in large commercial farms (especially the commercial large-scale flower farms that have quickly taken off in the past decade), there is also increasing uptake by smallholder farmers. For example, Tamru et al. (2017) show herbicides-applied cereals area, mostly by smallholders, doubled during 2004–2014 to more than a quarter of the cereal area in the country. One of the changes that typically happens during this agricultural transformation process is the increasing use of machines in agriculture. This is discussed in more detail in Chapter 3.

We also note some important changes in agricultural production, although it remains largely a smallholder economy. For example, Taffesse (2019) shows, pooling data from the annual agricultural sample surveys, that during the period 2003–2016 the average farm size was slightly less than a hectare. Half

FIGURE 7.1 **Imports of chemical fertilizers and agrochemicals in Ethiopia (million US$)**

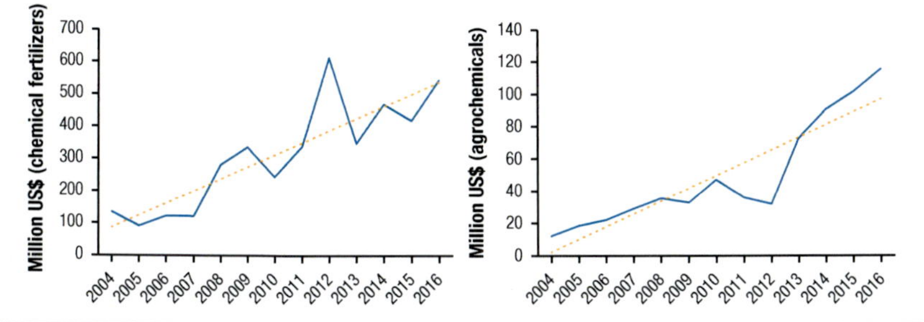

Source: UN Comtrade (https://comtrade.un.org/data/).
Note: The dotted line is a linear trend over the period examined.

of these smallholders operated less than 0.65 hectare of cropland, while the bottom tenth of the distribution manage less than 0.13 hectare. Only the top 10 percent of holders operated more than 2 hectares and 0.75 percent had more than 5 hectares. As shown in Chapter 6, these smallholders have over time become even smaller, and there is also increasing evidence that landlessness is on the rise (Vaughan 2018). Yet in recent years Ethiopia has pursued a policy aiming to increase the role of large commercial farms (those cultivating more than 25 hectares).[4] There are different estimates on the land area allocated to commercial farms. Based on large farm census data collected by the CSA in 2013, Ali, Deininger, and Harris (2017) estimated that 1.3 million hectares were allocated to 6,612 commercial farmers in 2013/2014. This compares to 1.4 million hectares estimated by Rulli, Saviori, and Odorico (2013) and 2 million hectares by Bekele (2016). However, it has been observed that a substantial part of the land allocated to commercial farms is not under cultivation (Ali, Deininger, and Harris 2017). It was also estimated that about 1 million hectares were cultivated by large commercial farms in 2013. It appears that the area cultivated by commercial farms increased significantly, coming from 0.46 million hectares in 2007/2008 (Ethiopia, CSA 2015). However, Ali, Deininger, and Harris (2017) indicate their growth has been slowing more recently as the number of new licenses that are given out had peaked in 2007/2008, and there is evidence that an increasing number of these investors are abandoning agriculture altogether (Fikade 2018).

---

4 An important change for potential land acquisitions by domestic and foreign investors was done in 2009 through the approval of "Proclamation 29/2001" giving federal authorities more leeway in large land allocations (those over 5,000 hectares).

Table 7.4 summarizes the most recently available CSA data (2017/2018) on the relative contribution of commercial farms. We compare in that year the production of smallholder farms—in the *belg* and *meher* seasons and in the irrigated dry season—with overall area allocation and production on commercial farms. The table shows that commercial farms made up 5.5 percent of overall area cultivated in that year. Commercial farms focus more on export crops than smallholder farms. For example, area under cereals made up 30 percent of the area cultivated by commercial farms but 68 percent of that cultivated by smallholders (Table 7.4). The most important crops cultivated by commercial farms are sesame, cotton, and coffee. These crops are spatially clustered for these commercial farms with sesame mostly grown in the northwest, maize mostly in the west, and wheat in the south and southeast of the country (cotton and sorghum are more scattered in different areas).

Ali, Deininger, and Harris (2017) further found that commercial farms have higher yields (about double) than smallholders. Table 7.4 also suggests higher yields as the share of production of commercial farms is higher than the share in area for all crop categories except oilseeds. Ali, Deininger, and Harris (2019) further assessed the spillovers from large to small farms, but they found that there was little job creation and that the benefits in terms of technology and input markets were modest. The upshot is that the role of commercial farms in national agricultural production—in particular for local markets—is still rather small. Moreover, despite the fiscal advantages that were given to these farms, their performance has been lower than anticipated (Fikade 2018).

We also find that smallholder farmers on average are becoming older. Figure 7.2 (left) indicates that the share of farmers under age 35 years has declined from 36 percent in 2004 to 30 percent in 2015, driven by the expanding life experience of more mature farmers as well as lower entry into farming by younger farmers. The CSA data also show that young farmers have smaller and declining farm sizes. We depict average land sizes cultivated by young and by mature farmers in Figure 7.2 (right). The decline in agricultural area has occurred in both age groups, but it is relatively more important for young farmers given their smaller farm sizes to start with. The absolute differences in farm sizes between the different age groups remained rather similar across years.

As most rural residents depend on agriculture as their main source of livelihood (Schmidt and Bekele 2016; Bachewe et al. 2017), access to land is crucial for agricultural activities. This is especially an issue for the younger population given that they have less access to agricultural land. In response

TABLE 7.4 Area allocation and production by smallholders and commercial farms in Ethiopia, 2017/2018

Crop	Total		Smallholders				Commercial farms			
	Area (1,000 hectares)	Production (million tons)	Area (1,000 hectares)	Production (million tons)	Area (%)	Production (%)	Area (1,000 hectares)	Production (million tons)	Area (%)	Production (%)
Cereals	11,976	29.62	11,680	28.65	97.5	96.7	296	0.97	2.5	3.3
Pulses	2,127	3.49	2,068	3.38	97.2	96.8	59	0.11	2.8	3.2
Oilseeds	1,165	1.10	879	0.86	75.5	78.2	286	0.24	24.5	22
Vegetables	439	3.64	431	3.54	98.2	97.3	8	0.1	1.8	2.9
Root crops	850	13.58	849	13.55	99.9	99.8	1	0.03	0.1	0.2
Permanent crops	1,504	9.75	1,157	2.86	76.9	29.3	347	6.89	23.1	70.7
Total	**18,061**	**61.18**	**17,064**		**94.5**		**997**		**5.5**	

Source: Ethiopia, CSA 2018.

FIGURE 7.2 Share of farmers in age categories and farm size (three-year moving average), 2004/2005–2016/2017

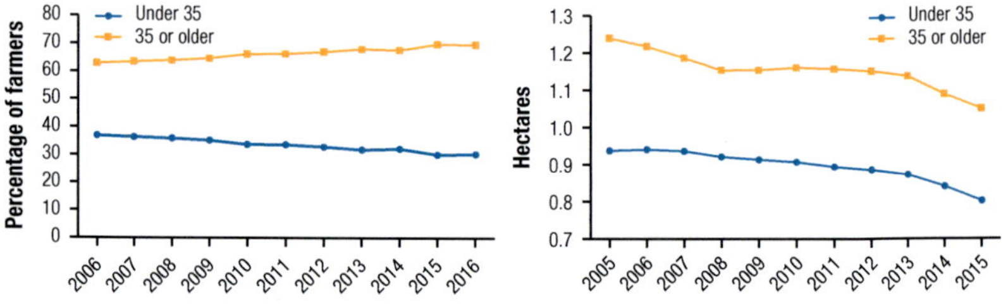

Source: Authors' calculations based on AgSS 2004/2005–2016/2017 (Ethiopia, CSA 2005–2016).

to more binding land constraints, rental markets are seemingly becoming more widespread, with the youth increasingly relying on these rental markets to access land. In Figure 7.3 we show the share of rented-in land out of the total land that is operated by youth and mature farmers, respectively. The share of rented-in land operated by youth farmers was nearly twice that operated by mature farmers in an average year during the 2004/2005–2015/2016 period, with the share of rented-in land especially increasing from 2009 forward. Vandercasteelen et al. (2018) illustrate that these formal rental markets are especially taking off in areas with better market access. For an assessment of the share of farm households that hire in or rent out labor and the evolution of wage rates, see Chapter 11.[5]

Supply Chain Transformation

Local Markets

EXPANSION

The quantities traded and the total value of agricultural commercial surpluses are rapidly increasing. To illustrate these changes, we use production data of all reported crops from the AgSS of CSA and value those at real yearly prices by crop.[6] We only rely on smallholder data as no sales data on commercial

5 It would also have been interesting to analyze trends in land rental rates and fertilizer prices to assess if the farming system is moving toward more land-saving, labor-saving, and/or capital-using forms of technology as seen in other African countries (Jayne et al. 2019). However, we are not aware of easily available datasets in Ethiopia that would capture these dynamics.

6 Including all grain crops, vegetables, fruit crops, root crops, chat, coffee, hops, and sugarcane.

FIGURE 7.3 Share of total land operated by farmer that is rented-in, by age category, 2004/2005–2016/2017

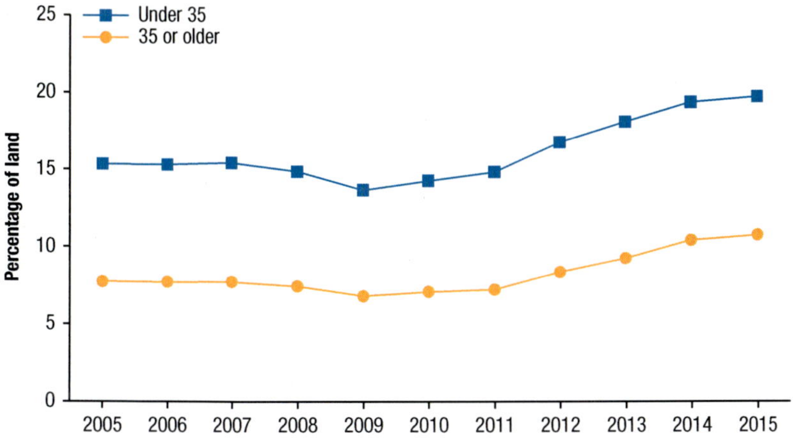

Source: Authors' calculations based on AgSS 2004/2005–2016/2017 (Ethiopia, CSA 2005–2016).

farms (those with more than 25 hectares) are available. We also only rely on *meher* data. Note that these omissions will only influence overall trends marginally, given the low importance of large farms and *belg* production compared to *meher* smallholder production. We see in Figure 7.4 that the overall real value of agricultural sales in Ethiopia has more than doubled over the past decade, from 8 billion birr in 2006 to 17.4 billion birr in 2017. We also note relatively small changes in the crop composition of these sales. Teff is by far the most important crop, making up about 23 percent of the value of agricultural sales in the country. The other important crops are coffee (13 percent), wheat (9 percent), chat (7 percent), and sesame (6 percent). Maize—despite being on top in quantities produced in the country—is relatively less important as a component of agricultural sales (at 6 percent), consistent with consumption patterns in rural areas noted above.

The increased agricultural commercialization patterns also show up in the purchase behavior of households. Respondents during the National Consumption Survey of 2011 were asked how they paid for specific food purchases. As expected, there are large differences between rural and urban areas. Consumption of households' own agricultural products accounts for 42 percent of total food expenditures in rural areas (Worku et al. 2017), but that share is much higher in calorie terms (two-thirds), indicating that

FIGURE 7.4 Real value of agricultural sales (*meher*) in Ethiopia, 2006–2017

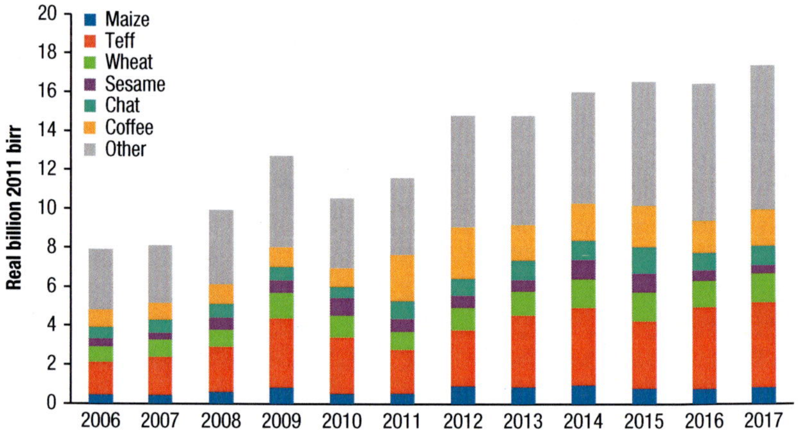

Source: Authors' calculations based on CSA smallholder data.

relatively more expensive foods are acquired through purchase (WFP and CSA 2014). These numbers reflect the high level overall of auto-subsistence of the rural Ethiopian economy (Worku et al. 2017).

However, the value and share of auto-consumption is considerably lower than is often assumed. Even rural households depend in important ways on commercial food markets. This is a consistent finding with other countries in eastern and southern Africa. For example, Dolislager, Tschirley, and Reardon (2015) show rural households bought 44 percent (in value terms) of the food they consume. As might be expected, in urban areas only 3 percent of food expenditures comes from own production. Towns and cities therefore are important commercial food markets. Despite the low urbanization share in the total population of Ethiopia, urban markets made up about one-third of the total commercial food market in 2011.

Change in the Structure of Supply Chains

We note important growth and transformation in the trading and transport, processing, and food service sectors, and in retailing and distribution (discussed below consecutively). It is estimated that 4 percent of the population was employed in agricultural trade or transport in 2011 per Ethiopia's social accounting matrix (SAM) and is one of the most important areas of employment in the service sector. This sector has shown rapid change. Minten, Tamru, and Stifel (2014) reported that the number of trucks in urban wholesale markets rose by about 75 percent over a decade. They also illustrate a shift to larger trucks being used for the transportation of agricultural products. Trader focus groups indicated considerable growth over time in the number of traders in these markets as well (Minten, Tamru, and Stifel 2014). With the number of traders perceived to be growing by almost 150 percent over the past decade, and the number of brokers growing by more than 250 percent, competition in these markets appears to have become keener and turnover per trader and broker lower. Bachewe et al. (2016) also show that agricultural trade is one of the main nonfarm activities in rural areas in Ethiopia, especially so for the relatively rich.

The increased consumption of processed food is a pattern that is seen globally with increases in income. Benfica and Thurlow (2017) estimate that processed food accounted for 17 percent of the food budget in 2011 in Ethiopia. We rely on the most updated data from two manufacturing enterprises surveys conducted by CSA to assess the food processing sector. In 2010/2011, CSA surveyed large and medium-scale manufacturing firms, defined as those that employ 10 or more people and use electricity-driven machinery. Based on this survey, 686 firms were identified as being involved in the manufacturing of food products and beverages, employing more than 67,000 people (Table 7.5). Of these employees, two-thirds are men and one-third are women. By subsector the most important subsectors in terms of employment in the food processing sector are firms involved in grain milling, baking, and the production of sugar and sugar confectionery.

In 2013/2014, CSA surveyed small-scale manufacturing firms, defined as those that employ fewer than 10 people. Such firms are significantly more important in terms of employment than the larger firms: they employ 1.7 million people nationally, of which 52 percent is involved in food processing (Table 7.5). A full 21 percent of these firms manufacture beverage or food products, with the production of bakery products being particularly important, employing more than 220,000 people. Grain milling accounts for 31 percent of people employed in small-scale manufacturing industries with more than 540,000 people engaged. In contrast to larger firms, more females are employed in these smaller firms, making up almost half of permanent employees. Overall, it is estimated that almost one million people are engaged in food processing in Ethiopia (bottom Table 7.5), or around 2 percent of the economically active population of the country defined as those ages 15–64 years. While food processing is important within the manufacturing sector, employment in food processing is overall small given the small share of the population employed in manufacturing in Ethiopia. Unfortunately, good

TABLE 7.5 Food processing sector in Ethiopia, by scale

Industry	Number of establish-ments	Number of persons engaged	Of which permanent	
			Male	Female
Large and medium-scale manufacturing (2010/2011)				
Total	2,170	175,698	95,211	52,037
Manufacturing of food products and beverages	686	67,471	38,134	18,612
Share (%)	32	38	40	36
Production, processing and preserving of meat, fruits, and vegetables	10	2,716	1,890	579
Manufacture of...				
vegetable and animal oils and fats	34	1,198	743	198
dairy products	24	1,867	1,165	509
grain mill products	197	10,077	5,590	2,419
prepared animal feeds	11	601	375	136
bakery products	247	14,917	6,619	7,696
sugar and sugar confectionery	31	15,273	8,897	1,303
macaroni and spaghetti	20	1,855	1,230	569
other food products	29	2,522	973	624
distilling, rectifying, and blending of spirits	18	1,886	1,079	739
wines	2	524	353	171
malt liquors and malt	10	6,049	4,398	1,021
soft drinks and production of mineral water	53	7,986	4,822	2,648
Small scale manufacturing industry (2013/2014)				
Total	116,604	1,744,544	532,859	445,209
Manufacturing of food products, except milling services	25,430	373,259	116,458	110,102
Share (%)	22	21	22	25
Production, processing and preserving of meat, fruits, and vegetables	269	3,271	1,162	1,238
Manufacture of...				
vegetable and animal oils and fats	720	7,487	3,082	2,640
bakery products	14,218	221,301	69,754	65,760
cocoa, chocolate, and sugar confectionery	29	320	134	115
other food products	10,195	140,881	42,326	40,349
Grain mill services	35,430	540,539	152,896	129,843
Share (%)	30	31	29	29
Total food processing sector	61,546	981,269	307,488	258,557

Source: Authors' calculations based on Ethiopia, CSA (2012a, 2015).

and consistent time-series data are lacking for the processing sector as a whole. However, important changes are happening in the sector, as illustrated by two examples. First, in the case of dairy processing plants, there were eight companies active in the country in 2007. By 2017 this had more than tripled to 25 (Figure 7.5, left side).[7] While there now is significant overcapacity in dairy processing, this example illustrates significant change in the subsector.

Second, data were obtained from the Ministry of Trade on the number of licenses given out to flour factories in Ethiopia. Beginning in the 2000s, about 15 licenses were given out annually, which tripled to more than 50 in 2016 (Figure 7.5, right side). Further illustrating large changes in the milling sector, Comtrade data indicate that the three-year moving average of the value of imports of machinery for mills (HS code 8437) quadrupled from US$7 million in 2006 to US$28 million in 2015. Minten, Tamru, and Stifel (2014) further provide evidence of the much greater presence of mills in urban areas, their evolving roles in moving toward one-stop retail shops, increasing competition, and lower milling margins over time.

In Ethiopia the share of food eaten away from home is increasing—a development that is seen globally as incomes rise.[8] In 2011 the share of food eaten away from home in the urban food budget was estimated to be 16 percent—that is, twice as high as expenditures on vegetables and fruits. Moreover, a strong gradient over income is seen. The poorest quintile spends 6 percent of its food budget on food eaten away from home, compared to 25 percent for the richest quintile. Associated with this, we see the increasing emergence of a rapidly growing food service sector. For example, in the case of commercial enjera markets, Minten et al. (2016a) show how these markets are quickly transforming with a large number of people, especially in urban areas, now buying enjera instead of preparing it themselves. They estimate that more than 100,000 people in urban centers in Ethiopia make their living in enjera-making enterprises or through the retailing of the enjera produced. This sector therefore provides a high level of employment in Ethiopia, comparable to the much publicized flower export sector (see, for example, Oqubay 2015; Schaefer and Abebe 2015).

---

7 Data from the Ethiopian Meat and Dairy Industry Development Institute.

8 For example, Smith, Dupriez, and Troubat (2014) note that food eaten away from home increased from 10 percent to 49 percent of total food expenditures in the United States between 1900 and 2010 and that similar fast changes are seen in this area in a number of quickly transforming economies such as China, India, and Mauritius.

FIGURE 7.5 Change in number of agro-processing firms in Ethiopia

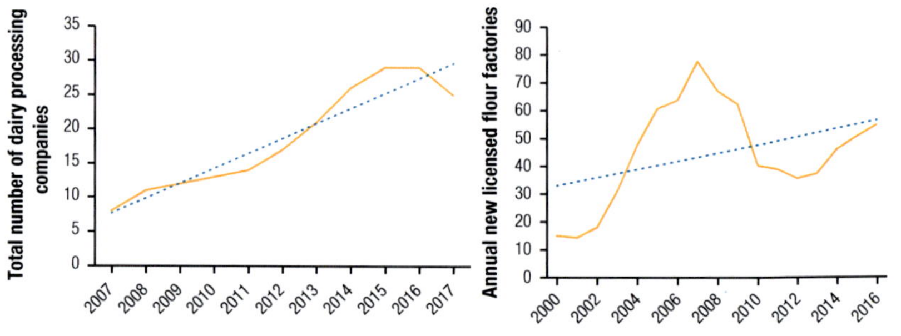

Source: Data from Ethiopian Meat and Dairy Industry Development Institute (left) and Ministry of Trade (right).
Note: The dotted lines are linear trends in number of agro-processing firms over this period.

Modern retail and different food distribution and retail systems are emerging. Assefa et al. (2016) analyze the urban food retail market in Ethiopia and find increasing differentiation in food retail markets in recent years. Despite the prohibition of foreign direct investment in food retail, a domestic modern private retail sector is quickly appearing. However, its share of the total urban food retail market is still very small and, in contrast to rollouts of modern retailing systems in other countries (for example, Reardon et al. 2003), it has not yet entered the cereal sector in a big way. The cereal sector remains overwhelmingly in the hands of local flour mills, cereal shops, and cooperative retail outlets. Another example of changes in distribution systems—in input markets—is the model of Ethio-chicken, where a private firm has engaged 2,500 agents in the country to distribute their chickens.

Innovations in Supply Chains

Three major innovations in agricultural markets in the past decade can be highlighted. First, a modern commodity exchange, the Ethiopian Commodity Exchange (ECX), was started in 2008. From December 2008 on, it became mandatory to sell coffee—followed by other export crops, such as sesame and other oilseeds—through the ECX. The ECX trades standard contracts, based on a warehouse receipt system, with standard parameters for grades, transaction size, payment, and delivery. Before the establishment of the ECX, there was no third-party quality control except when exported. All trade was centralized and sold through an auction system in Addis Ababa. While the ECX has had important impacts on the structure of value chains, it appears not to have led to important improvements in value chain performance (Hernandez et al. 2017). The ECX is only involved in the trade of export crops.

Second, mobile phones are used ubiquitously by agricultural traders. Minten, Tamru, and Stifel (2014) show that cell phone usage rates increased to 100 percent for traders and brokers of all the cereal crops in the various markets within an average of only four to five years after the introduction of cell phone coverage. This has contributed to behavioral changes such as the bypassing of the Addis Ababa wholesale markets as the clearinghouse for agricultural trade and the increasing use of mobile phones by traders to coordinate logistics and trade. However, surprisingly, no major effect of these improvements in communication on trade margins were found (Riera and Minten 2017). Also, farmers and pastoralists have increasingly adopted these mobile phones. They have, for example, been useful for pastoralists to find buyers as well as pastures to migrate to (Debsu et al. 2016). Despite these better communication possibilities, Tadesse and Bahiigwa (2015) did not find an effect of mobile phone ownership on the prices obtained by crop farmers.

Third, contract farming has been promoted by a number of institutions and firms in recent years. As argued by Reardon et al. (2019), contract farming might take off in transitional marketing systems but is more widespread in modern markets. Contracting schemes have the potential to solve a number of constraints that might exist in these markets, such as access to improved inputs, credit, extension, and lack of market access, which could lead to important spillover effects (Negash and Swinnen 2013). Holtland (2017) reviewed a number of implemented contracting schemes in recent years in Ethiopia. He showed that none of them was long-lived and that the only large-scale contracting scheme with smallholders that existed for a number of years, with the Heineken brewery, was subsidized by the company. Holtland identified a number of issues with these contract farming schemes, including firms not following up on their promises or farmers failing to deliver because of production problems or side-selling. Such issues have also been found to be major constraints to contract farming in other developing countries (Otsuka, Nakano, and Takahashi 2016).

Market Performance

We look at some measures of agricultural market performance over time. First, the degree to which cereal prices move together across markets throughout Ethiopia—that is, how well they are integrated—provides a measure of how well these markets function. Several authors have shown that markets have

TABLE 7.6 Degree of market integration of Addis Ababa with other cereal wholesale markets between 1999 and 2016

Indicator	Year	White teff	Mixed teff	Maize	White wheat	White sorghum
Total market pairs	1999	11	11	8	10	5
	2004	11	11	8	10	5
	2010	11	11	8	10	4
	2016	11	11	8	10	4
Percent integrated markets	1999	36	27	25	70	60
	2004	38	27	25	80	60
	2010	36	45	50	80	75
	2016	50	45	50	80	50
Half-life of adjustment to price changes (months)	1999	1.2	4.0	2.0	4.3	0.7
	2004	1.0	3.7	1.6	4.0	0.6
	2010	1.0	3.7	0.8	3.6	0.5
	2016	0.8	3.2	0.8	2.2	3.0
Transaction costs measured as percentage of average cereal prices in that period (%)	1999	16	8	13	10	11
	2004	10	7	13	10	10
	2010	7	6	12	9	5
	2016	6	6	10	8	10

Source: Authors' calculation based on data from EGTE.

become better integrated over time and are more resilient to drought (Minten et al. 2016b; Dercon 1995; Hill and Fuje 2017). We update that analysis and assess the integration of cereal wholesale markets from 1999 to 2016 by studying market integration between various market pairs for each of the major cereals using the methodology developed by Van Campenhout (2007). In particular, we pair Addis Ababa with the most important regional wholesale markets for each of the cereals, thus our analysis reflects the country's major cereal flows (Table 7.6).

Three important results can be highlighted from this market integration analysis. First, there has been an overall improvement in market integration over the past two decades. While wholesale markets for white teff, mixed teff, maize, and white wheat were well integrated at the end of 2016, this was less the case in 1999. About 50 percent of the regional white teff, mixed teff, and maize markets were integrated with respect to the Addis Ababa market in 2016, compared to 36 percent, 27 percent, and 25 percent, respectively, in 1999. In contrast, there seems to be no improvement in sorghum market integration over time, but integration was already high to start with. Second, the speed of price adjustments has improved. This is illustrated in the average half-life of adjustment to price changes declining from an average of three months in 1999 to a month and a half in 2016. Compared to the case in 1999, it now takes about half the time for prices between wholesale markets to return back to half of the long-run equilibrium after a given price shock. Third, estimated transaction costs between the wholesale markets considered have come down substantially over time (bottom row, Table 7.6). Transaction costs fell by about 50 percent for all of the cereals examined, except for white sorghum.[9]

Second, the analysis shows that marketing margins between agricultural markets are seemingly decreasing over time. Relying on EGTE price data from a number of wholesale markets in the country, Figure 7.6 shows that the standard deviation between wholesale market prices for cereals have all come down over time, expect for sorghum. Using the trend line for maize, wheat, and teff, standard deviations fell by 100 birr per quintal over the 15-year period, indicating that price differences between maize and wheat wholesale markets fell to one-third and half the differences at the beginning of the period. This decrease is likely a reflection of investments in road and communication infrastructure and the more competitive markets that have developed over time (Minten, Tamru, and Stifel 2014). However, there is still substantial volatility around this trend. This deviation was especially high in 2008 and 2009 during periods of global food price crises when there also were spikes in food prices in the country.

Third, seasonality is a major characteristic of agricultural markets. We assess with the EGTE wholesale market price data to what extent seasonal behavior is present in market prices and how this seasonal behavior is changing. To do this, we run a simple regression on the logarithm of the wholesale prices with market, year, and monthly dummies on the right-hand side. We then interpret the coefficients for the monthly dummies (Table 7.7). There is significant price seasonality, with prices low immediately after the main harvest and highest at the end of the rainy period before the new harvest starts coming in. Over the period considered, price seasonality was the highest for maize with a price amplitude of 16 percent, while it was lowest for teff with a price amplitude of 7 percent. This difference between crops might reflect

9 In the latter case sorghum is not widely consumed in major markets such as in Addis Ababa. As such, the thin sorghum markets with limited flows among them appear to be reflected in these estimation results.

FIGURE 7.6 Standard deviation of cereal market prices between wholesale markets in Ethiopia, 2001–2016 (birr per quintal)

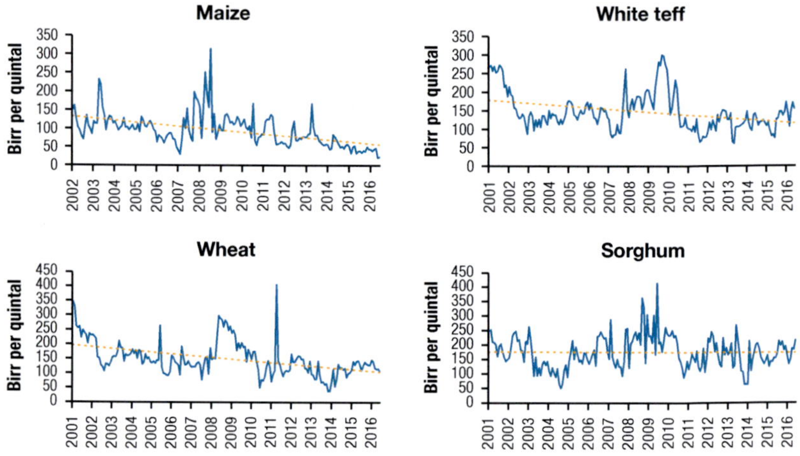

Source: Authors' calculations based on EGTE price data.
Note: Dotted lines are linear trends in prices over the period considered.

aptitude for storage as maize is less suited for long storage periods compared to teff.

Interestingly, we see a significant decline in price seasonality over time, comparing seasonal amplitudes for 2002–2008 and 2009–2016 periods. Seasonal price amplitudes were reduced over these two periods by between 8 percentage points for teff and 10 percentage points for wheat. For example, while the price difference for wheat between the highest and the lowest-priced month in the first period was 17 percent, this had declined to 7 percent in the second period. Several explanations could account for this reduced-price seasonality. First, as shown above, agricultural production during the main *meher* season has increased rapidly. Thus there might be more food available for storage and sale later in the year. Second, markets are better integrated and different supply regions can be used to smooth the lack of seasonal supplies in some areas. Third, farm households are relatively richer now than before and are less cash constrained in the immediate period after the harvest. Consequently, they are not quite so obliged as previously to sell all of their harvest immediately. Fourth, storage conditions have improved due to infrastructural improvements, and due to easier access to agrochemicals used to reduce storage losses. However, quantifying the impacts of each of these factors is left for future research.

TABLE 7.7 Price seasonality in cereals in Ethiopia, 2002–2016

	Amplitude							
	All	Period 1	Period 2	Reduction	2002–2008		2009–2016	
	2002–2016	2002–2008	2009–2016	p1–p2	Lowest price	Highest price	Lowest price	Highest price
All wholesale markets								
Teff	7.34	13.05	5.48	7.57	February	August	December	September
Wheat	8.45	16.81	6.53	10.28	January	August	December	September
Barley	10.84	16.96	8.04	8.92	March	October	December	August
Sorghum	12.89	19.41	11.06	8.35	February	November	April	January
Maize	15.99	25.18	16.58	8.60	January	August	December	August
Addis Ababa								
Teff	9.22	17.17	6.33	10.84	March	August	December	September
Wheat	8.83	17.43	6.84	10.59	January	August	December	June
Barley	12.78	23.04	5.49	17.55	February	August	February	June
Sorghum	12.22	19.52	11.24	8.28	February	November	April	September
Maize	19.33	29.39	19.25	10.14	January	August	December	August

Source: Authors' calculations based on EGTE data.
Note: Seasonality calculated as dummies in regression of the form log (real price) = f (city, year, month).

Finally, different food groups have shown different price developments over time. Using a large-scale price dataset collected monthly in 116 urban and rural markets across the country, Bachewe et al. (2017) show that real prices of all nutritionally rich food groups increased between 19 percent and 62 percent over the past ten years. This contrasts with staple crops (grains, roots, and tubers), which did not show any real price increase, and with oils, fats, and sugar, the prices of which decreased substantially. Given the large influence of prices on consumer choices in countries like Ethiopia, these findings suggest that more investments and attention to the production of nutritious foods—combined with behavioral change messaging—is needed to improve their affordability for consumers.

Changes in International Trade

An important policy consideration is also Ethiopia's dependence on international food markets. Ethiopia is a consistent importer of wheat, mostly by the government or international agencies such as the World Food Programme (WFP) to be used for distribution in humanitarian activities or in its large social safety net, the Productive Safety Net Program (PSNP). Figure 7.7 (left) illustrates the quantities of wheat imported into the country over the past

FIGURE 7.7 Wheat imports, 2007–2016, and number of aid beneficiaries, 2005–2016

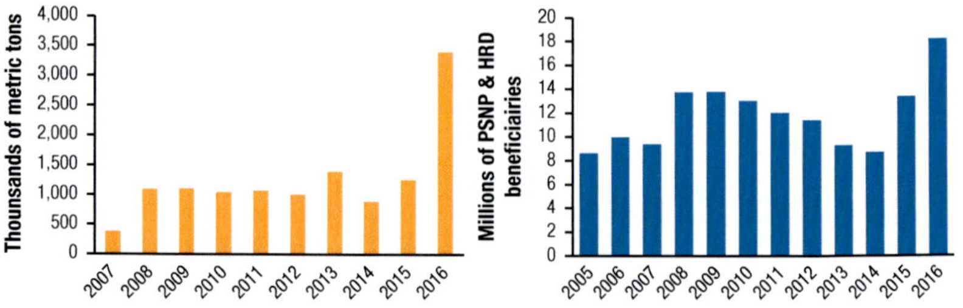

Source: UN Comtrade, AKLDP (2017), and World Bank (2017a).
Note: Aid beneficiaries from Productive Safety Net Program and Humanitarian Requirements Document.

decade. Wheat imports were typically about 1 million tons per year until 2016, during which imports more than tripled because of a large El Niño–induced drought in the country. Figure 7.7 (right) shows how the overall number of beneficiaries in safety net and aid programs has changed over the years. Imported quantities of wheat and the number of aid beneficiaries are, on average, not declining over time.

Ethiopia is also an exporter of a number of agricultural products, particularly coffee but also oilseeds, chat, flowers, and meat. Figure 7.8 shows to what extent agricultural exports have changed over the past 15 years. We see an important increase, partly driven by increasing commodity prices in international markets over the past decade for crops such as coffee and sesame, but also by rapidly increasing exported quantities. The value of agricultural exports overall rose six-fold, from US$0.4 billion in 2000/2001 to $US2.5 billion in 2016/2017. When we compare the value of these exports with agricultural imports, including those products that are major inputs in the agricultural production process such as chemical fertilizer, we find that both agricultural imports and exports have increased significantly since 2000.[10] However, Ethiopia in the past decade was a net agricultural exporter in all years except one. The annual value of net agricultural exports hovered around US$1 billion over the past five years. This increasing reliance on international trade is noteworthy, especially given complicated trade regimes in Ethiopia (World Bank 2015).

10 Relying on the National Bank of Ethiopia classification system, food and live animals, tobacco, fertilizer, and grains are included in agricultural imports. Only data on official imports and exports are included in this analysis. Given lack of data on informal trade, such as that of exports of livestock or imports of rice and pasta, these could not be included.

FIGURE 7.8 Ethiopia's agricultural imports (–) and exports (+), 2000/2001–2016/2017

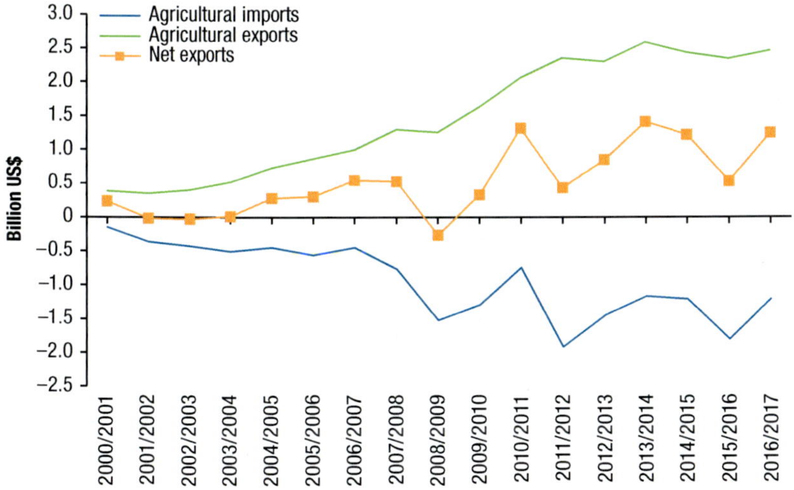

Source: Authors' calculations based on data from the National Bank of Ethiopia.

Structural Transformation and Future Food Value Chains

Structural Transformation

Structural transformation of economies is an essential process for poverty reduction and welfare improvements (Timmer 2014). In this process the share of agriculture in the overall economy declines while the services and, most important, the manufacturing sectors grow. As these sectors grow, lower labor productivity that is usually seen in the agricultural sector catches up with productivity levels in the other economic sectors. Ethiopia is still a relatively poor developing country where agriculture is important for the economy and for overall employment. However, the situation of Ethiopia's economy is rapidly changing. Figure 7.9 illustrates sectoral GDP changes over the past decade. While agriculture made up more than 50 percent of Ethiopia's GDP in 2006/2007, this share came down to 35 percent in 2015/2016. Based on share of GDP, agriculture is no longer the largest sector of the economy. Since 2014/2015, it has been overtaken by the services sector.

While the importance of agriculture in GDP has come down rapidly, the share of workers employed in agriculture has barely changed. Based on data

FIGURE 7.9 Share of different sectors in Ethiopia's economy, 2006/2007–2017/2018 (%)

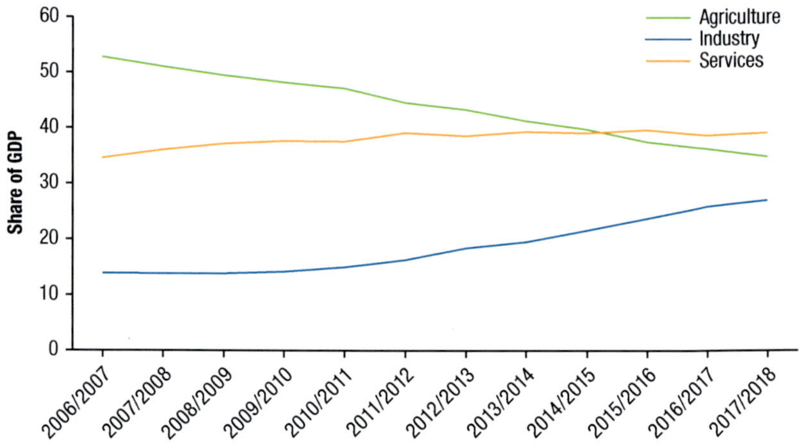

Source: NBE (2018).
Note: Base year = 2015/2016.

from the national labor survey, Schmidt and Bekele (2016) find that the share of the population employed in agriculture declined by only 4 percent between 2005 and 2013—from 80.2 percent in 2005 to 76.7 percent in 2013. The large share of employment in agriculture and the declining share of its economic weight is a typical pattern in developing countries and is reflective of low productivity levels in agriculture. Timmer (2014) uses a measure of the difference between the share of agriculture in GDP and employment as an "agricultural gap" to illustrate this phenomenon. We plot this measure for all countries for which World Bank data were available for the year 2016 (Figure 7.10). It shows to what extent the situation of Ethiopia is fairly typical for countries at its level of development. Timmer (2014) further argues that to achieve the desired closing of the gap, growth of all sectors—including agriculture—is required. Next we assess some of the trends that will likely shape the agricultural sector and food value chains in Ethiopia during this ongoing process of structural transformation.

Future Food Value Chains

To assess these future developments, we follow the guiding framework of the three transformations discussed above: dietary, agriculture, and supply chain transformation. To frame the discussion, we compare Ethiopia to other countries characterized by higher and lower levels of GDP per capita. We also

FIGURE 7.10 Relationship between GDP per capita, share of agriculture in GDP, share of employment in agriculture, and the "agricultural gap" at national level and globally, 2016

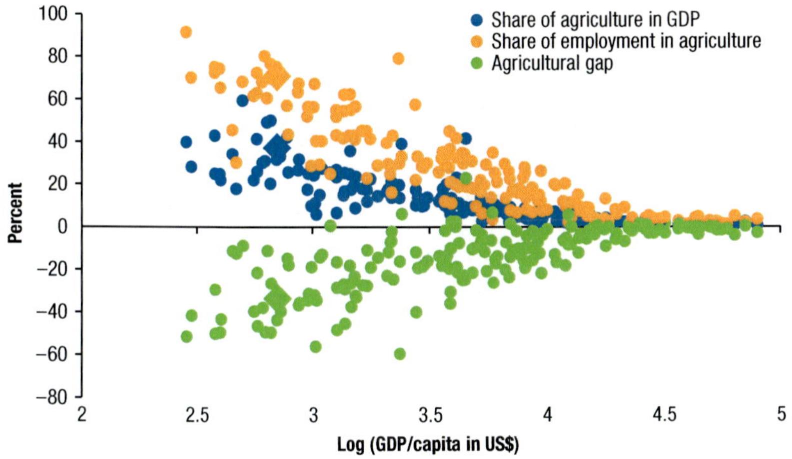

Source: Authors' calculations based on World Bank data.

simulate through simple empirical models the expected changes in food value chains based on income growth and urbanization projections.

CONTEXTUAL CHANGES

First, while the growth of the population is expected to slow down in Ethiopia, we will still see important growth in the total population through 2030. Figure 7.11 shows the expected growth for both the rural and the urban populations. Projections put the rural population at 106 million in 2030—26 million more than in 2016 (World Bank 2015). The share of the urban population is expected to further increase with especially the population in secondary cities growing in importance. The World Bank (2017b) estimates that the urban population will reach 40 million in 2030 and that one-third of Ethiopia's population would then be urban.

Second, Ethiopia is planning to continue on its high economic growth path. Several policy documents state that Ethiopia aims to reach middle-income status by 2025 (for example, World Bank 2015). While the outcome needs to be seen—especially given recent political unrest in the country as well as the increasing constraints on foreign capital (Dorosh et al. 2017)—we assess the impact of economic growth on Ethiopia's food value chains by looking at two scenarios: one with no income growth and a second one with annual income growth of 2 percent to 2030.

FIGURE 7.11 Projected population growth in Ethiopia, urban and rural, in millions of persons added, 2017–2030

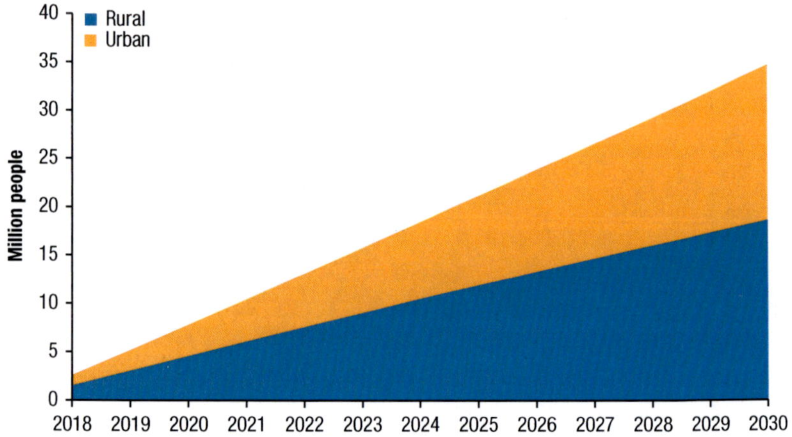

Source: Authors' calculations based on World Bank.

DIETARY TRANSFORMATION

Two important dynamics will shape future diets in Ethiopia: higher consumption levels and different types of foods. First, it can be expected that calories and proteins consumed will increase with increasing incomes in the country. Relying on data from multiple countries (FAOSTAT for the year 2013), Figure 7.12 illustrates to what extent calorie and protein consumption are associated with different GDP levels. A doubling of GDP per capita is associated with an increase in calorie consumption of 700 calories per capita, or an increase of 23 percent. However, protein consumption shows relatively larger differences between poorer and richer countries. A doubling of income leads to 30 grams more consumption per capita, or an increase of 60 percent for lower-income countries.

Second, we see changes in types of foods demanded with income growth, specifically further diversification in diets and an increase in the consumption of higher-value products, in particular fruits and vegetables, animal-sourced foods, and fish, and a decline in the consumption of starchy staples (Bennett 1941). Figure 7.12 illustrates, again using cross-sectional data from FAOSTAT for 2013, to what extent the consumption of these different food groups is associated with different GDP levels. It also shows the current situation of Ethiopia in the international context.

FIGURE 7.12 Global association of dietary patterns and per capita GDP (US$)

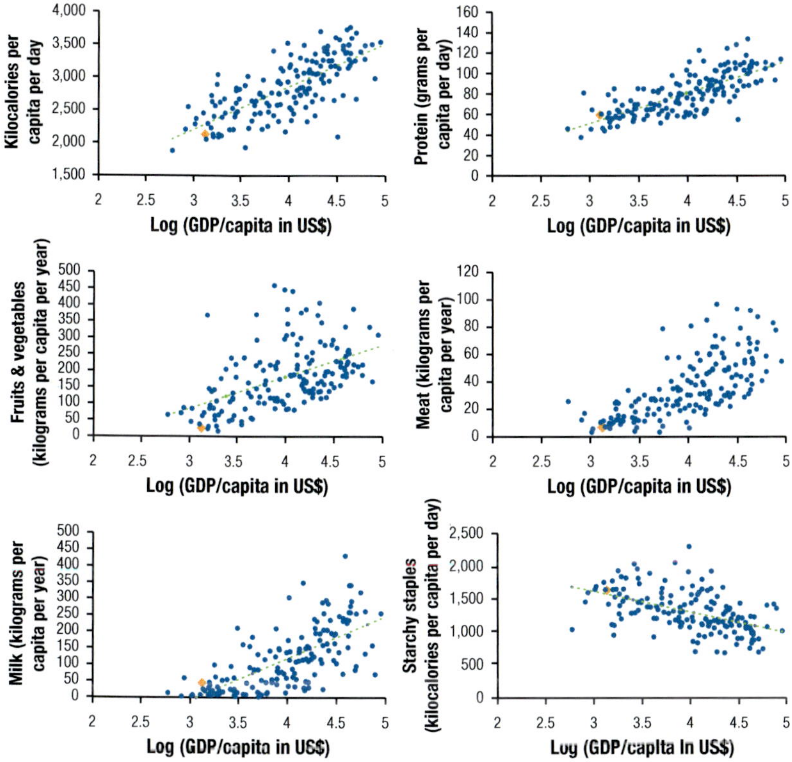

Source: Authors' calculations based on FAO Food Balance Sheets (www.fao.org/economic/ess/fbs/en/) and World Development Indicators (https://datacatalog.worldbank.org/dataset/world-development-indicators).

Note: Ethiopia data points are plotted with red diamonds. Gross domestic product (GDP) per capita is expressed in Purchasing Power Parity.

Agricultural Transformation

The rapid increase in population, income, and urbanization will further propel agricultural transformation and intensification (Vandercasteelen et al. 2018). To illustrate the expected changes necessary to feed the increasing population and the growing better-off part of the population, if planned income growth will happen, we simulate below some implications of this expected change. In the "no income growth" scenario, we illustrate what expected daily calorie requirements would be with no income growth and stable per capita calorie consumption. In this case, total national calorie consumption would go up by 92 percent by 2030 (Figure 7.13). In a second scenario, we assume

FIGURE 7.13 National food consumption expansion scenarios between 2011 and 2030 in Ethiopia: No income growth (scenario 1); 2 percent annual income growth (scenario 2)

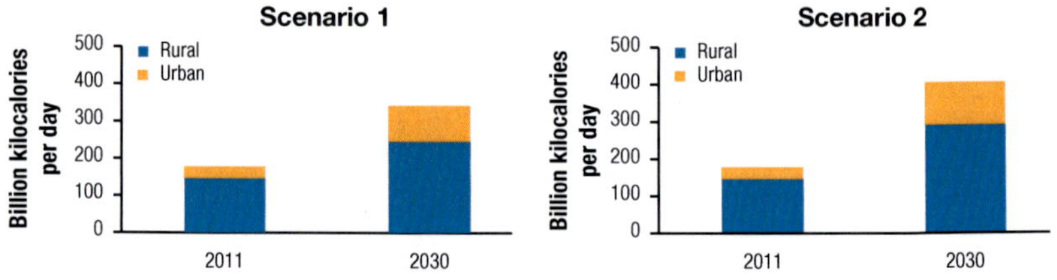

Source: Authors' calculations.

a 2 percent growth in income, a change in average calorie consumption from the current third quintile to the current fifth quintile levels as per the data of Ethiopia, MoFED (2012). Calorie consumption for the country as a whole would increase by 127 percent by 2030 in this scenario, 35 percent higher change than in the case of no income growth. Note that we did not account for likely price increases in the second scenario that might possibly dampen food consumption increases.

Providing this extra food can be met through different channels—that is, increased productivity in existing agricultural areas, bringing into production additional land, or food imports. For simplicity and illustrative purposes, we show in Figure 7.14 the required annual yield increases for the increased productivity and the land expansion options, respectively, assuming that food imports would not play an important role for local food provision.[11] In the case of no income growth and annual land expansion of 0.5 percent, annual yield growth would need to be 2.8 percent. In the same land expansion scenario but with income growth, yield would need to grow by 3.7 percent annually. If land expansion would be larger at 1.0 percent annually, required annual yield growth would be smaller, at 2.3 percent and 3.0 percent in the scenarios of no income growth (scenario 1) and 2 percent income growth (scenario 2), respectively. These simulations illustrate the further need for innovation and improved technology adoption so as to increase agricultural yields. They also show the need for further land investments and the opening up of areas previously not used for agriculture to make the provision of sufficient food supplies for the country easier to achieve.

11 Note that in 2011, food aid/imports (wheat) accounted for about 11 kilograms per capita per year—that is, 3 percent of the total quantity consumed.

FIGURE 7.14 Required productivity increases to assure food self-sufficiency scenarios: No income growth (scenario 1); 2 percent annual income growth (scenario 2)

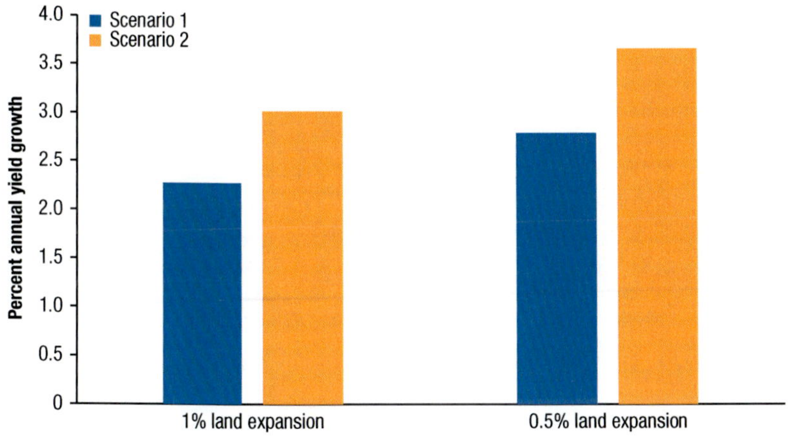

Source: Authors' calculations.

SUPPLY CHAIN TRANSFORMATION

A number of changes can be expected in supply chains. First, there will be increasing demand for processed products and convenient and ready-to-eat products. Figure 7.15 illustrates to what extent higher incomes—as measured by GDP per capita levels of countries—are related to a larger food processing sector, given increasing demand for such types of processed foods with a better-off population. We see in the right chart of Figure 7.15 an especially rapid increase after a logarithm of 4 (equivalent to US$10,000 per capita). However, even below that level, growth rates are considerable.

Second, different types of food distribution systems will emerge and grow. Modern retail is currently in its infancy in Ethiopia—partly because of the prohibition of foreign direct investment in retailing—but changes can happen very fast. For example, it is estimated in China that already half of the distribution of rice in urban markets is in the hands of modern retail (Reardon et al. 2014). Figure 7.15 (left) shows to what extent GDP levels are associated with the share of modern retail in food markets. This development has important implications on value chain performance given the limited number of supermarket chains that are often active in a country and the consequent power that they subsequently might have on agricultural and food value chains—for example, through imposition of standards and prices (Timmer 2014).

FIGURE 7.15 Global relationships of size of modern retail and value-added by food processing industry with GDP per capita (US$)

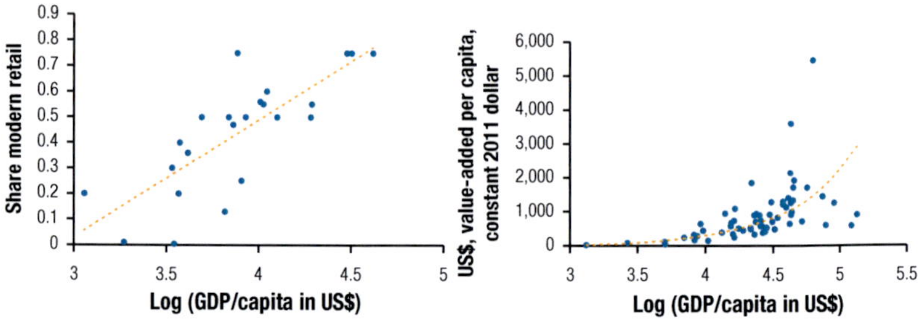

Source: Figure at left adjusted from World Bank (2008); figure at right from World Development Indicator database (https://data.worldbank.org/).

Note: Value-added of food, beverages, and tobacco in manufacturing sector.

Overall, commercial food markets are expected to expand rapidly. We look at two scenarios based on the consumption patterns observed in the national household survey of 2011. In a first conservative scenario, we build upon the differential food expenditures in urban and rural areas, assume urbanization and population growth rates as predicted by the World Bank, and assume no change in purchasing behavior—58 percent and 97 percent of the food obtained from the market in rural and urban areas, respectively. In this scenario we see overall a doubling of the commercial food market by 2030, with an increase in the value of rural expenditures by 67 percent and a tripling of the size of the urban food market (Figure 7.16, left).

In the second scenario, we assume that there is a growth of food expenditures by 2 percent annually and an increase in use of commercial markets in rural areas from 58 percent to 75 percent, driven by better infrastructure and by increasing incomes. Under this scenario we see almost a quadrupling (a growth of 264 percent) of the commercial food market size, with a 360 percent and 220 percent increase in urban and rural areas, respectively (Figure 7.16, right). Such large changes have also been found in other developing countries (Haggblade 2011). These numbers indicate that we should expect significant dynamism in food markets in Ethiopia, even with conservative assumptions. The scenarios also point to significant growth in related sectors associated with agricultural production and marketing, including input supply, logistics, trading, and distribution.

FIGURE 7.16 Commercial market expansion scenarios between 2011 and 2030 in Ethiopia: No income growth (scenario 1); 2 percent annual income growth (scenario 2)

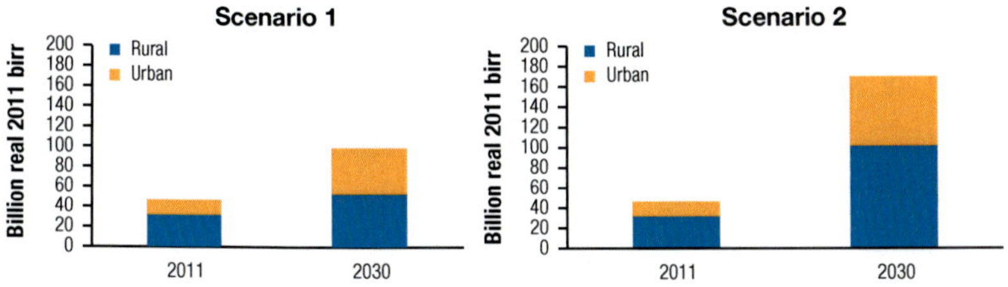

Source: Authors' calculations.

Conclusion

In this chapter we assess the changes that have happened in Ethiopia's food value chains in the past decade. We do so by documenting changes happening downstream at the consumption level, midstream with agricultural markets, and upstream with agricultural producers. We note major changes in dietary patterns, in agricultural production practices, and in agricultural supply chains driven by major contextual changes including high population growth, rapid urbanization, infrastructure investments, and income growth. To assess developments in the future, we simulate—relying on different assumptions of income growth, urbanization, and population growth—the impact of different income growth scenarios. We also put Ethiopia in an international context of countries with different GDP levels, as economic growth is an important associate of transformation in agricultural and food value chains (Timmer 2014).

At the consumption level we see important dietary change. Overall food consumption in Ethiopia, measured in calories, has been increasing over several decades. The relative share of cereals in food expenditures is declining and those of high-value products, including animal-sourced foods and fruits and vegetables, is rising. We note the increasing emergence of processed and convenience foods and out-of-home food consumption and a significantly different dietary pattern for urban households. Given fast growth in population, urbanization, and incomes, these diets are expected to evolve rapidly in the future. It is expected that there will be further increasing consumption of high-value products, such as meat, dairy products, and fruits and vegetables. Although this will be good for nutritional outcomes, a concern is the issue of the double-burden of nutrition. While food security will likely become less of an issue, at least at the national level, avoiding the obesity trends that have been noted in other transforming countries will likely become an important new challenge in the decades ahead (Ghebru et al. 2018).

To assure that this increased local demand for these high-value products can be met by local supply and not by imports, attention will need to be paid to their production with increased availability of seeds, agrochemicals, extension advice, and cold storages. Increasing the supply of animal-sourced foods will require increased livestock-related investments. These include broader adoption of improved animal husbandry and feeding practices, increased production of genetically superior breeds of livestock, the provision and use of appropriate veterinary health practices, and the facilitation of an enabling environment that will allow for efficient livestock markets. Changes in these high-value sectors has been slow in the past decade, as illustrated by the increasing real prices of high-value products—an indication that increasing demand is outstripping supply.

At the production level we have noted significant growth in agricultural production over the past decade. We have seen a process of intensification and modernization, but from a low base, as illustrated by the rapid change in the adoption of chemical fertilizers and agrochemicals. There has been increasing emphasis on large commercial farms as a way to stimulate agricultural production, but only a share of the land allocated to such large-scale farmers is effectively cultivated—amounting to 5 percent of all cultivated land in Ethiopia. Yields have not been much higher on these farms than under smallholder farming conditions. We also see important changes with smallholder agriculture: smallholder farmers are estimated to cultivate 95 percent of all agricultural land. We find over the past decade that

- Average farm sizes of smallholders are declining;
- Farmers are becoming older—the share of Ethiopian farmers under 35 years of age declined from 36 percent to 30 percent over the past decade;
- Young farmers have smaller and declining farm sizes, declining from 0.9 hectare to 0.8 hectare on average over the past decade; and
- Agricultural land rental markets are becoming more important, with 12 percent of cropland now being rented-in. Young farmers especially rely on the rental market to access land.

Given increasing land constraints across Ethiopia, agricultural innovations to increase productivity will be increasingly demanded and adopted. Access to land to supply these increasingly demanded products will be crucial, and better functioning land markets, in particular, will be needed to assure more efficient land allocations to achieve higher productivity levels.

There has been a substantial growth in agricultural commercial surpluses. In consequence, there have been a number of structural changes in supply chains as seen by evidence from the trading and transport, food service, and processing sectors, and from urban retail and distribution. We see more reliance on markets by consumers, better integrated markets, and smaller spatial and seasonal margins. However, an increase of prices of noncereals—especially of those products important to improve diet diversity and therefore nutritional outcomes—also is seen in the data, suggesting the demand for these foods is outstripping supply. While food imports are high, and the number of food aid beneficiaries has not come down in the past decade, nonetheless, Ethiopia exports significant amounts of agricultural goods. In a normal year the country is a net exporter of food in value terms. In the future, supply chains will need to cater to larger and different demands. We should expect to see rapid growth in the agro-processing sector and in the modern distribution sector. Under reasonable assumptions we expect that the value of the commercial food sector will almost quadruple over a 20-year period. This will have enormous implications on growth in agricultural input supply, logistics, trading, and distribution sectors, as well as on agricultural production across Ethiopia.

References

AKLDP (Agriculture Knowledge, Learning, Documentation and Policy Project). 2017. "El Nino in Ethiopia: Bridging the Gap—Cereal Production and Imports: October 2016 to September 2017." Unpublished, Cereal Availability Diagnostic Study, Addis Ababa.

Ali, D. A., K. Deininger, and A. Harris. 2017. "Using National Statistics to Increase Transparency of Large Land Acquisition: Evidence from Ethiopia." *World Development* 93: 62–74.

———. 2019. "Does Large Farm Establishment Create Benefits for Neighboring Smallholders? Evidence from Ethiopia." *Land Economics* 95 (1): 71–90.

Assefa, T., G. Abebe, I. Lamoot, and B. Minten. 2016. "Urban Food Retailing and Food Prices in Africa: The Case of Addis Ababa (Ethiopia)." *Journal of Agribusiness in Developing and Emerging Economies* 6 (2): 90–109.

Bachewe, F., G. Berhane, B. Minten, and A. S. Taffesse. 2016. *Non-farm Income and Labor Markets in Rural Ethiopia*. IFPRI-ESSP Working Paper 90. Addis Ababa: International Food Policy Research Institute (IFPRI)/Ethiopia Strategy Support Program (ESSP).

———. 2018. "Agricultural Transformation in Africa? Assessing the Evidence in Ethiopia." *World Development* 105: 286–298.

Bachewe, F., K. Hirvonen, B. Minten, and F. Yimer. 2017. *The Rising Costs of Nutritious Foods.* IFPRI-ESSP Research Note 67. Addis Ababa: IFPRI/ESSP.

Bekele, M. S. 2016. "Economic and Agricultural Transformation through Large-Scale Farming: Impacts of Large-Scale Farming on Local Economic Development, Household Food Security and the Environment in Ethiopia." PhD dissertation, Leiden University, Netherlands.

Benfica, R., and J. Thurlow. 2017. *Identifying Priority Value-Chains in Ethiopia.* IFPRI-ESSP Working Paper 110. Addis Ababa: IFPRI/ESSP.

Bennett, M. K. 1941. "International Contrasts in Food Consumption." *Geographical Review* 31: 365–374.

Braudel, F. 1982. *Civilization and Capitalism, 15th–18th Century: The Wheels of Commerce.* Berkeley, CA, US: University of California Press.

Deaton, A., and J. Drèze. 2009. "Food and Nutrition in India: Facts and Interpretations." *Economic and Political Weekly* 44 (7): 42–65.

Debsu, D. N., P. D. Little, W. Tiki, S. A. Guagliardo, and U. Kitron. 2016. "Mobile Phones for Mobile People: The Role of Information and Communication Technology (ICT) among Livestock Traders and Borana Pastoralists of Southern Ethiopia." *Nomadic Peoples* 20 (1): 35–61.

Dercon, S. 1995. "On Market Integration and Liberalization: Method and Application to Ethiopia." *Journal of Development Studies* 32: 112–143.

Dolislager, M. D., D. Tschirley, and T. Reardon. 2015. "Consumption Patterns in Eastern and Southern Africa." Report to USAID by Innovation Lab for Food Security Policy, Michigan State University, East Lansing, US.

Dorosh, P., J. Thurlow, F. Kebede, T. Ferede, and A. S. Taffesse. 2017. "Perspectives on the Future of Ethiopia's Agriculture." Presentation at the 15th International Conference on the Ethiopian Economy, Addis Ababa, July 20–22.

EGTE (Ethiopian Grain Trade Enterprise). 2017. "Monthly Wholesale Grain Prices." Accessed November 28, 2017. www.egte-ethiopia.com/en/.

Ethiopia, CSA (Central Statistical Agency). 1996. *Household Income and Consumption Expenditure Survey.* Addis Ababa.

———. 2000. *Household Income and Consumption Expenditure Survey.* Addis Ababa.

———. 2005. *Household Income and Consumption Expenditure Survey.* Addis Ababa.

———. 2005a–2016a. *Agricultural Sample Survey: Volume IV–Report on Land Utilization (Private Peasant Holdings,* Meher *Season).* Addis Ababa.

———. 2011. *Household Consumption and Expenditure Survey.* Addis Ababa.

———. 2012a. *Report on Large and Medium Scale Manufacturing and Electricity Industries Survey.* Addis Ababa.

———. 2012b. *Household Consumption and Expenditure Survey.* Addis Ababa.

———. 2015. *Report on the Small-Scale Manufacturing Industries Survey.* Addis Ababa.

———. 2018. *Key Findings of the 2017/18 (2010 E.C.) Agricultural Sample Surveys.* Addis Ababa.

Ethiopia, MoFED (Ministry of Finance and Economic Development). 2012. "Ethiopia's Progress towards Eradicating Poverty: An Interim Report on Poverty Analysis Study (2010/2011)." Photocopy, Addis Ababa.

Fikade, B. 2018. "Half a Million Hectare Commercial Farms Lay Bare." *Ethiopian Reporter.* March 24.Accessed September 9, 2019. www.thereporterethiopia.com/article/half-million-hectare-commercial-farms-lay-bare.

Gebru, M., R. Remans, K. Baye, M. B. Melesse, N. Covic, F. Habtamu, A. Abay, et al. 2018. *Food Systems for Healthier Diets in Ethiopia: A Research Agenda.* IFPRI Discussion Paper 1720. Washington, DC: International Food Policy Research Institute.

Haggblade, S. 2011. "Modernizing African Agribusiness: Reflections for the Future." *Journal of Agribusiness in Developing and Emerging Economies* 1 (1): 10–30.

Headey, D. 2014. *An Analysis of Trends and Determinants of Child Undernutrition in Ethiopia, 2000–2011.* IFPRI-ESSP Working Paper 70. Addis Ababa: IFPRI/ESSP.

Hernandez, M. A., S. Rashid, S. Lemma, and T. Kuma. 2017. "Institutions and Market Integration: The Case of Coffee in the Ethiopian Commodity Exchange." *American Journal of Agricultural Economics* 99 (3): 683–704. https://doi.org/10.1093/ajae/aaw101.

Hill, R., and H. Fuje. 2017. "What Is the Impact of Drought on Prices? Evidence from Ethiopia." Unpublished, Poverty and Equity Global Practice, Africa Region, World Bank, Washington, DC.

Holtland, G. 2017. *Contract Farming in Ethiopia: Concept and Practice.* Arnhem, Netherlands: AgriProFocus.

Humphries, D. L., J. R. Behrman, B. T. Crookston, K. A. Dearden, W. Schott, and M. E. Penny. 2014. "Households across All Income Quintiles, Especially the Poorest, Increased Animal Source Food Expenditures Substantially during Recent Peruvian Economic Growth." *PLoS ONE* 9 (11): p.e110961.

Jayne, T. S., S. Snapp, F. Place, and N. Sitko. 2019. "Agricultural Intensification in an Era of Rural Transformation in Africa." *Global Food Security* 20: 105–113.

Minten, B., T. Assefa, G. Abebe, E. Engida, and S. Tamru. 2016a. *Food Processing, Transformation and Job Creation: The Case of Ethiopia*. IFPRI-ESSP Working Paper 96. Addis Ababa: IFPRI/ESSP.

Minten, B., S. Tamru, E. Engida, and T. Kuma. 2016b. "Transforming Staple Food Value Chains in Africa: The Case of Teff in Ethiopia." *Journal of Development Studies* 52 (5): 627–645.

Minten, B., S. Tamru, and D. Stifel. 2014. "Structural Transformation in Cereal Markets in Ethiopia." *Journal of Development Studies* 50 (5): 611–629.

NBE (National Bank of Ethiopia). 2018. *Annual Report 2016/17.* Addis Ababa. Accessed February 8, 2018. https://nbebank.com/wp-content/uploads/pdf/annualbulletin/NBE%20Annual%20report%202016-2017/NBE%20Annual%20Report%202016-2017.pdf.

Negash, M., and J. F. Swinnen. 2013. "Biofuels and Food Security: Micro-evidence from Ethiopia." *Energy Policy* 61: 963–976.

Oqubay, A. 2015. *Made in Africa: Industrial Policy in Ethiopia*. Oxford, UK: Oxford University Press.

Otsuka, K., Y. Nakano, and K. Takahashi. 2016. "Contract Farming in Developed and Developing Countries." *Annual Review of Resource Economics* 8: 353–376.

Reardon, T., K. Chen, B. Minten, L. Adriano, T. A. Dao, J. Wang, and S. das Gupta. 2014. "The Quiet Revolution in Asia's Rice Value Chains." *Annals of the New York Academy of Sciences* 1331: 106–118.

Reardon, T., R. Echeverria, J. Berdegué, B. Minten, L. Liverpool-Tasie, D. Tschirley, and D. Zilberman. 2019. "Rapid Transformation of Food Systems in Developing Regions: Highlighting the Role of Agricultural Research." *Agricultural Systems* 172: 47–59.

Reardon, T., and B. Minten. Forthcoming. "Food Value Chain Transformation in Developed and Developing Regions." In *Agricultural Development: New Perspectives in a Changing World*, edited by K. Otsuka and S. Fan.

Reardon, T., and C. P. Timmer. 2014. "Five Inter-Linked Transformations in the Asian Agrifood Economy: Food Security Implications." *Global Food Security* 3 (2): 108–117.

Reardon, T., C. P. Timmer, C. B. Barrett, and J. A. Berdegué. 2003. "The Rise of Supermarkets in Africa, Asia, and Latin America." *American Journal of Agricultural Economics* 85 (5): 1140–1146.

Riera, O., and B. Minten. 2017. "Mobile Phones and Price Dispersion in Agricultural Markets in Ethiopia." Paper presented at the XVth EAAE Congress 2017, Parma, August 28–September 1, 2017.

Rulli, M. C., A. Saviori, and P. D'Odorico. 2013. "Global Land and Water Grabbing." *Proceedings of the National Academy of Science of the United States of America* 110 (3): 892–987.

Schaefer, F., and G. Abebe. 2015. *The Case for Industrial Policy and Its Application in the Ethiopian Cut Flower Sector*. EDRI Working Paper 12. Addis Ababa: EDRI (Ethiopian Development Research Institute).

Schmidt, E., and F. Bekele 2016. *Rural Youth and Employment in Ethiopia*. IFPRI-ESSP Working Paper 98. Addis Ababa: IFPRI/ESSP.

Schmidt, E., P. Dorosh, M. K. Jemal, and J. Smart. 2018. "Ethiopia's Spatial and Structural Transformation: Public Policy and Drivers of Change." Mimeo. Washington, DC, IFPRI.

Smith, L. C., O. Dupriez, and N. Troubat. 2014. *Assessment of the Reliability and Relevance of the Food Data Collected in National Household Consumption and Expenditure Surveys*. IHSN Working Paper 8. International Household Survey Network.

Tadesse, G., and G. Bahiigwa. 2015. "Mobile Phones and Farmers' Marketing Decisions in Ethiopia." *World Development* 68: 296–307.

Taffesse, A. S. 2019. "The Transformation of Smallholder Crop Production in Ethiopia (1994–2016)." In *Oxford Handbook of the Ethiopian Economy*, edited by F. Cheru, C. Cramer, and A. Oqubay, chapter 27. Oxford, UK: Oxford University Press.

Tamru, S., B. Minten, F. Bachewe, and D. Alemu 2017. "The Rapid Expansion of Herbicide Use in Smallholder Agriculture in Ethiopia: Patterns, Drivers, and Implications." *European Journal of Development Research* 29 (3): 628–647.

Timmer, C. P. 2014. *Managing Structural Transformation: A Political Economy Approach*. WIDER Annual Lecture 18. Helsinki, Finland: UNU-WIDER (United Nations University–World Institute for Development Economics Research).

Van Campenhout, B. 2007. "Modelling Trends in Food Market Integration: Method and an Application to Tanzanian Maize Markets." *Food Policy* 32 (1): 112–127.

Vandercasteelen, J., S. T. Beyene, B. Minten, and J. Swinnen. 2018. "Cities and Agricultural Transformation in Africa: Evidence from Ethiopia." *World Development* 105: 383–399.

Vaughan, S. 2018. "Land and Urbanization." WIDE Series III Discussion Brief 1 of 7. Accessed November 9, 2019. http://ethiopiawide.net/wp-content/uploads/WIDEBridge_DB1_Land_2018.pdf.

WFP (World Food Programme) and CSA (Central Statistical Agency). 2014. "Comprehensive Food Security and Vulnerability Analysis: Ethiopia." Unpublished, Addis Ababa.

Worku, I., M. Dereje, B. Minten, and K. Hirvonen. 2017. "Diet Transformation in Africa: The Case of Ethiopia." *Agricultural Economics* 48: 73–86.

World Bank. 2008. *2008 World Development Report: Agriculture for Development*. Washington, DC.

———. 2014. *Ethiopia: Poverty Assessment*. Report No. AUS6744, Poverty Global Practice, Africa Region, World Bank Group, Washington, DC.

———. 2015. *Ethiopia's Great Run: The Growth Acceleration and How to Pace It.* Report 99399-ET. Washington, DC.

———. 2017a. "Quantifying Costs of Drought Risk in Ethiopia: A Technical Note." Disaster Risk Financing and Insurance Program. Unpublished, Washington, DC.

———. 2017b. World Bank Open Data: Population Growth (Annual Percent). Accessed November 26, 2017. https://data.worldbank.org/indicator/SP.POP.GROW.

Chapter 8

EVOLVING ANIMAL-SOURCED FOODS AND LIVESTOCK MARKETS

Fantu Bachewe, Bart Minten, and Feiruz Yimer

Introduction

There is a strong desire by policymakers to improve nutritional outcomes in developing countries, given the often high levels of undernutrition, consequent high prevalence of stunted children, and the resultant high human and economic costs (Hoddinott et al. 2013). To address this problem, a strong emphasis is generally put on increasing diet diversity, on top of other interventions, given the well-established link between, for example, improved dietary quality and reduced stunting rates (Arimond and Ruel 2004). Recent research shows that animal-sourced foods (ASF) play an important part in the beneficial impact of higher dietary diversity, as a direct link has been shown worldwide between higher consumption levels of ASF and improved nutritional outcomes (Randolph et al. 2007; Leroy and Frongillo 2007; Weaver 2014; Black et al. 2008; Jin and Iannotti 2014; Iannotti et al. 2014; Givens et al. 2014). As ASF have more accessible crucial vitamins and minerals relative to plant-based foods and contain bioactive factors, they have been shown to be beneficial for physical growth and for brain development of children (Dror and Allen 2011, 2014). There also is increasing evidence that bone development is strongly associated with increased ASF consumption (Givens 2010).

In Ethiopia there is strong suggestive evidence on the link between the consumption of ASF (especially milk) and improved nutritional outcomes among children (Hoddinott, Headey, and Dereje 2015; Sadler and Catley 2009). However, ASF consumption in Ethiopia is low (Tafere and Worku 2012). Increasing consumption levels of these products is therefore desirable to improve nutritional indicators for the country's population. Previous research has shown that factors associated with low ASF consumption levels in Ethiopia are high prices (for example, Gordon, Tegegne, and Tadesse 2007; Tefera et al. 2010; Iannotti et al. 2012), shortage of ASF, particularly milk in urban areas (for example, Gordon, Tegegne, and Tadesse 2007; Tegene et al. 2013), fasting habits (FVI-Idele 2016), and lack of awareness regarding dietary diversity or the nutritional benefits of ASF (for example, Gordon, Tegegne, and Tadesse 2007; Tefera et al. 2010; Warren and Frongillo 2017). This work investigates the first issue in particular, examining price formation of ASF and livestock in Ethiopia.

Price formation of ASF is not well researched and therefore not well understood in Ethiopia. This is an important knowledge gap because the diversity and quantity of food consumed (and thereby the resultant nutritional outcomes) are strongly impacted by food and agricultural prices (for example, Brinkman et al. 2010). This is specifically so for ASF, which are often prohibitively priced for poor and vulnerable households to afford (for example, Iannotti et al. 2012). Although we are not aware of recent studies of ASF pricing in Ethiopia, a number of researchers have looked at the associated livestock markets in Ethiopia.[1] Livestock prices are linked to many factors, including the type of livestock (Ayele et al. 2006; Teklewold et al. 2009), market institutions and transaction costs (Jabbar et al. 2008; Bellemare and Barrett 2006), weather and rainfall patterns (De Waal 1988), periodical cash needs of farmers, particularly to purchase food (Little, Debsu, and Tiki 2014), rising feed costs that are linked with increasing land scarcity (Mekasha et al. 2014; Negassa et al. 2012), and international trade (Tadesse, Lanos, and MasAsparisi 2014).

This chapter is aimed at filling this knowledge gap with respect to ASF prices, while also adding to existing knowledge on livestock price formation. Relying on a large-scale nationally representative price dataset that has been consistently collected in 116 urban retail markets in the country (Ethiopia, CSA 2017b), we study patterns of ASF prices in Ethiopia over the past decade. We find that ASF products are expensive compared to basic staples. This is important given high poverty levels and the relatively high share of food expenditures in the overall budget of an average Ethiopian (Worku et al. 2016). For example, Warren and Frongillo (2017) show that households do not consume ASF—even when they are aware of their nutritional benefits—because of high prices. As seen in other developing countries, this seemingly explains why only the relatively richer part of the population regularly

1 Several researchers have looked at price formation in food and agricultural markets in Ethiopia. Strong seasonality and spatial patterns have been shown to exist in these markets (Minten, Stifel, and Tamru 2014). While a number of factors that constrain the well-functioning of the markets persisted in the past (Rashid and Minot 2010; Rashid and Negassa 2011), these markets are growing and their efficiency is improving over time, seemingly driven by improved road and communication infrastructure, rapid urbanization, and the increasing demands from urban residents, based on their income growth and consequently increases in their willingness to pay for food (for example, Minten, Stifel, and Tamru 2014).

consumes ASF. We also see significant seasonality in both prices and consumption of ASF. This is associated with large swings in demand, due to the timing of religious celebrations and the fasting periods that precede these religious events. We further find that there are significant spatial variations in ASF prices, with relatively higher prices in cities and in more commercially oriented livestock areas.

We find that there have been significant increases in real prices of ASF in the past decade, on the order of 33 percent, 36 percent, and 32 percent increases for beef, milk, and eggs, respectively. This is in contrast with cereal prices, which did not show such increases. While increases in ASF prices are good news for livestock producers, as their terms of trade improved, such price increases hamper the affordability of ASF, especially so for the poorer and more vulnerable populations in the country. As price elasticities are high for ASF products—Tafere and Worku (2012) estimate an elasticity of −0.73 for beef and −0.67 for dairy products—this implies that such price increases will have resulted in decreases in per capita consumption of beef and dairy products by almost 25 percent, holding other things constant, when we compare the end to the beginning of the decade.[2]

The results of these findings have important implications for agriculture and nutrition policy in Ethiopia. The Ethiopian government has a good track record with respect to improvement of cereal production in the country in the past decade. This improvement has been driven by a focus on a modernization of the cereal sector and on the increased adoption of modern technologies (Bachewe et al. 2015). It has been shown that changes in the cereal sector have brought about considerable improvements in poverty alleviation and calorie intake (World Bank 2014). However, in the further transformation process, ASF and "high-value" crops will have a more important role to play, given food preferences of consumers, but also because of their nutritional benefits.[3] To increase diet diversity and improve nutritional outcomes in the country, ASF and "high-value" crops need to be made available at more affordable prices. This, in turn, points to the needs for more investments to increase livestock, ASF, and high-value crop production as well as to achieve a livestock production system that can accommodate the growing demand in ASF. This is important since most recent efforts to achieve nutritional improvements

2 The increasing incomes in the country, however, might have compensated for these increases in prices, since these products also have high-income elasticities (Tafere and Worku 2012).

3 As has been noted in other growing economies, the relative importance of cereals in total food expenditures is decreasing in Ethiopia. We are beginning to see a shift toward more preferred but also more expensive foods, including animal-sourced foods (Worku et al. 2016).

in Ethiopia have focused on behavioral change communication (BCC) and improvements in other sectors, such as health, water, and sanitation. While important and successful (for example, Kim et al. 2016), improving access to and achieving low prices for these nutritious ASFs also have an important role to play in multidimensional efforts to improve nutritional outcomes.

The chapter gives background information on livestock and ASF consumption in Ethiopia. We discuss data and descriptive statistics, present findings on seasonal and spatial patterns in ASF and livestock prices, and look at changes over time in real prices and in terms of trade. We finish with the conclusions and implications of the findings.

Background on Livestock and Animal-Sourced Foods in Ethiopia

Production and Consumption

While numbers on livestock and livestock holdings in Ethiopia are uncertain and subject to debate, it is generally acknowledged that Ethiopia has the largest livestock population in Africa (Negassa et al. 2017). The Central Statistical Agency (CSA) of Ethiopia estimated that in 2014/2015 there were 56 million cattle, 29 million sheep, 29 million goats, and 57 million poultry (Ethiopia, CSA 2015a).[4] Figure 8.1 illustrates the spatial distribution of the cattle population and the goat and sheep population in the country based on CSA livestock population data. These livestock species are shown to be densely concentrated in the central highlands of Ethiopia. However, the reported estimates do not account for the significant number of pastoralists—mostly in the eastern parts of the country—as CSA's regular data collection operations only cover sedentary parts of the country.[5] Negassa et al. (2012) estimated that 20 percent of the cattle population and 40 percent of sheep and goats are held in these pastoralist areas.

On the consumption side, based on the nationally representative Household Consumption Expenditures Survey (HCES) conducted by CSA,

4 The Ethiopian Orthodox Christian Church and Islam—the two dominant religions in the country—do not permit pork consumption. Pig production, therefore, is rare in Ethiopia.

5 Livestock ownership per household in the Somali region is considerably higher and that in Afar is somewhat higher relative to average ownership in the central highlands. However, the map would likely still show a dense livestock population in the central highlands even if data were available for the eastern parts, because the central highlands are significantly more densely populated.

FIGURE 8.1 Estimated cattle, sheep, and goat population per woreda, 2014/2015

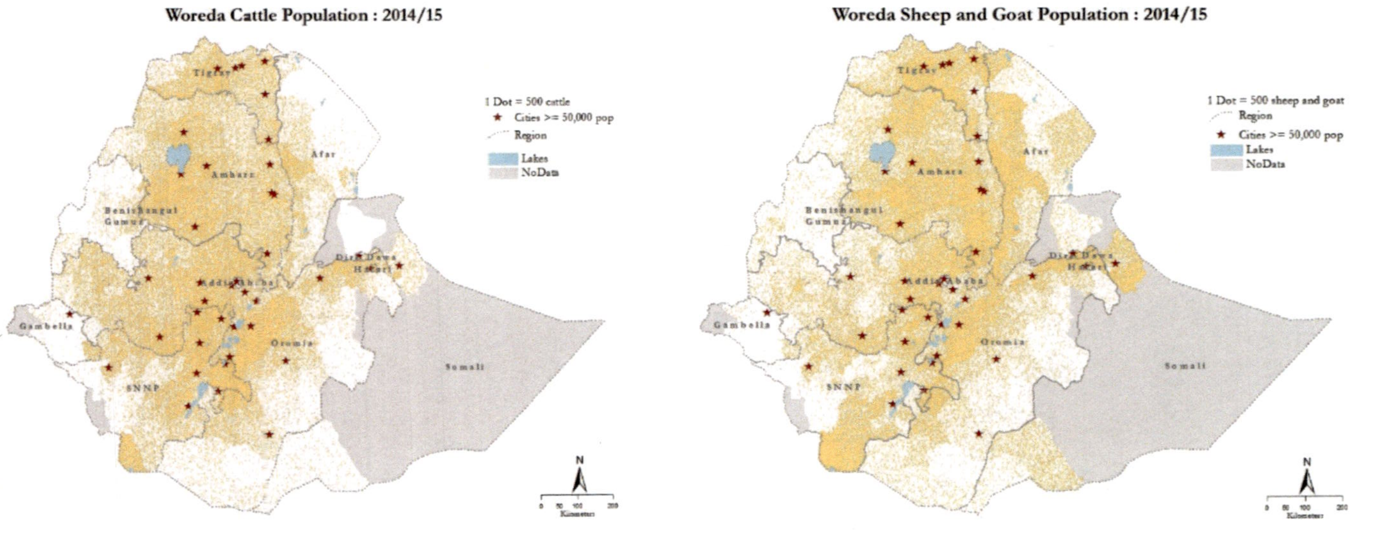

Source: Authors' calculations based on Ethiopia, CSA data, following the method of Tilahun and Schmidt (2012).

Note: The authors thank Helina Tilahun for this analysis.

it was estimated in 2011 that 13.4 percent of consumption expenditures of an average Ethiopian was devoted to ASF. Of this expenditure, 42 percent was used for dairy products, 40 percent for beef, and 12 percent for sheep and goat meat. Expenditures on other ASF products (including poultry) are relatively minor.[6] As seen in several other countries, ASF are mostly consumed by relatively better-off households (Tafere and Worku 2012). Worku et al. (2016) estimate that consumers in the richest quintile used 17.6 percent of their food budget for ASF. This compares to only 6.6 percent by the poorest quintile, about one-third the budget share level of the rich. Moreover, expenditure shares of ASF are significantly higher in urban areas (14.3 percent) relative to rural areas (9.8 percent). Within urban areas, differences are further noted with higher consumption in bigger cities (16.2 and 16.1 percent in Addis Ababa and secondary towns, respectively) compared to smaller ones (13.2 percent).

The high consumption of ASF in cities as well as rapid urbanization in Ethiopia is leading to rapid increases in the size of rural-urban livestock and ASF value chains. This is illustrated by the rapid increase in the number of animals slaughtered in Addis Ababa, where an estimated one-quarter of the meat consumption in the country occurs (FVI-Idele 2016).[7] The number of sheep and goats and number of cattle slaughtered in Addis Ababa, respectively, increased from 75,015 and 151,977 head in 2007 to 175,818 and 232,822 head in 2015, or an annual increase of 12.9 percent and 6.3 percent, respectively (based on data reported by FVI-Idele 2016).

International Trade

Ethiopia has increasingly become more integrated to international trading systems.[8] This trend is also seen in the livestock sector, as can be observed in

6 Average consumption levels of meat were around 5 kilograms per year per adult equivalent, based on these HCES data from 2011. This is relatively low but not out of line for a country with GDP levels of Ethiopia (Msangi 2016).

7 While slaughtering of sheep and goats is mostly done informally, this is not the case with cattle since beef is mostly sold through butcheries, which are required by law to buy from certified slaughterhouses. Sheep and goat meat is most often directly bought by consumers, so the animals are slaughtered informally in backyard settings. It is estimated that one-third of cattle slaughtering is informal or illegal in Addis Ababa (FVI-Idele 2016). Data on the number of cattle slaughtered in slaughterhouses (formal slaughter) might therefore provide a good indication of underlying trends in urban consumption.

8 The National Bank of Ethiopia reports that, measured as a share of GDP, real value of both exports and imports on average declined during 2004/2005–2014/2015. However, in absolute terms, exports and imports were 70 percent and 134 percent higher in 2014/2015 than in

increases in formal exports of livestock as well as in imports of ASF. On the import side, Ethiopia is increasingly importing dairy products, most importantly milk powder. The value of dairy products Ethiopia imported tripled from US$6 million in 2005 to $18 million in 2015 (Figure 8.2). Meat imports are also growing but are much lower in value ($1.5 million in 2015). Meanwhile, exports from the livestock sector are relatively much more important and have grown impressively over the past decade. The value of exports of live animals increased more than tenfold from $24 million in 2005 to $332 million in 2015; those of meat products increased from $18 million in 2005 to $107 million in 2015; and raw hides and skins increased by just below 50 percent, from $67 million to $98 million over the same period. Part of the increases in exports has been achieved through closer oversight by the government, leading to a shift from informal to formal markets (Farmer 2010). However, there are still significant informal exports of live animals (Negassa et al. 2012). Farmer (2010) estimated informal exports to be seven times the volume of formal exports in 2005/2006 (Farmer 2010), while USAID (2013) estimated more recently that informal exports made up 80 percent of total exports.

Given the increasing importance of live animal exports, we present further information on those in Figure 8.3. On the left we note that the cattle are the most important live animal exports as they made up around 60 percent of all live animal exports over the past decade. On the right we see the destination countries for these live animal exports. We see surprisingly large changes by year. However, the most important destinations in recent years are Djibouti, Egypt, and Somalia.

We also document to what extent the livestock sector is relying on international trade to obtain some of its modern inputs. The evolution of imports of feed and vaccines is illustrated in Figure 8.4. There have been significant increases, possibly showing important recent transformation and the increasing commercial orientation for some parts of the livestock subsector. For example, inputs of pre-mixed feeds have increased tenfold over the past ten years, with their value standing at more than US$5 million in 2015. The value of veterinary inputs also increased fivefold over the period considered. However, while increasing rapidly, the levels are still low. To put these

2004/2005, growing at average annual rates of about 7 percent and 9 percent, respectively. All dollars in this chapter are US dollars.

FIGURE 8.2 Ethiopia's import and export of livestock and animal-sourced foods, 2005–2015

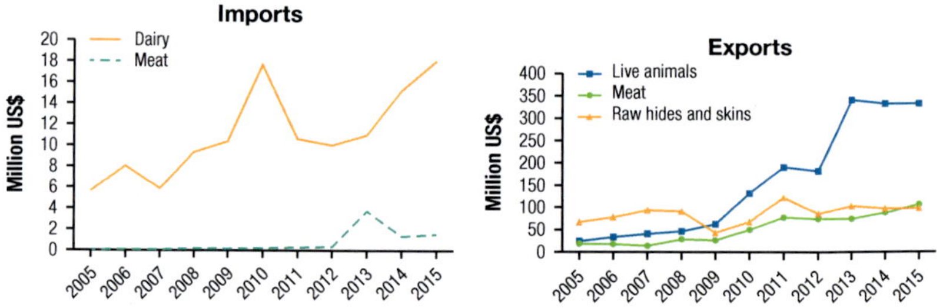

Source: Authors' calculations based on UN Comtrade data (UN Comtrade 2017).

FIGURE 8.3 Ethiopia's exports of live animals, composition, and destination countries, 2005–2015

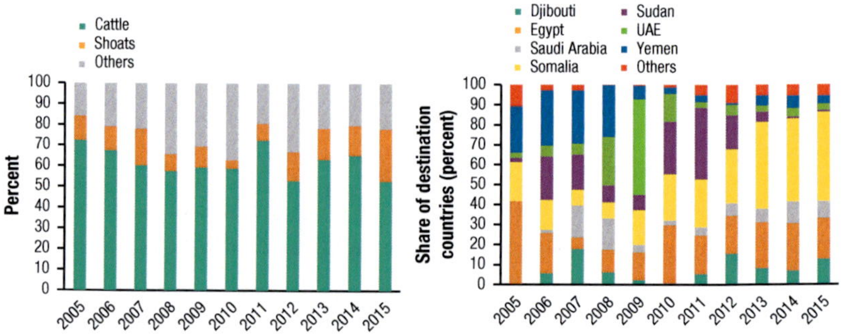

Source: Authors' calculations based on UN Comtrade data (UN Comtrade 2017).

Note: "Shoat" combines sheep and goats into a single category.

FIGURE 8.4 Imports of inputs used in livestock production in Ethiopia, 2007–2015

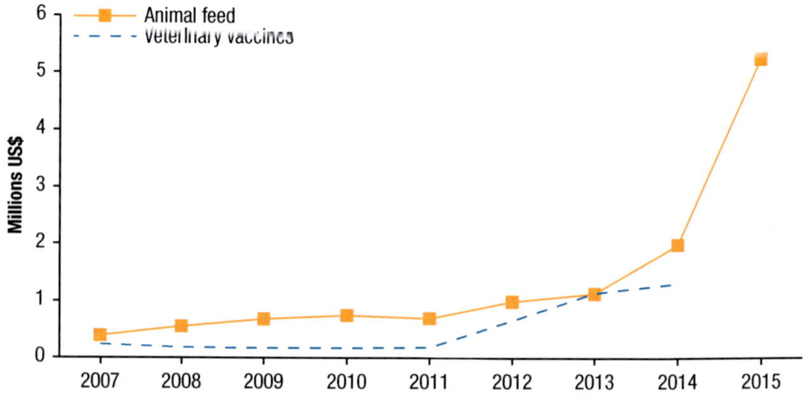

Source: Authors' calculations based on UN Comtrade data (UN Comtrade 2017).

imports in perspective, imports of chemical fertilizer used for crop agriculture amounted to more than $500 million in 2016 (UN Comtrade 2017).

Data and Descriptive Statistics

Data

The price data used in this study are collected as part of the CSA's Consumer Price Survey (Ethiopia, CSA 2017b). These data were collected in 116 urban retail markets in all regions of Ethiopia between January 2007 and December 2016. The number of markets in each region is approximately proportional to the region's share of the total urban population to ensure a sufficient degree of national representativeness (Figure 8.5). Thirty-two markets are surveyed in Southern Nations, Nationalities, and Peoples (SNNP) region, 24 in Oromia, and 20 in Amhara (the three biggest regions), while 12 markets are surveyed in Addis Ababa (by far the largest urban center with officially more than three million residents). The smaller regions include only a handful of markets.

CSA enumerators, who reside permanently in these markets, collect price and weight/volume data from traders, retailers, and consumers. For each item a maximum of three price quotations are collected from three different retailers in the first 15 days of each month in the European calendar, though enumerators are encouraged to survey the same retailers across months if possible. In the case of livestock, prices are collected on cattle (heifers, cows, bulls, and oxen), small ruminants (goats, sheep), and poultry (hens, cocks). Prices of 11 types of ASF are collected: beef, camel meat, milk (processed and unprocessed cow milk, goat milk, camel milk, and milk powder), eggs, and processed dairy products (yoghurt, cheese, and butter). Most of our descriptive analyses rely on price data of 8 of the 11 ASF items since data on camel meat, camel milk, and goat milk is unavailable in most months for most markets. We also study trends in real prices of cattle and ASF using CSA's Producer Survey Data (Ethiopia, CSA 2017c). The producer price surveys are also conducted monthly and mostly in the same manner as the retail price surveys, but they differ in that they include only primary crop and livestock outputs.[9]

9 As a result, four of the eight ASF items are excluded from the dataset: beef, pasteurized cow milk, powdered milk, and yoghurt. Although the producer price data leaves out one-quarter of the items considered in this study, its use, albeit briefly, is justified as most rural residents purchase ASF and livestock, particularly oxen for use in crop production, from these markets. For instance, the HCES (Household Consumption Expenditures Survey) (Ethiopia, CSA 2011) data

FIGURE 8.5 Administrative zones covered by Ethiopia Central Statistical Agency retail price data

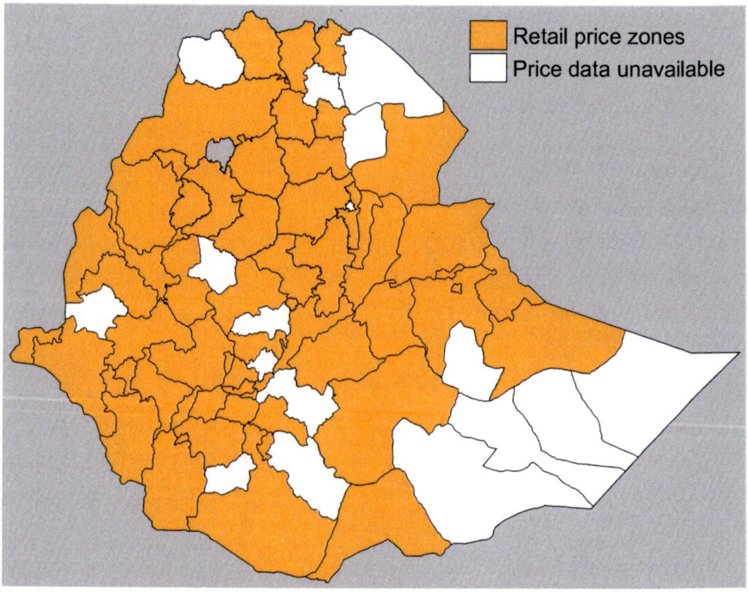

Source: Authors' mapping using CSA retail price data (Ethiopia, CSA 2017b).

We deflate prices using the regional general Consumer Price Index (CPI) calculated by CSA. This is done to express all prices in December 2011 birr (Ethiopia, CSA 2017a).[10] As the weight of individual products in the Consumer Price Index (CPI) is relatively low and given the lack of any reasonable alternative, we rely on the regional CPI to deflate the retail prices. We use a deseasonalized index, since one of our interests is studying seasonal patterns of prices that typically characterize the agricultural and food sectors.[11]

indicates that 78 percent of the dairy consumed in rural areas in 2010/2011 was own-produced while 20.3 percent was purchased. Similarly, 35 percent of the eggs and about 77 percent of the meat consumed was purchased.

10 Unless specified otherwise, all prices in this chapter are in real December 2011 birr. The US dollar-to-birr exchange rate in that month was US$1 = 17.19 birr.

11 To construct such an index, we first calculate a 12-month moving average of the real price for each item. Then we deflate/normalize the price series so constructed by dividing it to its January 2007 values. Accordingly, the price index has a value of one in January 2007 and it is greater/less than one in months with deseasonalized prices greater/less than January 2007 prices.

Descriptive Statistics

Table 8.1 provides summary statistics of the real prices of livestock and ASF over the period 2007–2016. The dataset used in our analyses comprises a large number of observations for livestock, ranging from 10,898 observations for cows and 11,479 for oxen to 12,260 for sheep and 12,676 for cocks. The high number of observations indicate the wide presence of this type of livestock in these retail markets. In the case of ASF, we see wide availability of prices for beef, unpasteurized milk, butter, and eggs. Yoghurt, cheese, powdered milk, and pasteurized milk are less available. The price data least available are for goat milk, camel milk, and camel meat, which are mostly consumed in pastoralist areas (Afar and Somali regions) and in Dire Dawa, Harari, and parts of Oromia regions.

As might be expected, we see large price differences across livestock categories. Poultry are the cheapest livestock category with median prices for a hen of about 43 birr (US$2.5). Prices for a hen are about 10 percent the price of a sheep, which had a median price of 438 birr (US$25). The price of a bull is more than five times the price of a sheep. While there are obvious differences in the quantity of meat that can be obtained from different livestock categories, the price of beef is the most expensive at 71 birr per kilogram, on average, while the average price of meat coming from sheep and goats is estimated at 45 birr per kilogram, and that of poultry ranged from 52 to 78 birr per kilogram.

In the case of ASF, beef is sold at slightly higher prices than camel meat. However, it is to be noted that camel meat is mostly found in eastern pastoralist areas, close to production areas and away from major urban centers, such as Addis Ababa, and its lower price might possibly be explained by this. Unpasteurized cow milk is the cheapest of all dairy products. The median prices of camel and goat milk are on average 22 percent and 26 percent higher than the median price of cow milk. Using medians, pasteurized cow milk is almost twice as expensive as nonpasteurized milk, implying that likely relatively richer households consume this type of milk. Looking at processed dairy products, butter has the highest price with a median price more than tenfold that of milk. The price of cheese is more than twice that of milk, and the price of yoghurt is 65 percent higher than that of milk.

We also present at the bottom of Table 8.1 prices per kilogram of the staple cereals in the country. When these prices are compared to prices of ASF, the latter have much higher prices. Compared to the prices of maize, the most important cereal in calorie terms in the country, median beef prices are 16 times more expensive over the period considered. Relative to the price of maize, unpasteurized cow milk prices are more than double, and a dozen of

TABLE 8.1 Prices in 2011 birr, descriptive statistics, 2007–2016

Category	Unit	Observations	Mean	Median	Standard deviation	Coefficient of variation
Livestock						
Heifer (2–4 years)	piece	10,905	2,074.3	1,973.7	703.7	0.34
Cow (4 years and above)	piece	10,898	3,117.4	2,953.0	993.0	0.32
Bull (2–4 years)	piece	11,142	2,607.4	2,403.8	992.3	0.38
Ox (4 years and above)	piece	11,479	5,029.1	4,750.4	1,629.4	0.32
Sheep (10–15 kilograms)	piece	12,260	460.9	438.5	165.7	0.36
Goat (10–15 kilograms)	piece	12,158	425.0	393.3	161.1	0.38
Hen (indigenous)	piece	12,593	44.7	43.1	15.3	0.34
Cock (indigenous)	piece	12,676	66.7	64.5	21.0	0.31
Animal-sourced foods						
Beef	kilograms	11,697	71.0	70.2	16.6	0.23
Camel meat	kilograms	1,018	63.1	65.1	19.6	0.31
Cow milk unpasteurized	liter	11,102	9.5	9.1	2.5	0.27
Cow milk pasteurized	liter	2,392	18.8	17.1	6.0	0.32
Camel milk	liter	1,438	11.8	11.1	4.1	0.35
Goat milk	liter	469	11.7	11.5	3.4	0.29
Powdered milk	450 grams	7,853	99.8	99.4	14.6	0.15
Yoghurt	liter	8,884	15.8	15.0	5.2	0.33
Cheese	kilograms	5,997	23.8	21.1	11.4	0.48
Butter	kilograms	11,694	98.1	96.3	22.0	0.22
Egg	dozen	12,837	20.3	20.1	4.7	0.23
Cereals						
Teff (mixed)	kilograms	12,592	10.5	10.2	2.3	0.22
Wheat (white)	kilograms	12,416	7.6	7.3	2.0	0.26
Barley (mixed)	kilograms	12,512	7.5	7.2	2.0	0.27
Maize (white)	kilograms	11,823	4.7	4.4	1.6	0.35
Sorghum (white)	kilograms	11,040	6.0	5.8	2.3	0.38

Source: Authors' computations using Ethiopia, CSA price data.
Note: Exchange rate December 2011: 1 US$ = 17.19 birr.

eggs was five times higher. On the other hand, milk is less expensive than teff. Based on the national household survey of 2011, Worku et al. (2016) compare the calorie prices of different food groups in average consumption baskets in Ethiopia and find that ASF calorie prices are on average 9.5 times more expensive than those of cereals. Finally, we compare the price ratios of ASF

TABLE 8.2 Cereal–ASF (animal-sourced foods) relative calorie price ratios, Ethiopia compared with other regions

Region	Cow's milk, fresh	Cow's milk, processed	Chicken eggs	Meat	Fish
High income	3.2	2.2	3.0	2.0	4.3
Latin America and Caribbean	3.9	3.0	4.9	3.2	3.4
Middle East and North Africa	10.1	3.1	6.1	6.2	6.0
South, Central, and Southeast Asia	7.8	3.8	6.2	6.5	5.3
Western and Central Africa	16.5	4.0	9.9	5.3	5.0
Eastern and Southern Africa	13.9	5.8	9.1	5.6	6.1
Ethiopia	10.4	—	19.6	52.9	38.1

Source: Numbers for all regions (except Ethiopia) from Table 9 in Headey, Hirvonen, and Hoddinott (2018); Ethiopia data are from Bachewe et al. (forthcoming).

Note: — = data not available.

in Ethiopia with other countries (Table 8.2). They indicate the significantly higher price ratios observed in Ethiopia compared to other countries. While the consumption of ASF obviously has benefits other than providing calories, these price differences indicate the challenge of assuring affordability of those products in Ethiopia where assuring food security remains a policy priority.

Seasonal and Spatial Patterns

Seasonality

Seasonal patterns are a defining characteristic of agriculture all over the world (for example, Sahn 1989). These seasonal movements are often linked to inherent constraints to consistent year-round supplies due to climatic and, most important, rainfall patterns. This is also the case in Ethiopia. Rainfall is unimodal in the west and north of the country with rainfall mostly falling between June and September and one main crop harvest (*meher*), while it is bi-modal in the east and southeast of the country, giving rise to a second season (*belg*) in those areas. Moreover, rainfall is more reliable in the west and south of the country compared to other parts. As the production of feed for livestock is linked to this rainfall pattern, it induces seasonal patterns in the sector. Farmer (2010) shows that seasonality is an important factor for pastoralist households as rangelands can support fewer animals during the dry season, leading owners to sell livestock during the dry season or migrate to areas with better feed availability, while holding on to livestock during the wet season. For farmers in the highlands, fewer sales of livestock occur during the rainy period as feed is more easily available and cattle are needed for crop cultivation.

We also note significant seasonality in demand. One important characteristic of ASF consumption, particularly in Ethiopia, is its link with religion. An estimated 43 percent of the population in Ethiopia are Orthodox Christians (Ethiopia, CSA 2009), and their religion is characterized by important constraints on food intake, especially related to ASF. Figure 8.6, based on the Household Consumption and Expenditures Survey (HCES) of 2011, shows how quantities of ASF consumed vary over the course of the year for rural and urban residents. We see peaks in the beginning of the year (for the festivities of Ethiopian Christmas and Timket), in April (for Easter), and in September (for the New Year and Meskel). We see similar patterns in rural as well as urban areas, although levels are lower in the former areas as discussed above. The major consumption peaks associated with the major religious events are preceded by troughs, which are linked with fasting periods that come before and culminate during these important celebrations. This is especially so for Christmas, which occurs on January 7 every year, and Easter, which mostly occurs in April. During the fasting periods of Orthodox Christians—the Lent leading up to Easter during which the fasting period lasts up to 56 days and the one before Christmas, which lasts for about 40 days—no ASF are consumed. While there are other fasting periods during the year, they are much shorter.[12] This reduction in consumption of ASF also shows up in slaughterhouse data in Addis Ababa. The number of head of cattle and of sheep and goats slaughtered in the "Christian hall" of the slaughterhouse in the month before Easter typically drops to one-quarter the level of other months (FVI-Idele 2016). No such seasonality is seen for the "Muslim hall" during the Muslim major fasting season of Ramadan, and there is no evidence of higher activity at the end of Ramadan or for the Eid festivities.

Seasonal changes in demand seemingly drive changes in prices of livestock and ASF. A summary of the monthly price indexes is reported in Table 8.3 for different livestock categories. The results show that all seasonal indexes for all livestock categories are higher than 1.0 in January, February, April, May, and September, illustrating to what extent these religious events are associated

12 On top of these long fasting periods, Orthodox Christians do not eat ASF on Wednesdays and Fridays throughout the year, except in the two months after Easter. It has been estimated that the total number of days that clergy (priests, nuns, monks) are not allowed to eat ASF amounts to 250 per year (Ayenew et al. 2009). For nonclergy practicing Orthodox Christians, the number of fasting days amounts to between 166 and 180.

FIGURE 8.6 Consumption of animal-sourced foods in Ethiopia, by month (2011)

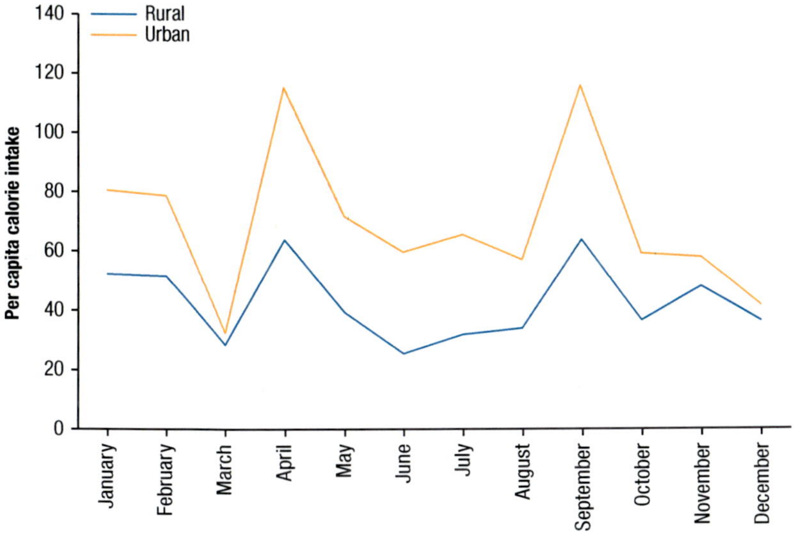

Source: Adjusted from Hirvonen, Taffesse, and Hassen (2016).

TABLE 8.3 Seasonal price indexes of livestock in Ethiopia, 2007–2016

	Livestock					
Month	Sheep	Goat	Cow	Bull	Ox	Hen
January	1.03	1.02	0.99	1.01	1.01	1.11
February	0.98	0.98	0.98	0.98	0.97	1.02
March	0.96	0.97	0.97	0.99	0.96	0.99
April	1.02	1.03	1.01	1.01	1.01	1.07
May	1.02	1.03	1.02	1.03	1.04	1.01
June	0.99	0.99	1.02	1.01	1.02	0.94
July	0.99	0.99	1.01	1.00	1.02	0.92
August	0.99	0.99	1.01	1.00	1.00	0.93
September	1.06	1.05	1.03	1.01	1.04	1.08
October	1.00	0.99	0.99	0.99	0.99	0.98
November	0.98	0.97	0.98	0.98	0.97	0.97
December	0.97	0.97	0.98	0.98	0.96	0.98
Amplitude	**0.10**	**0.08**	**0.06**	**0.05**	**0.08**	**0.19**

Source: Authors' calculations based on Ethiopia, CSA price data.

Note: Seasonal price indexes were calculated over the January 2007–December 2016 period using the percentage moving average method. In this method the original data values in the time-series (or real prices of the items in a given month) are expressed as a ratio of the 12-month moving average real prices of the items in that month.

with higher prices.[13] Slightly higher indexes for the prices of oxen/bulls/cows during June through August are also noted, possibly because of demand for cultivation during the *meher* season. The lowest prices are noted during August, when fodder is easily available, as well as in October, November, and December. Seasonal amplitudes are not that large and vary between 5 percent and 10 percent for cattle and for sheep and goats. It is highest for poultry, for which prices vary 19 percent over the year in the case of hens. Ayele and Rich (2010) also link the low prices of poultry with regular outbreaks of Newcastle disease that typically occur in the late spring.

In Table 8.4 seasonal indexes are reported for several ASF. A similar pattern emerges as in the case of livestock. We see increases of prices at the time of major religious celebrations and a decrease during July/August, when livestock feed is easily available, products are more plentiful, and milk yields are higher. Prices are also lower at the end of the year during or just after crops are harvested and when the stock of crop-residue, which is used as livestock feed, is at its maximum. The September peak for ASF is much less pronounced than in the case of livestock.

Spatial Patterns

We next look at livestock and ASF price variation by region, using average prices over the 2007–2016 period (Table 8.5). We note significant variation by region. For example, the price of oxen is more than 50 percent higher in Addis Ababa compared to the Afar region. Similar variations are noted for other types of livestock. Overall, prices for Addis Ababa are consistently above the national median. The difference varies from 23 percent for cocks to 98 percent for goats. Prices are also higher than the national median for the cities of Dire Dawa and Harari, except for poultry. Prices for livestock are also relatively higher in the western region of Benishangul-Gumuz (BG), seemingly linked with the hostile environment there for livestock production due to warm and humid conditions and the presence of tsetse fly (SID-Consult 2010).

Prices in the SNNP region, though, are consistently and significantly lower for all types of livestock.[14] Prices for most livestock are also low in the lowland and pastoralist areas of Somali. This is not the case for all types of livestock, however, as cattle prices are higher than the median. The breed sold in Afar markets is almost exclusively Danakil, which is among the lowest valued,

13 Except for cows in January.

14 The only exception is cock.

TABLE 8.4 Seasonal price indexes of animal-sourced food in Ethiopia, 2007–2016

	Animal-sourced food						
Month	Beef	Cow milk	Goat milk	Yoghurt	Cheese	Butter	Eggs
January	0.98	1.00	1.04	1.00	1.02	0.97	1.02
February	0.95	1.00	1.02	1.00	1.02	0.96	1.00
March	0.96	1.00	1.00	1.00	0.97	0.99	0.94
April	0.97	1.00	1.01	0.99	1.00	1.11	0.94
May	1.00	1.00	1.00	1.00	1.05	1.11	1.01
June	1.02	1.00	0.98	1.00	1.03	1.05	1.00
July	1.02	1.00	0.98	1.00	0.98	0.99	0.99
August	1.03	1.00	0.99	1.00	0.98	0.98	0.99
September	1.03	1.00	1.03	1.00	1.02	1.01	1.06
October	1.02	1.00	1.01	1.00	0.99	0.95	1.04
November	1.01	1.00	0.98	0.99	0.98	0.93	1.00
December	0.99	0.99	0.96	1.00	0.95	0.93	0.98
Amplitude	**0.07**	**0.01**	**0.08**	**0.01**	**0.09**	**0.18**	**0.12**

Source: Authors' calculations based on Ethiopia, CSA price data.

Note: Seasonal price indexes were calculated over the January 2007–December 2016 period using the percentage moving average method. In this method the original data values in the time-series (or real prices of the items in a given month) are expressed as a ratio of the 12-month moving average real prices of the items in that month.

while the breed sold in Harari region, the Harari breed, has the highest value. All sorts of breeds are sold in Addis Ababa markets, which is why cattle prices in Harari region are generally higher than those in Addis Ababa, but lower for other livestock.

The price of sheep is lowest in the Somali region, a major supply region for export markets. Despite the large herd of livestock and its remoteness, the Afar region shows surprisingly high prices for cattle overall. However, there are significant variations in breeds over the country with different quality premiums attached to it (Teklewold et al. 2009). Sheep and goats from lowland areas of Somali and Afar regions have a higher fat content, and their meat is perceived not to "darken" when prepared. Consequently, such meat is preferred in export markets. Sheep and goats from lowlands are therefore more destined to export markets compared to those of the highlands, which are more used by domestic consumers. Comparing livestock prices, without quality considerations, is therefore not straightforward. Doing this more detailed analysis is left for future research.

Median prices of ASF are presented by region in Table 8.6. We note again significant variation in prices across regions. Prices of beef and milk

TABLE 8.5 Prices of livestock in Ethiopia by region, 2007–2016

		Region											
Livestock type	Unit	Tigray	Afar	Amhara	Oromia	Somali	Benishangul-Gumuz	SNNP	Gambella	Harari	Addis Ababa	Dire Dawa	Total
Median prices (2011 birr)													
Heifer (2–4 years)	piece	2,074	2,050	2,162	1,925	2,039	2,323	1,677	1,889	2,683	2,915	2,767	1,976
Cow (4 years or more)	piece	3,334	3,338	3,130	2,786	3,417	3,396	2,633	2,942	4,660	4,142	3,156	2,983
Bull (2–4 years)	piece	2,661	2,411	2,624	2,425	2,466	2,516	2,098	2,077	4,167	3,384	3,100	2,406
Ox (4 years or more)	piece	4,511	4,520	4,642	4,919	5,926	4,614	4,484	4,222	7,562	7,506	7,089	4,779
Sheep (10–15 kilograms)	piece	548	412	452	419	369	441	371	584	697	713	415	437
Goat (10–15 kilograms)	piece	502	454	377	368	378	374	361	542	646	784	440	396
Hen (indigenous)	piece	56	56	46	41	31	51	37	48	26	58	30	43
Cock (indigenous)	piece	72	70	56	63	47	75	64	83	49	79	52	64
Comparison with national median (%)													
Heifer (2–4 years)	(%)	105	104	109	97	103	118	85	96	136	148	140	100
Cow (4 years or more)	(%)	112	112	105	93	115	114	88	99	156	139	106	100
Bull (2–4 years)	(%)	111	100	109	101	103	105	87	86	173	141	129	100
Ox (4 years or more)	(%)	94	95	97	103	124	97	94	88	158	157	148	100
Sheep (10–15 kilograms)	(%)	125	94	104	96	84	101	85	134	160	163	95	100
Goat (10–15 kilograms)	(%)	127	114	95	93	95	94	91	137	163	198	111	100
Hen (indigenous)	(%)	131	131	107	95	73	118	87	113	62	135	70	100
Cock (indigenous)	(%)	111	109	88	98	73	117	99	129	76	123	81	100

Source: Authors' computations using Ethiopia, CSA price data.

Note: Exchange rate December 2011: 1 US$ = 17.19 birr. SNNP = Southern Nations, Nationalities, and Peoples.

TABLE 8.6 Prices of animal-sourced foods by region, 2007 to 2016

		Region											
Animal-sourced food type	Unit	Tigray	Afar	Amhara	Oromia	Somali	B.G.	SNNP	Gambella	Harari	Addis Ababa	Dire Dawa	Total
Median real prices (2011 birr)													
Beef	kg	65	77	66	73	89	57	71	63	84	76	84	70
Cow milk unpasteurized	liter	9	11	8	9	11	8	9	8	12	11	12	9
Cow milk pasteurized	liter	21	25	15	18	25	15	18	17	24	16	23	18
Powdered milk	450 g	94	98	102	97	107	102	100	112	111	94	100	100
Yoghurt	liter	22	15	13	16	18	15	15	19	18	13	15	15
Cheese	kg	24	32	21	19	23	6	19	21	47	39	50	21
Butter	kg	105	109	90	96	109	74	96	98	116	113	107	97
Egg	dozen	20	25	18	20	26	23	19	26	23	24	22	20
Comparison with national median (percent)													
Beef	percent	93	110	95	104	126	82	101	90	120	109	120	100
Cow milk unpasteurized	percent	97	119	91	103	123	88	99	93	131	124	135	100
Cow milk pasteurized	percent	118	142	84	103	143	83	102	98	139	88	133	100
Powdered milk	percent	95	98	102	97	108	103	101	112	111	94	101	100
Yoghurt	percent	145	102	83	106	117	100	97	127	118	85	99	100
Cheese	percent	116	156	102	93	113	28	93	100	228	188	240	100
Butter	percent	109	113	93	100	112	77	99	102	120	117	110	100
Egg	percent	98	123	88	100	128	114	93	128	115	117	110	100

Source: Authors' computations using Ethiopia CSA price data.

Note: Exchange rate December 2011: 1 US$ = 17.19 birr. SNNP = Southern Nations, Nationalities, and Peoples. B.G. = Benishangul-Gumuz.

are relatively low in the northern part of the country—that is, Tigray and Amhara. Beef and milk prices in Tigray are 7 percent and 3 percent lower than the national median, respectively. This compares to 5 percent and 9 percent, respectively, in the Amhara region. Prices are also low for these products in the western regions of Benishangul-Gumuz and Gambella. Prices for beef and cow milk are high in the lowlands of Somali and Afar. Processed dairy product prices are also high in Somali and Afar, seemingly as there is little habit of processing milk in those settings. FAO (2017) notes that pastoralists and agro-pastoralists mostly consume milk fresh and as yoghurt and ghee.

ASF prices are also high in cities, particularly in the eastern cities of Harari and Dire Dawa compared to the capital Addis Ababa, which has the highest or second highest median prices for all livestock types and the lowest prices for some ASF items. Prices for pasteurized cow milk and yoghurt are among the lowest in Addis Ababa. This might possibly be explained by the considerable development of the dairy sector around Addis Ababa (Minten et al. 2018).[15] Improved dairy processing plants combined with high demand allows for economies of scale, which seems to have contributed to relatively low prices in the capital compared to other cities.

Egg prices are also high in cities. They are lower than the national median in the three major agricultural regions of the country (Tigray, Amhara, and SNNP) and at par with the median in the Oromia region. Prices are also very high in pastoralist areas, seemingly as holding chicken is less conducive for seminomadic lifestyles; in addition, chicken are susceptible to heat stress, and most breeds are not well adapted to these hot areas (Nyoni, Grab, and Archer 2019).

Semiurban and rural areas around cities are often suppliers of food to these cities and therefore are usually well integrated with the end markets in the cities (Minten, Stifel, and Tamru 2014). As a result, one would expect prices in these areas to be lower than in the cities, with the differences in prices reflecting marketing costs. We present supplying areas and urban prices in the case of eggs, milk, sheep, and oxen, comparing the median prices of Addis Ababa and the surrounding Shewa zones (east, west, and north Shewa in Oromia, and north Shewa in Amhara) over the 2007–2016 period (Figure 8.7). Prices in Addis Ababa are all above the prices in the surrounding areas. More important, these prices have a close relationship during the whole period, except for oxen prices, which appeared to move in different directions and were wider

15 Two-thirds of the current dairy processing plants in Ethiopia are based in Addis Ababa or its surrounding areas.

FIGURE 8.7 **Real prices of sheep, oxen, eggs, and milk in Addis Ababa and surrounding Shewa zones, 2007–2016**

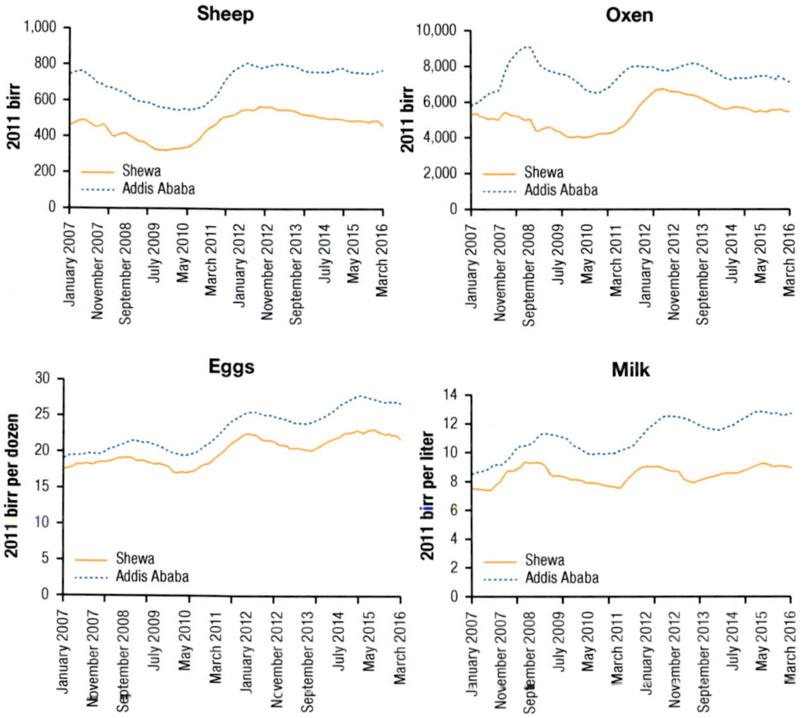

Source: Authors' calculations based on Ethiopia, CSA price data.
Note: Exchange rate December 2011: 1 US$ = 17.19 birr.

apart during mid-2007 to December 2008. Despite the wide differences in oxen prices in the beginning of the period, the differences are rather stable at the end of the period. Price differences are rather stable for sheep but increasing toward the end. In particular, Addis Ababa sheep prices were 37 percent higher than prices in Shewa in an average month in 2011, and this price gap increased to 56 percent in 2016. Differences between Addis Ababa and Shewa milk and egg prices have also slowly but consistently increased during the period.[16] More in-depth research would be needed to explain these increasing trends in marketing costs between rural and urban areas.

16 The gap between Addis Ababa and Shewa milk prices increased from 17 percent in an average month in 2007 to 40 percent in 2016. The oxen price gap increased from 23 percent to 33.4 percent, while the egg price gap increased from 8 percent to 20 percent during the same period.

Price Evolution over Time

Real Animal-Sourced Foods and Livestock Prices

We next look at the extent that livestock and ASF prices have changed over the past decade. Retail prices of cows, bulls, and oxen are presented for the years 2007, 2011, and 2016 in Table 8.7. Real prices in 2011 were at similar (or even lower) levels than in 2007. However, there has been a significant increase since 2011. By 2016 real prices of cattle had risen between 10 percent and 13 percent relative to 2007 prices. Sheep and goats show similar upward trends, with prices 22 percent and 26 percent higher in 2016, respectively, while they also were higher in 2011. Finally, chicken prices increased the highest of all—prices of cocks were 46 percent higher in 2016 than a decade earlier. In the case of ASF, we see similar increases in retail prices over time (Table 8.8). Beef prices increased by 33 percent between 2007 and 2016. Cow milk and egg prices increased by 36 percent and 32 percent, respectively. Important increases are also seen for processed dairy products and for imported powdered milk.

We also use CSA's producer price dataset (Ethiopia, CSA 2017c) to study trends in prices of livestock and four of the eight ASF items included in the rural producer price dataset. In general, similar price increases are noted as in urban retail prices. Table 8.7 indicates that real producer cattle prices increased by 5 percent to 10 percent, while the increase was 14 percent in sheep, 24 percent in goats, and by more than 18 percent in poultry. Producer livestock prices in 2016 were lower than retail prices ranging from 14 percent (for hens) to 30 percent (for oxen). Moreover, growth in producer prices was lower relative to retail prices for all livestock types. However, livestock types with rapidly/slowly growing retail prices generally showed rapid/slow growth in producer prices. Real producer prices of cheese were essentially the same as retail prices in 2016, while retail prices were at least 18 percent higher than producer prices for the remaining three items (Table 8.8). However, growth rates in producer prices of these ASF items was generally higher than growth in retail prices.

Figure 8.8 presents zonal maps indicating real prices at the beginning and at the end of the past decade for the three most important ASF—beef, milk, and eggs. Here, an increase in price is indicated by a change of colors from light to darker green and vice versa. Accordingly, the overall darker colors on the right illustrate the increase in prices of the three ASF items over the past decade. In the case of beef, we note relatively lower prices and a relatively lower change in the north of the country. However, we note a considerable increase

TABLE 8.7 Price evolution for livestock in Ethiopia, 2007, 2011, and 2016

	Urban retail prices			Rural producer prices		
Livestock type	2007	2011	2016	2007	2011	2016
Median real prices (2011 birr, per piece)						
Heifer (2–4 years)	1,866	1,811	2,059	1,636	1,407	1,713
Cow (4 years and above)	2,732	2,788	3,105	2310	2127	2536
Bull (2–4 years)	2,285	2,216	2,547	2,041	1,816	2,168
Ox (4 years and above)	4,405	4,250	4,992	3,606	3,258	3,858
Sheep (10–15 kilograms)	393	422	479	345	345	392
Goat (10–15 kilograms)	349	376	440	292	319	363
Hen (indigenous)	38	39	48	36	34	42
Cock (indigenous)	54	58	78	50	50	65
Comparison with national median in 2007 (%)						
Heifer (2–4 years)	n.a.	97	110	n.a.	86	105
Cow (4 years and above)	n.a.	102	114	n.a.	92	110
Bull (2–4 years)	n.a.	97	111	n.a.	89	106
Ox (4 years and above)	n.a.	96	113	n.a.	90	107
Sheep (10–15 kilograms)	n.a.	107	122	n.a.	100	114
Goat (10–15 kilograms)	n.a.	108	126	n.a.	109	124
Hen (indigenous)	n.a.	102	128	n.a.	96	118
Cock (indigenous)	n.a.	109	146	n.a.	101	132

Source: Authors' calculations based on Ethiopia CSA price data
Note: Exchange rate December 2011: 1 US$ = 17.19 birr. n.a. = not applicable.

in the east, central, and southern parts of the country. The map suggests that relative beef prices have especially gone up in those areas that are well integrated in commercial livestock circuits.

Similarly, cow milk prices are relatively low and seem to have stayed the same or changed little over the past ten years in the northern part of the country. Although milk prices were already higher in the south and east of the country in 2007, that difference seems to even have widened over the decade. Real milk prices around the capital Addis Ababa do not seem to have increased very much over time, possibly driven by the development of this active periurban dairy value chain. Eggs also show similar patterns, a worsening over time overall, a relatively stable-to-slow increase in the north, around Amhara and Tigray, and worsening in the most southern, eastern, and western zones of the country.

TABLE 8.8 Price evolution for animal-sourced foods in Ethiopia, 2007, 2011, and 2016

		Urban retail prices			Rural producer prices		
Animal-sourced foods type	Unit	2007	2011	2016	2007	2011	2016
Median real prices (2011 birr)							
Beef	kilograms	62	56	82			
Cow milk unpasteurized	liter	7	9	10	6	7	8
Cow milk pasteurized	liter	14	17	17			
Powdered milk	450 grams	77	97	100			
Yoghurt	liter	13	15	16			
Cheese	kilograms	18	18	22	15	16	22
Butter	kilograms	83	101	100	71	82	85
Egg	dozen	17	20	22	12	15	18
Comparison with national median in 2007 (%)							
Beef	(%)	n.a.	90	133			
Cow milk unpasteurized	(%)	n.a.	118	136	n.a.	130	141
Cow milk pasteurized	(%)	n.a.	122	122			
Powdered milk	(%)	n.a.	126	129			
Yoghurt	(%)	n.a.	115	122			
Cheese	(%)	n.a.	101	125	n.a.	105	146
Butter	(%)	n.a.	122	120	n.a.	116	120
Egg	(%)	n.a.	116	132	n.a.	123	142

Source: Authors' calculations based on Ethiopia, CSA price data.
Note: Exchange rate December 2011: 1 US$ = 17.19 birr. n.a. = not applicable.

As ASF are obtained from livestock, we assess to what extent price increases of ASF and livestock have been linked over the past decade. To do so, we plot different pairs of ASF and livestock prices (Figure 8.9) and compute correlation coefficients of paired monthly average price indexes and percentage changes in prices. As one would expect, we see strong correlations in prices, indicating the close link of ASF and livestock sector performance. Moreover, there are fairly high correlations in price changes of the items considered. Beef prices seem more volatile than bull prices—as prices of the latter went relatively lower and higher than bull prices over the period considered—but the correlation coefficient of price indexes is still high at 0.94 while correlation of price growth rates is 0.51. Beef prices in 2011 dropped significantly compared to bull prices, possibly because of price-fixing of beef by the government in the beginning of that year (Hassan 2011). The price-fixing led to less meat being brought to the market, and seemingly only those in need of cash were willing

FIGURE 8.8 Real prices of beef (birr per kilogram), milk (birr per liter), and eggs (birr per dozen), 2007 and 2016 by zone (2011 birr)

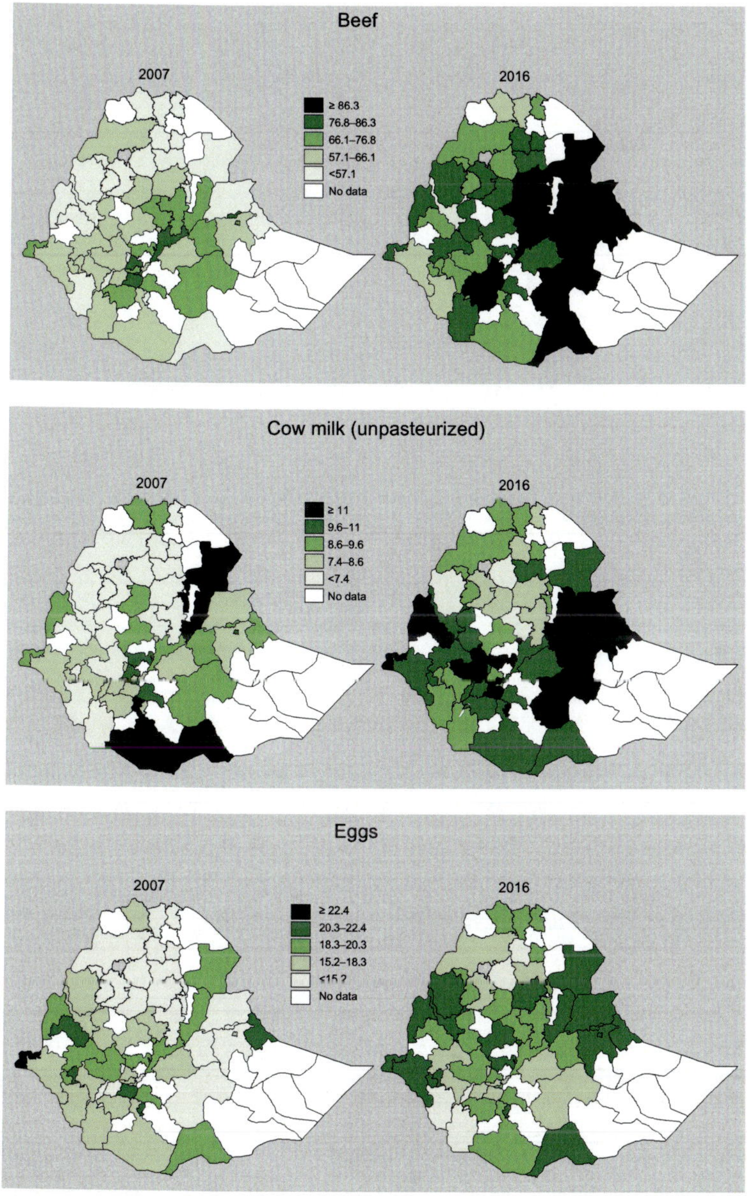

Source: Authors' calculations based on Ethiopia, CSA price data.

Note: Exchange rate December 2011: 1 US$ = 17.19 birr.

FIGURE 8.9 Correlations of animal-sourced foods and livestock price indexes (January 2007 = 1.0)

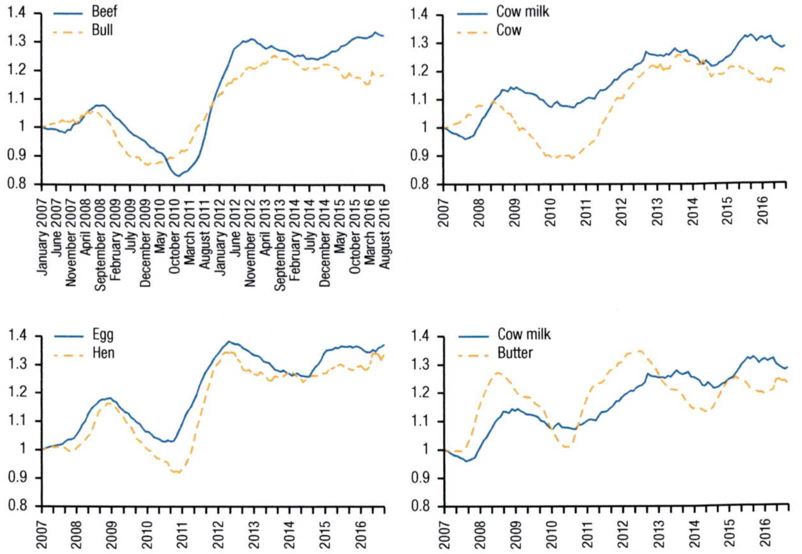

Source: Authors' calculations based on Ethiopia, CSA price data.

to sell their oxen.[17] Cow milk and cow prices are less correlated with a correlation coefficient of 0.78 for price indexes and 0.31 for price growth rates. Eggs and hens, however, are characterized by a very high correlation coefficient of 0.96 for price indexes and 0.68 for price growth rates. We note in this case the disruption of the price-fixing done in the beginning of 2011 as the prices of hens in that period were the lowest of the whole decade. Butter and milk are also less correlated (at 0.60 for prices and 0.30 for price growth rate). In any case, the figure suggests overall that the strong increases in prices of ASF are associated with increases in livestock prices.

Animal-Sourced Foods and Livestock versus Other Food Prices

Since increasing the consumption of ASF would imply at least a partial trade from other major products in the consumption basket, we compare to what

17 On January 6, 2011, the government announced a list of 18 products for which it fixed prices (Hassan 2011). However, as several products were coming to be in short supply, most of the price caps were lifted in late May 2011.

FIGURE 8.10 Annual per capita food consumption in 2011 in Ethiopia, by food type and region (kilograms per capita per year)

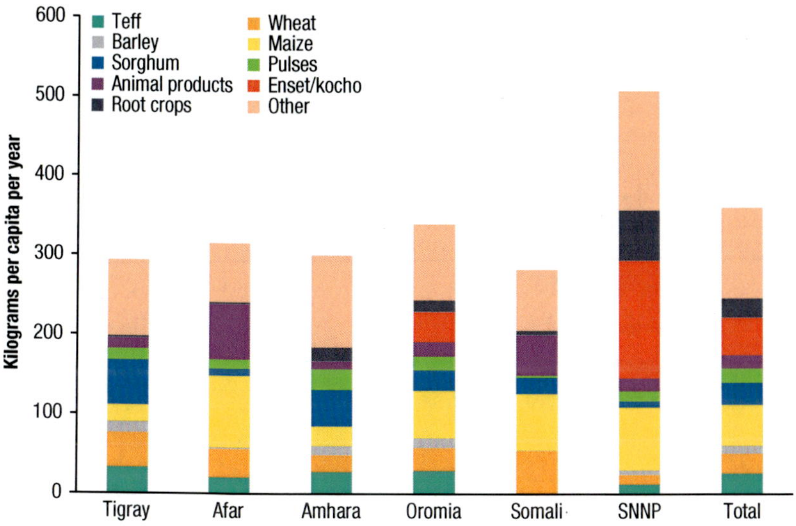

Source: Authors' calculations from CSA HCES (Ethiopia, CSA 2011).
Note: SNNP = Southern Nations, Nationalities, and Peoples.

extent the real prices of cereals changed relative to real prices of livestock and ASF products over the past decade.[18] To do so, we first examine the quantity of different crops and products consumed. Figure 8.10 shows these data for the four major crop-producing regions as well as for the Afar and Somali regions. We use the average consumption basket from the Household Consumption Expenditures Survey (HCES) in 2011. Figure 8.10 illustrates that cereals are, in quantity terms, very important in these regions. They comprise 60 percent of the average consumption basket in Tigray, decreasing to 53 percent in Afar and 56 percent in Somali, and still remain at 50 percent in the Amhara and Oromia regions. In contrast, they comprise only 24 percent of the quantity consumed in SNNP due to the high levels of consumption of enset and root crops there.

18 CSA data indicate that more than 80 percent of the population in 2015 was engaged in crop production; nearly 86 percent of crop farmers produced one or more type of cereals and consumed 66 percent of the cereals output (Ethiopia, CSA 2013, 2015b, 2015c). In the same year, 73 percent of the total crop area was used to produce cereals (Ethiopia, CSA 2015c).

TABLE 8.9 Terms of trade of livestock relative to cereals in Ethiopia, 2007, 2011, and 2016

		Year		
Livestock type	Unit	2007	2011	2016
Terms of trade (in terms of 100 kg of cereal)				
Heifer (2–4 years)	piece	3.3	2.9	3.8
Cow (4 years and above)	piece	5	4.4	5.6
Bull (2–4 years)	piece	4	3.6	4.8
Ox (4 years and above)	piece	8	7	9.2
Sheep (10–15 kg)	piece	0.7	0.7	0.9
Goat (10–15 kg)	piece	0.6	0.6	0.8
Hen (indigenous)	piece	6.5	6.1	8.9
Cock (indigenous)	piece	9.4	9.2	13.7
Comparison with national median in 2006 (percent)				
Heifer (2–4 years)	percent	n.a.	88	116
Cow (4 years and above)	percent	n.a.	89	113
Bull (2–4 years)	percent	n.a.	90	120
Ox (4 years and above)	percent	n.a.	88	115
Sheep (10–15 kg)	percent	n.a.	94	120
Goat (10–15 kg)	percent	n.a.	97	127
Hen (indigenous)	percent	n.a.	94	138
Cock (indigenous)	percent	n.a.	98	146

Source: Authors' calculations based on CSA price data.

Note: The following ratio is used to calculate terms of trade = (real price of a head of livestock/100 kilograms of per capita consumption weighted real cereal prices) for all livestock types except hens, for which the denominator is 1 kilogram of per capita consumption weighted real cereal prices. n.a. = not applicable.

We use these consumption data, together with CSA's retail price series, to compute the terms of trade of each ASF/livestock type as a ratio of the real prices of ASF/livestock to per capita consumption weighted real cereal prices. The results of this exercise are provided in Table 8.9 and Table 8.10.

To buy 1 kilogram of beef in 2016, one would have needed 15 kilograms of cereals (Table 8.10). This compares to 10.7 kilograms in 2007. This change in terms-of-trade corresponds to a 40 percent higher price of beef relative to cereals—a cereal farmer wanting to buy a kilogram of beef would need 4.3 kilograms more of cereals in 2016 than in 2007 to do so. A similar deterioration is seen in the case of milk, where prices expressed in kilograms of cereals deteriorated by 41 percent over the past decade. Egg prices also showed a deterioration of 39 percent. A similar exercise is done for livestock, indicating to what

TABLE 8.10 Terms of trade of animal-sourced foods relative to cereals in Ethiopia, 2007, 2011, and 2016

		Year		
ASF type	Unit	2007	2011	2016
Terms of trade (compared to 1 kg of cereal)				
Beef	kg	10.7	8.8	15.1
Cow milk unpasteurized	liter	1.3	1.4	1.9
Cow milk pasteurized	liter	2.4	2.9	3.5
Powdered milk	450 gr	14.1	15.5	18.1
Yoghurt	liter	2.4	2.4	3.2
Cheese	kg	3.1	3.1	4
Butter	kg	14.3	16.3	18.5
Egg	dozen	2.8	3.1	4
Comparison with national median in 2006 (percent)				
Beef	percent	n.a.	82	140
Cow milk unpasteurized	percent	n.a.	103	141
Cow milk pasteurized	percent	n.a.	122	146
Powdered milk	percent	n.a.	111	129
Yoghurt	percent	n.a.	101	134
Cheese	percent	n.a.	98	129
Butter	percent	n.a.	114	130
Egg	percent	n.a.	107	139

Source: Authors' calculations based on CSA price data.
Note: The following ratio is used to calculate terms of trade = (real price of a unit of ASF/1 kilogram of per capita consumption weighted real cereal prices). n.a. = not applicable.

extent livestock producers can buy more cereals with the sales of their livestock (Table 8.9) or how much more product cereal producers must sell to buy each type of livestock. The results show that cereal producers' terms-of-trade clearly deteriorated. The increase in the quantity of cereals needed to buy livestock ranged from 15 percent (1.2 quintals) in oxen to 46 percent in cocks, the latter of which sold for 9.4 kilograms and 13.7 kilograms of cereals in 2007 and 2016, respectively.

Local versus International Prices

Part of the changes in the domestic livestock and ASF prices is likely explained by changes in export and import prices of livestock and ASF. Using export data from the Ethiopian Revenue and Customs Authority (ERCA 2017),

FIGURE 8.11 Export (from Ethiopia) border prices and local prices of live animals/meat, 2007–2015 (US$ per kilogram)

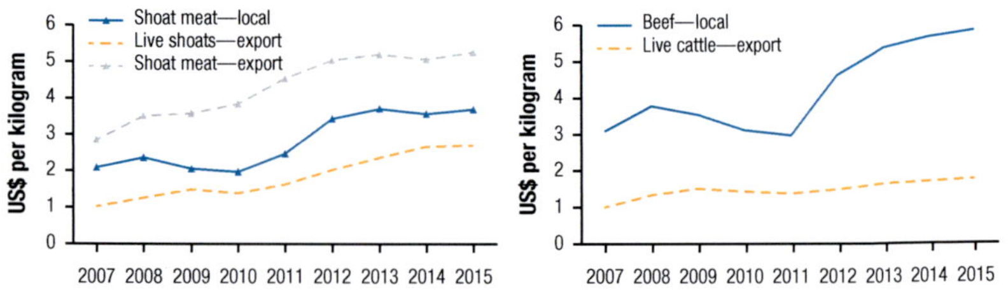

Source: Authors' calculations based on custom data (ERCA 2017) and CSA price data.
Note: "Shoat" combines sheep and goats into a single category.

Figure 8.11 shows to what extent prices, expressed in US dollars per kilogram, have changed over the past 10 years for both locally sold and exported live animals and meats. The data show important increases for live cattle and sheep and goats (shoats) as well as meat both sold locally and exported. Figure 8.12 shows how imported powdered milk prices have changed since 2007 and to what extent similar trends show up in local pasteurized and nonpasteurized milk prices. We find a strong correlation for these prices. The figure suggests that price increases in local markets have at least in part been linked to price increases in international markets.

We present the evolution of prices in international meat markets to understand to what extent the border prices that are noted in Ethiopia are synchronized with international patterns. We look at meat prices in Saudi Arabia, a major export country for Ethiopia; US import prices; and the international meat price index (Figure 8.13). The figure shows significant increases since 2007. However, we also see that meat prices have been coming down in the two most recent years, partly explaining why growth in meat prices in Ethiopia in recent years has been slower.

While it is likely that price increases of some of the livestock categories and ASF in the country have been influenced by international price increases, as Ethiopia is a major exporter of meat, such price increases were also seen for nontraded ASF, such as eggs and milk, and price increases in export destinations seemingly only explain part of these local price rises. In the case of milk, prices of powdered milk in comparison with local milk are very high.[19] There

19 Retail powdered milk prices in 2016 were around US$16 per kilogram. Assuming a typical 6.5 liters of liquid milk per 1 kilogram of powdered milk, this would imply a price of US$2.5 per liter, significantly above the local unpasteurized (at less than US$1 per liter) as well as pasteurized (at just more than US$1 per liter) milk prices.

FIGURE 8.12 Import (to Ethiopia) border prices and local prices of milk, 2007–2016

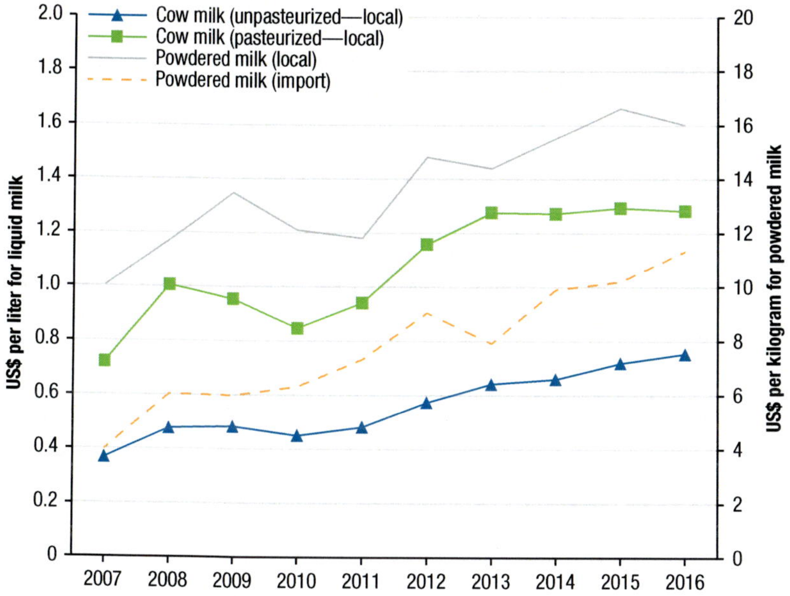

Source: Authors' calculations based on custom data (ERCA 2017) and CSA price data.

FIGURE 8.13 International meat price trends, 2007–2016

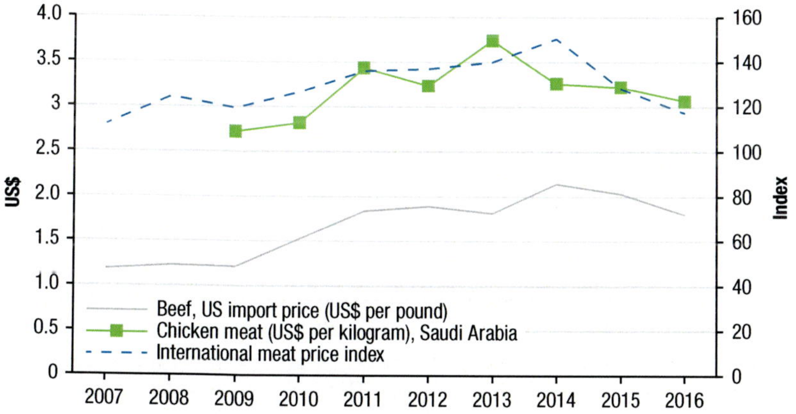

Source: Authors' calculations.

is likely limited substitution away from these local fresh milk products to imported sources, especially for poorer households, making local milk mostly a nontradable good. For example, the OECD-FAO (2016) estimates that fresh dairy products, with less than 1 percent of production traded, are one of the least traded agricultural commodities.

Conclusion

Increasing evidence shows that consumption of ASF have important beneficial nutritional impacts. However, only a minority of consumers in developing countries eat such foods, and poorer households especially forego ASF consumption seemingly as prices are perceived to be prohibitively high. Using a large national retail price dataset for Ethiopia that spans the past 10 years, we look at patterns in ASF prices, assess changes over time of ASF and associated livestock prices, and compare these prices with other food products, such as staple cereals. Several typical patterns in ASF and livestock pricing can be highlighted. First, there is price seasonality mostly driven by changes in demand due to religious festivities and fasting periods in the country. Second, there are significant spatial patterns with higher prices in cities and in commercial livestock areas. Third, ASF are relatively expensive. Average prices of ASF per calorie and per kilogram are about ten times as high as for staple cereals, an important consideration in these settings and especially so for the poorer part of the population.

Moreover, we find that real prices of ASF have been increasing rapidly in the past decade, on the order of 33 percent, 36 percent, and 32 percent for beef, milk, and eggs respectively. Such ASF price increases are strongly related with increases in livestock prices. These price trends are in contrast with staple cereal prices that do not show such increases. These are important findings as low prices of food are important in consumption decisions in these environments as shown by consistently high empirical estimates of price elasticities. For example, Tafere et al. (2010) used variation in prices to estimate to what extent consumption patterns change with changing prices, based on the national consumption survey of 2005. They found that most price elasticities were close to –1.0, suggesting that a 10 percent increase in prices would lead to a 10 percent decrease in consumption. Tafere and Worku (2012) followed a similar methodology to estimate price elasticities for ASF products. They found price elasticities for beef and dairy products as high as –0.73 and –0.67, respectively. These estimates imply that the price increases seen in the past decade in the country would have led to decreases in per capita consumption of beef and dairy products of almost 25 percent, holding other things constant, when we compare the end to the beginning of the decade.

The findings have important implications for policy. There is increasing emphasis on behavioral change communication to stimulate the adoption of more diverse diets in these settings. While these investments are important, households also should be able to afford these more nutritious, but also more expensive, foods. Although Ethiopia has done rather well in improving cereal production in the country and keeping prices of cereals stable, these higher priced foods that might be good for nutrition—as well as for farmers' income—have received relatively limited attention by government and development partners. Relative to the rapid growth in crop output and productivity recorded in the past decade, livestock and ASF output grew slower and productivity stagnated such that their share in agriculture production overall consistently declined (Chapter 5, Negassa et al. 2017). This neglect should be addressed to stimulate further agricultural and nutritional transformation in the country.[20]

Exploring the exact reasons for the increases of livestock and ASF prices should be fertile ground for further research. Several hypotheses can be forwarded. First, yield levels of livestock products are low and changed little during the period studied, and this appears to be because of the low adoption of modern inputs and methods, such as artificial insemination, improved breeds, feeds, and extension, as well as high unplanned deaths of livestock (Chapter 5). This is unlike the increase in crop productivity that resulted from the rapid growth in modern input adoption in the subsector (Chapter 3). Consequently, production has not kept up with the increasing demand for ASF from an increasingly better-off population in some pockets of Ethiopia, leading to such price increases. Second, cropland cultivation has expanded rapidly in Ethiopia. The increasing land pressure might have led to less land being available for (cheaper) grazing and subsequently to the adoption of more expensive purchased feeds (Chapter 5). These higher production costs are reflected in higher livestock and ASF prices. Third, wages are rapidly increasing in rural Ethiopia (Chapter 11). This combined with the lower availability

20 Unlike development policies in the past two decades, the current Growth and Transformation Plan II (GTP II) provides more focus on livestock production and includes reduction of undernutrition as one of its targets. Under GTP II (2015/2016–2019/2020), the Ethiopian government targeted the reduction of stunting levels in young children from 40 percent in 2014/2015 to 26 percent in 2019/2020 (National Planning Commission 2016). Furthermore, improving dietary diversity is one of the focus areas of the Agricultural Growth Program II (2015/2016–2019/2020) and is one of the four outcome indicators for the program (Ethiopia, Ministry of Agriculture and Natural Resources 2015).

of youth that often tend the animals of a household—as they are increasingly enrolled in schools—might also have led to higher costs for keeping animals.

Fourth, while improvements in road and information infrastructure in the country contributed to lower food prices and narrowing of urban-rural price gaps (Chapter 7), this appears to have not happened in the livestock subsector. Moreover, the rapid increase in live animal and meat exports following improvements in access to the rewarding export markets might have contributed to general price rises in the livestock sector. International trade and price movements in destination markets should be looked at as they might have driven some ASF price increases in Ethiopia. However, it should be noted that nontraded ASF showed similar increases. Unraveling the contribution of each of these factors through further research will help with better design of policies in this area with the purpose of making ASF more affordable—especially so for the poorer population.

References

Arimond, M., and M. T. Ruel. 2004. "Dietary Diversity Is Associated with Child Nutritional Status: Evidence from 11 Demographic and Health Surveys." *Journal of Nutrition* 134 (10): 2579–2585.

Assefa, T., G. Abebe, I. Lamoot, and B. Minten. 2016. "Urban Food Retailing and Food Prices in Africa: The Case of Addis Ababa, Ethiopia." *Journal of Agribusiness in Developing and Emerging Economies* 6 (2): 90–109.

Ayele, G., M. A. Jabbar, H. Teklewold, E. Mulugeta, and G. Kebede. 2006. "Seasonal and Inter-Market Differences in Prices of Small Ruminants in Ethiopia." *Journal of Food Products Marketing* 12 (4): 59–77.

Ayele, G., and K. M. Rich. 2010. *Poultry Value Chains and HPAI in Ethiopia*. Pro-Poor HPAI Risk Reduction Project Working Paper. Nairobi: International Livestock Research Institute (ILRI)

Ayenew, Y. A., M. Wurzinger, A. Tegegne, and W. Zollitsch. 2009. "Handling, Processing and Marketing of Milk in the North Western Ethiopian Highlands." *Livestock Research for Rural Development* 21 (7). www.lrrd.org/lrrd21/7/ayen21097.htm.

Bachewe, F., Y. Bai, D. Headey, and W. Masters. Forthcoming. "Measuring the Affordability of Nutrient Adequacy and Dietary Energy in Ethiopia, 2001–2017." Unpublished.

Bachewe, F., G. Berhane, B. Minten, and A. S. Taffesse. 2015. *Agricultural Growth in Ethiopia (2004–2014): Evidence and Drivers*. IFPRI-ESSP Working Paper 81. Addis Ababa: International Food Policy Research Institute/Ethiopia Strategy Support Program (IFPRI/ESSP).

Bellemare, M. F., and C. B. Barrett. 2006. "An Ordered Tobit Model of Market Participation: Evidence from Kenya and Ethiopia." *American Journal of Agricultural Economics* 88 (2): 324–337.

Black, R. E., L. H. Allen, Z. A. Bhutta, L. E. Caulfield, M. De Onis, M. Ezzati, C. Mathers, and J. Rivera. 2008. "Maternal and Child Undernutrition Study Group, Maternal and Child Undernutrition: Global and Regional Exposures and Health Consequences." *The Lancet* 371 (9608): 243–260.

Brinkman, H. J., S. de Pee, I. Sanogo, L. Subran, and M. W. Bloem. 2010. "High Food Prices and the Global Financial Crisis Have Reduced Access to Nutritious Food and Worsened Nutritional Status and Health." *Journal of Nutrition* 140 (1): 153S–161S.

De Waal, A. 1988. "Famine Early Warning Systems and the Use of Socio-Economic Data." *Disasters* 12 (1): 81–91.

Dror, D. K., and L. H. Allen. 2011. "The Importance of Milk and Other ASF for Children in Low-Income Countries." *Food and Nutrition Bulletin* 32 (3): 227–243.

———. 2014. "Dairy Product Intake in Children and Adolescents in Developed Countries: Trends, Nutritional Contribution, and a Review of Association with Health Outcomes." *Nutrition Reviews* 72 (2): 68–81.

ERCA (Ethiopian Revenues and Customs Authority). 2017. *Monthly Import-Export Data*. Accessed April 2017. www.erca.gov.et/index.php/import-export-information/.

Ethiopia, CSA (Central Statistical Agency). 2009. *Population Census of Ethiopia 2007*. Addis Ababa.

———. 2011. *Household Income and Consumption Expenditure Survey*. Addis Ababa.

———. 2013. *Population Projection of Ethiopia for All Regions at Wereda Level from 2014–2017*. Addis Ababa.

———. 2015a. *Agricultural Sample Survey 2014/15 [2007 E.C.]. Volume 2. Report on Livestock and Livestock Characteristics (Private Peasant Holdings)*. Addis Ababa.

———. 2015b. *Agricultural Sample Survey 2014/15 [2007 E.C.]. Volume 4. Report on Land Utilization (Private Peasant Holdings,* Meher *Season)*. Addis Ababa.

———. 2015c. *Agricultural Sample Survey 2014/15 [2007 E.C.]. Volume 1. Report on Area and Production of Crops (Private Peasant Holdings,* Meher *Season)*. Addis Ababa.

———. 2017a. *Country and Regional Level Consumer Price Indices*. Addis Ababa.

———. 2017b. *Consumer Price Survey*. Addis Ababa.

———. 2017c. *Producer Price Survey*. Addis Ababa.

Ethiopia, Ministry of Agriculture and Natural Resources. 2015. *Agricultural Growth Program 2 (AGP 2): Program Implementation Manual (PIM)*. Addis Ababa.

Ethiopia, National Planning Commission. 2016. *Growth and Transformation Plan II (GTP II) (2015/16–2019/20): Volume 1: Main Text*. Addis Ababa.

FAO (Food and Agricultural Organization of the United Nations). 2015. FAOSTAT database. Rome. http://faostat.fao.org/.

———. 2017. "Livestock and Nutrition." Mimeo. Rome.

Farmer, E. 2010. "End Market Analysis of Ethiopian Livestock and Meat." MicroREPORT #164. Prepared for USAID. www.value-chains.org/dyn/bds/docs/801/USAID%20 Ethiopia%20Livestock%20End%20Market%20Study.pdf.

FVI-Idele (France Vétérinaire Internationale—Institute de l'Elevage). 2016. *Complementary Feasibility Study for the Relocation and Modernization of Addis Ababa Abattoirs*. Study prepared for Addis Ababa Abattoirs Enterprise (AAAE). Paris.

Givens, D. I. 2010. "Milk and Meat in Our Diet: Good or Bad for Health?" *Animal* 4 (12): 1941–1952.

Givens, D. I., K. M. Livingstone, J. E. Pickering, Á. A. Fekete, A. Dougkas, and P. C. Elwood. 2014. "Milk: White Elixir or White Poison? An Examination of the Associations between Dairy Consumption and Disease in Human Subjects." *Animal Frontiers* 4 (2): 8–15.

Gordon, A., S. D. Tegegne, and M. Tadesse. 2007. *Marketing Systems for Fish from Lake Tana, Ethiopia: Opportunities for Improved Marketing and Livelihoods*. IPMS (Improving Productivity and Market Success) of Ethiopian Farmers Project Working Paper 2. Nairobi: ILRI.

Hassan, S. 2011. "Futility and Damaging Effects of Ethiopian Price Caps (ANALYSIS)." Accessed October 13, 2016. http://hornofafrica-abdikarim.blogspot.com/2011/01/futility and -damagingeffects-of.html.

Headey, D., K. Hirvonen, and J. Hoddinott. 2018. "Animal Sourced Foods and Child Stunting." *American Journal of Agricultural Economics* 100 (5): 1302–1319.

Hirvonen, K. 2016. "Rural-urban Differences in Children's Dietary Diversity in Ethiopia: A Poisson Decomposition Analysis." *Economics Letters* 147: 12–15.

Hirvonen, K., A. S. Taffesse, and I. W. Hassen. 2016. "Seasonality and Household Diets in Ethiopia." *Public Health Nutrition* 19 (10): 1723–1730.

Hoddinott, J., H. Alderman, J. R. Behrman, L. Haddad, and S. Horton. 2013. "The Economic Rationale for Investing in Stunting Reduction." *Maternal and Child Nutrition* 9 (S2): 69–82.

Hoddinott, J., D. Headey, and M. Dereje. 2015. "Cows, Missing Milk Markets, and Nutrition in Rural Ethiopia." *Journal of Development Studies* 51 (8): 958–975.

Iannotti, L. L., C. K. Lutter, D. A. Bunn, and C. P. Stewart. 2014. "Eggs: The Uncracked Potential for Improving Maternal and Young Child Nutrition among the World's Poor." *Nutrition Reviews* 72 (6): 355–368.

Iannotti, L. L., M. Robles, H. Pachón, and C. Chiarella. 2012. "Food Prices and Poverty Negatively Affect Micronutrient Intakes in Guatemala." *Journal of Nutrition* 142 (8): 1568–1576.

Jabbar, M., S. Benin, E. Gabre-Madhin, and Z. Paulos. 2008. "Market Institutions and Transaction Costs Influencing Trader Performance in Live Animal Marketing in Rural Ethiopian Markets." *Journal of African Economies* 17 (5): 747–764.

Jin, M., and L. L. Iannotti. 2014. "Livestock Production, Animal Source Food Intake, and Young Child Growth: The Role of Gender for Ensuring Nutrition Impacts." *Social Science and Medicine* 105: 16–21.

Kim, S. S., R. Rawat, E. M. Mwangi, R. Tesfaye, Y. Abebe, J. Baker, E. A. Frongillo, M. T. Ruel, and P. Menon. 2016. "Exposure to Large-Scale Social and Behavior Change Communication Interventions Is Associated with Improvements in Infant and Young Child Feeding Practices in Ethiopia." *PLoS ONE* 11 (10): e0164800.

Leroy, J. L., and E. A. Frongillo. 2007. "Can Interventions to Promote Animal Production Ameliorate Undernutrition?" *Journal of Nutrition* 137 (10): 2311–2316.

Little, P. D., D. N. Debsu, and W. Tiki. 2014. "How Pastoralists Perceive and Respond to Market Opportunities: The Case of the Horn of Africa." *Food Policy* 49: 389–397.

Mekasha, A., B. Gerard, K. Tesfaye, L. Nigatu, and A. J. Duncan. 2014. "Inter-connection between Land Use/Land Cover Change and Herders'/Farmers' Livestock Feed Resource Management Strategies: A Case Study from Three Ethiopian Eco-environments." *Agriculture, Ecosystems and Environment* 188: 150–162.

Minten, B., Y. Habte, S. Tamru, and A. Tesfaye. 2018. *Transforming Agri-Food Systems in Ethiopia: Evidence from the Dairy Sector.* ESSP Working Paper 129. Addis Ababa: International Food Policy Research Institute (IFPRI).

Minten, B., D. Stifel, and S. Tamru. 2014. "Structural Transformation of Cereal Markets in Ethiopia." *Journal of Development Studies* 50 (5): 611–629.

Msangi, S. 2016. "Results IMPACT Model IFPRI." Presentation FAO. Unpublished, FAO, Rome.

National Bank of Ethiopia. 2016. *Annual Report 2015/16.* Addis Ababa.

Negassa, N., F. Bachewe, A. S. Taffesse, and M. Dereje. 2017. *An Assessment of the Livestock Economy in Mixed Crop-Livestock Production Systems in Ethiopia.* IFPRI-ESSP Working Paper 101. Addis Ababa: IFPRI/Ethiopia Strategy Support Program (ESSP).

Negassa, A., S. Rashid, B. Gebremedhin, and A. Kennedy. 2012. "Livestock Production and Marketing." In *Food and Agriculture in Ethiopia: Progress and Policy Challenges*, edited by P. Dorosh and S. Rashid, 159–189. Philadelphia: University of Pennsylvania Press on behalf of IFPRI.

Nyoni, N. M. B., S. Grab, and E. R. Archer. 2019. "Heat Stress and Chickens: Climate Risk Effects on Rural Poultry Farming in Low-Income Countries." *Climate and Development* 11 (1): 83–90.

OECD (Organisation for Economic Co-operation and Development)–FAO. 2016. *OECD-FAO Agricultural Outlook 2016–2025*. Paris: OECD; Rome: FAO.

Randolph, T. F., E. Schelling, D. Grace, C. F. Nicholson, J. L. Leroy, D. C. Cole, M. W. Demment, A. Omore, J. Zinsstag, and M. Ruel. 2007. "Role of Livestock in Human Nutrition and Health for Poverty Reduction in Developing Countries." *Journal of Animal Science* 85 (11): 2788–2800.

Rashid, S., and N. Minot. 2010. "Are Staple Food Markets in Africa Efficient? Spatial Price Analyses and Beyond." Paper presented at the COMESA policy seminar "Food Price Variability: Causes, Consequences, and Policy Options," Maputo, January 25–26.

Rashid, S., and A. Negassa. 2012. "Policies and Performance of Ethiopian Cereal Markets." In *Food and Agriculture in Ethiopia: Progress and Policy Challenges,* edited by P. Dorosh and S. Rashid, 53–83. Philadelphia: University of Pennsylvania Press.

Sadler, K., and A. Catley. 2009. *Milk Matters: The Role and Value of Milk in the Diets of Somali Pastoralist Children in Liben and Shinile, Ethiopia*. Addis Ababa: Feinstein International Center, Tufts University and Save the Children.

Sahn, D. E. 1989. *Seasonal Variability in Third World Agriculture: The Consequences for Food Security*. London: IFPRI; Baltimore: Johns Hopkins University Press.

SID-Consult. 2010. "Market Assessment and Value Chain Analysis in Benishangul-Gumuz Regional State, Ethiopia." Mimeo. Accessed December 5, 2017. www.cangoethiopia.org/assets/docs/BG%20Market%20Assessment%20Final%20Report%20300510.pdf.

Tadesse, K., B. Lanos, and A. MasAsparisi. 2014. "Analysis of Price Incentives for Live Cattle in Ethiopia." Technical notes series, MAFAP (Monitoring and Analysing Food and Agricultural Policies), FAO, Rome.

Tafere, K., A. S. Taffesse, S. Tamiru, N. Tefera, and Z. Paulos. 2010. *Food Demand Elasticities in Ethiopia: Estimates Using Household Income Consumption Expenditure (HICE) Survey Data.* ESSP II Working Paper 11. Addis Ababa: IFPRI.

Tafere, K., and I. Worku. 2012. *Consumption Patterns of Livestock Products in Ethiopia: Elasticity Estimates Using HICES (2004/05) Data*. ESSP II. Addis Ababa: IFPRI and Ethiopian Development Research Institute.

Tefera, T. L., R. Puskur, D. Hoekstra, and A. Tegegne. 2010. *Commercializing Dairy and Forage Systems in Ethiopia: An Innovation Systems Perspective.* ILRI Working Paper 17. Nairobi: ILRI.

Tegegne, A., B. Gebremedhin, D. Hoekstra, B. Belay, and Y. Mekasha. 2013. *Smallholder Dairy Production and Marketing Systems in Ethiopia: IPMS Experiences and Opportunities for Market-Oriented Development.* IPMS (Improving Productivity and Market Success) of Ethiopian Farmers Project Working Paper 31. Nairobi: ILRI.

Teklewold, H., G. Legese, D. Alemu, and A. Negasa. 2009. *Determinants of Livestock Prices in Ethiopian Pastoral Livestock Markets: Implications for Pastoral Marketing Strategies.* Contributed paper prepared for presentation at the International Association of Agricultural Economists Conference, Beijing, China.

Tilahun, H., and E. Schmidt. 2012. *Spatial Analysis of Livestock Production Patterns in Ethiopia.* ESSP II. Addis Ababa: IFPRI.

UN Comtrade (United Nations–Comtrade). 2017. UN Comtrade Database. Accessed April 28, 2017. http://comtrade.un.org/.

USAID. 2013. "Agricultural Growth Program—Livestock Market Development. End Market Analysis for Meat/Live Animals, Leather and Leather Products, Dairy Products Value Chains." Expanding Livestock Markets for the Small-Holder Producers. Prepared by the AGP-Livestock Market Development Project of CFNA (Cultivating New Frontiers in Agriculture), Washington, DC.

Warren, A. M., and E. A. Frongillo. 2017. "Mid-level Actors and Their Operating Environments for Implementing Nutrition-Sensitive Programming in Ethiopia." *Global Food Security* 13: 66–73.

Weaver, C. M. 2014. "How Sound Is the Science Behind the Dietary Recommendations for Dairy?" *American Journal of Clinical Nutrition* 99 (5): 1217S–1222S.

Worku, I., M. Dereje, B. Minten, and K. Hirvonen. 2016. *Diet Transformation in Africa: Evidence from Ethiopia.* IFPRI-ESSP Working Paper 87. Addis Ababa: IFPRI/ESSP.

World Bank. 2014. *Ethiopia: Poverty Assessment.* Report No. AUS6744. Poverty Global Practice. Africa Region. World Bank. Washington, DC.

Chapter 9

DROUGHTS, CEREAL PRICES, AND PRICE STABILIZATION OPTIONS

Paul Dorosh, Jennifer Smart, Bart Minten, and David Stifel

Introduction

Increases in cereal prices can have adverse effects on poor net food buyers. Conversely, falling domestic prices of some cereals (especially maize), typically at harvest time, can be detrimental to producers who are net sellers. Price volatility is especially problematic for Ethiopia because of frequent weather shocks. The country is often hit by droughts that result in adverse effects on all farmers due to major crop failures, subsequent price instability, extremes in income streams, and severe hardship. The average rural household in Ethiopia is a net buyer when it comes to the seven most important commodities in the country (coffee, maize, beans, barley, wheat, teff, and sorghum) and hence is adversely affected by price spikes. However, price volatility more negatively affects the welfare of wealthier rural households, who are more likely to be producers, than poorer rural households (Bellemare, Barrett, and Just 2013). For surplus-producing farmers, net revenues are vulnerable to both production shocks and price volatility. Price stabilization efforts are therefore an important consideration for Ethiopian policymakers.[1] This chapter evaluates options for cereal price stabilization in Ethiopia drawing on the experiences of other developing countries.

Twice in the past fifty years, crop failures contributed to widespread famines in Ethiopia (1972–1974 and 1984–1985).[2] Because of their devastating effects, these famines received significant attention in the literature. In their analysis of the 1984–1985 famine, Webb and von Braun (1994) identified four major causal factors. The most important factor was drought-induced crop

1 Bellemare, Barrett, and Just (2013) find that the average household's willingness to pay to fully stabilize commodity prices at their means is about 18 percent of its income and that this may very well explain the government's frequent interest in price stabilization since, on average, households stand to benefit from it.

2 See Appendix Table 9A.1 for a chronology of major food shortages in Ethiopia over the past 50 years.

failure.[3] The other three factors reflect longer-term causes. In addition to misguided land policies and conflict, markets in Ethiopia were not well integrated.[4] The latter followed from market restrictions (particularly regulation and bans on interregional movement of grain and labor), requirements for licensed private traders to make half of their purchases available to the Agricultural Marketing Corporation at fixed prices, and poor market infrastructure.

This chapter assesses how Ethiopia tried to assure food security and to achieve food price stability during the major El Niño–induced drought of 2016. A series of droughts and crop losses in 2015 and 2016 threatened the food security of an estimated 10.2 million people who were declared in need of emergency food assistance in addition to the 7.9 million already covered by the Productive Safety Net Program (PSNP) (WFP 2016a-i). Fortunately, the experiences of 2016 were very different from the earlier famines for several reasons. First, major reforms in government policies and massive public investments in agricultural extension and road infrastructure have contributed to substantial increases in cereal production (Bachewe et al. 2018; Rashid, Dorosh, and Alemu 2018). Second, the functioning of agricultural markets has significantly improved with most markets now being much more integrated (Minten et al. 2014; Hill and Fuje 2017). Third, the introduction of the well-targeted PSNP in the mid-2000s provided an effective means of providing access to food for many of Ethiopia's poorest households (Berhane et al. 2014).

The structure of the chapter is as follows. First we review the international experiences with cereal price stabilization policies and the institutional context in Ethiopia; next is an assessment of cereal price variability in the past decade in Ethiopia and a comparison of that price variability with the experiences of other East African countries. We then analyze production shortfalls

3 Per capita production of cereals fell from 154.1 kilograms per capita in 1982/1983 to 122.6 in 1984/1985 and then 91.2 and 100.4 kilograms per capita in 1985/1986 and 1986/1987 (FAO data, including Eritrea: note that FAO reports 1983/1984 production as 1983 [Seyoum Taffesse, Dorosh, and Asrat 2013]). Even taking account of increased imports (including food aid) of approximately 800,000 metric tons per year in1985/1986 and 1986/1987, per capita cereal availability overall fell from 151.5 kilograms per person in 1982/1983 to 100.1 kilograms per person in 1985/1986 and 110.2 kilograms per capita in 1986/1987.

4 Government land reform policies, including abolishing private land ownership in 1975 and ceilings on private land access rights to 10 hectares per farm, were welcomed in much of central and southern Ethiopia, but subsequently government investment in agriculture was allocated mainly to state farms and producer cooperatives, instead of independent small farmers. Ongoing civil war between the Mengistu regime and both the Eritrean People's Liberation Front and the Tigray Peoples' Liberation Front resulted in loss of life and serious injuries, a reduction in labor availability for crop cultivation, and diversion of scarce public resources away from needed investments in agriculture, roads, and telecommunications.

and policy responses in the case of the major El Niño–induced drought of 2016. Price stabilization options for the Ethiopian government are presented before our final conclusions.

International Experience with Cereal Price Stabilization

In light of the sharp rise in international cereal prices in 2007 and 2008 and a rethinking on the part of many developing countries on their policies regarding food prices, public sector food stocks, and reliance on international trade, it is helpful to review the experiences of countries that have implemented price stabilization policies. There are good reasons why governments intervene in markets to prevent excess price spikes or price declines. These include protecting the welfare of the urban poor and of rural net buyers by dampening price increases, enhancing incentives for domestic production and farmer incomes by preventing large price falls, and promoting political and social stability.

While the overall record on price stabilization is mixed, there are some success stories. Examples from South and Southeast Asia include India, Pakistan, and Indonesia, where price stability objectives have been achieved. However, price stabilization has often come at considerable cost to the public purse, along with distributional and leakage costs. Pressure from domestic farmer interest groups has led to sustained high levels of procurement (and subsequent distribution). For example, India's government procures 20 percent to 25 percent of total domestic wheat and rice production annually.

In East and Southern Africa, several countries, including Kenya, Madagascar, and Zambia, have enjoyed moderate success in raising price levels for farmers and reducing price volatility through their pricing, marketing, and trade policies. For example, the Kenyan National Cereals and Produce Board's purchases and sales of maize at administratively determined prices alongside competition from the private sector contributed to average prices for maize being 20 percent higher than they would have otherwise been between 1989 and 2004 (Jayne, Myers, and Nyoro 2008). Similarly, interventions by the Food Reserve Agency (FRA) in Zambia contributed to average maize prices being roughly 18 percent higher than they otherwise would have been between 2003 and 2008, while price volatility was more than 30 percent lower (Chapoto and Jayne 2009). In both of these cases, however, there is also evidence that these countries could have done better if their policies had been characterized by transparency and predictability.

The importance of transparency, predictability, and ultimately more certainty is directly related to the price-stabilizing benefits of private-sector international trade. When domestic prices rise above international prices (or the effective cost of imports as measured by the import parity price), there is an incentive for private traders to import. This results in a suppression of exceedingly high price spikes. For example, following severe flooding and a rice production shortfall in Bangladesh in 1998, private-sector rice imports, made possible by earlier trade liberalization and investments in infrastructure and market development, effectively stabilized market supplies, as imported rice filled the gap, and prices, since international prices provided an upper price bound (Dorosh 2001, 2008). In relying on the international market in this manner, price stabilization can be achieved in a low-cost way without relying on government expenditures.

For international private-sector trade flows to play a price-stabilizing role, however, governments need to minimize uncertainty related to their policies and interventions. For example, uncertainty about government intentions to import or about when or whether governments will alter import duties can result in traders worrying that they will incur financial losses if they do import and the government subsequently changes its policy. The consequence of this can be a temporary underprovision of imports when there is a domestic shortfall, with resulting food shortages and price spikes above the cost of imports. On the face of it, the lack of the private-sector imports may appear to reflect market failures and a weak private sector. However, a good case can be made that the behavior of private-sector traders is a rational response to uncertainty.

This type of private-sector response to uncertain policies has been documented in such countries as Kenya, Madagascar, Malawi, and Zambia. When Kenya was hit by maize shortfalls in 2008, private traders were reluctant to import maize because the 50 percent import duty made private importation uneconomical. The government's history of making sudden changes to the tariff (including zero-ratings), however, meant that grain traders also expected that a change was imminent. This indeed occurred in 2009 when the government eliminated the duty. The consequence of delays in government importation and in government's decision to maintain the 50 percent tariff on imports throughout 2008 was that maize prices stayed at very high levels in late 2008 despite the tumbling of world prices starting in October 2008 (Jayne and Tschirley 2009).

In Malawi, in response to what is now believed to be overstated official estimates of maize production in 2008 and 2009, the government purchased maize, encouraged exports, and restricted imports despite the private sector's difficulties in sourcing maize for government tenders and despite soaring prices. Suggesting that private maize traders had orchestrated the price rise, the government initially banned private maize trade and then required traders to operate within an official price band. But since market prices were well above the ceiling, traders stopped buying maize. In the end the government eventually arranged with one large trader to supply maize to the government at prices well above the official ceiling price (Jayne and Tschirley 2009).

In Madagascar, when domestic production and imports were low in 2004, the government offered tenders for imports below the tariff-inclusive import parity price for rice while leaving open the possibility that the 45 percent tariff might be waived. In this case government policy—in particular, its lack of transparency, the harassment of importers (including detaining ships with imported rice), uncertainty about tariff levels and enforcement, and an official selling price that made private-sector imports paying full tariffs unprofitable—ultimately made the situation worse than would have been the case if there had been no intervention. Total rice imports in 2004 were 100,000 metric tons below the 2003 level despite domestic market prices rising above the import parity level (Minten and Dorosh 2006; Coady, Dorosh, and Minten 2009).

Finally, Zambia has experienced unintended price-instability effects of lags between policy announcements and implementation. For example, three times in the period studied (2002, 2003, and 2008/2009) announcements of intentions to import maize resulted in market prices skyrocketing as private traders stayed out of the market for maize because they realized they could not compete against subsidized imports. Due to lags in implementing the imports, government stock levels dwindled, which contributed to panic, rationing, and rapidly rising maize prices, the opposite of the policy's intended effect. Simulations based on econometric estimates show that prices in Zambia have been more volatile with government marketing and trade policies than without them, but that this result was not due to the government's purchases/sales. Rather, it was due to uncertainty over government policy and timing of implementation (Chapoto and Jayne 2009).

The message coming out of international experiences in price-stabilization policies is that rules-based approaches to marketing and trade policies may reduce levels of uncertainty and the price instability associated with them. But, ultimately, the approach that countries take in addressing price stabilization depends on the visions that policymakers have with regard to food markets. Jayne and Tschirley (2009) classify these visions into three models of

TABLE 9.1 Competing visions of staple food market development

Model 1	Model 2	Model 3
Rely on markets with limited role of state	Primary reliance on markets with rules-based role of state	Role for markets with discretionary state intervention
Role of state limited to • Public goods investment • Regulatory framework • Strengthening of institutions and property rights • Policies supportive of private sector entry and competition	Role for rules-based state operations • Buffer stock release to defend stated ceiling price • Marketing board purchases at stated floor price announced in advance • Transparent rules for initiating state imports • Public goods investments	• Based on premise that the private sector cannot ensure adequate food supplies in response to production shortfalls • Justification for unconstrained role for state intervention in markets to correct for market failures

Source: Jayne and Tschirley (2009).

how markets work and how the private and public sectors interact. These are described in Table 9.1.

In Model 1 the role of the state is confined to providing public goods to strengthen markets. This vision relies on the private sector to carry out the main direct marketing functions and is close to the "Washington Consensus" around the desirability of an expansion of market forces within domestic economies, which is now generally out of favor. In Model 2 the role of the state is expanded to include direct marketing operations based on the premise that markets do fail in certain circumstances and that direct rules-based nondiscretionary state operations are necessary to maintain good prices within reasonable bounds. The price bounds are typically defined by long-run import and export parity prices. Finally, Model 3 is a vision in which discretionary state intervention provides the state with the flexibility needed to achieve state policy objectives. The staple food market policies of most governments in East and Southern Africa are characterized by this vision in that they adopt highly unpredictable and discretionary trade policies.

While Model 3 is the most common vision underlying the approaches of governments in food markets, this discretionary approach to food price policy creates risks for the private sector and hampers the private sector from performing the functions envisioned of it in Models 1 and 2. Transitioning from Model 3 to Model 2 is likely to promote market predictability and lead to greater supplies and price stability in food markets during times of domestic production shortfalls. The challenge in moving from policies motivated by Model 3 to those consistent with Model 2, however, is convincing people of the credibility of governments' commitments to a rules-based approach given long histories of discretionary state intervention in food markets.

Approaches to Price Stabilization and Safety Nets in Ethiopia

Although the role of the state has been substantially less than in the 1980s and early 1990s (Rashid and Negassa 2012; Rashid et al. 2018), Ethiopia's price-stabilization and safety net policies in the past two decades have involved substantial discretionary state interventions (Model 3). For example, the government's price-stabilization and safety net programs typically involve importing several hundred thousand tons of grain per year. In addition, private-sector international trade, including wheat imports (Dorosh and Ahmed 2011), has been severely hampered by foreign exchange and trade restrictions (for instance, import licensing requirements).[5] The fact that this international grain trade takes place almost exclusively through the major seaport in neighboring Djibouti means that high transport and marketing costs make cross-border trade even less attractive to private-sector traders.

Three main institutions govern and implement Ethiopia's price stabilization and safety nets: the Ethiopian Grain Trade Enterprise (EGTE), the Emergency Food Security Reserve Agency (EFSRA), and the National Disaster Risk Management Commission (NDRMC). EGTE is tasked with stabilizing grain prices by buying and selling grain in the market. EFSRA maintains relatively small food reserves (averaging about 145.1 thousand tons of cereals—139.4 thousand tons of wheat and 5.6 thousand tons of maize—from calendar years 2009 to 2015) that are available for emergency loans of working cereal stocks to nonprofit and public relief and rehabilitation agencies, prior to the arrival of imported food aid. The NDRMC, which reports directly to the prime minister, coordinates and implements all disaster risk management and humanitarian assistance at both the international and national levels (Rashid et al. 2018).

The EGTE's grain distributions are used by the government of Ethiopia to influence the supply of wheat and flour in urban areas. Millers are entitled to buy wheat from the EGTE at prices about 30 percent lower than the price the EGTE paid for it. Flour made from subsidized wheat is then subject to a price cap. EGTE sales are concentrated in urban areas and limited to licensed enterprises. For example, in 2012 only 59 bakeries and flour mills in Addis Ababa

5 In recent years exporters have been able to retain some of their foreign exchange earnings and sell them to importers at parallel market rates. This trade has thus reduced the adverse effects of an appreciation of the real exchange rate on the real price of coffee and incentives for coffee producers (Tamru, Minten, and Swinnen 2019).

were permitted to purchase EGTE subsidized wheat in 2012 (USDA FAS 2012).[6]

The Ethiopian Commodity Exchange (ECX), which is jointly owned by the private sector and the Ethiopian government, is another institution that was established (in 2008) to increase grain trade and to help stabilize markets. However, in part because of foreign exchange restrictions that limited the ability of private traders to import grain or compete with the EGTE subsidized wheat distribution system, the ECX was unable to attract substantial trade in cereals. Thus the ECX quickly changed its focus to coffee, for which all exports were required to pass through the exchange.[7]

An Assessment of Cereal Price Variability in Ethiopia

The above institutions have not been particularly successful in stabilizing Ethiopia's cereal prices, at least in comparison with cereal price variability in two other East African countries in recent years. In particular, we compare the variability of maize prices in Ethiopia with those in Uganda and Kenya as well as in the broader international (that is, world) market over the 2009–2018 period.

World wheat prices (measured in US dollars) from 2009 to 2018 were on average 28.5 percent higher than world maize prices and were also considerably more stable, with a coefficient of variation (CV) of 0.196 as compared to 0.287 for maize. Comparing across countries, Ethiopia's maize prices were more stable than both Kenya's and Uganda's in real terms, although in nominal terms they were less stable than Kenya's but more stable than Uganda's. Overall, real prices of maize and wheat in Ethiopia declined over time (growth rates of −0.31 and −0.21 percent per month), as did the real price of maize in Kenya (with a growth rate of −0.25 percent per month). Real prices of maize in Uganda were, on average, almost unchanged (a growth rate of −0.03 percent per month). (For a more detailed comparison of the variability of maize prices in Ethiopia with those in Uganda and Kenya, as well as in the broader international market from 2009 to 2018, see Appendix 9B).

6 EGTE also maintains relatively small stocks of wheat and maize, which averaged 139.4 thousand tons and 5.6 thousand tons, respectively, from 2009 to 2015 (authors' calculations from EGTE data).

7 The ECX later included sesame, another major export crop. The history of the ECX and its effects on market prices of coffee are described in Rashid et al. (2018); USDA FAS (2012); Minten et al. (2014); and Minten et al. (2019).

FIGURE 9.1 Cereal prices in Ethiopia, Kenya, and Uganda, 2006–2018 (US$ per metric ton)

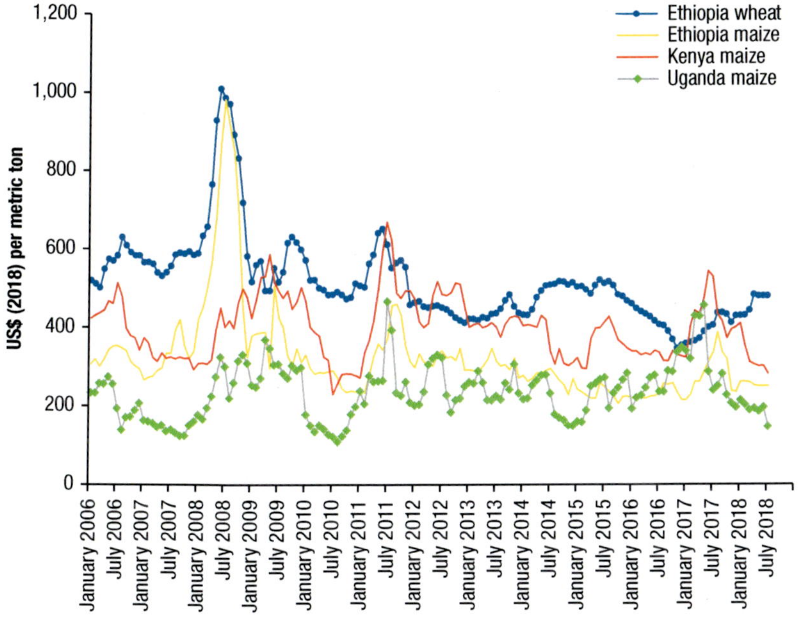

Source: Authors' calculations from Dorosh, Minten, and Stifel (2015); EGTE (2018); IMF (2018); and Ethiopia, CSA (2018) data.

Much of the variation in Ethiopian prices occurred in 2007 and 2008, a period of substantial domestic inflation. As shown in Figure 9.1, US dollar prices of maize and wheat in Ethiopia spiked in mid-2008 but have been relatively much more stable since then. A faster decline in the dollar prices of wheat relative to those of maize in Ethiopia, however, has lowered the gap between these prices over time. Unlike in Ethiopia, there has been no spike in US dollar prices of maize in Kenya and Uganda during this period, although, as indicated by the numerical measures of variance described earlier, there has been substantial variation in these prices as well.

Figure 9.2 shows the evolution of the nominal wholesale and import parity prices of wheat in Addis Ababa over the 2000–2018 period.[8] In well-functioning integrated markets, domestic prices would not exceed the import parity prices of a product for an extended period of time if there are no

8 The import parity price is calculated as the import price of wheat at the port of Djibouti plus transport and other marketing costs to the Addis Ababa market.

FIGURE 9.2 Ethiopia wheat prices, 2000–2018 (birr per kilogram)

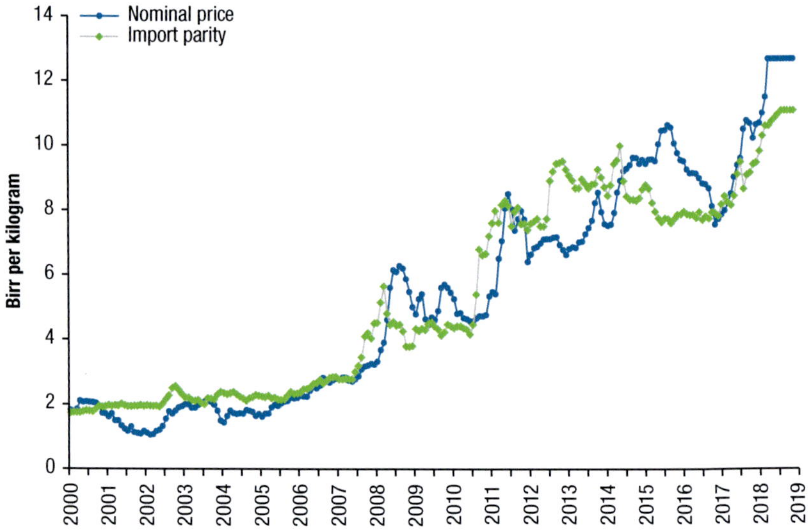

Source: Authors' calculations from Ethiopia, EPAU (2018); IMF (2018); and Ethiopia, CSA (2018) data.
Notes: Import parity in Figure 9.2 is based on international wheat prices (US Gulf, Hard Red Winter) and international shipping costs.

restrictions on trade. Nominal prices of wheat were, in fact, equal to or lower than import parity from mid-2010 to mid-2014. However, from early 2008 to mid-2010, domestic wholesale prices were significantly higher than import parity prices because of a relatively poor harvest along with foreign exchange and wheat import restrictions that limited imports (Dorosh and Ahmed 2011). Moreover, domestic wheat prices were substantially higher than import parity prices from mid-2014 through mid-2016. As discussed below, if private-sector import flows had been unhindered in this period, domestic prices could have fallen to import parity levels, thereby reducing the adverse effects of the drought that happened during that period on net wheat consumers.

Figure 9.3 presents a similar graph for maize. As in the case of wheat, domestic wholesale prices of maize were significantly above the import parity levels of maize coming from international markets through Djibouti to Addis Ababa in 2008 and 2009 and from mid-2017 to mid-2018. Nonetheless, there were no imports of maize in these periods, suggesting that foreign exchange and trade restrictions (such as import tariffs and import licensing) may have

FIGURE 9.3 **Ethiopia maize prices, 2000–2018 (birr per kilogram)**

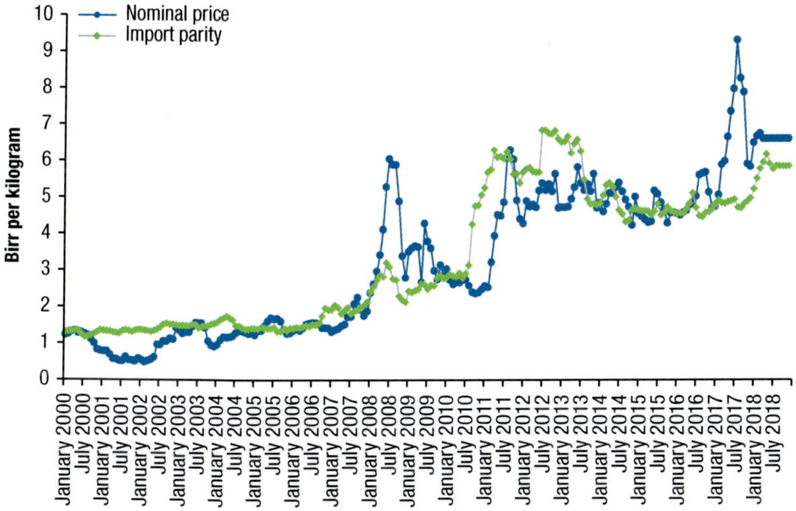

Source: Authors' calculations from Ethiopia, EPAU (2018); IMF (2018); and Ethiopia, CSA (2018) data.

prevented these import inflows.[9] Note also that in early 2017 there were opportunities for exports of maize to northern Kenya, where prices were extremely high due to poor harvests.[10]

In summary, several key aspects of cereal price movements in Ethiopia should be noted. First, since the 2008 price spikes, wheat and maize prices have been relatively stable in Ethiopia, and overall price variability of maize is similar to that in Kenya and considerably less than in Uganda. Second, for wheat (for which there have been substantial, mostly public, imports), international dollar prices have been considerably more stable than domestic prices in Ethiopia, implying that macroeconomic instability (movements in nominal exchange rates and periods of overall price inflation) played a large role in domestic wheat price fluctuations. Third, similar price movements of wheat and maize in Ethiopia suggest that the markets for these two cereals are closely linked (Rashid and Lemma 2015) and that, as discussed next, interventions in wheat markets may help to stabilize maize prices and therefore overall cereal

9 High transaction costs because of a lack of established trade links may also have hindered flows.

10 The export parity price in western Ethiopia, measured on the basis of prices in Kenya, was somewhat higher than the import parity price of maize measured on the basis of international market prices.

consumption. Fourth, there have been extended periods during which maize and wheat prices in Ethiopia have been above import parity prices, indicating restrictions on foreign exchange and trade.

Droughts, Production Shortfalls, and Policy Response: The Case of the 2015/2016 Drought

We present a case study of the 2015/2016 drought, its effect on cereal production and prices, and what government policy measures, regarding both imports and safety nets were put into place. We then discuss the price effects of alternative policy responses to the drought, before concluding with a discussion of effective price-stabilization options for Ethiopia.[11]

Drought Chronology and Cereal Production

Cereal production in Ethiopia is concentrated in the *meher* season with the harvest between November and January. *Meher* crops accounted on average for 96 percent of Ethiopia's annual total production (24.4 million metric tons) from 2013/2014 to 2015/2016 (Figure 9.4). Cereal production in the *belg* season, harvested between June and August, averaged only 1.04 million metric tons in this period but was highly variable, ranging from only 0.74 million metric tons in 2014/2015 to 1.48 million metric tons in 2015/2016. The poor *belg* harvest in 2013/2014 was due to below-average March to May short-season rains leading to water shortages and affecting food security, especially in the eastern part of the country. According to the July 2014 report of the World Food Programme (WFP), in these areas "pastures have not adequately regenerated, water sources are poorly replenished, and crops planted during the season have suffered huge losses from water stresses" (WFP 2014a).

The *belg* season rains were also late and erratic in 2015, marking the start of the most severe drought Ethiopia had experienced in 50 years (AKLDP 2016). Table 9.2 and Figure 9.5 show rainfall totals in Dessie (a cereal deficit area in eastern Amhara) and Nekemte (a cereal surplus area in western Oromia) from January 2013 to June 2017. For both locations a significantly lower level of rainfall was seen in 2015. Crops were heavily damaged in Afar, Amhara, eastern and central Oromia, and pockets of Southern Nations, Nationalities, and Peoples (SNNP) regions. The drought worsened in the *meher* growing season (July/September), and harsh conditions spread to

11 For further details on cereal production and prices in the 2013/2014–2015/2016 period, see Dorosh et al. (2018).

FIGURE 9.4 Ethiopia cereal harvests by season, 2013/2014–2015/2016 (million metric tons)

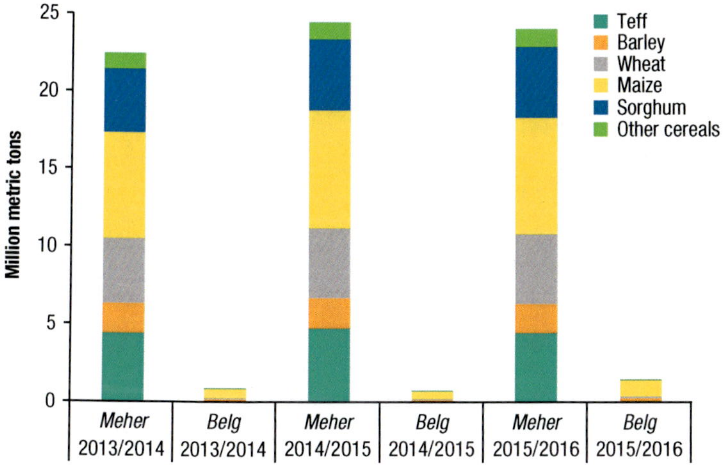

Source: Authors' calculations from Ethiopia, CSA data (2016).

Note: The *meher* season includes commercial farms annual production.

TABLE 9.2 Annual rainfall in Dessie and Nekemte, January 2013–June 2017 (millimeters)

Year	Dessie	Nekemte	Dessie, % change to year earlier	Nekemte, % change to year earlier
2013	1,021	1,749	n.a.	n.a.
2014	944	2,008	−7.5	14.9
2015	444	1,276	−56.5	−27.0
2016	905	2,003	−11.3	14.6

Source: National Oceanographic and Atmospheric Administration (NOAA) data (Arguez et al. 2010).

Note: n.a. = not applicable.

southern and central Tigray as well as to the northern Somali region. In August the government of Ethiopia made its first calls for emergency assistance, and the WFP increased the number of people it was assisting from two million to six million by November 2016 (Table 9.3).

Ultimately, the total smallholder *meher* cereal harvest was only 23.1 million metric tons in 2015/2016, 2.0 percent less than the 23.7 million metric tons harvested in 2014/2015. Production effects were felt far less in western Ethiopia than in the drought-prone highlands, where maize

FIGURE 9.5 Monthly rainfall in Dessie and Nekemte, January 2013–June 2017 (millimeters)

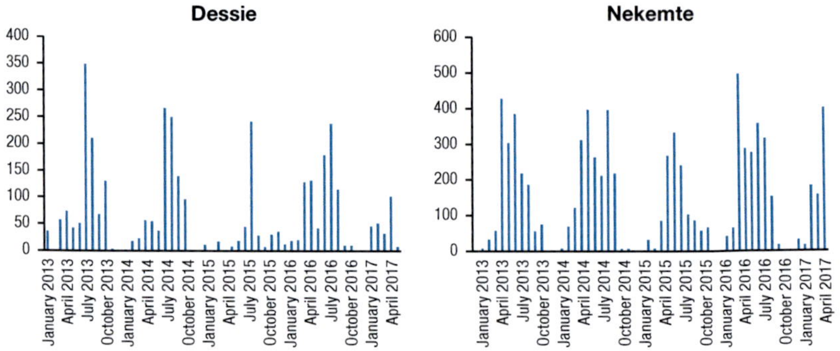

Source: National Oceanographic and Atmospheric Administration (NOAA) data (Arguez et al. 2010)

TABLE 9.3 Drought chronology for Ethiopia, 2014–2017

Period	Summary	Size of population declared as in need of emergency relief (millions)
March–October 2014 (*belg* 2013/2014)	Below-average *belg* rains lead to water stresses and affect food security primarily in the east.	3.2
March–May 2015 (*belg* 2014/2015)	*Belg* rains are late and erratic, causing failed harvest.	2.9
August 2015	The government of Ethiopia makes its first calls for emergency assistance.	4.5
November 2015 (*meher* 2015/2016)	The main *meher* harvest is reduced by 50 percent to 90 percent in the drought-affected highlands. Farmers' incomes decrease up to 60 percent compared to the year prior.	8.2
March–May 2016 (*belg* 2015/2016)	As in 2015, late and unevenly distributed *belg* rains cause flooding in the Somali region, and displacement, the disruption of public services, and the spread of disease continue.	10.2
June–September 2016	Drought and food insecurity lessens in the northeast after improved *belg* rains but worsens in the southeast pastoralist areas of Somali and SNNP regions.	9.7
February 2017	Distress migration, drought-induced school drop-outs and border tensions increase in primarily Somali pastoralist areas.	5.6
May 2017	The new 2016–2017 drought continues to get worse as the year progresses.	7.8 (projected)

Source: WFP reports (WFP 2014a, 2014b, 2015a, 2015b, 2016a, 2016b, 2016c, 2016d, 2016e, 2016f, 2016g, 2016h, 2016i, 2017a, 2017b, 2017c).

Note: SNNP = Southern Nations, Nationalities, and Peoples.

production was worst affected. *Meher* maize production in 2015/2016 was down 5.8 percent in the drought-prone highlands from the year prior, compared to only 1.2 percent for Ethiopia as a whole. *Meher* season teff and barley production shortfalls in the highlands were also larger: −13.1 percent and −10.1 percent in the highlands, respectively, compared to −5.9 percent and −4.9 percent for the country as a whole.

Six regions of the country were reported to be severely affected by the drought, and 10.2 million people were reported in need of food assistance by April 2016 (on top of the people that were already covered by the Productive Safety Net Program [PSNP]). The drought significantly weakened coping capacities, further extensive floods caused displacement, diseases broke out, basic public services were disrupted, and more than 1 million livestock died with 1.7 million more at risk. Combined, this amounted to approximately 3 percent of Ethiopia's total livestock population (see Table 9.4 for a chronology of the livestock losses during the drought). At the time of the 2016 drought (April 2016), WFP reported that "the poorest 20 percent of the population can meet, on average, only 15 percent of their food needs through their own means, while 76 percent eat, on average, only one meal per day. Eighty percent of the affected population are consuming fewer calories than the daily minimum recommended by WHO" (WFP 2016b). These findings are broadly consistent with national household survey data presented in Chapter 10 that show (1) cereals in general are consumed by nearly all households; (2) cereals are the most important calorie source in the country, making up just more than 60 percent of total calories for the average household; and (3) 25.6 percent of Ethiopians are food energy deficient even outside of the specific effects of a given drought period, with poor diet quality only exacerbated by the losses in livestock and associated animal-sourced foods that contribute to more diverse and nutritious diets, discussed in Chapter 8 (Chapter 10 elaborates more on these dietary quality trends).

April and May 2016 were marked by late and unevenly distributed spring rains. Nonetheless, *belg* maize production in 2016 was twice that of 2015: 1.06 million metric tons versus 0.49 million metric tons. In the drought-prone highlands, production increased from 0.04 million metric tons to 0.21 million metric tons between the two years. Overall, total cereal production from the *meher* and *belg* seasons combined decreased by 2.0 percent between 2014/2015 and 2015/2016. In the drought-prone highlands, total cereal production considering both seasons combined fell very little.

By July 2016 the regions with the worst food consumption levels shifted to Somali and SNNP. May floods during the main wet season *(gu)* in the

TABLE 9.4 Livestock losses chronology for Ethiopia, 2014–2017

Period	Summary
April and May 2014	There are 4,000 livestock deaths in Afar region due to flooding, 2,000 livestock deaths in Oromia region due to drought.
May 2015	Abnormal livestock deaths are on the rise in Somali region.
March 2016	Over one million livestock die over the course of the poor 2015/2016 *meher* season, largely in Afar and Somali regions, with 1.7 million more at risk.
March–May 2016	*Belg* rains are good, even in pastoral areas, greatly ameliorating dire conditions.
November 2016	There are 6,000 livestock deaths reported in Oromia just from October to November.
January 2017	Livestock death counts are in the tens of thousands again by January. In parts of Somali, 35 percent to 55 percent of total livestock holdings die.
May 2017	Hundreds of thousands of livestock owned by 1.9 million households critically require survival and supplementary feed as the 2017 drought gets worse.

Source: FEWSNET reports including June 2014, May 2015, Outlook June 2016–2017, and FAO reports from March 2016, November 2016, January 2017, and May 2017.

Note: In 2014 there was estimated to be a total of 54 million cattle, 26 million sheep, and 24 million goats in the country (Leta and Mesele 2014). This would make the one million deaths at the end of the 2015–2016 drought almost 1 percent of the total livestock population, with closer to 2 percent at risk of death by March 2016.

Somali region again left homes, fields, and more livestock destroyed. This was followed by a prolonged *hagaa* dry season. *Hagaa* is the lesser of the two dry seasons of the agricultural year, typically lasting from July to September, and precedes the *deyr* season of short rains. The difficult *hagaa* season in 2016 quickly deteriorated food security and led to distress migration as pastoralists trekked further afield with their animals looking for water and grazing. By late 2016 the prolonged drought and food insecurity lessened in the northeast after improved *belg* rains, and good rains enabled a successful *meher* harvest throughout most of Ethiopia, even while hundreds of thousands of livestock were still left in critical condition primarily in Somali pastoralist areas.

Market Prices and Government Policy Measures: Imports and Safety Nets

The main policy instrument used to counter the adverse effects of the 2015/2016 drought and crop shortfalls was to increase cereal imports for humanitarian aid and expanded food distribution through PSNP.[12] Ethiopia imported 2.46 million metric tons of wheat from July 2015 through

12 Ethiopia's PSNP transfers (1) reduce poverty (the food gap) by half a month; (2) reduce vulnerability (the expected food gap) given a drought has occurred, from 4.14 months to 1.8 months (57 percent decline); and (3) increase resilience by reducing the food gap by a further 1.75 months after a drought and reducing the time to recover, from four years to two (Knippenberg and Hoddinott 2017).

FIGURE 9.6 Ethiopia's wheat imports by import channel, 2014/2015–2016/2017 (thousand metric tons)

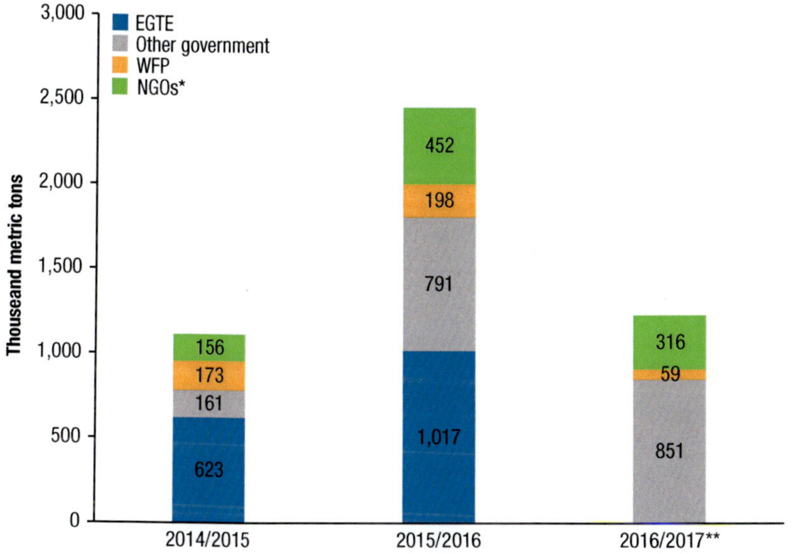

Source: EGTE and Ethiopia customs data (ERCA).

Note: Each set of years represents the months from July to June. EGTE = Ethiopian Grain Trade Enterprise; WFP = World Food Programme; NGO = nongovernmental organization. *Includes private trade equal to 25,000 metric tons in 2014/2015 and 2,000 metric tons in 2015/2016. **Data from July 2016 through March 2017.

June 2016, 1.35 million metric tons more than in the same period in 2014 to 2015 (Figure 9.6). The EGTE imported more than 40 percent of this wheat (1.02 million metric tons) but only distributed 556 thousand metric tons through its own sales channels. The other 1.11 million metric tons were imported by the World Food Programme, other NGOs, and other government channels (Figure 9.6).

Most of these imports (including some of those imported by the EGTE) were distributed through PSNP in return for labor for households with able-bodied members, or as direct support free of charge elsewise (Wiseman, Van Domelen, and Coll-Black 2010; Coll-Black et al. 2013), which benefited 8 million people in 2015 (although more than 10 million additional people were reported to be in need of food aid due to the drought). Data from nationally representative surveys from before (2014) and after (2016) the drought indicate that the share of households residing in drought-exposed areas were 7.5 percentage points more likely, or more than twice as likely, to have received humanitarian assistance in 2016 compared to the year prior (2015). The

FIGURE 9.7 Nominal wholesale Addis Ababa cereal prices and national wheat distribution, 2014–2017

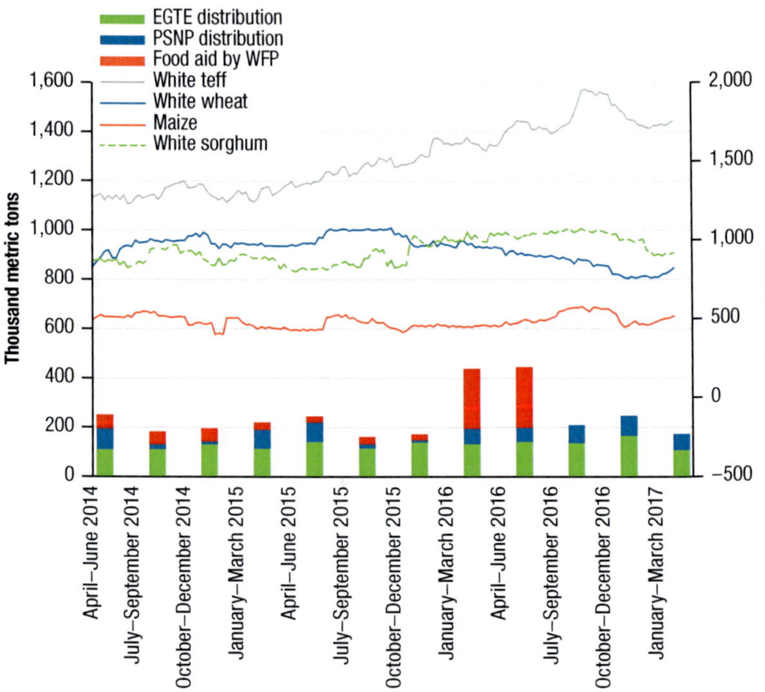

Sources: Authors' calculations from Ethiopia, CSA data (2018); EGTE (2018) data; Ethiopian Revenues and Customs Authority (ERCA) data; the IFPRI PSNPIV 2013 baseline survey (2017); FAOSTAT data for recent years on food aid shipments (WFP); USAID (2016); and Tesfanews (2016).

Note: EGTE = Ethiopian Grain Trading Enterprise; PSNP = Productive Safety Net Program; WFP = World Food Programme

share of households enrolled in PSNP also increased between 2014 and 2016, increasing from 10 percent to 12 percent among the surveyed households (Hirvonen, Sohnesen, and Bundervoet 2018).

Monthly price movements during this period reflect the massive wheat import flows and overall changes in supply and demand. As wheat imports increased sharply in early 2016, the price of wheat (as recorded in the Addis Ababa retail market) fell below that of sorghum (Figure 9.7). The real price of wheat in Addis Ababa fell by 13.9 percent between 2014/2015 and 2015/2016 (Figure 9.8), while in Dessie real wheat prices fell by 13.1 percent from 2014/2015 to 2015/2016. By contrast, the prices of teff, maize, and sorghum in Dessie rose by 11.3 percent, 2.1 percent, and 16.1 percent, respectively, during the *meher* season due to the drought's adverse effects on cereal

FIGURE 9.8 Real wholesale cereal prices in Addis Ababa, 2012–2018

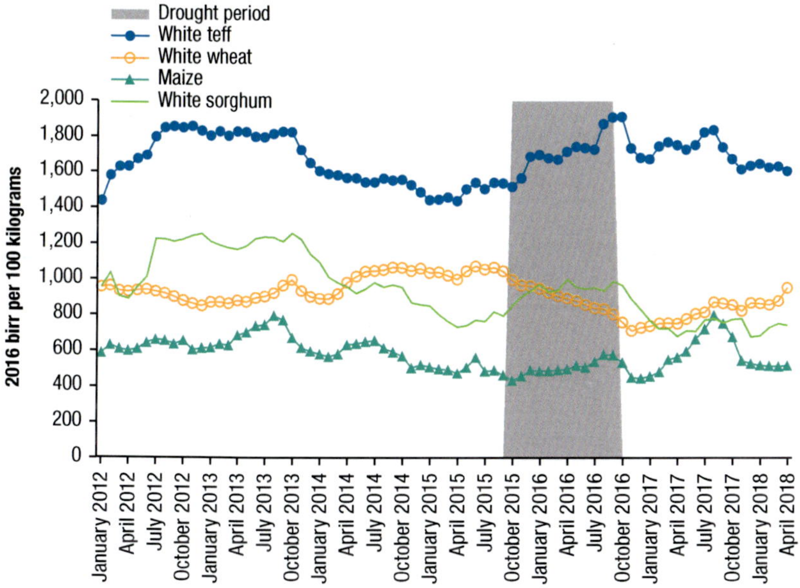

Source: Authors' calculations from Ethiopia, CSA data (2018).

production in that part of the country. The relatively small increase in real maize prices (only 2.1 percent as compared with the 11.3 percent and 16.1 percent real price increases for white teff and sorghum, respectively) reflected the relatively good maize harvest in 2015/2016 in western Ethiopia.

There is further evidence of improved functioning of markets and market integration among cereal products across regions, even in drought years (Minten et al. 2014; Hill and Fuje 2017; Bachewe, Yimer, and Minten 2017). Remarkably, there was not a widespread increase in chronic or acute child undernutrition rates in Ethiopia due to this drought, as there had been in the famine of the early 1980s, *except* in areas characterized by limited road networks (Hirvonen, Sohnesen, and Bundervoet 2018). This highlights the important role of market integration and public infrastructure for improved resilience to future droughts in the country. Chapter 12 discusses how urbanization, when supported by appropriate infrastructure, has the potential to boost total factor productivity and economic output through positive agglomeration effects. Mann, Warner, and Malik (2018) show that the impacts of the droughts were focused on the drought-prone highlands and therefore areas that typically produce less cereals, leading to lower impacts on production and prices than initially feared.

The 2015/2016 Drought: Price Effects of Alternative Policies

To estimate the effects of the drought and alternative policies on wheat consumption, imports, and market prices, we use a basic partial equilibrium framework with exogenous domestic production (Table 9.5). Supply of each cereal is calculated assuming 17 percent total losses for seed, feed, and wastage, a rate in range of the typical 15 percent to 20 percent of food availability assessed by agencies such as the World Food Programme (WFP and Ethiopia, CSA 2019; FAO and WFP 2007, 2019a, and 2019b). Household demand is modeled as a function of prices and (exogenous) income.[13] In simulations where imports are exogenous, the domestic price adjusts to equate supply and demand. In Simulation 4 in which the model liberalized private-sector trade, the domestic price is set equal to the exogenous import parity price and imports are endogenous.

Simulation 1 models the effects of the historically observed production shocks to the four major cereals in Ethiopia—teff, wheat, maize, and sorghum.[14] According to the official data, wheat production was almost unchanged between 2014/2015 and 2015/2016, leading to a per capita drop in production of 2.1 percent. If wheat imports would have remained the same in 2015/2016 as the previous year (0.948 million metric tons in 2014/2015), then wheat supply per capita would also have fallen by 2.1 percent. At the same time, demand for wheat per capita is estimated to have risen by 4.0 percent—that is, a 5 percent increase in per capita incomes multiplied by an income-elasticity of demand of 0.8. Given an own-price elasticity of demand of −0.95, domestic wheat prices would have risen by 6.2 percent.

Including production and imports of teff, maize, sorghum, and rice that are assumed to equal their actual 2015/2016 levels, the total supply of cereals falls by only 0.1 percent since maize production actually increased in this

13 The model is a simplified version of the models used for Ethiopian wheat markets (Dorosh and Ahmed 2011) and Bangladesh rice markets (Dorosh and Rashid 2013). Own-price elasticities of demand for wheat, teff, maize, and sorghum are −0.952, −0.920, −0.735, and 0.743, respectively (weighted averages of urban and rural elasticities reported in Tafere et al. [2009] and Berhane et al. [2013]). Per capita incomes are assumed to have increased by 5 percent between 2014/2015 and 2015/2016. We use an average income elasticity of demand of 0.8.

14 The model does not capture potential cross-price effects of a change in price of one commodity (for example, teff) on the production or demand for another crop (for example, wheat). Note that Tafere et al. (2009) estimated cross-price elasticities very close to zero. The simulation results for "all cereals" implicitly assume perfect substitution in demand across all cereals and thus provide broad sensitivity analysis.

TABLE 9.5 Simulated price effects of production shocks and increased imports in Ethiopia, 2015/2016 versus 2014/2015

Simulation:	S1	S2	S3	S4
	Wheat production shock	Wheat production shock with higher imports	All cereals production shock with higher imports	All cereals production shock with free trade
Wheat production[a] (millions of metric tons)	4.550	4.550	3.849	3.849
Wheat production per capita (% change)	−2.1	−2.1	−17.2	−17.2
Wheat imports (millions of metric tons)	0.948	2.483	2.483	3.103
Wheat supply[b] (millions of metric tons)	4.724	6.259	5.677	6.297
Wheat supply per capita (% change)	−2.1	29.7	17.7	30.5
Wheat price (% change)	6.2	−26.0	−13.8	−26.8
All cereals supply[c] (millions of metric tons)	19.586	21.460	20.140	20.760
All cereals price (% change)	6.8	−4.4	3.2	−0.5

Source: Model simulations.

Notes: a. Production includes commercial farms' production.

b. Supply is calculated as production minus a 17 percent adjustment for seed, feed, and wastage plus imports.

c. Supply of all cereals includes supply of wheat, maize, sorghum, teff, and rice. Supply of these nonwheat cereals is fixed at the 2015/2016 level in Simulations S1 and S2. In Simulations S3 and S4 production of teff, maize, and sorghum are each reduced by 5 percent relative to their 2015/2016 levels.

Simulation S3: The wheat production shock is set at the level at which wheat (real) price change matches historical value of −13.8 percent.

Simulation S4: Wheat imports are adjusted so that wheat (real) price change matches historical value of the percentage difference between 2014/2015 wholesale price and estimated 2015/2016 import parity price.

period. However, given a sharp 6.6 percent fall in per capita supply of teff and an 11.0 percent increase in its simulated price, the weighted average simulated real price increase of all four major cereals is 6.8 percent.

Simulation 2 models the effects of increased distribution of wheat from wheat imports (2.483 million metric tons, an increase of 1.535 million metric tons over 2014/2015).[15] Given this increase in imports, net wheat supply per capita increases by 29.7 percent relative to 2014/2015. The simulated wheat price falls by 26.0 percent relative to 2014/2015, far greater than the historical real price decline of only 13.8 percent (Simulation 2). Again, assuming that the levels of production and imports of teff, maize, and sorghum are unchanged relative to their actual 2015/2016 levels, the average price of cereals falls 4.4 percent.

In Simulation 3 we reduce wheat production another 15.4 percent, for a total decline in per capita production of 17.2 percent, so that the simulated

15 These wheat import figures are the estimates of wheat imports actually released by EGTE for these years. See Table 4.6 in Dorosh et al. (2018) for the change in wheat supply due to imports.

price fall (13.8 percent) matches the historical price fall. Net wheat supply per capita is then 17.7 percent greater than in 2014/2015. We also model a 5 percent reduction in production of teff, maize, and sorghum in this simulation. As a result, the average price of all cereals rises by 3.2 percent.

Finally, in Simulation 4 we simulate the counterfactual policy of allowing private-sector wheat trade at the import parity price (without tax), which was substantially below domestic prices from mid-2014 through mid-2016 (Figure 9.9). In this scenario the government of Ethiopia and NGOs procure wheat for PSNP and other programs in the wholesale market at the prevailing (import parity) price. Here, we maintain the same reduced level of wheat production as in Simulation 3 and fix the domestic price level at the estimated import parity price, which was 26.8 percent below the historical value of the wholesale domestic price of wheat in 2014/2015.[16]

In the case of Simulation 4 total wheat imports rise to 3.1 million metric tons and net wheat supply per capita is 30.5 percent higher than in 2014/2015. Compared to Simulation 3, in which net wheat supply per capita is only 17.7 percent higher than in 2014/2015, wheat supply per capita in Simulation 4 is 10.9 percent higher (12.8 percentage points higher). In consequence, the wheat price is 15.1 percent lower, benefiting net wheat consuming households purchasing from domestic markets. As in Simulation 3, we again model a 5 percent reduction in production of teff, maize, and sorghum. Given the sharp reduction of the wheat price due to increased wheat imports, the average price of cereals actually falls by 0.5 percent, as compared to the 3.2 percent rise in Simulation 3.

Using more inelastic demand parameters, which may better reflect the behavior of consumers and markets in the short-run, produces broadly similar results (Table 9.6). Relatively high wheat production with 2.48 million metric tons of imports results in an even greater price decline (57.1 percent) than with the base parameters (26.0 percent) (Simulation 2a). To achieve a price reduction that matches the historical change (–13.8 percent) requires a slightly larger reduction in production (–28.4 percent instead of –17.2 percent) (Simulation 3a). Finally, with the alternate parameters, liberalized private-sector trade leads to a smaller increase in imports (306,000 versus 620,000 metric tons) (Simulation 4a).

16 7.68 (2016) birr per kilogram import parity in 2015/2016 as compared to 10.49 (2016) birr per kilogram average wholesale white wheat price in Addis Ababa in 2014/2015.

FIGURE 9.9 Wheat domestic wholesale and import parity prices, 2010–2018

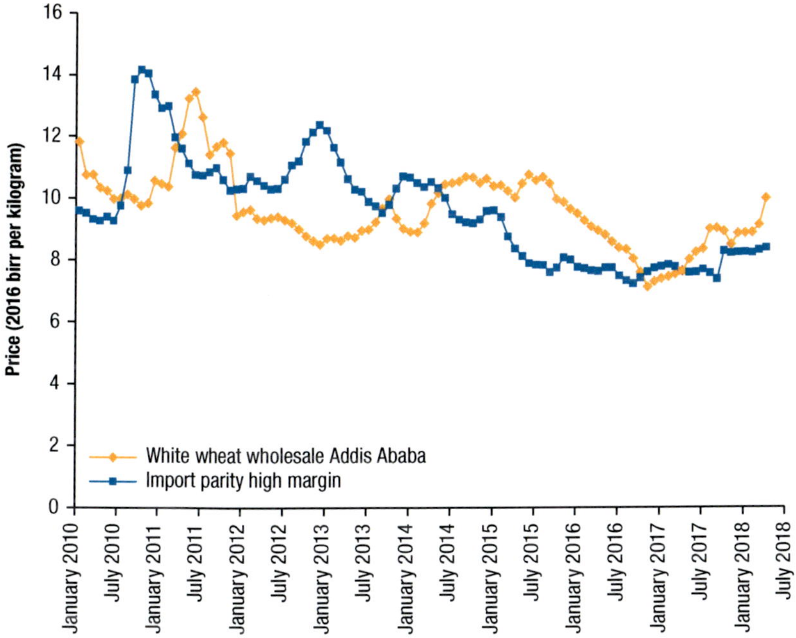

Source: Authors' calculations from Ethiopia, CSA (2018) and Ethiopia, EPAU (2018) data.

Notes: Import parity in this figure is based on wheat originating from Black Sea ports, reflecting the major source of wheat in recent years. Note that the import parity prices in Figure 9.2, which covers a longer period, are based on wheat shipped from US Gulf ports (Hard Red Winter).

Thus, liberalizing private-sector wheat imports in 2015/2016 could have provided even greater stability of supply and price. To the extent that the private sector could import wheat and bring it to wholesale markets in Addis Ababa or other major cities at a lower cost than the Ethiopian Grain Trading Enterprise (EGTE) or international agencies, the government of Ethiopia could also have reduced costs of its distribution programs.[17] However, it is not always the case that the import parity price is less than a target domestic price level. When international prices are very high, stabilizing domestic prices requires either a drawdown of public stocks, food aid inflows, or subsidized sales of government commercial imports.

17 There was no explicit trade restriction in 2015/2016, but private traders were not granted import licenses and foreign exchange to import wheat.

TABLE 9.6 Simulated price effects of production shocks and increased imports in Ethiopia, inelastic parameters, 2015/2016 versus 2014/2015

Simulation:	S1a	S2a	S3a	S4a
	Wheat production shock	Wheat production shock with higher imports	All cereals production shock with higher imports	All cereals production shock with free trade
Wheat production[a] (millions of metric tons)	4.550	4.550	3.327	3.327
Wheat production per capita (% change)	−2.1	−2.1	−28.4	−28.4
Wheat imports (millions of metric tons)	0.948	2.483	2.483	2.789
Wheat supply[b] (millions of metric tons)	4.724	6.259	5.244	5.550
Wheat supply per capita (% change)	−2.1	29.7	8.7	15.0
Wheat price (% change)	8.4	−57.1	−13.8	−26.8
All cereals supply[c] (millions of metric tons)	19.586	21.460	20.297	20.603
All cereals price (% change)	9.3	−13.6	0.5	−3.3

Source: Model simulations.

Notes: a. Production includes commercial farms' production.

b. Supply is calculated as production minus a 17 percent adjustment for seed, feed, and wastage, plus imports.

c. Supply of all cereals includes supply of wheat, maize, sorghum, teff, and rice. Supply of these nonwheat cereals is fixed at the 2015/2016 level in Simulations S1a and S2a. In Simulations S3a and S4a, production of teff, maize, and sorghum are each reduced by 1 percent relative to their 2015/2016 levels.

Simulation S3a: The wheat production shock is set at the level at which wheat (real) price change matches historical value of −13.8 percent.

Simulation S4a: Wheat imports are adjusted so that wheat (real) price change matches historical value of the percentage difference between 2014/2015 wholesale price and estimated 2015/2016 import parity price.

Price Stabilization Options for Ethiopia

Effective Price Stabilization Policy

The above review of the international experience with price stabilization and the case of the policy response to the 2016 cereal production shortfall and resulting drought suggests several important principles in designing an efficient price-stabilization strategy for Ethiopia. The most important elements for such a strategy are discussed below.

INTERNATIONAL MARKETS

Access to international markets to add to domestic supplies and lower prices through imports or to reduce domestic surpluses and support higher prices through exports can greatly reduce the cost of a price-stabilization policy. The alternative is to build up large stocks through domestic procurement in excess of distribution.[18] Likely due to policymakers' high perceptions of the risk of a food crisis, the food reserve agency is mandated to hold 407 thousand metric tons of grain in its stock. However, the actual stock had not exceeded 350 thousand metric tons from 2004 to 2015, and in some quarters stocks were well below 100 thousand metric tons (Rashid et al. 2018).

International trade in wheat through food aid and government commercial imports already constitutes a major source of wheat supplies for Ethiopia. In contrast, Ethiopia's current levels of international trade in teff, sorghum, and maize are small. The lack of an international market for teff—Ethiopia is essentially the only major producer and consumer of teff globally—would make it very difficult and costly to stabilize teff prices, as it would require very large stocks. International trade in sorghum is much larger. Sudan has exported several hundred thousand metric tons in some years to countries in the Middle East. The United States is also a major exporter and a potential source of sorghum supplies for Ethiopia. Nonetheless, the international market for sorghum is rather thin. For maize there is a broad international market. Any maize exports or imports by Ethiopia would represent only a small fraction of total international trade.

Currently, high transport costs relative to the value of the grain make the import parity–export parity price band for Ethiopia very wide. As described earlier, Ethiopia's domestic prices for maize have generally been well above export parity and mostly (but not always) below import parity. This has made both exports and imports of maize unprofitable for the private sector in most years.

PROMOTING PRIVATE TRADE

The international experience (such as in Bangladesh the late 1990s) suggests that facilitating and promoting private international trade is often the least costly, most efficient, and quickest means of enhancing and maintaining market supply and promoting price stability. The private sector can usually react much more quickly to potential domestic supply shortfalls than can public institutions (EGTE or donor food aid). Moreover, private-sector imports involve no direct costs to the government. Ethiopia's private-sector

18 To stabilize domestic prices with grain from security stocks, India holds stocks equal to about 24 percent of wheat production and 19 percent of rice production, and Pakistan holds stocks equal to about 6 percent of wheat production from 2001 to 2007 (Dorosh 2009). Ethiopia's cereal stocks averaged only 188.3 (145.1) million metric tons in 2001–2008 (2009–2015), equivalent to 1.8 (0.7) percent of cereal production (emergency food reserve agency data).

international trade in cereals remains very small, however, mainly limited to imports of rice and high-gluten and other specialty wheat types suited for pasta or for baking purposes.

Private-sector wheat imports could substantially reduce the cost of price stabilization in Ethiopia. To a large extent, private-sector wheat imports that would otherwise have been profitable have been hindered by foreign exchange rationing, displaced by public sector imports, or made too risky because of uncertainties over government trade, exchange rate, and credit policies (Dorosh and Ahmed 2011; World Bank 2016; Dorosh and Ahmed 2009; Ferrand 2018).

TRANSPARENCY AND PREDICTABILITY

Uncertainty over government policy is a major disincentive to private-sector investment and both international and domestic trade. Yet price-stabilization regimes with frequent and abrupt policy changes are common.[19] Such regimes characterize what is often the least efficient price-stabilization alternative, when compared to the promotion of international trade using rules-based price-stabilization policies (see Table 9.1). Unfortunately, a lack of transparent and predictable policy has characterized price-stabilization efforts in much of East Africa and resulted in unnecessary price instability in Kenya, Malawi, and Zambia, with important fiscal costs as well as adverse effects on the food security of poor consumers.

One option for promoting transparency and predictability of policies and for fostering trust between the private sector and government is to establish a platform for discussions of current market conditions, current and planned government policies, and ideas for promoting more efficient markets, such as investments in key infrastructure.

ADMINISTRATIVE AND MARKET PRICES

Administratively setting prices can seriously weaken the ability of markets to stabilize food prices. This was seen during Madagascar's rice price crisis of 2004/2005, when official prices were set below import parity levels, effectively ending incentives for private-sector imports. Price targets are best achieved through market mechanisms by either adding to supply or making purchases from the market to increase demand. Thus it is important to communicate

19 Examples include periodic bans of maize exports in Zambia and Malawi (Diao and Kennedy 2016; Gondwe and Baulch 2017) and Ethiopia's seizure of private coffee stocks in 2009 (Minten et al. 2014).

that the target price (or better, a target price band) is a policy objective, but not a binding restriction on market transactions.

Guidelines for Implementation of Cereal Price Stabilization

Setting clear objectives and specific rules for interventions could contribute to successful cereal price stabilization in Ethiopia. The following elements as part of such a strategy should be considered.

PRICE TARGETS AND MARKET INTERVENTIONS

Annual target ceiling and floor prices for wheat, maize, and sorghum could be set. Current policy involves a government sales price and an implicit target (ceiling or floor) price that is not necessarily equal to the government sales price. The implicit goal of the interventions is often to reduce private market prices to the target ceiling price levels. Setting the target ceiling price for imported commodities at long-term import parity levels will ensure that the marginal cost of domestic production is equal to the long-term opportunity cost of alternative supplies from the international market, thus promoting an efficient use of resources. In years of relatively low import parity prices, keeping the target ceiling price at the higher long-run import parity price will provide incentives for private-sector imports and make public interventions unnecessary. In years of high import parity prices, subsidized sales of government imports may be required to maintain market prices at the target long-run import parity level.

The margin between floor and ceiling prices should be wide enough to give incentives for private-sector storage and trade. If the band is too narrow, the government may crowd out most private-sector trade, buying substantial amounts of grain to defend the floor price and then later selling grain to defend the ceiling price. This pattern of government interventions in cereal markets is seen in Pakistan where public procurement and subsequent sales, usually within the same crop marketing year, accounts for more than 20 percent of total wheat production, crowding out the private sector and entailing huge fiscal costs. Setting a low price with a wide margin between the floor and ceiling prices can avoid the need for any domestic procurement.

MANAGEMENT OF PUBLIC STOCKS

Public stocks are needed for an effective government price-stabilization program. About 60 percent of the grain stocks in Ethiopia are held by the Ethiopian Emergency Strategic Food Reserve Agency, while the remaining stocks are in the hands of the EGTE, in a few mills, and a small amount in private trader storage (USDA-FAS 2012). Stock levels should be kept to a minimum, however, to avoid excessive costs of storage, which include implicit interest, stock management, quality deterioration, and stock rotation costs. Food security reserve stocks are typically set according to the volume of likely distribution needs, allowing for a three- to four-month lag between the decision to import (if necessary) and the arrival of grain at distribution points. Ethiopia's emergency food reserve stocks have generally been small relative to domestic production and consumption and have been effectively managed (Graham, Rashid, and Malek 2012). From 2009 to 2015, Ethiopia's average public cereal stocks (145 thousand metric tons) were less than 1 percent of domestic cereal production. Incorporating price-stabilization objectives to this system would imply a need for larger stocks.

Price stabilization need not require huge stocks, however. As shown in Table 9.7, total stocks for Bangladesh averaged 2.9 percent of total wheat and rice production. Given that Bangladesh has two major rice harvests (and the risk of a production shortfall in both seasons in the same year is very small), this ratio of stocks to production may be rather small. India's stock to production ratio of 21.1 percent is likely excessively large, even given the fact that India is such a large country that it would have difficulty obtaining massive imports if it had a major shortfall. A better guideline might be the stock-to-production ratio of Pakistan with stocks at 5.7 percent of production.[20]

ENHANCING MARKET EFFICIENCY THROUGH TRANSPARENCY AND PREDICTABILITY OF INTERVENTIONS

There is a need for clear announcements of overall policy objectives and instruments to promote efficient private markets. This includes, first, clear annual announcements in advance of the (*meher*) planting season of target ceiling and floor prices for wheat, maize, and sorghum, and the potential interventions planned to achieve these targets. Second, clear announcements of quantities of planned public imports and exports are required so as not to discourage private trade (both imports and exports). Although, as noted above, when import parity prices are lower than the target ceiling price, private imports are more cost-effective, and they generally involve no costs to the government.

20 Note that in Bangladesh, India, and Pakistan, public stocks serve dual purposes: price stabilization and to provide grain for public food distribution programs. Since the early 1990s, Bangladesh has used trade policy as well as market interventions and stocks to stabilize prices.

TABLE 9.7 Cereal stock levels in selected Asian countries, average 2001–2007

	Pakistan	India			Bangladesh		
Indicator	Wheat	Wheat	Rice	Total	Rice	Wheat	Total
Production (millions of metric tons)	20.33	70.51	87.23	157.7	25.68	1.21	26.89
Stocks (millions of metric tons)	1.16	17.06	16.4	33.5	0.54	0.23	0.77
Stocks/Production (%)	5.7	24.2	18.8	21.2	2.1	19.0	2.9
Per capita stocks (kilograms)	7.6	16.2	15.4	31.6	1.7	3.8	5.5

Source: Dorosh (2009).

Conclusion

Increases in cereal prices can have adverse effects on poor net food buyers. Conversely, falling domestic prices of some cereals (especially maize), typically at harvest time, can be detrimental to producers who are net sellers. Price-stabilization efforts are therefore an important consideration for Ethiopian policymakers. This chapter sheds light on options for cereal price stabilization in Ethiopia drawing on experiences of other developing countries. The international experience in food price stabilization shows that, while some countries have successfully stabilized prices at generally high fiscal cost, the efforts of many others have actually destabilized market prices at great fiscal cost. Ethiopia's interventions in wheat markets (there have been no major interventions in maize markets in Ethiopia in recent years) have met with limited success.

In this chapter we assess the extent of cereal price variation in Ethiopia. For the period 2009–2018, coefficients of variation (CV) of nominal wheat prices (which benefited from international trade) and maize (which did not) were both 0.280. The coefficients of variation for real prices were only 0.132 for wheat and 0.214 for maize. While real maize price volatility is similar in Kenya (CV = 0.233), it is more variable in Uganda (CV = 0.291). The upshot is that a significant part of the variation in cereal prices in Ethiopia is explained by general inflation.

One of the major causes of price variability is also weather shocks, which lead to lower production. In this chapter we assess the extent to which Ethiopian stabilization efforts were effective during the major El Niño–induced drought of 2015/2016. The consequent crop losses threatened the food security of an estimated 10 million people who were declared in need of emergency food assistance in addition to the 8 million already covered by the Productive Safety Net Program. Ultimately, a major food crisis was averted through a combination of a timely large-scale government and donor response along with substantial private market flows of cereals from surplus to deficit regions. Nonetheless, opportunities were missed to enhance food security and consumer welfare further by allowing private-sector imports of cereals to minimize the rise in cereal prices that occurred in 2016 and to reduce the fiscal costs to the government and donors.

In general, Ethiopia's policy options to further enhance cereal price stability vary by crop. Because Ethiopia imports wheat (either through food aid or through government commercial channels) in most years, liberalization of private-sector wheat imports can often provide a low-cost mechanism for providing a price ceiling at the import parity price. Enhancing price stability in other years through direct domestic market interventions, however, may be costly in fiscal terms and in terms of discouraging private-sector storage and trade. Given that Ethiopia's maize price is typically far below import parity (and far above export parity), private external trade in maize will generally not contribute to price stability. Moreover, interventions in domestic maize markets may be even more problematic for maize than for wheat because storage losses for maize are typically greater than those for wheat. Interventions to stabilize teff prices may be especially difficult, since very little teff is traded on the international market. Nonetheless, as urban markets for teff grow, private-sector commercial storage may increase over time and help stabilize market prices.

Appendix 9A: Historical Account of Major Food Shortages in Ethiopia

TABLE 9A.1 Historical account of major food shortages in Ethiopia, 1971–2017

Date	Region affected	Attributed causes and severity
1971–1975	Ethiopia	Sequence of rain failures. Estimate 250,000 dead. 50 percent livestock lost in Tigray and Wollo.
1978–1979	Southern Ethiopia	Failure of *belg* rains.
1982	Northern Ethiopia	Late *meher* rains.
1983–1985	Ethiopia	Sequence of rain failures. Eight million affected. Estimated one million dead. Much livestock loss.
1987–1988	Ethiopia	Drought of undocumented severity in peripheral regions.
1990–1992	Northern, eastern, and southwestern Ethiopia	Rain failure and regional conflicts. Estimated four million people suffering food shortage.
1993–1994	Tigray, Wollo, Addis Ababa	Four million people requiring food assistance, including demobilized army and Somali refugees. New droughts.
1997–2000	Eritrea, northern Tigray	Localized food shortages due to conflict.
2002–2003	Ethiopia	Drought-induced crop shortages.
2009	Southern Ethiopia	Localized drought.
2014–2015	Eastern Highlands, drought-prone lowlands	Localized droughts lead to *belg* harvest shortfalls; losses of livestock.
2015–2017	Highlands, drought-prone lowlands	Localized drought reduces *meher* and *belg* harvests; 10.2 million people in need of relief in mid-2016.

Source: 1971–1994: Webb and von Braun (1994: 21); Graham, Rashid, and Malek (2012), Table 9.1.

Appendix 9B: Assessing Cereal Price Variability in Ethiopia

World wheat prices in Ethiopia (measured in US dollars) from 2008 to 2018 were considerably more stable than the world maize prices with a coefficient of variation (CV) of 0.196 as compared to 0.287 for maize (column 1). Column 2 of Table 9B.1 shows that there was a relative difference in the coefficients of variation of nominal exchange rates across countries. Thus the coefficients of variation of border prices of maize measured in local currency units are rather similar for the three countries (column 3). In all three countries domestic maize prices (column 4) were on average slightly above border prices; wheat prices in Ethiopia were on average 46.9 percent above border prices, though, as discussed in the chapter regarding price effects of alternative policies, there were considerable periods in which domestic wheat prices were somewhat below border prices. Domestic prices measured in US dollars (column 5) were

TABLE 9B.1 Average prices for wheat and maize in Ethiopia and maize in Kenya and Uganda, 2009–2018

	World price	Exchange rate	Border price	Domestic price	Domestic price	Border price/CPI	Domestic price/CPI
	US$ per metric ton (1)	lcu /$ (2)	lcu per kilogram (3)	lcu per kilogram (4)	$ per metric ton (5)	lcu (2018) per kilogram (6)	lcu (2018) per kilogram (7)
Ethiopia wheat							
Mean	262.12	19.12	5.53	8.07	418.74	8.41	11.86
Standard deviation	51.42	4.16	1.31	2.26	51.42	2.29	1.56
CV	0.196	0.217	0.237	0.280	0.123	0.272	0.132
Ethiopia maize							
Mean	204.02	19.12	4.32	4.86	254.38	2.57	2.60
Standard deviation	58.54	4.16	1.12	1.36	40.65	0.84	0.56
CV	0.287	0.217	0.260	0.280	0.160	0.328	0.214
Kenya maize							
Mean	204.02	90.81	21.05	31.42	348.28	27.92	40.73
Standard deviation	58.54	9.49	4.74	7.78	67.33	8.61	9.50
CV	0.287	0.104	0.225	0.248	0.193	0.211	0.233
Uganda maize							
Mean	204.02	2841.45	651.14	720.64	255.08	818.58	876.81
Standard deviation	58.54	590.90	148.22	250.65	75.04	191.42	255.13
CV	0.287	0.208	0.228	0.348	0.294	0.234	0.291

Source: Authors' calculations from Dorosh, Minten, and Stifel (2015); EGTE (2018); IMF (2018); and Ethiopia, CSA (2018) data.

Note: CPI = Consumer Price Index; lcu = local currency unit; CV = coefficent of variation.

considerably more stable than domestic prices measured in local currency. Finally, real prices of both wheat and maize in Ethiopia in local currency units (calculated using the domestic CPIs as deflators) were much more stable than the nominal domestic prices of these commodities.

Over this same period (2009–2018), Ethiopia's overall domestic inflation (measured by the CPI), as well as the growth rate of domestic maize prices, were higher than those in Kenya and Uganda.[21] The average increase in the CPI in Ethiopia was 0.98 percent per month, compared to only 0.60 percent per month in Kenya and 0.53 percent per month in Uganda. Likewise, the average growth rates in maize prices in Ethiopia was considerably higher

21 For more details on the exchange rates and cereal prices across these three countries, see Dorosh et al. (2018).

than those in Kenya or Uganda (0.66 percent, 0.35 percent, and 0.49 percent, respectively). Over this period, Ethiopia's nominal exchange rate (relative to the US dollar) increased by 0.64 percent per month. This was considerably less than the average monthly growth rate in the CPI, so the ratio of the exchange rate to the CPI (a crude measure of the real exchange rate that does not take into consideration changes in world prices), appreciated by 0.20 percent per month, almost identical to the change in this measure for Kenya (0.19 percent per month). This appreciation of the real exchange rate encouraged imports and reduced export competitiveness. In contrast, Uganda's real exchange rate depreciated over this period.

References

AKLDP (Agriculture Knowledge, Learning, Documentation and Policy Project). 2016. *Food Security in Ethiopia in 2016: Analysing Crop Production and Market Function after the Main* Meher *Agricultural Season.* Addis Ababa: AKLDP, USAID (United States Agency for International Development).

Arguez, A., I. Durre, S. Applequist, M. Squires, R. Vose, X. Yin, and R. Bilotta. 2010. "NOAA's U.S. Climate Normals (1981–2010)." NOAA National Centers for Environmental Information. Accessed July 5, 2017. www.ncdc.noaa.gov/data-access/land-based-station-data/land-based-datasets/climate-normals/1981-2010-normals-data.

Bachewe, F., G. Berhane, B. Minten, and A. S. Taffesse. 2018. "Agricultural Transformation in Africa? Assessing the Evidence in Ethiopia." *World Development* 105: 286–298.

Bachewe, F., F. Yimer, and B. Minten. 2017. *Agricultural Price Evolution in Drought versus Non-Drought Affected Areas in Ethiopia: An Updated Assessment Using National Producer Data (January 2014 to January 2017).* IFPRI-ESSP Working Paper 106. Addis Ababa: IFPRI/ESSP (International Food Policy Research Institute/Ethiopia Strategy Support Program).

Bellemare, M. F., C. B. Barrett, and D. R. Just. 2013. "The Welfare Impacts of Commodity Price Volatility: Evidence from Rural Ethiopia." *American Journal of Agricultural Economics* 95 (4): 877–899.

Berhane, G., D. O. Gilligan, J. Hoddinott, N. Kumar, and A. S. Taffesse. 2014. "Can Social Protection Work in Africa? The Impact of Ethiopia's Productive Safety Net Programme." *Economic Development and Cultural Change* 63 (1): 1–26.

Berhane, G., L. McBride, K. Tafere, and S. Tamiru. 2013. "Patterns in Foodgrain Consumption and Calorie Intake." In *Food and Agriculture in Ethiopia: Progress and Policy Challenges*, edited by P. Dorosh and S. Rashid, 190–216. Philadelphia: University of Pennsylvania Press.

Chapoto, A., and T. Jayne. 2009. *Effects of Maize Marketing and Trade Policy on Price Unpredictability in Zambia*. Food Security Research Project Working Paper 38. East Lansing, US: Michigan State University.

Coady, D., P. Dorosh, and B. Minten. 2009. "Evaluating Alternative Policy Responses to Higher World Food Prices: The Case of Increasing Rice Prices in Madagascar." *American Journal of Agricultural Economics* 91 (3): 711–722.

Coll-Black, S., D. Gilligan, J. Hoddinot, N. Kumar, A. Taffesse, and W. Wiseman. 2013. "Targeting Food Security Interventions in Ethiopia: The Productive Safety Net Programme." In *Food and Agriculture in Ethiopia: Progress and Policy Challenges*, edited by P. Dorosh and S. Rashid, 280–317. Philadelphia: University of Pennsylvania Press.

Diao, X., and A. Kennedy. 2016. "Economywide Impact of Maize Export Bans on Agricultural Growth and Household Welfare in Tanzania: A Dynamic Computable General Equilibrium Model Analysis." *Development Policy Review* 34 (1): 101–134.

Dorosh, P. A. 2001. "Trade Liberalization and National Food Security: Rice Trade between Bangladesh and India." *World Development* 29 (4): 673–689.

———. 2008. "Food Price Stabilization and Food Security: The International Experience." *Bulletin of Indonesian Economic Studies* 44 (1): 93–114.

———. 2009. "Price Stabilization, International Trade and National Cereal Stocks: World Price Shocks and Policy Response in South Asia." *Food Security* 1 (2): 137–149.

Dorosh, P., and H. Ahmed. 2009. *Foreign Exchange Rationing, Wheat Markets and Food Security in Ethiopia*. IFPRI-ESSP Working Paper 4. Addis Ababa: IFPRI/ESSP.

———. 2011. "Foreign Exchange Rationing and Wheat Markets in Ethiopia." *Ethiopian Journal of Economics* 20 (2): 83–104.

Dorosh, P., B. Minten, and D. Stifel. 2015. "Cereal Price Stabilization in Ethiopia: Implications from International Experience." In *Cereals Price Policy and Analysis for Ethiopia in the Context of Rapid Production Growth*, edited by D. Chanyalew, J. W. Mellor, B. Minten, and T. Kuma. Unpublished 2014 conference proceedings, ESSP/IFPRI, Addis Ababa.

Dorosh, P., and S. Rashid. 2013. "Trade Subsidies, Export Bans and Price Stabilization: Lessons of Bangladesh—India Rice Trade in the 2000s." *Food Policy* 41: 103–111.

Dorosh, P., J. Smart, B. Minten, and D. Stifel. 2018. *Droughts, Cereal Prices, and Price Stabilization Options in Ethiopia*. IFPRI-ESSP Working Paper 126. Addis Ababa: IFPRI/ESSP.

EGTE (Ethiopian Grain Trade Enterprise). 2018. *Commodity Statistics*. Accessed July 5, 2019. www.egte-ethiopia.com/en.

Ethiopia, CSA (Central Statistical Agency). 2016. *Annual Agricultural Sample Survey Data Bulletins.* Accessed July 5, 2017. www.csa.gov.et/survey-report/category/58-meher-main-season-agricultural-sample-survey.

———. 2018. Unpublished market price data.

Ethiopia, EPAU (Ethiopian Policy Analysis Unit). 2018. Unpublished data.

Ethiopia, ERCA (Ethiopian Revenues and Customs Authority). Unpublished imports data.

FAO (Food and Agriculture Organization of the United Nations). 2016a. *2015–2016 El Niño: Early Action and Response for Agriculture, Food Security and Nutrition Report.* Update #6. Accessed July 5, 2019. http://reliefweb.int/sites/reliefweb.int/files/resources/a-i5855e.pdf.

———. 2016b. *Ethiopia Situation Report—November 2016.* Accessed July 5, 2019. www.fao.org/emergencies/resources/documents/resources-detail/en/c/456019/.

———. 2017a. *Ethiopia Situation Report—January 2017.* Accessed July 5, 2019. www.fao.org/emergencies/resources/documents/resources-detail/en/c/463347/.

———. 2017b. *Ethiopia Situation Report—May 2017.* Accessed July 5, 2019. www.fao.org/emergencies/resources/documents/resources-detail/en/c/886170/.

———. 2017c. *Food Aid Shipments (WFP).* Accessed July 5, 2019. www.fao.org/faostat/en/#data/FA.

———. 2017d. *Production Indices.* Accessed July 5, 2019. www.fao.org/faostat/en/#data/QI.

FAO and WFP (World Food Programme). 2007. *Special Report: FAO/WFP Crop and Food Supply Assessment Mission to Ethiopia.* Accessed August 30, 2019. www.fao.org/3/a-j9325e.pdf.

———. 2019a. *Special Report: FAO/WFP Crop and Food Security Assessment Mission to South Sudan.* Accessed August 30, 2019. www.fao.org/3/ca3643en/ca3643en.pdf.

———. 2019b. *FAO/WFP Joint Rapid Food Security Assessment—Democratic People's Republic of Korea.* Accessed August 30, 2019. www.fao.org/3/ca4447en/ca4447en.pdf.

Ferrand, A. 2018. *Exchange Rate Management and Export Growth: Lessons for Ethiopia. K4D Helpdesk Report.* Brighton, UK: Institute of Development Studies.

FEWS NET. 2014. "Ethiopia Food Security Outlook Update: Food Security Likely to Deteriorate in Afar and Southern Somali." Accessed July 5, 2017. www.fews.net/east-africa/ethiopia/food-security-outlook-update/june-2014.

———. 2015. "Ethiopia Food Security Outlook Update: *Belg* Production Likely to Be Far Below Average in June/July." Accessed July 5, 2017. www.fews.net/east-africa/ethiopia/food-security-outlook-update/may-2015.

———. 2017. "Ethiopia Food Security Outlook: *Meher* Harvests to Improve Food Security in October, but *Deyr* Rains May Be Below Average." Accessed July 5, 2017. www.fews.net/sites/default/files/documents/reports/Ethiopia_OL_2016_06_final.pdf.

Gondwe, A., and B. Baulch. 2017. *The Case for Structured Markets in Malawi.* IFPRI-ESSP Policy Note 29. Addis Ababa: IFPRI/ESSP.

Graham, J., S. Rashid, and M. Malek. 2012. "Disaster Response and Emergency Risk Management in Ethiopia." In *Food and Agriculture in Ethiopia: Progress and Policy Challenges,* edited by P. Dorosh and S. Rashid, 256–279. Philadelphia: University of Pennsylvania Press.

Hill, R., and H. Fuje. 2017. "What Is the Impact of Drought on Prices? Evidence from Ethiopia." Unpublished, Poverty and Equity Global Practice, Africa Region, World Bank, Washington, DC.

Hirvonen, K., T. P. Sohnesen, and T. Bundervoet. 2018. *Impact of Ethiopia's 2015 Drought on Child Undernutrition.* IFPRI-ESSP Working Paper 114. Addis Ababa: IFPRI/ESSP.

IFPRI (International Food Policy Research Institute). 2017. "Ethiopia Productive Safety Net Program (PSNP) Baseline Community Survey, 2013." Harvard Dataverse, V1. Accessed July 5, 2018. https://doi.org/10.7910/DVN/SNL11W.

IMF (International Monetary Fund). 2018. International Financial Statistics. Accessed July 5, 2018. http://data.imf.org/regular.aspx?key=61545850.

Jayne, T., R. Myers, and J. Nyoro. 2008. "The Effects of NCPB Marketing Policies on Maize Market Prices in Kenya." *Agricultural Economics* 38: 313–325.

Jayne, T., and D. Tschirley. 2009. "Food Price Spikes and Strategic Interactions between the Public and Private Sectors: Market Failures or Governance Failures?" Paper presented at the FAO meeting on Institutions and Policies to Manage Global Market Risks and Price Spikes in Basic Food Commodities, Rome, October 26–27.

Knippenberg, E., and J. Hoddinott. 2017. *Shocks, Social Protection, and Resilience: Evidence from Ethiopia.* IFPRI-ESSP Working Paper 109. Addis Ababa: IFPRI/ESSP.

Leta, S., and F. Mesele. 2014. "Spatial Analysis of Cattle and Shoat Population in Ethiopia: Growth Trend, Distribution and Market Access." *SpringerPlus* 3 (310). https://link.springer.com/article/10.1186/2193-1801 3-310#citeas.

Mann, M. L., J. M. Warner, and A. S. Malik. 2018. *Predicting High-Magnitude Low-Frequency Crop Losses Using Machine Learning: An Application to Cereal Crops in Ethiopia.* IFPRI-ESSP Working Paper 120. Addis Ababa: IFPRI/ESSP.

Minten, B., M. Dereje, E. Engida, and T. Kuma. 2019. "Coffee Value Chains on the Move: Evidence in Ethiopia." *Food Policy* 83 (February): 370–383.

Minten, B., and P. Dorosh. 2006. *Rice Markets in Madagascar in Disarray: Policy Options for Increased Efficiency and Price Stabilization.* World Bank Africa Region Working Paper 101. Washington, DC: World Bank.

Minten, B., D. Stifel, and S. Tamru. 2014. "Structural Transformation of Cereal Markets in Ethiopia." *Journal of Development Studies* 50 (5): 611–629.

Rashid, S., P. Dorosh, and D. Alemu. 2018. "Grain Markets, Disaster Management, and Public Stocks: Lessons from Ethiopia." *Global Food Security* 19: 31–39.

Rashid, S., and S. Lemma. 2014. *Public Wheat Imports into Ethiopia since 2008: The Rationales and Cost-Effectiveness.* REAP (Research for Ethiopia's Agriculture Policy)/IFPRI Discussion Paper. Washington, DC: IFPRI.

Rashid, S., and A. Negassa. 2012. "Policies and Performance of Ethiopia Cereal Markets." In *Food and Agriculture in Ethiopia: Progress and Policy Challenges,* edited by P. Dorosh and S. Rashid, 53–83. Philadelphia: University of Pennsylvania Press.

Tafere, K., P. Zelekawork, A. S. Taffesse, T. Nigussie, and T. Seneshaw. 2009. *Food Demand Elasticities in Ethiopia: Estimates Using 2004/05 Household Income Consumption Expenditure (HICE) Survey Data.* Ethiopia Strategy Support Program II, Discussion Paper 11. Addis Ababa: IFPRI.

Taffesse, A. S., P. A. Dorosh, and S. Asrat. 2013. "Crop Production in Ethiopia: Regional Patterns and Trends." In *Food and Agriculture in Ethiopia: Progress and Policy Challenges,* edited by P. Dorosh and S. Rashid, 53–83. Philadelphia: University of Pennsylvania Press.

Tamru, S., B. Minten, and J. Swinnen. 2019. *Trade, Value Chains, and Rent Distribution with Foreign Exchange Controls: Coffee Exports in Ethiopia.* Ethiopia Strategy Support Program Working Paper 136 (September). Addis Ababa: IFPRI/ESSP.

TesfaNews. 2016. "Food Aid Destined for Ethiopia's Hungry Stuck in Djibouti Port." March 24. Accessed July 5, 2017. www.tesfanews.net/ethiopia-food-aid-stuck-djibouti-port/.

USAID (United States Agency for International Development). 2016. *Ethiopia-Complex Emergency Fact Sheet #8, Fiscal Year (FY) 2016.* Accessed July 5, 2017. https://scms.usaid.gov/sites/default/files/documents/1866/ethiopia_ce_fs08_04-13-2016.pdf.

USDA–FAS (United States Department of Agriculture–Foreign Agricultural Service). 2012. "Ethiopia Grain and Feed Annual Report." Global Agricultural Information Network Report ET 1201. Accessed August 30, 2019. https://gain.fas.usda.gov/Recent%20GAIN%20Publications/Grain%20and%20Feed%20Annual_Addis%20Ababa_Ethiopia_4-17-2012.pdf.

Webb, P., and J. von Braun. 1994. *Famine and Food Security in Ethiopia: Lessons for Africa.* Washington, DC: IFPRI.

WFP (World Food Programme). 2014a. "WFP Ethiopia July Monthly Situation Report #6." Accessed July 5, 2017. http://reliefweb.int/sites/reliefweb.int/files/resources/WFP%20Ethiopia%20Monthly%20Report%20of%20July%202014.pdf.

———. 2014b. "WFP Ethiopia October Monthly Situation Report #7." Accessed July 5, 2017. http://reliefweb.int/report/ethiopia/wfp-ethiopia-october-monthly-situation-report-7.

———. 2015a. "WFP Ethiopia May Monthly Situation Report #1." Accessed July 5, 2017. https://reliefweb.int/report/ethiopia/wfp-ethiopia-may-monthly-situation-report-1.

———. 2015b. "WFP El Niño: Preparedness and Response, Situation Report #1, 06 November 2015." Accessed July 5, 2017. http://reliefweb.int/report/ethiopia/wfp-ethiopia-may-monthly-situation-report-1.

———. 2016a. "WFP Ethiopia Drought Emergency Household Food Security Bulletin #1." Accessed July 5, 2017. http://reliefweb.int/report/ethiopia/wfp-ethiopia-drought-emergency-household-food-security-bulletin-1.

———. 2016b. "WFP Ethiopia Drought Emergency Situation Report #1." Accessed July 5, 2017. www.humanitarianresponse.info/en/operations/ethiopia/document/wfp-ethiopia-drought-emergency-situation-report-1-13-april-2016.

———. 2016c. "WFP Ethiopia Drought Emergency Situation Report #2." Accessed July 5, 2017. http://reliefweb.int/report/ethiopia/wfp-ethiopia-drought-emergency-situation-report-2-3-may-2016.

———. 2016d. "WFP Ethiopia Drought Emergency Situation Report #3." Accessed July 5, 2017. http://reliefweb.int/report/ethiopia/wfp-ethiopia-drought-emergency-situation-report-3-20-may-2016.

———. 2016e. "WFP El Niño 2015–2016: Preparedness and Response, Situation Report #4." Accessed July 5, 2017. http://reliefweb.int/report/world/wfp-el-ni-o-2015-2016-preparedness-and-response-situation-report-4-24-may-2016.

———. 2016f. "WFP Ethiopia Drought Emergency Household Food Security Monitoring Bulletin #3." Accessed July 5, 2017. https://reliefweb.int/report/ethiopia/wfp-ethiopia-drought-emergency-household-food-security-bulletin-3.

———. 2016g. "WFP Ethiopia Situation Report #8." Accessed July 5, 2017. http://reliefweb.int/report/ethiopia/wfp-ethiopia-drought-emergency-situation-report-8-4-august-2016

———. 2016h. "WFP Ethiopia Situation Report #9." Accessed July 5, 2017. http://reliefweb.int/report/ethiopia/wfp-ethiopia-drought-emergency-situation-report-9-15-august-2016

———. 2016i. "WFP Ethiopia Situation Report #14." Accessed July 5, 2017. http://reliefweb.int/report/ethiopia/wfp-ethiopia-drought-emergency-situation-report-14-1-30-november-2016.

———. 2017a. "WFP East and Central Africa Horn of Africa Drought Crisis Situation Report #02." Accessed July 5, 2017. https://reliefweb.int/report/somalia/wfp-east-central-africa-horn-africa-drought-crisis-situation-report-2-20-february.

———. 2017b. "WFP Ethiopia 2017 Emergency Situation Report #1." Accessed July 5, 2017. http://reliefweb.int/report/somalia/wfp-east-central-africa-horn-africa-drought-crisis-situation-report-2-20-february.

———. 2017c. "WFP Ethiopia 2017 Emergency Situation Report #3." Accessed July 5, 2017. http://reliefweb.int/report/ethiopia/wfp-ethiopia-drought-emergency-situation-report-3-may-2017.

WFP and Ethiopia, CSA (Ethiopia, Central Statistical Agency). 2019. "Comprehensive Food Security and Vulnerability Analysis (CFSVA)." Accessed August 30, 2019. https://reliefweb.int/sites/reliefweb.int/files/resources/wfp_ethiopia_cfsva_report_june_2019.pdf.

Wiseman, W., J. Van Domelen, and S. Coll-Black. 2010. *Designing and Implementing a Rural Safety Net in a Low Income Setting: Lessons Learned from Ethiopia's Productive Safety Net Program 2005–2009.* Washington, DC: World Bank.

World Bank. 2016. *Devaluing the Exchange Rate in Ethiopia: Why, When and How.* June. Addis Ababa.

Chapter 10

FOOD SECURITY

David Stifel and Ibrahim Worku Hassen

Introduction

Ethiopia has made remarkable progress in reducing poverty and malnutrition in the presence of overall economic growth since the turn of the century (Stifel and Woldehanna 2016). Furthermore, effective safety nets and targeted food and nutrition interventions have provided important sources of social insurance (Gilligan, Hoddinott, and Taffesse 2009; Coll-Black et al. 2011; Berhane et al. 2012). Nonetheless, food security remains an important concern in the country as the quantity and quality of diets, and related nutritional outcomes, are at risk (Ethiopia, CSA and WFP 2016). Despite the enormous economic progress in the country, it is still unclear to what extent food security has improved and what changes we are seeing in diet diversity and nutritional improvements.

To understand where the country currently stands, this chapter provides an assessment of food security and diet quality in Ethiopia using data from the most recent Ethiopian Household Consumption Expenditures Survey (HCES) dataset (2015/2016)—data that have only been recently released. The HCES are repeated nationally representative cross-sectional surveys collected by the Central Statistical Agency (CSA) of Ethiopia and serve as the official source for poverty statistics in the country. An extensive consumption-expenditure module allows us to estimate household food consumption patterns to assess food security in 2015/2016 and to compare these patterns to those observed in the 2010/2011 HCES. A word of caution must be noted, however. Temporal comparisons over this time period may not indicate long-term trends in food security since much of the country experienced an extended drought in 2015/2016 (Hirvonen, Sohnesen, and Bundervoet 2018) but not in 2010/2011. This drought, which especially affected the eastern part of the country during the main cropping season, appears not to have affected child undernutrition (Hirvonen, Sohnesen, and Bundervoet 2018). Furthermore, drought predictors from 2015 fail to show an impact on household consumption (Sohnesen 2018), and rainfall shocks in our regression analysis do not have an effect on household energy consumption and diet quality. Nonetheless, we remain cautious in our interpretation of long-term food security trends based on comparisons of the 2010/2011 and 2015/2016 HCES data.

Ethiopia continues to face high levels of food insecurity. Daily energy consumption is low (3,055 kilocalories per adult on average), and diet quality is poor (starchy staples account for 71.6 percent of calories) and monotonous (6.1 food groups are consumed per week out of a possible 12). Between 2011 and 2016 the diets of the poorest half of the population improved slightly (less reliance on starchy staples and greater dietary diversity), although this was mostly in rural areas. For the rest of the population, the changes were mixed (less reliance on starchy staples but less diverse diets). Ethiopian diets are also affected by seasonality. Energy consumption tends to be highest in both urban and rural areas during postharvest periods and lowest during lean periods; reliance on cereals, roots, and tubers (starchy staples) tends to vary little seasonally in urban areas but accounts for most of the decline in energy consumption during lean periods in rural areas; dietary diversity varies little seasonally in urban areas but is generally lower during lean periods in rural areas; and the poorest households in urban areas experience more seasonal variation in starchy staple consumption and dietary diversity than less poor households, whereas richest households in rural areas experience more seasonal variation than less poor households in terms of energy consumption, reliance on starchy staples, and dietary diversity.

In both rural and urban areas, smaller, more educated households with higher overall expenditure levels are generally more food secure. Those who have members working as legislators, professionals, technicians, and clerks are also more food secure. Agricultural households in urban areas are among the least food secure. In rural areas, although average energy intake is reasonably high for agricultural households, the quality of their diets is low. In rural areas male-headed households consume fewer calories per adult than female-headed households but have slightly higher-quality diets. In urban areas there is virtually no distinction between male- and female-headed households in terms of food security. We find little evidence of diets being affected by rainfall shocks. This is particularly salient since Ethiopia had just experienced its worst drought in decades.

At first glance the results of this chapter appear to be inconsistent with the evidence presented in Chapter 1 that stunting rates for children under five years of age fell from 44 percent to 38 percent from 2011 to 2016 (see Figure 1.2). This apparent disconnect, however, may follow from the different units of analysis, where food security in this chapter is measured at the household level, while stunting is measured at the individual child level. Since child nutrition outcomes depend on how resources are allocated within households (Thomas 1990; Sahn and Stifel 2002), as well as on other factors such as general feeding practices, mother's health and education, and household-level sanitation (Danaei et al. 2016), the link between average household-level consumption/diet quality and individual-level child nutrition is not automatic. Where this link may be observed, however, is among the poorest half of the population, where improved diet quality may have contributed to less growth faltering in these households. In addition, improved healthcare and access to improved water sources may have played an important role in improving child nutrition outcomes (Headey et al. 2019) despite the persistence of household food security.[1] Finally, it is worth noting that despite the improvements over time, the high level of child stunting in 2016, especially in rural areas (40 percent; Ethiopia, CSA and ICF 2016), is nonetheless consistent with the high levels of food insecurity described in this chapter.

The structure of the chapter is as follows: we begin with a description of the data and methods for measuring food security. After presenting a temporal comparison of the state of food security in 2010/2011 and 2015/2016, we turn to 2015/2016 by analyzing the seasonal nature of food security and providing a profile of the food insecure. We end with concluding remarks.

Methodology and Data

Methodology

> Food security . . . [is achieved] when all people, at all times, have physical and economic access to sufficient, safe and nutritious food to meet their dietary needs and food preferences for an active and healthy life.
>
> (FAO 1996)

In the spirit of Amartya Sen's notion of entitlements (Sen 1981), the statement from the 1996 World Food Summit makes clear that food security depends on peoples' abilities to acquire nutritious food. This in turn depends on food

1 For example, according to the 2011 and 2016 Demographic and Health Surveys (Ethiopia, CSA and ICF 2012; Ethiopia, CSA and ICF 2016) the percentage of children born in health facilities rose (10 percent to 26 percent), as did the share of deliveries assisted by skilled providers (11 percent to 28 percent). In addition, the percentage of children with four or more antenatal care visits rose from 19 percent to 32 percent, while antenatal care received from a nurse or midwife rose from 28 percent to 42 percent. Finally, households with access to improved water sources increased from 54 percent to 65 percent from 2011 to 2016.

being sufficiently available, and people being able to regularly acquire adequate quantities of it, and to store and prepare it in a manner that results in a positive nutritional impact (FAO 1996). As such, food-secure households are those that have diets that are sufficiently high in quantity and in quality. Using this framework as a guide for measuring food security in the context of a household consumption and expenditure survey, we employ two sets of indicators that are measured at the household level: diet quantity and diet quality.

The two diet-quantity indicators that we use are meant to capture the quantity of food that people consume. The first, food energy consumption, is a measure of the total daily number of calories per adult equivalent acquired by the household (see Table 10A.1 for the equivalence scales that we used). It is calculated by adding the energy contents of the edible portions of all foods (purchased, produced, or gifted) acquired by the household over the seven-day recall period in the survey and dividing this sum by seven and by the number of adult equivalents in the household. We computed energy content using the energy conversion factors reported in the Ethiopian food composition table (EHNRI 1997). This measure represents food acquired by the household, not necessarily the food consumed by the household. Nonetheless, the mean of this measure over a random sample of households provides a good estimate of mean food energy consumed by a population (Smith and Subandoro 2007). As such, and for ease of exposition, we refer to food energy acquisition in this report as food energy consumption.

The second diet-quantity measure, food energy deficiency, is an indicator of whether members of a household consume sufficient dietary energy. The idea is that members of a household are food energy deficient if the total household acquisition of dietary energy is not sufficient to meet the dietary energy requirements of all of the household members. These requirements are for basal metabolic function (state of rest) and light activity, and depend on the sex and age of each household member (Smith and Subandoro 2007). In practice, because we divide total household food energy consumption by the number of adult equivalents in the household, it is sufficient to compare the total daily calories per adult equivalent to the energy requirements for an adult male. As such, we record household members as being food energy deficient if their household's daily caloric consumption is less than 2,200 kilocalories per adult equivalent, the minimum caloric requirement that is used to estimate the national poverty line (Ethiopia, MoFED 2013).

We use three diet-quality measures in this analysis in recognition that living "active and healthy lives" requires more than just meeting energy requirements. Macronutrients such as proteins, and micronutrients such as iron, vitamin A, and iodine, are also essential for nutrition and health outcomes (FAO 1996; Welch 2004). In addition, there is a growing literature showing that insufficient energy consumption is becoming less of a binding constraint than inadequate diet quality (Ruel 2003; Graham, Welch, and Bouis 2001; Smith and Subandoro 2007) and that diet-quality measures are a better gauge of the impacts of shocks (Headey and Ecker 2013).

The first diet-quality measure is the percentage of food energy from starchy staples, which is measured as the percentage of total household dietary energy consumption (availability) derived from food staples (for example, maize, teff, wheat, sorghum, barley, roots, and tubers). Because energy-dense starchy staples have limited amounts of proteins and micronutrients, households that consume more of these staples compared to other foods are more vulnerable to protein and micronutrient deficiencies (Smith and Subandoro 2007). As such, a larger percentage of food energy from starchy staples represents a lower-quality diet.

Our second diet-quality measure is related to the first. We indicate that when households have a high proportion of calories from starchy staples, that is, when 75 percent or more of total household food energy consumption derives from energy-dense starchy staples, they have low-quality diets. We report percentages of individuals living in households that acquire a high proportion of their calories from starchy staples.

Our third diet-quality measure is a Household Dietary Diversity Score (HDDS), the number of food groups consumed by household members during the seven-day recall period. Dietary diversity indicators tend to be highly correlated with more complex measures of food and nutrient intake (Ruel, Harris, and Cunningham 2013). As Headey and Ecker (2013) note, there are two basic reasons that these indicators are effective measures of food and nutrition security. First, a diverse diet is one that is more likely to capture the consumption of both macro- and micronutrients (Ruel 2003). Second, there is evidence that individuals will only diversify into higher-value micronutrient-rich foods (such as meats, fish, eggs, dairy products) when they have satisfied their basic caloric needs. Thus, as their incomes grow, poor households substitute out of monotonous energy-dense diets into more diversified nutrient-rich diets (Jensen and Miller 2010). For our HDDS we follow Swindale and Bilinsky (2006) and categorize the 377 food items in the survey into twelve food groups: (1) cereals; (2) roots and tubers; (3) vegetables; (4) fruits; (5) meat, poultry, and offal; (6) eggs; (7) fish and seafood; (8) pulses, legumes, and nuts; (9) milk and milk products; (10) oils and fats; (11) sugar and honey; and (12) miscellaneous food items. Thus the HDDS ranges from

1 (all food consumption derives from one food group) to 12 (household consumed items from all 12 food groups), with higher values representing higher-quality diets.

For each of these quantity and quality measures, we calculate weighted averages of the household indicators to attain sample population means. To do this, we use as weights the household sampling weights multiplied by the household size. This approach results in diet-quantity estimates that are qualitatively similar but slightly different from those reported by the CSA (Ethiopia, CSA 2018). This follows for food energy consumption, for example, because the CSA divides their estimate of all calories consumed in the relevant strata by the estimate of the total number of adults in the strata to get an estimate of calorie consumption per adult equivalent in the strata. Our approach instead results in an estimate of the mean calorie consumption per adult equivalent in the strata.

Data

The primary data sources used for the analysis are the Household Consumption and Expenditures Surveys (HCES) for 2010/2011 (hereafter 2011) and 2015/2016 (hereafter 2016). The HCES data are collected by the CSA and are used to construct household consumption aggregates to calculate the official poverty statistics in Ethiopia. Both surveys were conducted over the course of a year for a similar time period (for example, July 8, 2010–July 7, 2011 and July 8, 2015–July 7, 2016). A total of 27,835 households were interviewed in the 12-month period in the 2011 survey, while 30,229 were interviewed in the 2016 survey. To minimize recall error on consumption, each survey household was visited twice within a single week. The survey covered all 11 regions of the country and included 864 rural and 1,104 urban enumeration areas in 2011, and 864 rural and 1,242 urban enumeration areas in 2016. The HCES did not cover the pastoralist communities in Afar (three zones) and Somali (six zones). The sampling began by stratifying the country into rural and urban areas. After that, the enumeration areas were selected using the probability-proportional-to-size approach, where more populated units have a higher probability of being selected into the sample. The final household sample was formed of households that were randomly selected from these enumeration areas. We use sampling weights, which are based on selection probabilities and provided by the CSA, to compute representative estimates for rural and urban areas of the country.

Table 10.1 shows the sources of calories for households in Ethiopia as measured by the 2011 and 2016 HCES data. Cereals in general are consumed

TABLE 10.1 Shares of calories and shares of households consuming food items in Ethiopia, 2011 and 2016 (%)

	Calorie shares		Share of households consuming	
Food item	2011	2016	2011	2016
All Cereals	63.3	61.1	99.0	99.0
Teff	11.0	12.1	67.5	68.5
Wheat	10.4	11.7	78.7	77.2
Barley	4.1	3.7	19.5	23.4
Maize	20.2	20.5	51.9	53.6
Sorghum	12.4	9.9	25.2	21.0
Other cereals	3.2	3.2	20.7	27.0
Processed cereals	0.2	0.1	4.2	2.1
Roots and tubers	3.6	3.7	5.1	56.9
Pulses	7.3	7.1	90.5	89.5
Oil seeds	0.2	0.1	7.0	3.4
Animal-sourced foods	1.8	2.4	59.0	61.7
Oil and fat	4.3	7.7	82.7	89.2
Vegetables and fruits	2.1	2.8	88.2	90.9
Pepper	1.7	1.0	5.5	73.0
Enset/kocho	8.8	6.8	11.1	10.6
Coffee/tea/chat	1.9	1.5	87.0	92.9
Sugar and salt	2.0	2.6	95.3	91.0
Other foods	3.0	3.2	58.2	59.9
Total	100.0	100.0	100.0	100.0

Source: Authors' calculations from Ethiopia, CSA 2011 and 2016.

by nearly all households and are the most important calorie source, making up just more than 61 percent of total calories for the average household. Among cereals, maize is the largest source (20.5 percent) and is consumed by slightly more than half of all households (53.6 percent). Although wheat is the most ubiquitous (77.2 percent of households consume it), it accounts for a smaller percentage of total calories (11.7 percent). Teff is also consumed by a large percentage of households (68.5 percent) and accounts for an average of 12.1 percent of calories. Other important sources of calories consumed by nearly 90 percent of households are oils and fats (7.7 percent) and pulses (7.1 percent). Although animal-sourced foods (ASF) account for only 2.4 percent of calories, nearly 62 percent of households consume them. This low share of calories attributable to ASF is not surprising given the high prices of these food items, as described in Chapter 8.

Changes in the dietary composition over time were not large (Table 10.1). While still making up the bulk of diets for nearly all Ethiopians, cereals accounted for a smaller share (2.1 percentage points) of calorie consumption in 2016 than in 2011. This was offset by more households consuming more oils and fats (82.7 percent in 2011 and 89.2 percent in 2016), and deriving a larger share of calories from them (4.3 percent to 7.7 percent, respectively). While teff and wheat consumption as a share of total calories increased by 1 percentage point between 2011 and 2016, sorghum consumption fell by over 2 percentage points. Finally, although 2.7 percentage points more households consumed ASF in 2016 compared with 2011, the share of calories derived from ASF increased by only 0.6 percentage point. This may be due to the rapid rise in ASF prices discussed in Chapter 8.

Table 10.2 illustrates how households acquire their calories. Interestingly, the majority of calories are acquired through market purchases (52.2 percent), while 44.3 percent come from consumption of own production, and the remainder (3.6 percent) from other sources such as gifts and in-kind transfers from NGOs and the government. Even though own production provides the largest source of calories from staples (49.9 percent), market purchases are important (45.7 percent), especially for teff (63.8 percent). It is noteworthy that 13.2 percent of calories from wheat are acquired from other sources. This is not surprising given that food aid is provided in the form a wheat. For non-cereals, households rely on own production for root crops (59.7 percent) and enset (76.4 percent). Conversely, 90.3 percent of calories from oils and fats and 80.3 percent of calories from fruits and vegetables are acquired through market purchases.

Food Security, 2011–2016

Ethiopia continues to face high levels of food insecurity. As illustrated in Table 10.3, the quantity of food consumed is low. The average daily energy consumption per adult in 2016 was 3,055 kilocalories.[2] Using a threshold of 2,200 kilocalories, we find that 25.6 percent of Ethiopians are food energy deficient. Furthermore, the quality of diets is poor as indicated by (1) starchy

2 This is larger than the Central Statistical Agency's (Ethiopia, CSA 2018) estimate of 2,468 kilocalories per person per day because the estimates reported here are kilocalories per adult instead of per person.

TABLE 10.2 Shares of calories by sources of acquisition in Ethiopia (%)

Food item	Own production	Purchased	Other†
All cereals	49.9	45.7	4.4
Teff	30.7	63.8	5.4
Wheat	40.5	46.3	13.2
Barley	78.8	20.2	1.0
Maize	55.6	42.6	1.8
Sorghum	62.1	37.1	0.8
Other cereals	50.5	48.2	1.3
Processed cereals	13.1	75.8	11.1
Root crops	59.7	40.0	0.4
Pulses	30.5	66.2	3.3
Oil seeds	45.7	54.0	0.4
Animal-sourced foods	45.5	50.8	3.7
Oil and fat	6.8	90.3	2.8
Vegetables and fruits	18.8	80.3	0.9
Pepper	7.1	92.3	0.6
Enset/kocho	76.4	23.6	0.0
Coffee/tea/chat	29.3	68.9	1.8
Sugar and salt	1.4	97.7	0.9
Other foods	26.6	63.8	9.6
All food	44.3	52.2	3.6

Source: Authors' calculations from Ethiopia, CSA 2016.
Note: † Includes gifts and in-kind transfers from NGOs and government.

staples accounting for an average of 71.6 percent of total household energy consumption, (2) 46.3 percent of the population living in households in which starchy staples account for more than 75 percent of energy consumption, and (3) individuals consuming an average of only 6.1 food groups per week out of a possible 12 food groups (Table 10.4).

Despite the high levels of food insecurity, there is evidence of some uneven improvements between 2011 and 2016. Average daily energy consumption rose by 1.0 percent, from 3,024 kilocalories per adult equivalent to 3,055 kilocalories per adult equivalent, while the percentage of the population that is energy deficient fell by 2.1 percentage points, from 27.7 percent to 25.6 percent (Table 10.3). This change in average levels, however, masks a rather mixed picture throughout the distribution of energy consumption

TABLE 10.3 Diet quantity in Ethiopia by region, 2011 and 2016

	Energy consumption†				Energy deficiency††			
Region	2011	2016	% difference		2011	2016	difference	
National	3,024	3,055	1.0	***	27.7	25.6	−2.1	***
Rural/urban								
Rural	3,091	3,144	1.7	***	26.9	24.6	−2.3	***
Urban	2,688	2,678	−0.4		31.8	29.8	−2.0	***
Region								
Tigray	2,870	2,759	−3.8	***	32.4	28.2	−4.2	***
Afar	2,834	2,870	1.3		31.9	35.0	3.2	*
Amhara	2,669	2,462	−7.8	***	36.6	44.7	8.1	***
Oromia	3,085	3,260	5.7	***	26.6	20.5	−6.1	***
Somali	2,879	2,875	−0.1		29.7	23.0	−6.7	***
Benishangul-Gumuz	3,210	2,826	−12.0	***	21.3	23.9	2.6	
SNNP	3,455	3,660	5.9	***	17.1	10.9	−6.1	***
Gambella	3,225	3,340	3.6	**	26.6	14.7	−11.9	***
Harari	3,054	2,814	−7.8	***	24.2	31.2	7.0	***
Addis Ababa	2,509	2,338	−6.8	***	35.6	33.9	−1.7	
Dire Dawa	2,799	2,648	−5.4	**	28.2	37.4	9.2	***
Rural								
Tigray	2,878	2,770	−3.8	**	33.7	28.7	−5.0	***
Afar	2,853	3,011	5.6	**	33.1	29.4	−3.7	
Amhara	2,667	2,389	−10.4	***	37.4	49.0	11.6	***
Oromia	3,147	3,346	6.3	***	25.6	18.9	−6.7	***
Somali	2,924	2,915	−0.3		29.3	21.7	−7.6	***
Benishangul-Gumuz	3,234	2,802	−13.4	***	21.1	24.3	3.2	
SNNP	3,517	3,811	8.3	***	16.1	7.5	−8.6	***
Gambella	3,466	3,570	3.0		20.4	8.6	−11.8	***
Harari	3,503	3,416	−2.5		13.6	15.9	2.3	
Dire Dawa	3,330	3,237	−2.8		11.9	13.5	1.5	
Urban								
Tigray	2,836	2,727	−3.8	**	27.4	26.8	−0.6	
Afar	2,789	2,255	−19.2	***	28.9	59.7	30.7	***
Amhara	2,686	2,853	6.2	***	31.3	21.8	−9.5	***
Oromia	2,673	2,742	2.6	***	33.2	30.0	−3.1	***
Somali	2,688	2,648	−1.5		31.0	29.9	−1.0	
Benishangul-Gumuz	3,063	2,924	−4.6	*	22.8	22.2	−0.6	

Region	Energy consumption†				Energy deficiency††			
	2011	2016	% difference		2011	2016	difference	
Urban (continued)								
Gambella	2,713	2,892	6.6	***	39.7	26.5	−13.1	***
Harari	2,554	2,310	−9.5	***	36.0	44.1	8.1	**
Addis Ababa	2,509	2,338	−6.8	***	35.6	33.9	−1.7	
Dire Dawa	2,547	2,327	−8.6	***	35.9	50.4	14.5	***

Source: Authors' calculations from Ethiopia, CSA 2011 and 2016.

Note: SNNP = Southern Nations, Nationalities, and Peoples. † Average kilocalories per adult equivalent per day. †† Percentage of population in households consuming less than 2,200 kilocalories per adult per day. *, ** and *** indicate significance levels of 10 percent, 5 percent, and 1 percent, respectively.

(Figure 10A.1). Indeed, there was virtually no change in the level of energy consumption for the lowest 30 percent of the population (distributions overlap), while there was a slight improvement for those in the 30th to 50th percentile and a slight deterioration for those in the 50th to 80th percentile. Changes in diet quality at the national level were also mixed. Although average energy consumption from starchy staples fell by 4.1 percentage points (and throughout the distribution; Figure 10A.2), and although the share of people deriving a high proportion of their energy from starchy staples fell by 11.5 percentage points, dietary diversity worsened slightly from 6.3 food groups to 6.1 food groups on average (Table 10.4). Here too the average masks different experiences throughout the distribution. Although Ethiopian diets became slightly more monotonous on average, those at the lower end of the distribution consumed more food groups in 2016 than in 2011 (Figure 10A.3).

Consistent with Berhane et al. (2012), the Central Statistical Agency (Ethiopia, CSA 2018), and Hirvonen, Tafesse, and Worku Hassen (2016), urban-rural comparisons indicate that energy consumption in rural areas (3,144 kilocalories per adult) is 17 percent higher than in urban areas (2,678 kilocalories per adult), and 34 percent higher than in Addis Ababa (2,338 kilocalories per adult). Similarly, the share of the rural population that is energy deficient (24.6 percent) is 17 percent lower than in urban areas in general (29.8 percent deficient) and 27 percent lower than in Addis Ababa (33.9 percent deficient). These differences, however, more likely represent differences in caloric energy needs than differences in levels of food security (Deaton and Drèze 2009; Headey and Ecker 2013). This follows from the need to expend less energy in urban areas due to less physically demanding labor, increased use of motorized transportation, and reduced energy losses through improved health.

While urban residents consume fewer calories than their rural counterparts, the quality of their diets is higher. The average urban resident in 2016 acquired 60.2 percent of her or his daily calories from starchy staples, which was 19 percent lower than for the average rural resident (74.3 percent). Moreover, the proportion of urban residents relying heavily on starchy staples (15.6 percent) was 70 percent lower than in rural areas (53.5 percent). This was even lower in Addis Ababa, where only 5.1 percent acquired more than 75 percent of their calories from starchy staples. As they shift out of staples, urban residents shift into more diverse diets. This is illustrated by urban residents consuming food from an average of 6.9 different food groups over the course of a week, compared with 5.9 food groups in rural areas.[3]

Diet quantity generally improved in both urban and rural areas between 2011 and 2016. For example, energy deficiency fell in each by roughly 2 percentage points. Interestingly, while average daily energy consumption increased by 1.7 percent in rural areas, from 3,091 kilocalories per adult to 3,144 kilocalories per adult, the level in urban areas in 2011 was not statistically different from the 2,678 kilocalories per adult observed in 2016. As was observed at the national level, changes in the quality of diets in urban and rural areas were mixed. The reliance of households on low-quality starchy staples fell considerably. The average share of starchy staples in urban and rural diets dropped by 6.6 percentage points and 3.2 percentage points, respectively. Similarly, the share of households consuming 75 percent or more of their calories from starchy staples fell by 11.4 percentage points in urban areas and by 10.3 percentage points in rural areas. This decline in the share of staples in total food consumption accompanying the increases in total calories, however, appears to represent shifts away from staples to other food groups that were already being consumed, rather than to new food groups. In fact, dietary diversity appears to deteriorate somewhat, from 6.0 to 5.9 food groups in rural areas and from 7.4 to 6.9 food groups in urban areas.

A regional analysis of rural areas reveals a mixed picture for 2016 with (a) energy consumption highest in SNNP and Gambella, (b) reliance on staples lowest in Somali, Benishangul-Gumuz and Amhara, (c) and dietary diversity highest in Harari and Oromia. Conversely, we find that (a) energy consumption was lowest in Amhara, (b) reliance on staples is highest in Dire Dawa and Harari, and (c) dietary diversity is lowest in Amhara.

---

3 Relatively large differences in dietary quality between urban and rural areas have also been observed in several other studies, including for Ghana (Ecker and Fang 2016), Malawi (Jones, Shrinivas, and Bezner-Kerr 2014), Mozambique (Smart, Tschirley, and Smart 2018), South Africa and Kenya (Steyn et al. 2012), and all of Africa south of the Sahara (Hirvonen 2016).

Compared to the average daily energy consumption in all rural areas in 2016 (3,144 kilocalories per adult), SNNP (3,811 kilocalories per adult), Gambella (3,570 kilocalories per adult), Harari (3,416 kilocalories per adult), and Oromia (3,346 kilocalories per adult) fared well. This translates into fewer than 10 percent of the rural population in SNNP and Gambella suffering from energy deficiency, and less than 20 percent being energy deficient in Harari and Oromia. At the other extreme, however, the low level of daily energy consumption in Amhara (2,389 kilocalories per adult) means that nearly half of the rural population there consumes fewer than 2,200 kilocalories per day on average. This is considerably higher than the all-rural average of 24.6 percent. Although rural food energy consumption fell between 2011 and 2016 in Benishangul-Gumuz (13.4 percent) and Tigray (3.8 percent), the 10.4 percent decrease in Amhara left an already low food-quantity region with even lower consumption. In fact, Amhara was the only province to observe an increase in rural dietary deficiency over this time period.

The low average quality of diets in rural areas masks considerable regional variation. The average share of total calories derived from starchy staples, for example, ranges from 66.1 percent in Somali to 83.7 percent in neighboring Dire Dawa. It is not surprising that rural residents in Somali are less reliant on staples (only 29 percent consume more than 75 percent of total calories from starchy staples) than elsewhere in the country (53 percent of all rural individuals consume a high proportion of staples) given the predominance of pastoralism and the higher consumption of animal-sourced foods consumed there. Interestingly, although residents in Amhara have the lowest levels of daily energy consumption among rural populations, they are less reliant on staples (70 percent of total calories) than those in all other regions except for Somali and Benishangul-Gumuz. Although most regions experienced declines, residents in rural Afar became more dependent on starchy staples as a share of total calories (3.6 percentage point increase to 77.6 percent).

Dietary diversity further adds to the mixed picture. While two regions with higher than average levels of rural daily energy consumption (Harari and Oromia) are also those with the highest levels of dietary diversity (6.4 and 6.3 food groups, respectively), another region with highest energy consumption (SNNP) has the lowest level of dietary diversity (5.8 food groups). So while rural households in SNNP appear to be satisfying their basic caloric needs with energy-dense diets, they are less likely to be acquiring essential macronutrients and micronutrients that accompany more diversified diets. In Amhara, though, low dietary diversity (5.5 food groups) is consistent with low daily caloric intake (2,389 kilocalories).

A regional analysis of urban areas reveals an equally mixed picture for 2016 with (a) energy consumption highest in Benishangul-Gumuz, Gambella, and Amhara, (b) reliance on staples lowest in Afar, and (c) dietary diversity highest in Dire Dawa and Harari. Conversely, we find that (a) energy consumption was lowest in Afar and Harari, (b) reliance on staples is highest in Tigray, Gambella, and SNNP, and (c) dietary diversity is lowest in Amhara and Afar.

In terms of quantity measures of food security, whereas urban residents in Benishangul-Gumuz had the highest levels of daily energy consumption in 2016 (2,924 kilocalories per adult) despite experiencing a 4.6 percent decrease between 2011 and 2016, urban residents in Amhara (2,853 kilocalories per day) and Gambella (2,892 kilocalories per day) consumed more than the urban average (2,678 kilocalories per day) due to growth over this period. Not surprisingly, these three regions also experienced the lowest levels of urban energy deficiency in 2016 (22.2 percent, 21.8 percent, and 26.5 percent, respectively).

An indication that calorie consumption may not be a good indicator of food security in urban areas is that the four regions in which urban food energy consumption is lowest and in which food energy deficiency is highest (Afar, Harari, Dire Dawa, and Addis Ababa) are also regions characterized by high-quality diets. For example, the proportion of urban residents in these regions whose diets include a high proportion of starchy staples are well below the urban average of 15.6 percent (for example, 4.0 percent, 6.1 percent, 5.1 percent, and 2.9 percent, respectively). In addition, although dietary diversity in urban Afar is among the lowest in the all urban areas (6.6 food groups), it is highest in Dire Dawa (7.5), Harari (7.3), and Addis Ababa (7.1).

Seasonality

Levels of food security in Ethiopia vary seasonally due to weather patterns, crop cycles and religious observances. General patterns observed over the 12 months from July 2015 through June 2016 during which data were collected for the HCES include the following: (1) energy consumption was highest in both urban and rural areas during postharvest periods and lowest during lean periods; (2) reliance on cereals, roots, and tubers (starchy staples) varied little in urban areas but appears to account for most of the decline in energy consumption during lean periods in rural areas; (3) dietary diversity varied little in urban areas but was generally lower during lean periods in rural areas; and (4) the poorest households in urban areas experienced more seasonal variation in starchy staple consumption and dietary diversity than less poor households, whereas the richest households in rural areas experienced more seasonal variation than less poor households in terms of energy consumption, reliance on starchy staples, and dietary diversity.

As illustrated in Figure 10.1a, while energy consumption in rural areas is higher and slightly more variable than in urban areas, the seasonal patterns are generally similar. Average calorie intake per adult equivalent was highest at the beginning of the *meher* harvest in September (rural: 3,763 kilocalories; urban: 2,929 kilocalories) and the *belg* harvest in April (rural: 3,281 kilocalories; urban: 2,791 kilocalories), and was low in the lean months just prior to these harvests (rural: 3,121 kilocalories for June–August and 3,027 kilocalories for January–March; urban: 2,558 kilocalories in August and 2,524 kilocalories in March).

Figure 10.1b and 10.1c show that although energy consumption varied over the course of the year in urban areas, the quality of diets there remained stable. Throughout the course of the year, urban diets consisted of roughly seven food groups, with starchy staples accounting for about 60 percent of calories consumed. In rural areas, however, the quality did change over time. For example, during the lean season months, the decline in rural energy intake appears to follow from the lower consumption of starchy staples. This is illustrated by the percentage of calories from starchy staples falling from more than 75 percent to 70 percent just prior to the *belg* harvest, and from 75 percent to 71 percent just prior to the *meher* harvest.

Turning to household dietary diversity, we find that rural diet quality is lowest in the *belg* lean period months of February and March (5.8 food groups on average) and highest in September (6.3 food groups) and July (6.4 food groups). The high dietary diversity score in July is puzzling since this is in the midst of the *meher* lean period, when energy consumption is low. Furthermore, this is nearly 0.5 food groups higher than in June and August. Interestingly, Hirvonen, Tafesse, and Worku Hassen (2016) find a similar result using the 2011 HCES. One partial explanation may be that this follows from an increase in vegetable consumption since harvests from homestead garden crops reach their peak in July in Ethiopia (Hirvonen and Headey 2018). This is consistent with our finding that the percentage of people consuming fruits and vegetables rose from 86 percent in June to 96 percent in July (Figure 10.2).

To get a better understanding the seasonal composition of diets in Ethiopia, we illustrate in Figure 10.3 the monthly shares of daily energy intake from noncereal sources. Consistent with the stability of the quality of urban diets observed in Figure 10.1, we find that the composition of urban diets generally did not change substantively over the course of the survey

FIGURE 10.1 Seasonality and food security in Ethiopia

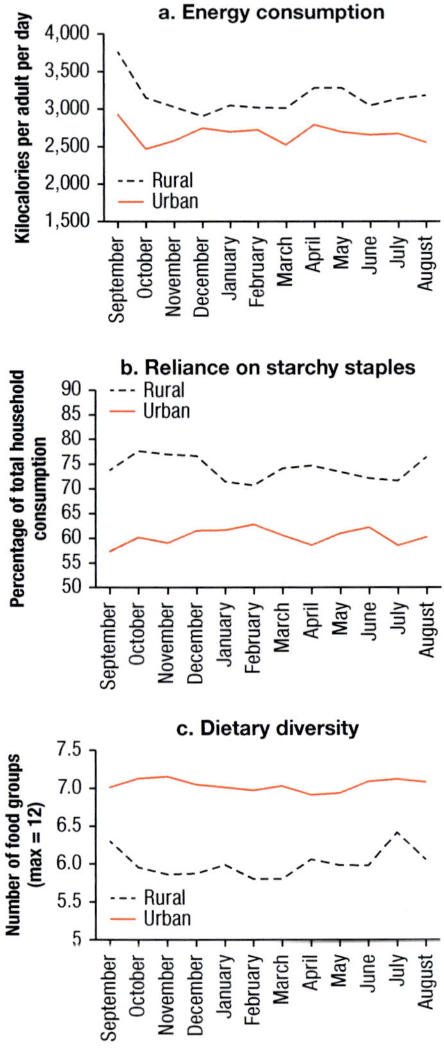

Source: Authors' calculations from Ethiopia, CSA 2016.

year (Figure 10.3a). The main source of calories was cereals, which constituted 56 percent of energy intake throughout the year. Other foods (oils, fats, sugars, and so on) were the next most important source of energy, making up a stable quarter of urban diets. Pulses, legumes, and nuts made up about 7 percent of urban energy consumption, while fruits and vegetables and roots

FIGURE 10.2 Percentage of households consuming items from noncereal sources in Ethiopia, by month

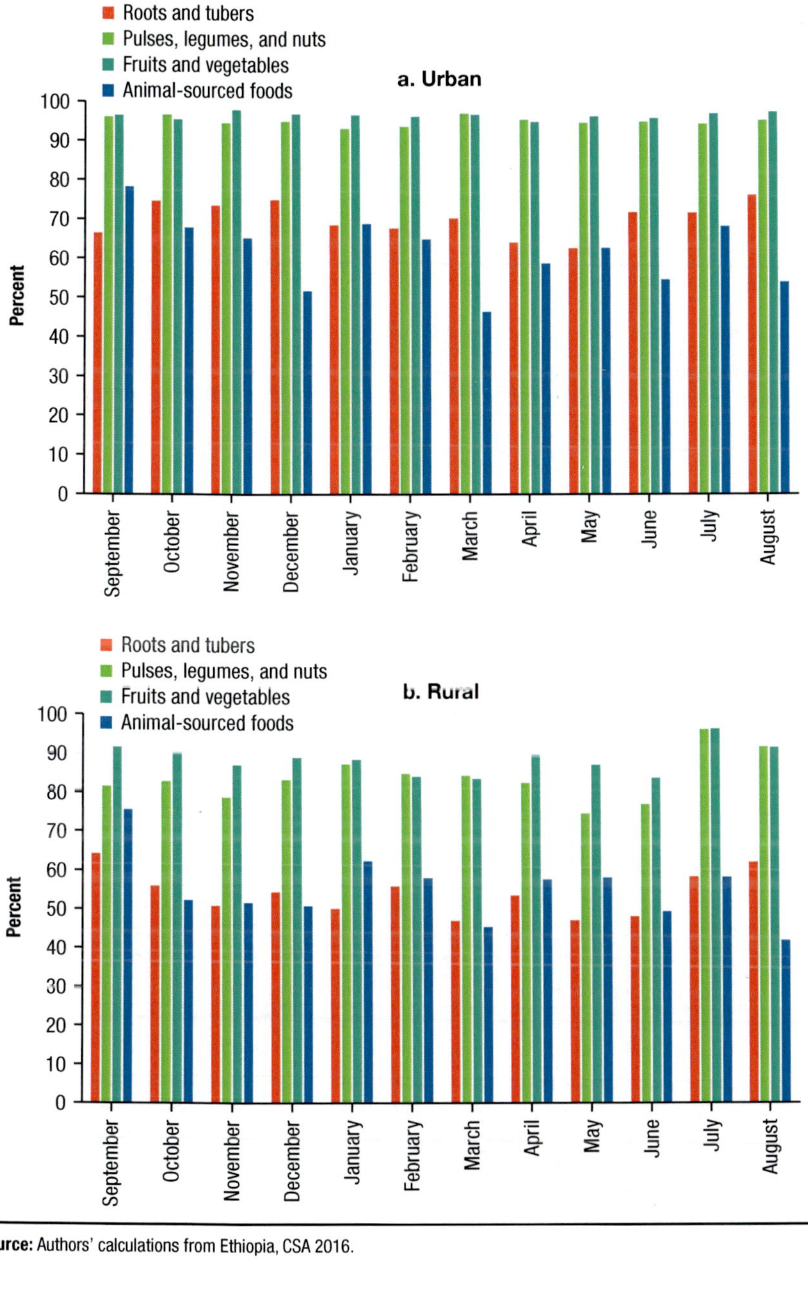

Source: Authors' calculations from Ethiopia, CSA 2016.

FIGURE 10.3 Percentage of daily energy intake from noncereal sources in Ethiopia, by month

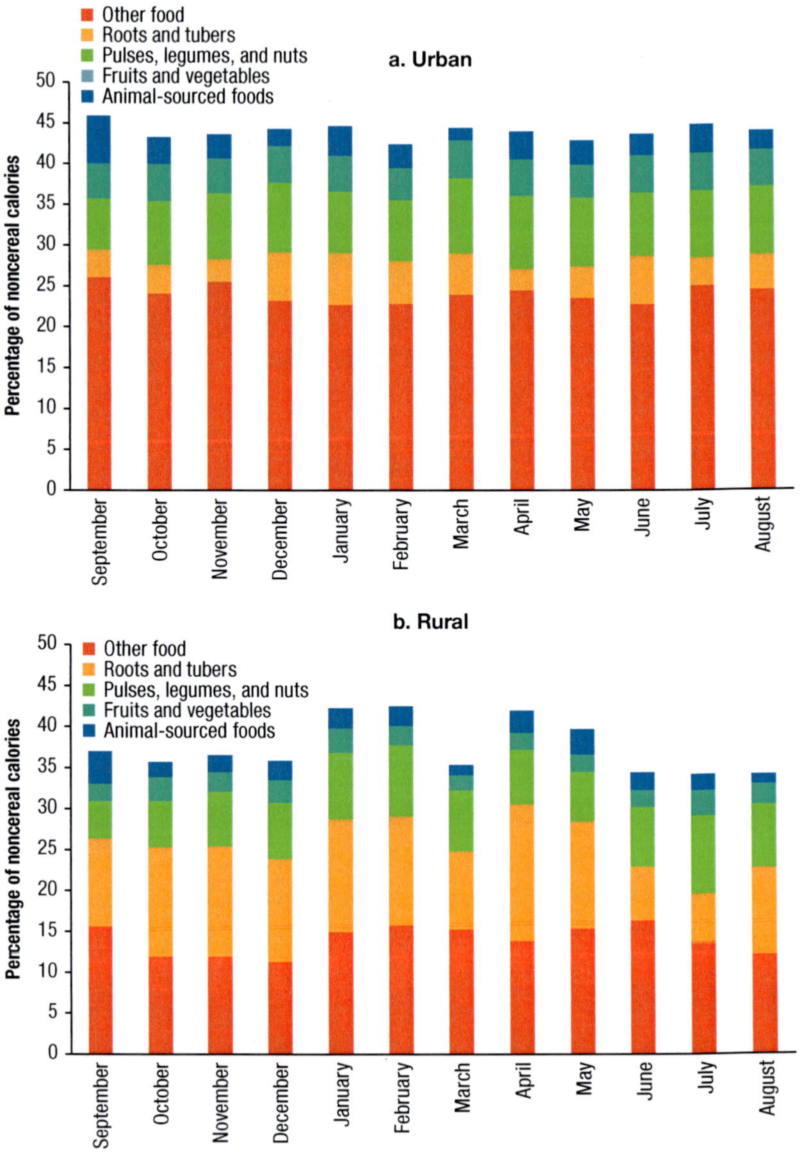

Source: Authors' calculations from Ethiopia, CSA 2016.

and tubers each accounted for 4 percent. Finally, animal-sourced foods (meat, poultry, fish, and dairy products) account for only 3 percent of calorie intake, but this does vary from a high of 6 percent in September to a low of 2 percent during the fasting months of December and March.

Rural diets (Figure 10.3b) are more heavily weighted toward cereals (63 percent) than urban diets (56 percent) and experience greater fluctuations over the course of the year. The largest sources of seasonal variation in rural diets come from cereals and roots and tubers. For example, cereals fell from 64 percent of rural diets in December to 57 percent in February. This was offset by a 2 percentage point rise of roots and tubers from 12 percent to 14 percent, as well as a 5 percentage point rise in other foods, to the extent that calorie intake over this period did not change (Figure 10.1a). In a similar manner the 6 percentage point decline in the share of roots and tubers between May and June was entirely offset by a 6 percentage-point increase in the share of cereals.

Figure 10.4 illustrates how the seasonal variation in our diet-quantity and -quality measure differs by total household expenditure quintile for urban and for rural areas. Note that the quintiles are calculated over the full sample.[4] So rural households in the second quintile, for example, have a similar range of expenditure levels as urban households in the second quintile. Several patterns emerge from these figures. First, the poorest households in both urban and rural areas have the lowest energy consumption, the highest reliance on starchy staples, and the most monotonous diets. Second, the fact that the richest urban households consume fewer calories on average than rural households in the middle quintile provides further evidence that we should not put too much emphasis on this measure of food security when making urban-rural comparisons. Third, aside from energy consumption, where seasonal variability was modest across all expenditure quintiles, poorer households in urban areas experienced more seasonal variation in their diets than households in higher quintiles. Conversely, in rural areas, richer households experienced more seasonal variation in their diets compared to poorer households. This is especially the case for energy consumption where daily energy consumption for households in the richest quintile ranges from 6,125 to 4,416 kilocalories per adult, compared to a range from 2,288 to 2,086 kilocalories per adult for those in the poorest quintile.

4 Using total household expenditure quintiles calculated separately within urban and rural areas does reduce some of the noisiness of the highest rural quintile and of the lowest urban quintile. But it does not qualitatively change the messages of Figure 10.4.

FIGURE 10.4 Seasonality and food security in Ethiopia, by expenditure quintile

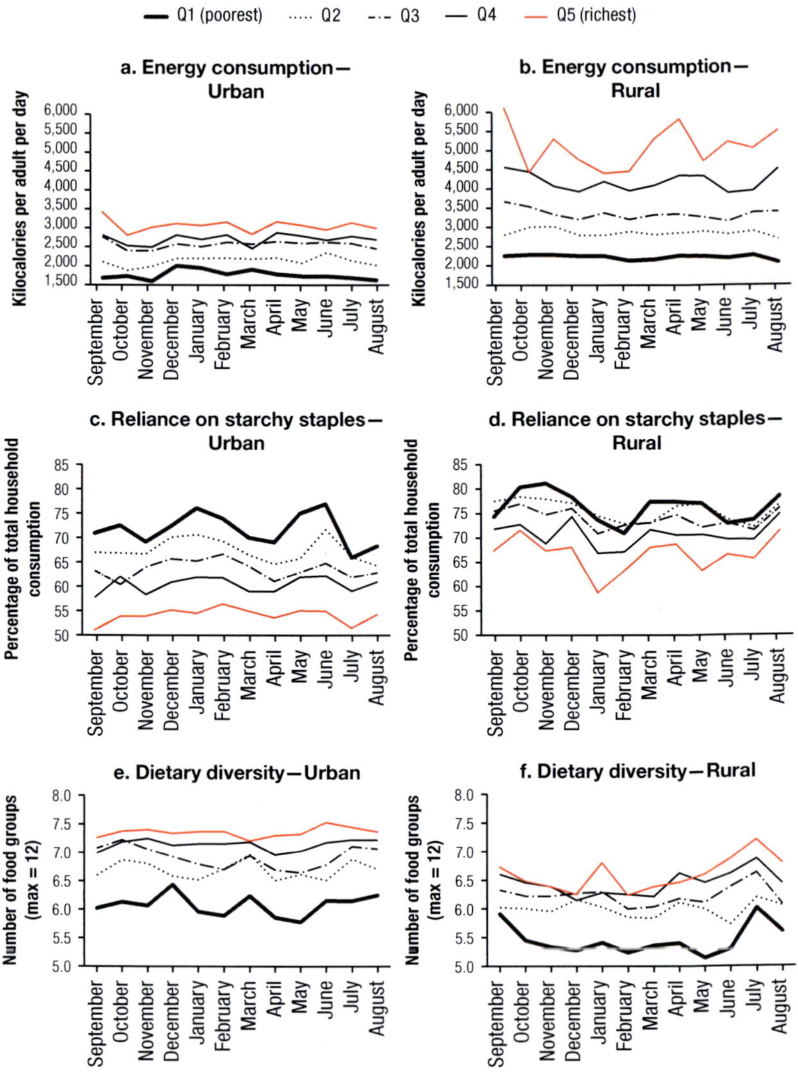

Source: Authors' calculations from Ethiopia, CSA 2016.

Correlates of Food Security

To better understand the mechanisms that affect household food security in Ethiopia in 2016, we present household-level correlates of food security in rural (Table 10.4) and urban (Table 10.5) areas. The major takeaways are that in both rural and urban areas, smaller, more educated households with higher overall expenditure levels are generally more food secure (Figures 10.A4 and 10.A5). Those who have members working as legislators, professionals, technicians, and clerks are also more food secure. Agricultural households in urban areas are among the least food secure. In rural areas, although average energy intake is reasonably high for agricultural households, the quality of their diets is low. In rural areas male-headed households consume fewer calories per adult than female-headed households but have slightly higher-quality diets. In urban areas there is virtually no distinction between male- and female-headed households in terms of food security.

Smaller households have higher energy consumption per adult equivalent than larger ones. For example, in rural areas one-person households consume an average of 4,568 kilocalories per day, compared with 2,829 kilocalories per day for households with six or more members. The gap is not as large in urban areas but remains significant nonetheless, with one-person households consuming 3,443 kilocalories, compared with 2,326 kilocalories for households with six or more members. While energy deficiency levels are higher in urban areas, as noted previously, the gap between small and large households is larger in urban areas (23.8 percentage points) than in rural areas (21.9 percentage points) as well. In terms of the quality of diets, households with fewer members generally rely less on staples and have more diverse diets. While the share of total calories from staples increases monotonically with household size (65.2 percent to 75.7 percent in rural areas, and 52.3 percent to 64.0 percent in urban areas), the number of food groups plateaus after four members. In rural areas one-person households consume an average of just under five food groups in a week, while households with four or more members consume about six food groups. In urban areas one-person households consume 1.5 fewer food groups (5.6) than those with four or more members (7.1).

Better educated households are generally more food secure. Using the education level of the household head as our measure of household education, we find that more educated households in both rural and urban areas tend to have higher energy intake and better quality diets. The exception is households whose head has adult or informal education (that is, literate but with no formal education). In urban areas the food security status of these

TABLE 10.4 Food security and household characteristics in Ethiopia, rural areas (2016)

	Diet quantity		Diet quality		
Characteristic	Energy consumed[a]	Energy deficiency[b]	Share from staples[c]	High portion staples[d]	Dietary diversity[e]
Sex of household head					
Female	3,451	19.2	75.1	57.5	5.8
Male	3,082	25.7	74.1	52.7	6.0
Household size					
1	4,568	8.7	65.2	27.7	4.9
2	4,202	9.8	69.8	41.4	5.7
3	3,898	13.3	71.8	43.1	5.8
4	3,522	16.5	72.8	46.4	6.0
5	3,180	21.5	73.6	51.2	5.9
6+	2,829	30.6	75.7	58.7	6.1
Education (household head)					
None	3,069	26.1	74.9	55.3	5.8
Grades 1–4	3,216	21.3	75.0	56.4	6.2
Grades 5–8	3,510	15.8	74.2	53.1	6.4
Grades 9–12	3,614	10.3	72.8	48.8	6.6
Postsecondary	4,112	6.1	67.3	29.3	7.0
Adult/Informal	2,557	45.2	70.4	41.3	5.8

households is not substantively different from households with no or little education. In rural areas these households are substantially worse off than all others in terms of energy intake. The following discussion thus focuses on the 93 percent of households who either have no education or have at least some formal education.

In rural areas diet quantity increases monotonically with each level of education. For example, households with no education consume an average of 3,069 kilocalories per adult per week, while those with postsecondary education consume 4,112 kilocalories per adult per week. Similarly, energy deficiency declines monotonically and dramatically with schooling. The share of individuals who consume fewer than 2,200 kilocalories per adult per day is 26.1 percent for those in households with no education, compared to 6.1 percent for those in households with postsecondary education. Diet quality for rural households tends to increase for households with five or more years of formal schooling. For example, while the share of calories derived from starchy staples is about 75 percent for households without any education

Characteristic	Diet quantity		Diet quality		
	Energy consumed[a]	Energy deficiency[b]	Share from staples[c]	High portion staples[d]	Dietary diversity[e]
Expenditure quintiles					
Q1 (poorest)	2,230	55.8	76.7	61.1	5.4
Q2	2,863	17.0	75.8	58.4	6.0
Q3	3,347	12.7	74.0	52.9	6.2
Q4	4,207	5.6	70.6	41.2	6.4
Q5 (richest)	5,126	1.7	66.7	31.1	6.6
Profession					
Agriculture	3,298	24.8	75.2	57.3	5.6
Legislator	3,375	18.0	65.5	12.9	6.6
Professional	4,059	6.0	65.3	20.5	7.2
Technician	3,673	8.9	65.0	33.4	7.0
Clerk	3,226	13.8	60.6	16.4	6.6
Service	3,100	28.3	72.6	49.5	6.0
Skilled	3,128	24.5	74.5	54.1	6.0
Craft	3,089	28.1	72.4	47.0	6.0
Plant/machine operator	2,922	24.9	68.4	20.1	6.2

Source: Authors calculations from HCES 2015/2016 (Ethiopia, CSA 2016).

Note: a. Average kilocalories per adult per day.

b. Percentage of population in households consuming less than 2,200 kilocalories per adult per day.

c. Average share of total calories from starchy staples.

d. Percentage of population in households consuming more than 75 percent of total calories from starchy staples.

e. Average number of food groups (max = 12) consumed by household during previous week.

and for those with one to four years of schooling, the reliance on staples falls for households that acquired more than four years of education (for example, 74.2 percent for those with five to eight years, 72.8 percent for those with nine to twelve years, and 67.3 percent for those with more than twelve years). This translates to a similar decline in the percentage of those living in households with a high proportion (75 percent) of their calories from starchy staples. For example, 56 percent of those in households with less than five years of schooling consume a high proportion from starchy staples, compared with 29.3 percent of those in households with postsecondary education. Dietary diversity also increases with education, rising from an average of 5.8 food groups for households with no education to 7.0 food groups for those with postsecondary education.

In urban areas diet quantity improves with education for those with more than eight years of schooling. For example, individuals in urban households

TABLE 10.5 Food security and household characteristics in Ethiopia, urban areas (2016)

	Diet quantity		Diet quality		
Characteristic	Energy consumed[a]	Energy deficiency[b]	Share from staples[c]	High portion staples[d]	Dietary diversity[e]
Sex of household head					
Female	2,712	30.7	60.0	15.4	7.0
Male	2,661	29.4	60.3	15.8	7.1
Household size					
1	3,443	16.4	52.3	8.5	5.6
2	3,163	20.9	54.7	9.5	6.7
3	2,933	22.5	56.9	12.0	7.0
4	2,792	24.9	59.8	13.8	7.1
5	2,575	28.6	61.1	14.7	7.2
6+	2,326	40.2	64.0	21.2	7.2
Education (household head)					
None	2,597	34.1	63.6	21.3	6.7
Grades 1–4	2,579	33.9	63.4	24.3	6.9
Grades 5–8	2,568	34.7	60.0	15.6	7.0
Grades 9–12	2,699	27.7	58.1	11.5	7.2
Postsecondary	2,915	20.2	56.7	8.3	7.5
Adult/informal	2,621	30.0	63.4	19.8	6.7

with eight years of schooling or less consume approximately 2,600 kilocalories per adult on average, while those with nine to twelve years consume 2,699 kilocalories per adult, and those with postsecondary education consume 2,915 kilocalories per adult. This pattern is reflected in energy deficiency levels, which fall from about 34 percent for those with eight years of schooling or less, to 27.7 percent for those with nine to twelve years, and further to 20.2 percent for those with postsecondary education. Diet quality similarly improves with education but starts with five to eight years of schooling, rather than nine to twelve years. For example, the average share of starchy staples in the diets of urban households falls monotonically from 63.4 percent for those with one to four years of schooling, to 56.7 percent for those with postsecondary education. Similarly, the share of those with a high proportion of starchy staples in their diets falls monotonically from 24.3 percent for households with one to four years of schooling to 8.3 percent for those with postsecondary education. Finally, dietary diversity in urban areas increases monotonically

	Diet quantity		Diet quality		
Characteristic	Energy consumed[a]	Energy deficiency[b]	Share from staples[c]	High portion staples[d]	Dietary diversity[e]
Expenditure quintiles					
Poorest	1,768	72.2	71.3	44.4	6.1
Q2	2,132	53.9	67.3	30.9	6.7
Q3	2,562	28.8	63.5	19.8	6.9
Q4	2,711	25.2	60.4	13.1	7.1
Richest	3,094	15.2	53.9	3.9	7.4
Profession					
Agriculture	2,570	33.3	60.8	15.7	7.0
Legislator	2,743	23.5	56.9	10.0	7.6
Professional	2,918	19.4	57.0	8.4	7.5
Technician	2,940	18.9	57.6	9.2	7.3
Clerk	2,789	24.1	56.7	8.3	7.3
Service	2,577	33.9	59.3	14.4	7.0
Skilled	2,850	26.9	68.7	33.5	6.7
Craft	2,655	31.0	62.8	18.9	6.8
Plant/machine operator	2,609	29.2	56.8	11.6	7.4

Source: Authors' calculations from HCES 2015/2016 (Ethiopia, CSA 2018).
Note: a. Average kilocalories per adult per day.
b. Percentage of households consuming less than 2,200 kilocalories per adult per day.
c. Average share of total calories from starchy staples.
d. Percentage of households consuming more than 75 percent of total calories from starchy staples.
e. Average number of food groups (max = 12) consumed by household during previous week.

throughout the formal education distribution, rising from 6.7 food groups for those without any education to 7.5 food groups for those with postsecondary education.

To explore the avenues through which education may affect household food security, we regress our energy consumption, share of starchy staples, and dietary diversity variables on household education and progressively add such variables as expenditure quintiles, household demographics, rainfall shocks, and profession of the household head. Comparisons of the estimated coefficients on education across the models yields insights into the possible mechanisms linking household education and food security.[5]

5 This approach, referred to as mediation or suppression effects, is common in the public health literature (MacKinnon, Krull, and Lockwood 2000).

TABLE 10.6 Regressions of household energy consumption in Ethiopia per adult equivalent

	Rural			
Variables	Model 1	Model 2	Model 3	Model 4
Education of household head (base = none)				
Grades 1–4	147.1** (45.57)	97.31* (38.09)	118.8** (37.74)	5.216 (36.11)
Grades 5–8	440.4*** (57.93)	202.5*** (48.47)	216.2*** (47.61)	98.38* (45.88)
Grades 9–12	544.3*** (108.20)	107.7 (85.10)	124.7 (88.29)	−34.62 (87.76)
Postsecondary	1042.7*** (152.60)	−67.26 (153.30)	−91.22 (158.50)	80.44 (222.90)
Adult/informal	−512.5*** (45.54)	−442.8*** (38.36)	−350.4*** (38.85)	−46.21 (38.68)
Expenditure quintile (base = poorest)				
Q2		610.6*** (24.15)	577.4*** (24.12)	415.0*** (23.92)
Q3		1078.3*** (39.05)	1034.8*** (37.87)	924.1*** (36.64)
Q4		1946.3*** (51.12)	1893.6*** (53.15)	1736.3*** (51.31)
Richest		2862.1*** (82.63)	2764.9*** (85.45)	2584.2*** (81.47)
Household head is male (dummy)			−53.01 (38.00)	−52.28 (37.10)
Age of household head			−5.322*** (1.03)	−6.208*** (1.02)
Household size			−97.80*** (7.67)	−122.6*** (7.80)
Share of dependents in household			1010.4*** (68.07)	979.3*** (63.45)
Meher rain deviation from average				0.992 (1.95)
Additional controls	No	No	No	Yes
R2	0.034	0.348	0.383	0.475
Number of observations	10,366	10,366	10,366	10,366

Source: Authors' calculations from HCES 2015/2016 (Ethiopia, CSA 2016).

Note: Additional controls include occupation, region, and agroecological zone dummies. Robust standard errors in parentheses. * $p < 0.05$, ** $p < 0.01$, *** $p < 0.001$.

Variables	Urban			
	Model 1	Model 2	Model 3	Model 4
Education of household head (base = none)				
Grades 1–4	−18.16 (38.71)	−77.67* (34.37)	−76.32* (32.55)	−65.94* (31.72)
Grades 5–8	−29.78 (30.97)	−194.4*** (27.77)	−187.1*** (27.85)	−112.5*** (26.81)
Grades 9–12	101.4*** (30.79)	−208.0*** (28.49)	−196.0*** (28.95)	−95.29*** (28.10)
Postsecondary	318.0*** (30.62)	−111.1*** (30.06)	−104.1*** (30.66)	−95.56** (32.60)
Adult/informal	23.89 (46.25)	−79.3 (41.42)	−10.07 (38.63)	−39.78 (37.83)
Expenditure quintile (base = poorest)				
Q2		383.1*** (34.08)	377.6*** (32.88)	431.9*** (32.08)
Q3		816.4*** (33.92)	789.6*** (33.19)	848.9*** (32.91)
Q4		978.2*** (31.28)	943.6*** (31.23)	1047.5*** (31.61)
Richest		1374.6*** (30.38)	1325.9*** (32.57)	1471.3*** (33.12)
Household head is male (dummy)			39.46* (18.58)	−63.04*** (19.15)
Age of household head			−3.276*** (0.64)	−0.321 (0.66)
Household size			−99.04*** (5.70)	−97.58*** (5.64)
Share of dependents in household			844.4*** (39.80)	726.0*** (37.51)
Meher rain deviation from average				0.537 (0.47)
Additional controls	No	No	No	Yes
R2	0.016	0.175	0.242	0.316
Number of observations	19,856	19,856	19,856	19,856

These models suggest that in rural areas the effects of education on energy consumption and the share of starchy staples are almost entirely mediated by households' abilities to acquire goods (that is, household expenditure quintile). Household expenditures, profession of the household head, and other controls account for roughly half of the effect of education on dietary diversity, suggesting that education still has a direct effect on the diversity of diets in rural areas. To see this, we note first that Model 1 for both rural (column 1) and urban (column 5) areas in Tables 10.6, 10.7, and 10.8 show the differences in energy intake, share of starchy staples, and dietary diversity by household education levels that are illustrated in Tables 10.4 and 10.5. When we add expenditure quintiles to the model (Model 2), the magnitudes of the education coefficients for energy consumption (Table 10.6) and share of starchy staples (Table 10.7) fall substantially and in most cases become insignificant.

Given the positive association between education and household expenditures (rural households with postsecondary education have expenditure levels that are 60 percent higher than those with no education in this sample), and the positive and monotonic relationship between household expenditure quintiles and all of our food security indicators (Table 10.4), the fall in the magnitude of the coefficients on education suggests that the positive relationship between education and energy consumption and share of starchy staples is due to the effect of education on household expenditure. While the magnitudes of the education coefficients also fall for the rural dietary diversity models (Table 10.8), they remain statistically and economically significant once household expenditure quintiles are added to the model. They fall further once household demographics, rainfall shocks, and profession of the household head are added but remain statistically and economically significant. Thus, although expenditures and profession mediate part of the effect of education on dietary diversity, education appears to have an independent effect as well.

For urban areas these models suggest that although household expenditure and profession mediate some of the effect, education continues to have an independent effect on diet quality (coefficients in urban Model 4 in Table 10.8 are positive and statistically significant). Interestingly, the mediated effect of education on energy consumption through household expenditures appears to be so large that education becomes negatively associated with energy consumption when we control for expenditures. The decline in magnitude of the education coefficients when we further add occupation suggest that the negative relationship between education and energy consumption may be due to less demanding work-energy requirements for those with more education. Nonetheless, even when the full controls are included in the model, the education coefficients in the energy requirement models remain statistically significant.

Finally, controlling for agroecological zones in Model 4, we find no evidence that rainfall shocks had an impact on food security in rural areas. We do find evidence of an impact on diet quality in urban areas, but the magnitudes are small. For example, one additional millimeter of rain during the *meher* season above the 30-year *meher* season average is associated with a modest increase of 0.002 food groups. This result is particularly salient since Ethiopia experienced in 2015 one of the worst droughts in decades (Sohnesen 2018). This lack of an impact is consistent with the findings of Hirvonen, Sohnesen, and Bundervoet (2018) that the drought did not lead to widespread undernutrition. Hirvonen, Sohnesen, and Bundervoet (2018) credit this resilience to the role of the improved road infrastructure and to the effective targeting of aid and social safety nets. We are unable to conduct inference on these possible sources of resilience. Nonetheless, the absence of a negative effect of rainfall shocks in the models, as well as the fact that we did not observe a widespread deterioration of food security between 2011 and 2016, suggests that the transformation and integration of food markets (Chapters 7 and 9) may have had a dampening effect on shocks and that responses to the drought by the Ethiopian government and the international community were successful.

Conclusion

This chapter provides an updated assessment of food security in Ethiopia using data from the most recent Ethiopian Household Consumption and Expenditure Survey dataset (2015/2016). The HCES are repeated nationally representative cross-sectional surveys collected by the Central Statistical Agency and serve as the official source for poverty statistics in Ethiopia. An extensive consumption–expenditure module allows us to estimate household food consumption patterns to assess food security in 2016 and to compare these patterns to those observed in the 2011 HCES.

Ethiopia continues to face high levels of food insecurity. Daily energy consumption is low (3,055 kilocalories per adult on average) and diet quality is poor (starchy staples account for 71.6 percent of calories) and monotonous (6.1 food groups are consumed per week out of a possible 12). Moreover, energy consumption is especially low for the poorest total household expenditure quintile (2,200 kilocalories per adult), compared with the richest quintile (3,916 kilocalories per adult). Between 2011 and 2016 the diets of the poorest

TABLE 10.7 Regressions of share of starchy staples in household diets in Ethiopia

	Rural			
Variables	Model 1	Model 2	Model 3	Model 4
Education of household head (base = none)				
Grades 1–4	0.0944 (0.44)	0.225 (0.43)	0.755 (0.45)	−0.618 (0.44)
Grades 5–8	−0.652 (0.48)	0.113 (0.47)	0.876 (0.49)	−0.256 (0.46)
Grades 9–12	−2.020* (0.92)	−0.534 (0.88)	0.654 (0.88)	−0.706 (0.84)
Postsecondary	−7.532*** (1.28)	−3.539** (1.20)	−2.061 (1.17)	−0.38 (1.53)
Adult/informal	−4.424*** (0.62)	−4.620*** (0.60)	−4.501*** (0.60)	−1.774** (0.56)
Expenditure quintile (base = poorest)				
Q2		−1.066* (0.44)	−0.824 (0.44)	−2.529*** (0.42)
Q3		−2.931*** (0.47)	−2.729*** (0.47)	−4.129*** (0.47)
Q4		−6.207*** (0.50)	−5.614*** (0.53)	−7.263*** (0.54)
Richest		−9.899*** (0.64)	−8.786*** (0.68)	−11.14*** (0.68)
Household head is male (dummy)			−2.164*** (0.42)	−1.810*** (0.41)
Age of household head			0.0695*** (0.01)	0.0431*** (0.01)
Household size			0.480*** (0.09)	0.308*** (0.09)
Share of dependents in household			1.365 (0.79)	0.593 (0.77)
Meher rain deviation from average				−0.0275 (0.02)
Additional controls	No	No	No	Yes
R^2	0.013	0.057	0.071	0.193
Number of observations	10,366	10,366	10,366	10,366

Source: Authors' calculations from HCES 2015/2016 (Ethiopia, CSA 2016).

Note: Additional controls include occupation, region, and agroecological zone dummies. Robust standard errors in parentheses. * $p < 0.05$, ** $p < 0.01$, *** $p < 0.001$.

Variables	Urban			
	Model 1	Model 2	Model 3	Model 4
Education of household head (base = none)				
Grades 1–4	−0.169 (0.61)	0.566 (0.56)	0.774 (0.56)	−0.38 (0.54)
Grades 5–8	−3.636*** (0.45)	−1.548*** (0.42)	−1.085* (0.45)	−1.307** (0.44)
Grades 9–12	−5.509*** (0.46)	−1.494*** (0.44)	−0.811 (0.47)	−0.879 (0.47)
Postsecondary	−6.893*** (0.42)	−1.236** (0.42)	−0.722 (0.45)	−1.615** (0.53)
Adult/informal	−0.156 (0.68)	1.133 (0.64)	0.406 (0.65)	−1.16 (0.62)
Expenditure quintile (base = poorest)				
Q2		−3.840*** (0.70)	−3.622*** (0.70)	−2.894*** (0.66)
Q3		−7.594*** (0.68)	−7.148*** (0.69)	−5.956*** (0.67)
Q4		−10.59*** (0.62)	−9.888*** (0.63)	−8.238*** (0.60)
Richest		−16.75*** (0.60)	−15.45*** (0.63)	−13.22*** (0.62)
Household head is male (dummy)			0.385 (0.29)	−0.303 (0.30)
Age of household head			0.0675*** (0.01)	0.0599*** (0.01)
Household size			0.399*** (0.08)	0.618*** (0.08)
Share of dependents in household			2.186*** (0.61)	0.976 (0.58)
Meher rain deviation from average				−0.0195** (0.01)
Additional controls	No	No	No	Yes
R^2	0.032	0.149	0.161	0.244
Number of observations	19,856	19,856	19,856	19,856

TABLE 10.8 Regressions of dietary diversity in household diets in Ethiopia (number of food groups out of 12)

	Rural			
Variables	Model 1	Model 2	Model 3	Model 4
Education of household head (base = none)				
Grades 1–4	0.418*** (0.04)	0.384*** (0.04)	0.276*** (0.04)	0.219*** (0.04)
Grades 5–8	0.587*** (0.04)	0.473*** (0.04)	0.366*** (0.04)	0.325*** (0.04)
Grades 9–12	0.800*** (0.08)	0.626*** (0.07)	0.540*** (0.07)	0.463*** (0.07)
Postsecondary	1.176*** (0.10)	0.806*** (0.09)	0.720*** (0.09)	0.398*** (0.12)
Adult education	0.0457 (0.06)	0.0926 (0.05)	0.0783 (0.05)	0.127* (0.05)
Expenditure quintile (base = poorest)				
Q2		0.538*** (0.04)	0.620*** (0.04)	0.594*** (0.04)
Q3		0.746*** (0.04)	0.826*** (0.04)	0.771*** (0.04)
Q4		0.966*** (0.04)	1.188*** (0.04)	1.136*** (0.05)
Richest		1.058*** (0.06)	1.414*** (0.06)	1.324*** (0.06)
Household head is male (dummy)			0.0287 (0.04)	−0.00677 (0.04)
Age of household head			−0.00376*** (0.00)	−0.00350*** (0.00)
household Size			0.118*** (0.01)	0.112*** (0.01)
Share of dependents in household			0.104 (0.07)	0.116 (0.07)
Meher rain deviation from average				−0.00058 (0.00)
Additional controls	No	No	No	Yes
R^2	0.05	0.138	0.18	0.226
Number of observations	10,366	10,366	10,366	10,366

Source: Authors' calculations from HCES 2015/2016 (Ethiopia, CSA 2016).

Note: Additional controls include occupation, region, and agroecological zone dummies. Robust standard errors in parentheses. * $p < 0.05$, ** $p < 0.01$, *** $p < 0.001$.

Variables	Urban			
	Model 1	Model 2	Model 3	Model 4
Education of household head (base = none)				
Grades 1–4	0.201*** (0.04)	0.152*** (0.04)	0.157*** (0.04)	0.127** (0.04)
Grades 5–8	0.355*** (0.04)	0.236*** (0.03)	0.264*** (0.04)	0.187*** (0.04)
Grades 9–12	0.489*** (0.04)	0.272*** (0.04)	0.322*** (0.04)	0.236*** (0.04)
Postsecondary	0.779*** (0.03)	0.492*** (0.03)	0.534*** (0.04)	0.441*** (0.04)
Adult education	0.012 (0.06)	−0.0664 (0.06)	−0.0981 (0.05)	0.0116 (0.05)
Expenditure quintile (base = poorest)				
Q2		0.553*** (0.06)	0.631*** (0.05)	0.641*** (0.05)
Q3		0.800*** (0.06)	0.942*** (0.06)	0.942*** (0.05)
Q4		0.951*** (0.05)	1.165*** (0.05)	1.162*** (0.05)
Richest		1.092*** (0.05)	1.477*** (0.05)	1.461*** (0.05)
Household head is male (dummy)			−0.136*** (0.02)	−0.151*** (0.02)
Age of household head			−0.00068 (0.00)	−0.00107 (0.00)
Household size			0.141*** (0.01)	0.133*** (0.01)
Share of dependents in household			0.596*** (0.05)	0.608*** (0.05)
Meher rain deviation from average				0.00153** (0.00)
Additional controls	No	No	No	Yes
R^2	0.051	0.112	0.201	0.236
Number of observations	19,856	19,856	19,856	19,856

half of the population improved slightly (less reliance on starchy staples and greater dietary diversity), although this was mostly in rural areas. For the rest of the population the changes were mixed (less reliance on starchy staples but less diverse diets). Ethiopian diets are also affected by seasonality. Energy consumption tends to be highest in both urban and rural areas during postharvest periods and lowest during lean periods; reliance on cereals, roots, and tubers (starchy staples) tends to vary little in urban areas but accounts for most of the decline in energy consumption during lean periods in rural areas; dietary diversity varies little in urban areas but is generally lower during lean periods in rural areas; and the poorest households in urban areas experience more seasonal variation in starchy staple consumption and dietary diversity than less poor households, whereas richest households experience more seasonal variation than less poor households in terms of energy consumption, reliance on starchy staples, and dietary diversity.

In both rural and urban areas, smaller, more educated households with higher overall expenditure levels are generally more food secure. Those who have members working as legislators, professionals, technicians, and clerks are also more food secure. Agricultural households in urban areas are among the least food secure. In rural areas, although average energy intake is reasonably high for agricultural households, the quality of their diets is low. In rural areas male-headed households consume fewer calories per adult than female-headed households but have slightly higher quality diets. In urban areas there is virtually no distinction between male- and female-headed households in terms of food security. We find little evidence of diets being affected by rainfall shocks. This is particularly salient since Ethiopia had just experienced its worst drought in decades.

Appendix 10A: Diet Consumption Distributions over Time and Expenditure Quintiles

FIGURE 10A.1 Distributions of energy consumption in Ethiopia, 2011–2016

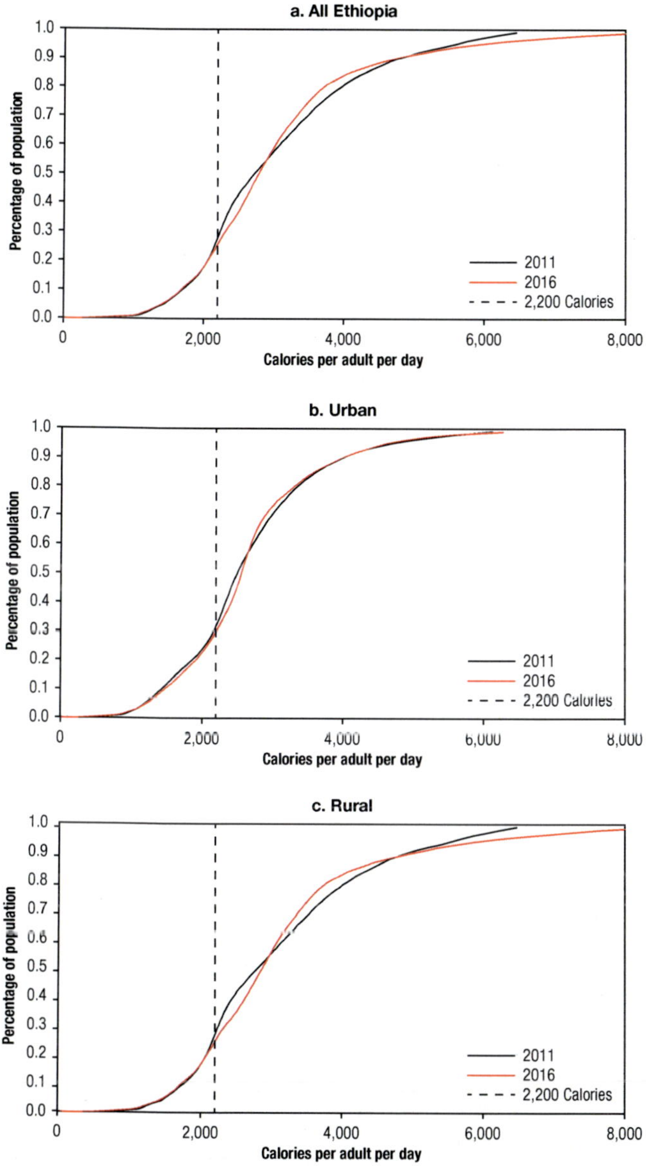

Source: Authors' calculations from HCES 2010/2011 and 2015/2016 (Ethiopia, CSA 2011 and 2016).

FIGURE 10A.2 Distributions of starchy staple consumption in Ethiopia, 2011–2016

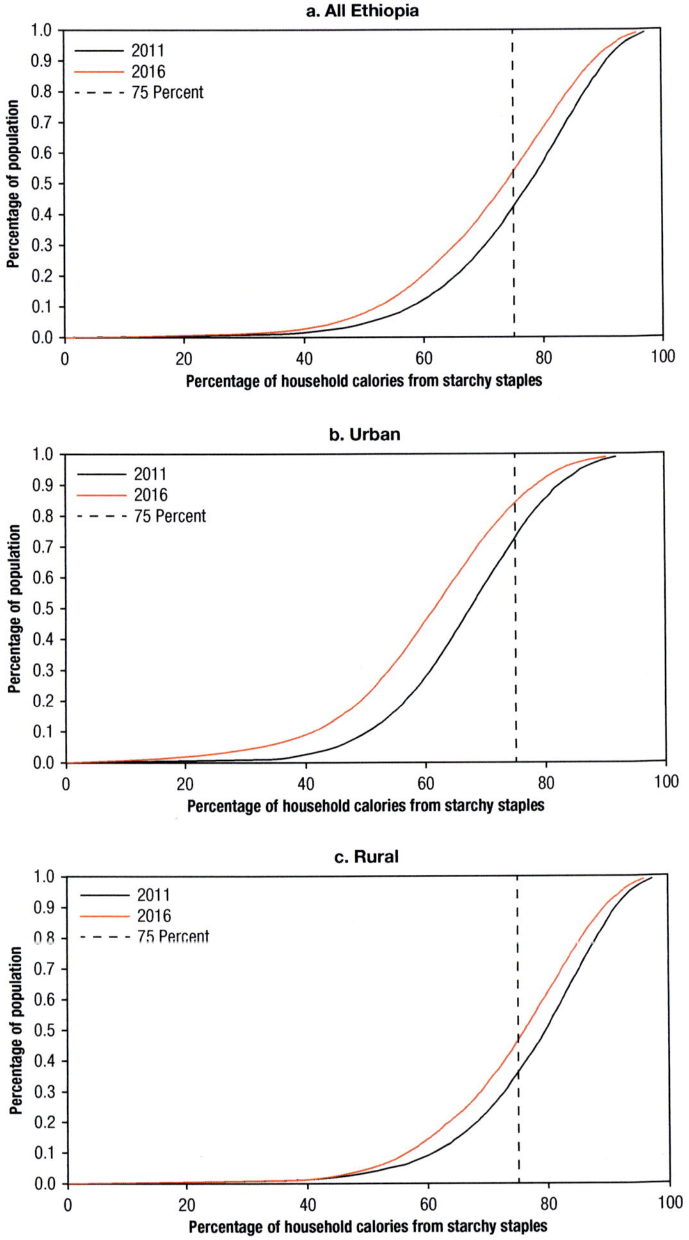

Source: Authors' calculations from HCES 2010/2011 and 2015/2016 (Ethiopia, CSA 2011 and 2016).

FIGURE 10A.3 Distributions of number of food groups consumed in Ethiopia, 2011–2016

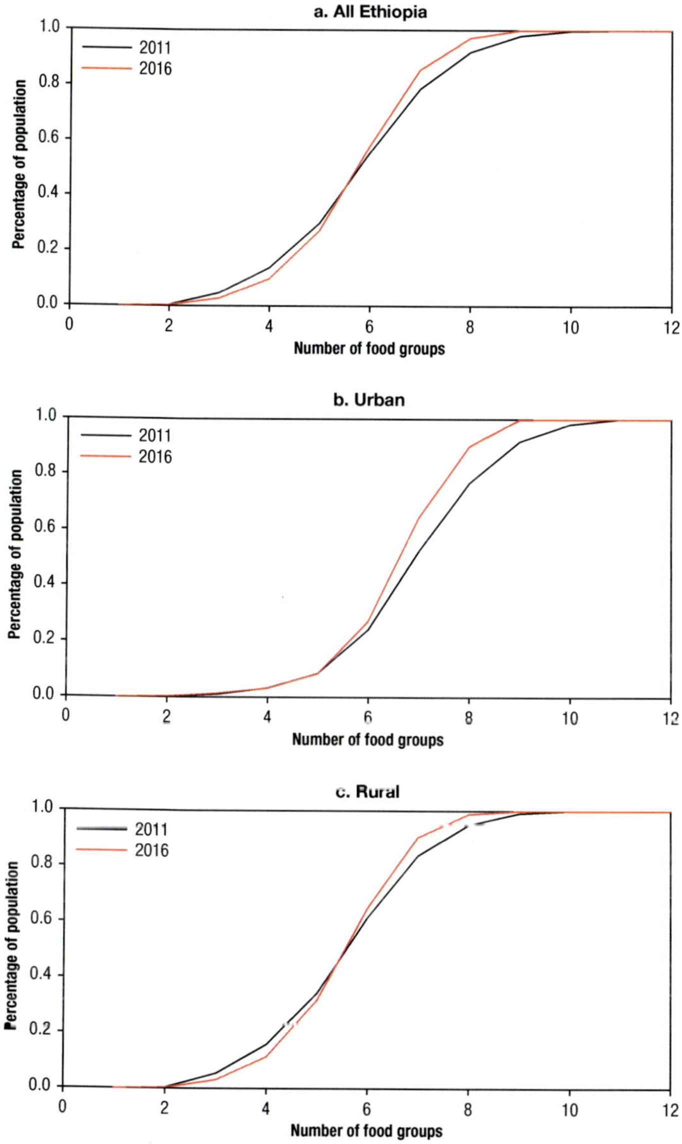

Source: Authors' calculations from HCES 2015/2016 (Ethiopia, CSA 2018)

FIGURE 10A.4 Distributions of energy intake in Ethiopia, by expenditure quintile (2016)

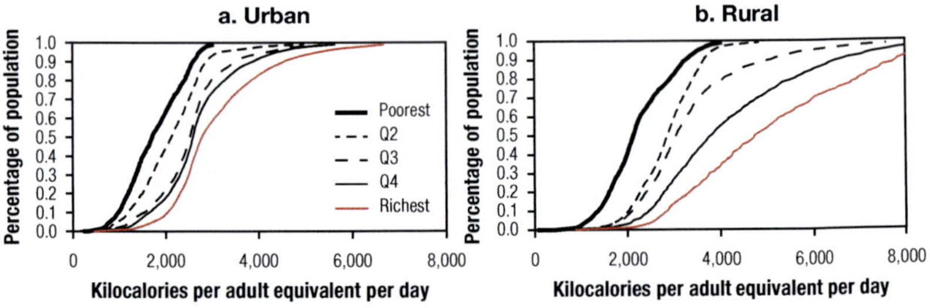

Source: Authors' calculations from HCES 2010/2011 and 2015/2016 (Ethiopia, CSA 2011 and 2016).

FIGURE 10A.5 Distributions of number of food groups consumed in Ethiopia, by expenditure quintile (2016)

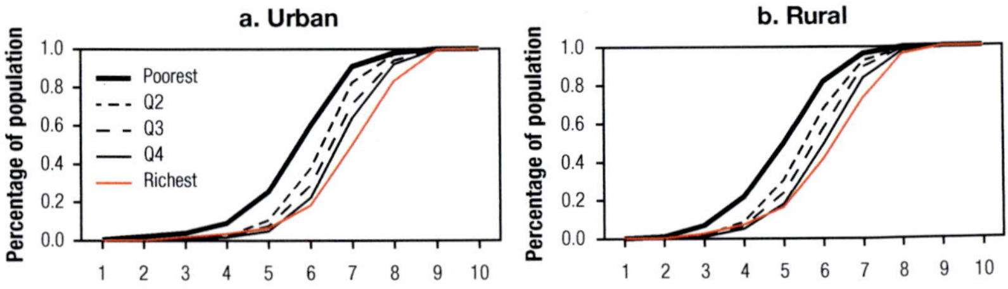

Source: Authors' calculations from HCES 2010/2011 and 2015/2016 (Ethiopia, CSA 2011 and 2016).

TABLE 10A.1 Adult Equivalence Scales

Age group (years)	Male	Female
< 1	0.30	0.30
1–2	0.46	0.46
2–3	0.54	0.54
3–5	0.62	0.62
5–7	0.74	0.70
7–10	0.84	0.72
10–12	0.88	0.78
12–14	0.96	0.84
14–16	1.06	0.86
16–18	1.14	0.86
18–30	1.04	0.80
30–60	1.00	0.82
> 60	0.84	0.74

Source: Ethiopia, CSA (2018).

Note: The numbers in the table are the ratios of food requirements of men and women of various ages to the food requirements of an average adult male between the ages of 30 and 60 years.

References

Berhane, G., L. McBride, K. T. Hirfrfot, and S. Tamiru. 2012. "Patterns in Foodgrain Consumption and Calorie Intake." In *Food and Agriculture in Ethiopia: Progress and Policy Challenges,* edited by P. Dorosh and S. Rashid. Philadelphia: University of Pennsylvania Press.

Coll-Black, S., D. Gilligan, J. Hoddinott, N. Kumar, A. S. Taffesse, and W. Wiseman. 2011. *Targeting Food Security Interventions When "Everyone Is Poor": The Case of Ethiopia's Productive Safety Net Programme.* IFPRI-ESSP II Working Paper 24. Addis Ababa: International Food Policy Research Institute/Ethiopia Strategy Support Program II (IFPRI/ESSP II).

Danaei, G., K. Andrews, C. Sudfeld, G. Fink, D. McCoy, E. Peet, A. Sania, and M. Smith. 2016. "Risk Factors for Childhood Stunting in 137 Developing Countries: A Comparative Risk Assessment Analysis at Global, Regional, and Country Levels." *PLoS Medicine* 13 (11): e1002164.

Deaton, A., and J. Drèze. 2009. "Nutrition in India: Facts and Interpretations." *Economic and Political Weekly* 44 (7): 42–65.

Ecker, O., and P. Fang. 2016. "Economic Development and Nutrition Transition in Ghana: Taking Stock of Food Consumption: Patterns and Trends." In *Achieving a Nutrition Revolution for Africa: The Road to Healthier Diets and Optimal Nutrition, ReSAKSS Annual Trends and Outlook Report 2015,* edited by N. Covic and S. Hendriks, 28–50. Washington, DC: IFPRI.

EHNRI (Ethiopian Health and Nutrition Research Institute). 1997. *Food Composition Table for Use in Ethiopia, Part III.* Addis Ababa.

Ethiopia, CSA (Central Statistical Agency). 2011. *Household Consumption Expenditures Survey.* Addis Ababa.

———. 2016. *Household Consumption Expenditures Survey.* Addis Ababa.

———. 2018. *The 2015/16 Ethiopian Household Consumption-Expenditure (HCE) Survey: Statistical Report.* Statistical Bulletin 585. Addis Ababa.

Ethiopia, CSA (Central Statistical Agency) and ICF. 2012. *Ethiopia Demographic and Health Survey 2011.* Addis Ababa: CSA; Calverton, MD, US: ICF International.

———. 2016. *Ethiopia Demographic and Health Survey 2016.* Addis Ababa: CSA; Rockville, MD, US: ICF.

Ethiopia, CSA, and WFP (World Food Programme). 2016. *Comprehensive Food Security and Vulnerability Analysis (CFSVA): Ethiopia Report.* Addis Ababa: CSA and WFP.

Ethiopia, MoFED (Ministry of Finance and Development). 2013. "Development and Poverty in Ethiopia. 1995/96–2010/11." Report. Addis Ababa.

FAO (Food and Agriculture Organization of the United Nations). 1996. *Rome Declaration on World Food Security and World Food Summit Plan of Action*. Document prepared at the World Food Summit, Rome. November 13–17.

Gilligan, D., J. Hoddinott, and A. S. Taffesse. 2009. "The Impact of Ethiopia's Productive Social Safety Net Programme and Its Linkages." *Journal of Development Studies* 45 (10): 1684–1706.

Graham, R., R. Welch, and H. Bouis. 2001. "Advancing Micronutrition Through Enhancing the Nutritional Quality of Staple Foods: Principles, Perspectives and Knowledge Gaps." *Advances in Agronomy* 70: 77–142.

Headey, D., and O. Ecker. 2013. "Rethinking the Measurement of Food Security: From First Principles to Best Practice." *Food Security* 5: 327–343.

Headey, D., D. Stifel, L. You, and Z. Guo. 2019. "Remoteness, Urbanization and Child Nutrition in Sub-Saharan Africa." *Agricultural Economics* 49 (6): 765–775.

Hirvonen, K. 2016. "Rural–Urban Differences in Children's Dietary Diversity in Ethiopia: A Poisson Decomposition Analysis." *Economics Letters* 147: 12–15.

Hirvonen, K., and D. Headey. 2018. "Can Governments Promote Homestead Gardening at Scale? Evidence from Ethiopia." *Global Food Security* 9: 40–47.

Hirvonen, K., T. P. Sohnesen, and T. Bundervoet. 2018. *Impact of Ethiopia's 2015 Drought on Child Undernutrition*. IFPRI-ESSP Working Paper 114. Addis Ababa: IFPRI/ESSP.

Hirvonen, K., A. S. Tafesse, and I. Worku Hassen. 2016. "Seasonality and Household Diets in Ethiopia." *Public Health Nutrition* 19 (10): 1723–1730.

Jensen, R., and N. Miller. 2010. *A Revealed Preference Approach to Measuring Hunger and Undernutrition*. NBER Working Paper 16555. Cambridge, MA, US: National Bureau of Economic Research.

Jones, A. D., A. Shrinivas, and R. Bezner-Kerr. 2014. "Farm Production Diversity Is Associated with Greater Household Dietary Diversity in Malawi: Findings from Nationally Representative Data." *Food Policy* 46:1–12.

MacKinnon, D. P., J. L. Krull, and C. M. Lockwood. 2000. "Equivalence of the Mediation, Confounding and Suppression Effect." *Prevention Science: The Official Journal of the Society for Prevention Research* 1: 173–181.

Ruel, M. 2003. "Operationalizing Dietary Diversity: A Review of Measurement Issues and Research Priorities." *Journal of Nutrition* 133: 3911S–3926S.

Ruel, M., J. Harris, and K. Cunningham. 2013. "Diet Quality in Developing Countries." In *Diet Quality: An Evidence-Based Approach*, edited by V. R. Preedy, 239–261. New York: Springer.

Sahn, D., and D. Stifel. 2002. "Parental Preferences for Nutrition of Boys and Girls: Evidence from Africa." *Journal of Development Studies* 39 (1): 21–45.

Sen, A. 1981. *Poverty and Famines: An Essay on Entitlement and Deprivation.* Oxford, UK: Clarendon Press.

Smart, J. C., D. Tschirley, and F. Smart. 2018. "Food System Transformation in Mozambique: An Assessment of Changing Diet Quality in the Context of a Rising Middle Class Development." IFPRI Discussion Paper 1769 (November). Washington, DC: IFPRI.

Smith, L., and A. Subandoro. 2007. *Measuring Food Security Using Household Expenditure Surveys.* Washington, DC: IFPRI.

Sohnesen, T. P. 2018. "Droughts, Measurement and Impact: The Case of Ethiopia in 2015." Unpublished, World Bank, Washington, DC.

Steyn, N. P., J. H. Nel, W. Parker, R. Ayah, and D. Mbithe. 2012. "Urbanisation and the Nutrition Transition: A Comparison of Diet and Weight Status of South African and Kenyan Women." *Scandinavian Journal of Public Health* 40: 229–238.

Stifel, D., and T. Woldehanna. 2016. "Poverty in Ethiopia, 2000–2011: Welfare Improvements in a Changing Economic Landscape." In *Growth and Poverty in Sub-Saharan Africa,* edited by C. Arndt, A. McKay, and F. Tarp, 43–68. Oxford, UK: Oxford University Press.

Swindale, A., and P. Bilinsky. 2006. *Household Dietary Diversity Score (HDDS) for Measurement of Household Food Access: Indicator Guide (v.2).* Food and Nutrition Technical Assistance (FANTA) Project. Washington, DC: Academy for Educational Development.

Thomas, D. 1990. "Intra-Household Resource Allocation: An Inferential Approach." *Journal of Human Resources* 25 (4): 535–664.

Welch, R. 2004. "Micronutrients, Agriculture and Nutrition: Linkages for Improved Health and Well Being." USDA-ARS, U.Sl Plant, Soil and Nutrition Laboratory, Ithaca, NY. Accessed June 20, 2018. www.css.cornell.edu/FoodSystems/Micros%26AgriMan1ref.html.

PART 3

Economywide Perspectives

Chapter 11

NONFARM INCOME AND RURAL LABOR MARKETS

Fantu Bachewe, Guush Berhane, Bart Minten, and Alemayehu Seyoum Taffesse

Introduction

Stimulating the off-farm economy in rural areas of developing economies is considered an important step toward structural transformation of such economies. In particular, the development of well-functioning rural labor markets is increasingly seen as crucial for economic growth and for the creation of livelihood opportunities for young people (for example, Fox et al. 2013). It is therefore important to understand how the off-farm sector and labor markets operate and to what extent their transforming influence is being felt. Moreover, rural wage increases are strongly linked with poverty reduction, since the majority of the population who are poor reside in rural areas and regularly depend on such wages for their livelihood (Ravallion 2000; Lanjouw and Lanjouw 2001).

We examine this issue in Ethiopia where the economy is changing fast. Having started from a low base, the Ethiopian economy still remains dominated by the agricultural sector, and the majority of the population still makes a living in this sector. While rapid transformation has occurred in agriculture (Chapter 3; Bachewe et al. 2015; World Bank 2015a), the implications of these changes on the rural off-farm economy are unclear. It has been shown in other settings that changes in off-farm income increasingly become important in rural areas when agriculture grows because of multiplier effects due to increased local spending by farmers (Haggblade, Hazell, and Reardon 2007).

This chapter provides new insights on the off-farm sector and labor markets in rural Ethiopia using recent large-scale datasets. While researchers have explored this issue in the past, the studies were based on relatively small surveys (Holden, Shiferaw, and Pender 2004; Dercon and Krishnan 1996; Block and Webb 2001). In particular, this chapter addresses the following four research questions: First, what is the importance of the off-farm sector in rural Ethiopia, and within the off-farm sector in particular, how important are labor markets? Second, what are the factors associated with rural income diversification? Third, are rural wages changing over time? Fourth, what are the drivers and the implications of these changes?

We define off-farm income as income generated from noncrop and nonlivestock production activities, or it is the sum of enterprise, agricultural wage, and nonagricultural wage incomes. Off-farm income makes up 18 percent of total rural household income in high-potential agricultural areas of the country. Agricultural and nonagricultural wage income accounted for 7 percent and 3 percent, respectively, of total household income, while enterprise income accounted for 8 percent. In these rural areas wage income is as important as livestock income, the latter accounting for 10.7 percent of household income. Hired labor is especially prevalent during specific periods of activity in the year—that is, during plowing, weeding, and harvesting. We find that off-farm income is more important in these high-potential agricultural areas than in other rural areas and that rural wages, driven by agricultural growth, are on the rise. These developments have important implications on poverty and on agricultural production practices. In particular, they affect incentives for the adoption of labor-saving technologies, such as herbicides and mechanization, as well as improve access to income in the lean period of the year when household food stocks may be depleted, possibly leading to improved agricultural productivity in the main season.

The chapter is organized as follows. First, we briefly describe the datasets used. Next we deal with the four research questions posed, analyzing rural off-farm income, looking specifically at agricultural hired labor use and agricultural wages and their associated factors, and studying trends in rural wages between 2004 and 2018 in Ethiopia. The results of our analyses on the drivers of change of real rural wages as well as the implications of these changes are discussed before our conclusion.

Data

For the descriptive and econometric analyses, we mainly use two data sources, complemented by a number of other datasets. First, we use price data collected by the Central Statistical Agency of Ethiopia (Ethiopia, CSA 2018a). CSA has been collecting monthly retail price data and data on prevailing wages since 1996 from about 120 markets, which were sampled to represent almost all administrative zones of the country.[1] From this dataset we use nominal wages

1 CSA datasets leave out three of the five zones of Afar and six of the nine zones in Somali regions.

of casual laborers and prices of a number of consumer food and nonfood items. We compute real wages using the Regional General Price Index data from CSA (Ethiopia, CSA 2018b) as well as a Poor-Persons' General Prices index that we created. The latter index is constructed using retail prices for 26 food and non-food categories (334 items) coupled with data from CSA's Household, Income, Consumption, and Expenditure Survey (HICES) dataset of 2004/2005 (Ethiopia, CSA 2007), which provides information on the expenditure shares of the 26 food and non-food categories (Headey et al. 2012). We define "the poor" as the bottom two quintiles of households based on their aggregate expenditure.

Second, we analyze data collected through the Agricultural Growth Program of Ethiopia (AGP) baseline survey. This dataset was collected in four regions of the country: Tigray, Amhara, Oromia, and Southern Nations, Nationalities, and Peoples (SNNP). The survey was conducted in May 2011 and covered close to 8,000 households, who were sampled to represent more than 9 million households in 93 woredas or districts (Berhane et al. 2013). Information on agricultural production in the AGP baseline survey dataset pertains to the 2010/2011 main agricultural season.

The AGP baseline survey dataset is complemented with data from several other sources, including IFPRI's Ethiopia Strategy Support Program (ESSP) Teff Producers dataset; the ESSP Coffee Producers dataset; the Feed the Future (FtF) midline impact evaluation survey; and from the Ethiopia Rural Socioeconomic Survey (ERSS). The Teff Producer dataset was collected in 2012 from 1,200 households in five major commercial teff-producing administrative zones of Amhara and Oromia regions. The Coffee Producer dataset, which includes about 1,600 households in five major coffee-producing areas of Oromia and SNNP, was collected in 2014. The FtF impact evaluation dataset, collected in 2015, and the ERSS dataset, collected in 2014, are two large-scale datasets that are used to provide additional insight into the importance of off-farm income in rural areas. We also use the World Development Indicators data (World Bank 2014) on real national GDP and value-added in the agriculture, manufacturing, and services sectors in our analyses of the drivers of rural wages.[2] Similarly, we use Zonal Poverty Head Count Index data obtained from Hill and Tsehaye (2018) in our investigation of the welfare implications of real wages.

In all of the surveys considered, data were gathered at the household level through comprehensive survey instruments on crop production and sales for a complete agricultural year, which includes two cropping seasons (*meher* and

2 For exact definitions, see World Bank (2014).

belg); on livestock production and sales; on agricultural and nonagricultural wage income; and on enterprise income for the 12 months preceding the survey period. While in some of the surveys, information on rental income and income transfers was collected, these data were not used, as these types of income constitute a relatively small share of total income and were not collected in all surveys. Therefore, for the purpose of this study, *total household income* is defined as the sum of net crop, livestock, and enterprise income, plus total wages earned from agricultural and nonagricultural labor.

Crop income is computed as the total value of crop output less the variable costs of production. Variable costs include money spent to purchase inputs such as fertilizer and improved seeds; to rent farm machinery or draft animals; to hire farm workers or other crop production-related costs. Similarly, *livestock income* is computed as the value of livestock sold or slaughtered and the value of livestock products, such as milk, honey, butter, and other products, less the variable costs of livestock production. Livestock production variable costs include wages paid to hire herding labor and costs incurred for veterinary and other related services. In each of the surveys price data collected at the community level are used to compute the value of crop and livestock production. *Net enterprise income* is computed as the difference between gross enterprise income for a business and total costs incurred for purposes of running the business. The definition of income for our purposes is different from monetary income, and in particular, incomes from crop and livestock production include both the value of home-consumed products and the value of sold quantities. Data on family labor used in enterprise and livestock production activities as well as services of nonpurchased inputs, such as own animals used for traction, were not collected. As a result, the values of these services are omitted from these calculations.

Off-Farm Income in Rural Areas

Here we investigate the importance of off-farm income in rural areas. The contribution of these income sources is described in both an international context and across regions in Ethiopia for different categories of households. We also present the results of our analyses of factors that are associated with participation in different income-generation activities.

Off-Farm Income in Ethiopia

Table 11.1 summarizes the data from the AGP baseline survey on the proportion of households with members employed in off-farm (agricultural and

TABLE 11.1 Importance of off-farm income in rural areas (%)

Income source	All regions	Tigray	Amhara	Oromia	SNNP
Contribution of source to total income					
Crop	71.4	59.8	69.6	72.2	73.3
Livestock	10.7	13.7	13.6	10.1	8.3
Agricultural wage income	6.6	7.8	9.1	5.9	5.1
Nonagricultural wage income	3.1	9.4	2.4	3.5	2.5
Enterprise income	8.1	9.3	5.3	8.3	10.9
Households earning some income by type					
Crop income	94.1	87.1	93.6	94.2	95.5
Livestock income	60.2	63.1	74.4	57.5	48.9
Agricultural wage income	21.4	16.3	29.0	20.2	15.9
Nonagricultural wage income	8.1	21.5	6.9	8.6	7.0
Enterprise income	25.4	20.2	18.0	28.8	28.1

Source: Authors' computation using the Agricultural Growth Program of Ethiopia baseline survey data (2011).
Note: SNNP = Southern Nations, Nationalities, and Peoples.

nonagricultural wage labor and enterprise) activities and those engaged in crop and livestock production. The table also summarizes the contribution of each source to total household income.[3]

Several interesting findings are seen. First, the share of off-farm income is relatively low compared to shares observed in other settings (Reardon et al. 2006), and off-farm income is significantly related to the agricultural sector. Crop income makes up 71 percent of total household income. Wage income makes up 10 percent of total income, which is about the same share that households derive from livestock and livestock products. Enterprise income accounts for 8 percent of household income. If we divide incomes between agricultural and nonagricultural sources, however, only 11 percent of total household income is not related to agriculture—nonagricultural wage income accounts for only 3 percent of total household income.

Second, examining the share of households that obtain some income from each source, the number engaged in crop or livestock production is high, 94 percent and 60 percent, respectively. About 21 percent of the rural

3 In this chapter we focus on agricultural wage labor, which we view as distinct from other rural wage labor and business activities, thereby eliminating discrepancies that may arise due to differences in definitions of rural and agricultural.

households earned income from agricultural wage labor, 25 percent were engaged in business enterprises, and 8 percent in nonagricultural wage labor.

Third, we see relatively little variation across different regions. Crop income is relatively more important in Amhara, Oromia, and SNNP regions, at 70, 72, and 73 percent, respectively, compared to Tigray (60 percent). Rural households in Tigray are somewhat more diversified with respect to income sources. Agricultural wage income as a share of total income is highest in the Amhara region (9 percent), possibly linked to seasonal migration to areas that produce cash crops, such as sesame.

To see how the results on income sources from the AGP survey compare to other recent large-scale datasets in the country, we examine information on income sources for rural households in the FtF and ERSS datasets (Table 11.2). Overall, we note the same orders of magnitude by income source: crop income makes up 80 percent and 71 percent of total household income in the FtF and ERSS dataset, respectively. However, in both cases, livestock income is evaluated to be more important than was observed in the AGP dataset, accounting for 11 percent and 20 percent of total income in the FtF and ERSS datasets, respectively. Off-farm income accounts for 7 percent and 9 percent, respectively, which is significantly smaller than in the AGP dataset (18 percent). Overall, these results confirm that the off-farm economy is small in rural Ethiopia. We use the Feed the Future impact evaluation survey datasets collected in 2015 and 2018 (as consistent income data were collected in both surveys) to see whether there were changes in the importance of off-farm income during that period. This analysis indicates that changes were relatively small: the importance of crop income in total household income declined by 4 percent, while livestock and transfer incomes increased by 1 percent and 3 percent, respectively, during that period.

We explore the factors associated with the importance of off-farm income using simple graphs and a regression analysis. First, we look at how the share of income sources varies by income level. To do so, we divide all households into five income quintiles and calculate shares of total income by income source for each (Figure 11.1).

Income from crop production contributes the lowest (64 percent) for households in the poorest income quintile (quintile 1 in Figure 11.1), and the contribution increases with total income. The contribution of livestock and nonagricultural wage income is higher in the bottom two income quintiles and declines with increases in total income. Agricultural wage income also contributes the highest in the lowest income quintile, while its contribution declines with increases in total income. Finally, enterprise income is more

TABLE 11.2 Importance of off-farm income in rural areas, alternative estimates (%)

Contribution of source to total income	Four AGP regions	Five regions (including Somali)	Tigray	Amhara	Oromia	SNNP	Somali
Feed the Future (FtF) dataset							
Crop	82.0	80.4	74.6	82.1	82.1	84.4	28.3
Livestock	11.3	12.7	19.1	12.0	11.3	7.2	61.2
Agricultural wage	1.5	1.5	1.3	1.7	1.7	1.0	0.8
Nonagricultural wage	1.6	1.6	1.5	1.5	1.8	1.4	3.7
Enterprise	3.7	3.8	3.6	2.7	3.2	5.9	6.0
Contribution of source to total income	**Four AGP regions**	**All of Ethiopia, excluding Addis Ababa**	**Tigray**	**Amhara**	**Oromia**	**SNNP**	**All other regions**
Ethiopia Rural Socioeconomic Survey (ERSS) dataset							
Crop	71.3	69.0	65.0	69.7	71.7	74.4	31.6
Livestock	19.6	21.1	22.2	20.9	19.4	17.3	45.9
Casual labor	2.0	2.0	3.3	1.9	1.8	2.1	2.5
Salaried workers	3.6	4.1	5.6	3.4	4.1	2.5	11.2
Enterprise	3.5	3.8	4.0	4.0	2.9	3.8	8.8

Source: Authors' computation using the FTF (2013) and the Ethiopia Rural Socioeconomic Survey (ERSS) datasets.
Note: AGP = Agricultural Growth Program of Ethiopia; SNNP = Southern Nations, Nationalities, and Peoples.

FIGURE 11.1 Contribution of different income sources to overall income in Ethiopia, by income quintile

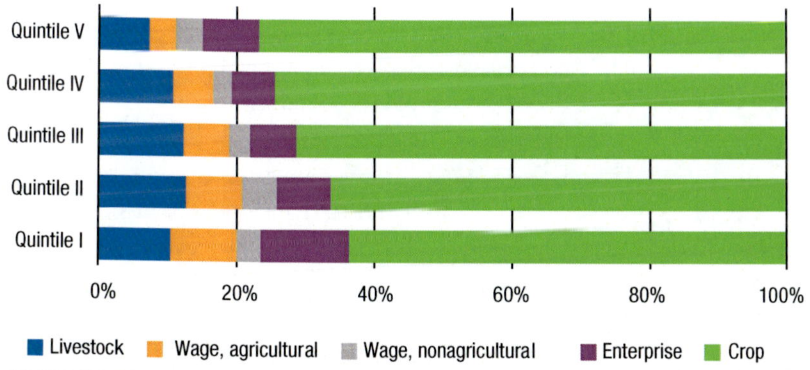

Source: Authors' computation using the 2011 AGP baseline survey data (Berhane et al. 2013).

important in the lowest and highest income quintiles. The higher contribution of enterprise income in the lowest income quintile may appear to contradict the assumption that establishing businesses requires considerable initial investments. However, the businesses in which households within the highest and lowest income quintiles are engaged partly explains the apparent contradiction. For instance, brewing and selling local liquor and ale, which require relatively lower initial investments, are the two most important business activities in the lowest income quintile, while trading activities in grain and livestock are important for households in the highest income quintile. Overall, the off-farm sector is more important for the poor than for the rich, suggesting that push factors might still be relatively more important in rural areas for engagement by households in the rural off-farm income sector than are pull factors (Reardon et al. 2006; USAID 2011).

Second, we consider the contribution of income sources across land endowments as an alternative measure of wealth. Similar to the previous exercise, we look at the share of income sources for five quintiles of landownership, ranked from poor to rich. Figure 11.2 shows that the share of crop and livestock income significantly increases with land endowment. The share of income from the three off-farm income sources (agricultural wage, nonagricultural wage, and enterprise income) generally declines with land size. In particular, the contribution of each of the three income sources is nearly twice as high for households in the lowest land quintile (quintile 1) relative to those in the highest quintile. When we look at absolute levels, agricultural wage income of households in land quintile 1 and quintile 2 are higher than in both quintile 4 and quintile 5. The same holds for nonagricultural wage income of quintile 1 households. Enterprise income increases with land size, except that enterprise income of quintile 1 households is 37 percent higher than quintile 2 households. Crop and livestock income increase linearly with land size.

Third, we look at female- and youth-headed households (under 35 years of age).[4] Table 11.3 shows the contribution of different sources of income to the overall income of households with heads in four gender–age categories. Off-farm income sources are relatively more important for youth-headed households. Off-farm income makes up 21 percent and 23 percent of total income for male-youth– and female-youth–headed households, respectively. This compares to 15 percent and 17 percent of male-mature– and female-mature–headed households, respectively. The younger households might have to rely

4 As these households were an explicit target of the AGP, they were oversampled to detect impact of the program on these particular households.

FIGURE 11.2 Contribution of different income sources to overall income in Ethiopia, by land endowment quintile

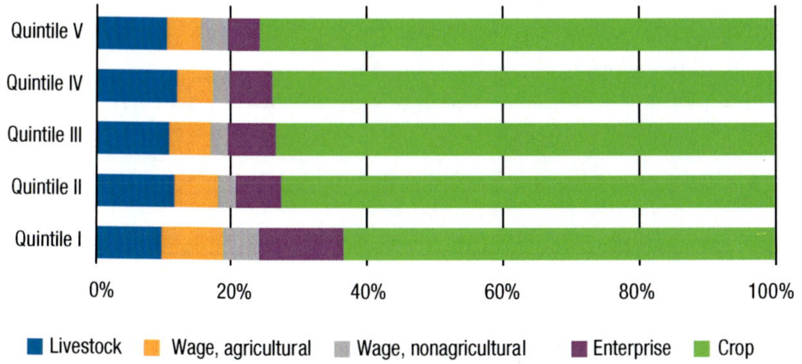

Source: Authors' computation using the 2011 AGP baseline survey data (Berhane et al. 2013).

TABLE 11.3 Importance of wage income in rural areas in Ethiopia, by gender and age of household heads (%)

Income source	Male-youth	Male-mature	Female-youth	Female-mature
Contribution of source to total income				
Crop	68.0	74.3	65.0	72.9
Livestock	10.7	11.0	12.2	9.7
Agricultural wage income	7.4	5.4	9.9	6.9
Nonagricultural wage income	3.5	2.8	3.2	3.1
Enterprise income	10.4	6.6	9.7	7.3
Households earning some income by type				
Crop income	93.6	96.0	88.6	93.2
Livestock income	61.2	62.6	60.6	54.2
Agricultural wage income	24.1	20.3	21.5	19.7
Nonagricultural wage income	9.5	8.1	6.7	6.7
Enterprise income	32.0	21.7	27.7	22.2

Source: Authors' computation using the 2011 AGP baseline survey data (Berhane et al. 2013).

more on off-farm sources of income as they often own less land than the more established households (Headey et al. 2012). The enterprise income and agricultural wage labor income especially are relatively important for youth-headed households. Notably, the female-headed households overall rely more on off-farm income than do male-headed households.

Fourth, Table 11.4 indicates the different types of enterprises that rural households are involved in. We see significant differences by income quintile. Selling homemade local liquor, producing handicrafts, trade in crop outputs, selling local ale, and selling prepared food items are important for households in the lowest income quintile in the order given. A higher proportion of households in the highest income quintile are engaged in selling crop outputs, livestock and livestock products, local liquor, retail nonagricultural items, and handicrafts. Several of the activities that a larger proportion of households in the highest income quintile are engaged in, such as trade in grains and livestock and livestock products, require substantial investment, while the production and sales of local liquor, handicrafts, and food items, enterprises in which poorer households are engaged, require a relatively lower investment.

Finally, we look at the factors associated with involvement in off-farm income using a regression framework. Using Escobal's (2001) methodology, we run Tobit models that show the factors associated with diversification (defined as the share of income in percent from sources other than crop agriculture and livestock production). In particular, we look at agricultural and nonagricultural wage income and enterprise income. We conduct the analysis using as dependent variable an aggregate measure of income diversification that accounts for both the number and share of income sources: the Herfindahl Diversification Index (HDI), which is defined as:

$$HDI_i - 1 - HI_i - 1 - \Sigma_j^K \left(\frac{\mathrm{Y_{ij}}}{\Sigma_{\mathrm{j}=1}^{\mathrm{K}} Y_{ij}}\right)^2$$

in which the term in parentheses is the contribution of income from source j (Y_{ij}) to total income ($\Sigma_{\mathrm{j}=1}^{\mathrm{K}} Y_{ij}$).[5] Results of these analyses are provided in Table 11.5.

First, the results show a strong association between income sources and the age and education of the head of household. Crop and livestock income are positively associated with the age of the household head, while agricultural wage and enterprise income are negatively related to age. Diversification

5 HDI ranges between 0 and 1. HDI = 0 for households that generate their entire income from only one source, and it increases with number and balancedness in share of income sources.

TABLE 11.4 Percentage of households engaged in different business enterprises in Ethiopia, by income quintile (%)

Business activity	Quintile 1	Quintile 2	Quintile 3	Quintile 4	Quintile 5	All
Weaving/spinning	5.6	9.0	1.8	0.5	1.6	3.2
Milling	0.1	1.2	0.1	0.2	3.3	1.2
Handicrafts	16.5	16.2	17.2	14.2	6.6	13.5
Trade in grains and other crop items	13.3	12.2	11.9	17.1	22.3	16.0
Trade in livestock and livestock products	6.4	3.7	10.3	13.1	15.8	10.8
Trade in nonagricultural merchandise	4.6	9.4	9.2	10.1	12.4	9.8
Transport (by pack animal)	0.0	0.1	0.2	0.2	2.6	0.8
Tella (local ale)	11.8	4.6	4.4	4.1	4.5	5.1
Araqi (local liquor)	19.7	18.3	22.0	21.3	15.1	19.1
Prepared food (bread/*enjera*)	6.8	2.5	4.9	0.9	2.8	3.2
Other businesses	15.2	22.8	17.8	18.3	13.0	17.3
Total	100.0	100.0	100.0	100.0	100.0	100.0

Source: Authors' computation using the 2011 AGP baseline survey data (Berhane et al. 2013).

(HDI) is negatively associated with the age of the household head. This suggests that households with younger heads are more likely to rely on off-farm income sources to assure their livelihoods. Diversification is also positively related to the education of the household head and to the number of members in the household. Enterprise and nonagricultural wage income, which require more skills than agricultural labor, are positively associated with the heads' education, while the latter is negatively associated with self-employed and hired agricultural labor income.

Second, there is a strong gender component to off-farm income. The proportion of females among total household members of working age (16–65 years) is negatively associated with both categories of hired labor. This could be because female members of households in rural areas spend more time on household chores and have limited time to engage in hired labor. This is in addition to the fact that females, particularly younger girls, are generally discouraged from working as hired labor because in many communities it is socially unacceptable for these young girls to work outside of the household. The number of females engaged in hired labor is about half (56 percent) the number of males.

Third, a greater quantity and better quality of agricultural assets lead to less diversification. As implied in the discussion above, total household land

TABLE 11.5 Factors associated with the contribution of farming, business enterprise, and wage employment to total income in Ethiopia

Variables	Self-employment, agriculture (crop + livestock)	Self-employment, nonagriculture (enterprise)	Wage employment, agriculture	Wage employment, nonagriculture	Herfindahl Diversification Index
Household variables					
Gender of household head (=1 if male)	0.932	–3.439	1.914	2.434	0.028*
Age of household head (years)	0.407***	–0.543***	–0.378***	–0.110	–0.003***
Education of head	–3.002***	5.742***	–2.283**	5.276***	0.018***
Household size	–0.335	0.122	0.435	2.113**	0.011***
Proportion of females in working age	5.370	10.295	–16.570***	–17.640*	–0.030
Farm characteristics					
Purchased at least one input on credit	–0.239	0.268	–3.561	7.820*	0.026*
Total cultivated area (hectares)	4.149***	–0.426	–2.741**	–7.684***	–0.011
Total cultivated area-squared	–0.239***	0.102	0.066	0.481***	0.000
Land quality index	1.138***	–1.400***	–0.500	–0.255	–0.006**
Tropical livestock units	2.161***	–1.370***	–1.951***	–5.416***	–0.012***

area is positively associated with agricultural income and negatively with agricultural and nonagricultural wage income. Higher land quality, computed from self-reported soil fertility and slope measures by the respondent, is positively associated with agricultural income and generally negatively associated with diversification. The number of tropical livestock units (livestock measured in camel units) is positively associated with agricultural income, while it is negatively related to the remaining income sources (wage and enterprise income) as well as to diversification.

Fourth, consistent with other settings and countries, proximity to cities is often an important factor associated with income diversification in rural Ethiopia (Deichmann, Shilpi, and Vakis 2009; Fafchamps and Shilpi 2005, 2003; Jacoby and Minten 2009). Diversification and enterprise income decline with distance from both the capital Addis Ababa and the nearest urban center of 50,000 or more inhabitants, while agricultural income increases with distance from urban centers. Agricultural and nonagricultural

Variables	Self-employment, agriculture (crop + livestock)	Self-employment, nonagriculture (enterprise)	Wage employment, agriculture	Wage employment, nonagriculture	Herfindahl Diversification Index
Location					
Travel time to nearest 50,000 town (minutes)	0.032***	−0.018**	−0.006	−0.087***	−0.000***
Distance from Addis Ababa (hundred kilometers)	1.524*	−5.482***	2.996***	−6.132***	−0.014**
Total district population[a]	0.024*	−0.048**	−0.003	0.046	−0.000
Zonal poverty head count index[b]	−0.144**	0.565***	−0.283***	0.016	0.001*
Amhara	18.840***	−24.408***	21.050***	−73.010***	−0.093***
Oromia	17.810***	−8.727*	14.460***	−68.950***	−0.090***
SNNP	15.050***	1.749	3.295	−72.620***	−0.086***
Constant	50.950***	−13.316	−20.860**	−1.588	0.195***
Chi-squared	389	314	154	464	318
Number of observations			7,178		
Turning point for total cultivated area	8.692	n.a.	n.a.	7.981	n.a.

Source: Authors' analysis using the 2011 AGP baseline survey data (Berhane et al. 2013) except as noted: a. Ethiopia, CSA (2014); b. Hill and Tsehaye (2014).

Note: SNNP = Southern Nations, Nationalities, and Peoples. Coefficients with superscripts ***, **, and * are significant at 1 percent, 5 percent, and 10 percent levels, respectively. n.a. = not applicable.

wage employment are positively and negatively associated with distance from Addis Ababa. The size of the coefficients indicates that households located 100 kilometers from Addis Ababa have 11 percent less income from the non-agricultural sector (5 percent less from self-employment and 6 percent less from wages) than households living close to Addis Ababa. Finally, the Zonal level Poverty Head Count index is also positively related to diversification overall, suggesting again that push factors are important for household income diversification in rural Ethiopia.

Off-Farm Income in Ethiopia in an International Context

Comparing off-farm income patterns for rural households in Ethiopia with those in other countries provides a broader perspective. First, we compare the importance of the rural off-farm sector with other countries. Reardon et al. (2006) find that off-farm income made up 36 percent of total income in Africa, 50 percent in Asia, and 43 percent in Latin America (Figure 11.3).

FIGURE 11.3 Off-farm income as share of total income in rural areas in Ethiopia, from local sources and from migration income (%)

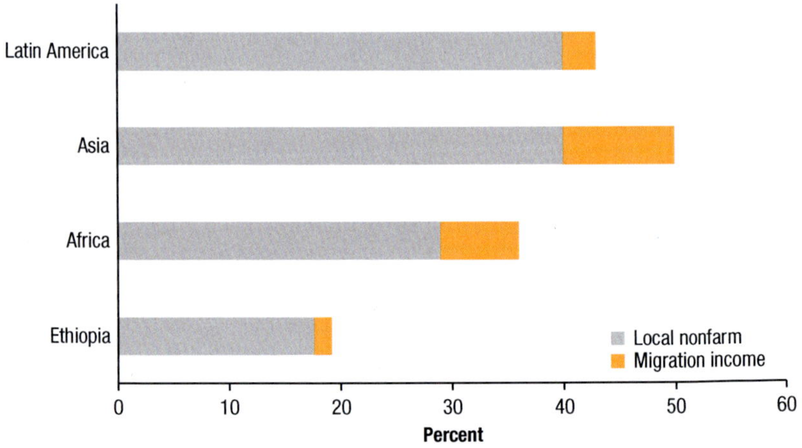

Source: Reardon et al. (2006), and for Ethiopia the migration income (transfer) and local nonfarm numbers are from ERSS and AGP baseline survey datasets, respectively.

They show that most off-farm income comes from local off-farm sources and less from wage income obtained through migration. In Ethiopia the off-farm sector accounts for 18 percent of total rural household income, which is significantly smaller than the African average. As shown in Figure 11.3 in more developed continents, the share of this off-farm sector is expected to increase with anticipated growth in Ethiopia's economy.

Second, we look at wage levels using the AGP data. The distribution of agricultural wages paid in US dollars (US$) during the survey period are shown in Figure 11.4 (left). An average wage of agricultural workers in these high potential areas was 19.5 birr (US$1.27) per day. Average wages of all rural workers, including nonagricultural labor, are slightly lower at 19 birr (US$1.24) per day. Wage data from other countries, obtained from Wiggins and Keats (2014) and Zhang et al. (2014), are also shown in Figure 11.4 (right). Unfortunately, these data provide only an indication of wages paid in Asian countries, and data from the African continent are sparse. In any case, we find that wages in Ethiopia are significantly lower than in Asian countries for which data are available. It is estimated that agricultural wages in Ethiopia were only 57 percent, 44 percent, and 56 percent of the average agricultural wages paid in Nepal, Bangladesh, and Myanmar, respectively.

FIGURE 11.4 Frequency distribution of agricultural wages in Ethiopia (left), and agricultural wages in Ethiopia compared to a number of other countries in US$ per day (right)

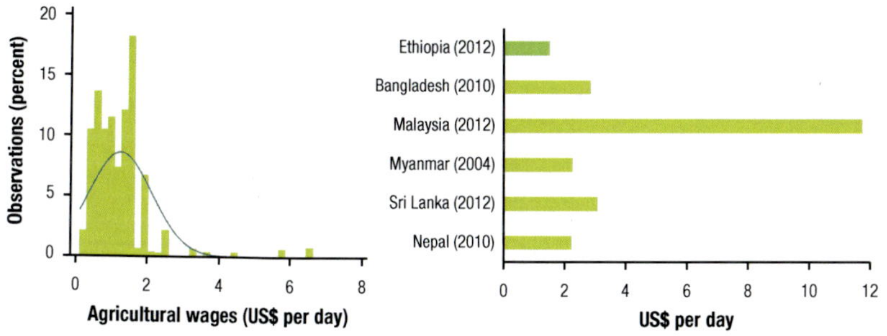

Source: Wiggins and Keats (2014); Zhang et al. (2014).

Agricultural Wages and Agricultural Labor Market Use

Here we investigate the contribution of hired-in labor in total agricultural labor; seasonality in agricultural labor use; and factors associated with agricultural wages. We study factors associated with participation in rural labor markets, particularly from the demand side.

Agricultural Wage Labor

Table 11.6 summarizes the share of crop-producing households that relied entirely on family or on hired-in labor, and those that used both family and hired-in labor. Three important observations can be made. First, family labor is by far the most important contributor to agricultural work—hired labor contributes only 7 percent to all agricultural work. Second, there is important regional heterogeneity in terms of the share of hired-in labor. The share of hired-in labor is highest in Tigray region, where it makes up 14 percent of total labor used in crop production, and lowest in Amhara, constituting only 4 percent. The low levels of hired-in labor in Amhara is surprising given the relatively larger importance of agricultural wage income in the region. This result possibly indicates the importance of seasonal migration in the region to work on commercial farms. Third, the proportion of farms that exclusively rely on hired-in labor is low at only 1 percent. The vast majority of households rely exclusively on family labor—more than three-quarters of these farms exclusively use family labor for crop production. Nearly 23 percent of households used a combination of family and hired-in labor. We disaggregate the

TABLE 11.6 Share of hired-in labor and households using hired-in labor for crop production in Ethiopia (%)

	Share of total labor use		Proportion of households using		
Region	Family labor	Hired-in labor	Only family labor	Only hired labor	Family and hired labor
All regions	93.0	7.0	76.1	1.1	22.8
Tigray	86.1	13.9	62.2	3.5	34.3
Amhara	96.1	3.9	81.4	0.3	18.3
Oromia	93.0	7.0	72.5	0.4	27.1
SNNP	90.6	9.4	78.8	3.0	18.2

Source: Authors' computation using the 2011 AGP baseline survey dataset (Berhane et al. 2013).
Note: SNNP = Southern Nations, Nationalities, and Peoples.

data by the type of crop cultivated. The results reveal surprisingly little differences in the share of hired-in labor use across crops, ranging from 4.5 percent of total labor used for pulses to 7.5 percent for oilseeds.

We triangulate this issue using the Teff Producer and Coffee Producer surveys. These surveys hold more detailed information on the type of labor use for the different agricultural activities during the year as well as information on the importance of exchange labor that is prevalent in some regions. The latter information was not collected in the AGP surveys. The results are presented in Table 11.7. Note that the larger farms are oversampled in both these surveys to ensure that a representative picture for the total quantity produced and commercialized would emerge.

Family labor accounted for 63 percent and 68 percent of all labor used in major teff- and coffee-producing areas, respectively. In addition, exchange labor or other forms of social labor accounted for 22 percent and 18 percent of the total labor used in the respective producing areas. The share of exchange labor is even more important than hired-in labor in these contexts, the latter standing at 14 percent. While exchange labor is invariably important, there are key variations in its use over space. Figure 11.5 depicts the information on the share of hired-in, family, and exchange labor by distance to Addis Ababa as measured in terms of the cost of transporting a quintal of grain to the capital city. The figure shows that households in areas closer to Addis Ababa rely more on hired-in labor than they do on exchange labor, while the reverse holds in areas farther away from Addis Ababa. This figure suggests that households

TABLE 11.7 Share of labor arrangements for different activities in Ethiopia (%)

	Teff			Coffee		
Practice	Family	Hired	Exchange	Family	Hired	Exchange
Tree management	n.a.	n.a.	n.a.	88	4	9
Mulching	n.a.	n.a.	n.a.	83	7	10
Tilling	85	7	8	63	12	25
Manure and organic input use	93	4	3	92	3	5
Sowing and fertilizer use	86	6	7	85	6	9
Weeding	73	10	16	70	11	19
Herbicide and pesticide application	76	23	1	n.a.	n.a.	n.a.
Harvesting	41	32	27	68	19	13
Postharvesting activities	69	7	24	98	1	1
Threshing and winnowing	57	11	32	n.a.	n.a.	n.a.
Total	**63**	**14**	**22**	**68**	**14**	**18**

Source: Authors' computation using IFPRI and EDRI (2019, 2020).
Note: n.a. = not applicable.

FIGURE 11.5 Labor arrangements and remoteness in Ethiopia

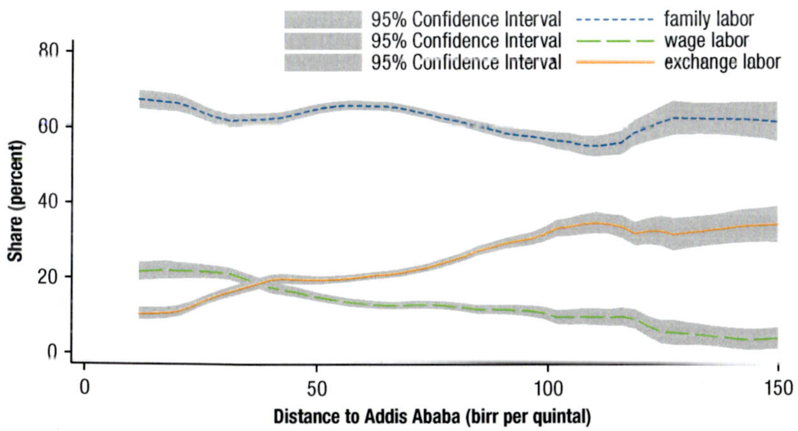

Source: Authors' computation using IFPRI and EDRI (2019, 2020).

shift to wage and market labor arrangements in more commercialized and better-connected areas.[6]

Table 11.7 shows that there also exist significant seasonal variations in hired-in labor use. Hired-in labor is much more important during harvest periods in both major teff- and coffee-producing areas. Hired-in labor was also relatively important during the tilling and weeding periods. Of the total labor used in teff and coffee harvesting, 32 percent and 19 percent was accounted for by hired-in labor, respectively.

Factors Associated with Agricultural Wages

Next we investigate factors associated with wage levels and test if and to what extent agricultural workers' wages vary with type of agricultural work, worker characteristics, and locational factors. In the econometric analyses we use the AGP baseline survey dataset. We complement this data with an urban dummy variable constructed from CSA's woreda population predictions (Ethiopia, CSA 2014) and zonal poverty head count (HC) index data from Hill and Tsehaye (2018).[7] The variables used in the regression analyses are defined and the results provided in Table 11.8.

The results indicate that workers that hired-out labor for planting earn lower wages (13 percent lower) relative to those that worked on land preparation, which is the omitted category, while wages of laborers engaged in weeding and harvesting earned 12 percent and 17 percent more, ceteris paribus. Wages for livestock herding were not statistically different from land preparation. The results also indicate that male workers are paid 8 percent more than females, controlling for the type and location of the work. Moreover, wages decline with the age of the worker, possibly indicating that the productivity of older workers is presumed to be lower.

The results further indicate that wages increase with proximity to urban centers. A further 100 kilometers away from Addis Ababa reduces agricultural wages by 7 percent. Remoteness from the local market is also associated with the payment of lower wages, but these effects are small. Furthermore,

6 A similar picture shows up for a switch to land rental agreements from sharecropping as households live closer to Addis Ababa.

7 To address the endogeneity problem that could arise if the poverty HC index is endogenously determined with wages, we use the predicted value of the poverty HC index obtained from a first-stage regression of the latter variable on factors associated with poverty but not used in the second-stage analysis. The dependent variables used in the first stage (zonal level) regression are: household-level average values of cultivated area, value of crop and livestock outputs, and proportion of households that accessed credit to purchase inputs. We thank an anonymous reviewer for indicating this problem.

TABLE 11.8 Region, activity, and gender factors associated with rural wages in Ethiopia

Dependent variable: log of agricultural wages (birr per day)		
Variable	Coefficient	Standard Error
Planting	−0.120**	0.048
Weeding	0.127***	0.037
Harvesting	0.181***	0.032
Herding	−0.114	0.153
Gender of worker (=1 if male)	0.081***	0.027
Worker age (years)	−0.002*	0.001
Distance to daily/periodic market (kilometers)	−0.001***	0.000
Distance from Addis Ababa (hundred kilometers)	−0.120***	0.015
Total district population	0.002***	0.000
Zonal poverty head count index (predicted)	−0.010***	0.003
Amhara	−0.179***	0.057
Oromia	−0.581***	0.063
SNNP	−0.264***	0.051
Constant	3.369***	0.123
F-statistics	28	
Number of observations	2,365	

Source: Authors' analysis using the 2011 AGP baseline survey data (Berhane et al. 2013), Ethiopia, CSA (2014), and Hill and Tsehaye (2018). Coefficients with superscripts ***, **, and * are significant at 1 percent, 5 percent, and 10 percent levels, respectively.

Note: SNNP = Southern Nations, Nationalities, and Peoples.

agricultural wages decline with the Poverty Head Count index in the administrative zone, even after controlling for location differences, indicating the strong link of agricultural wage levels with poverty measures. This suggests that in areas where a higher proportion are poor, the resulting lower wages from a higher labor supply means that the poor might rely relatively more on these labor markets for their livelihoods. Region dummy variables used in the analysis imply that agricultural wages are relatively higher in Tigray.

Factors Associated with Agricultural Labor Use

We look at factors associated with hired-in agricultural labor use. For this, we use a Tobit model to investigate whether household and location-specific factors explain the proportion of hired-in labor out of total labor. Variables used in the regression analysis and results obtained are provided in Table 11.9.

The results in Table 11.9 indicate that household characteristics are important for decisions on labor markets. First, the share of hired-in labor

TABLE 11.9 Factors associated with hired-in labor use (Tobit regression) in Ethiopia

Dependent variable: share of hired-in labor (%)		
Variables	Coefficient	Standard error
Household variables		
Gender of household head (=1 if male)	−4.10	2.624
Age of household head (years)	−0.08	0.062
Education of head	4.60***	0.840
Dependency ratio	0.15***	0.041
Household size	−3.57***	0.490
Proportion of females in working age	14.79***	5.115
Farm characteristics		
Total cultivated area (hectares)	12.47***	1.088
Total cultivated area-squared	−0.70***	0.100
Land quality index	1.82***	0.431
Tropical livestock units	1.30***	0.224
Household obtains production information	5.01***	1.324
Model farmer in last five years? (=1 if yes)	7.01**	3.295
Visited by extension agent (=1 if yes)	0.94	1.959
Location		
Travel time to nearest 50,000 town (minutes)	−0.02***	0.006
Distance from Addis Ababa (hundred kilometers)	−4.14***	0.943
Urban (=1 if woreda population ≥ 50,000)	73.03***	12.680
Zonal poverty head count index	0.22***	0.073
Amhara	−30.29***	3.365
Oromia	−38.69***	3.746
SNNP	−20.61***	3.212
Constant	−105.30***	15.190
Chi-squared	576.1	
Number of observations	7,257	
Turning point for total cultivated area	8.891	

Source: Authors' analysis using the 2011 AGP baseline survey data (Berhane et al. 2013), except those with superscripts a and b, which are from Ethiopia, CSA (2014) and Hill and Tsehaye (2018). Coefficients with superscripts ***, **, and * are significant at 1 percent, 5 percent, and 10 percent levels, respectively.

Note: SNNP = Southern Nations, Nationalities, and Peoples. "Tropical livestock units" normalizes the number of livestock a household owns in camel units. "Model farmer" takes a value of 1 if the household head was selected as a model farmer in the last five years. "Production information" takes a value of 1 if the head obtained production information from radio, newspaper, or information boards. "Extension" takes a value of 1 if the household was visited by extension agents at least once in the last 12 months. The "Land Quality index" is computed by multiplying perceived fertility (three categories) and slope (three categories) of each plot of land; whereby the index ranges from the poorest (1) to the best (9) land quality. Household-level land quality index is computed as a plot area weighted sum of land quality indexes of plots cultivated.

decreases with the dependency ratio and vice versa for household size. As household size increases, there is more labor available and the likelihood of hiring in labor is therefore reduced. A higher share of elder and younger people—and therefore a higher dependency ratio—reduces labor supply, ceteris paribus, leading to more hiring-in of labor. The education levels of household heads also show a significant association with reliance on labor markets. More educated household heads are more likely to hire-in labor, since they are more likely to be engaged in alternative activities outside of their farms.

Second, farm characteristics also determine how households will participate in agricultural labor markets. The size of the farm is an important factor associated with labor market participation. The bigger the farm, the more likely the household will use hired-in labor. An increase in the size of the farm by one hectare leads to an increase of the share of expected hired-in labor by 12 percent. Households that cultivate better quality land are more likely to hire-in laborers, possibly to cultivate the better-quality land more intensively. Moreover, the proportion of hired-in labor increases with livestock ownership, probably from the need to use more labor to care for livestock as well as due to the wealth effect of owning a larger herd.

Third, we look at the effect of location variables. Holding other factors constant, farmers in zones with a higher incidence of poverty are more likely to hire agricultural workers, possibly because of the lower wage in such zones. The proportion of hired-in labor increases in urban areas and with proximity to Addis Ababa. The latter is consistent with better functioning labor markets in areas closer to large population centers, along with incentives for more intensive cultivation of land. Better functioning labor markets also imply employment opportunities, other than agricultural work. All regional dummies are significant and negative, which indicates that the proportion of hired-in labor is lower in Amhara, Oromia, and SNNP compared with Tigray.

Rural Wages: Changes, Drivers, and Implications

Changes in Rural Wages

Here we describe trends in real wages of unskilled or casual labor in both rural and urban areas. This is important for at least three reasons. First, since rural wages benchmark the wages paid in the manufacturing and service sectors, trends in rural-urban wage gaps over time will have important implications for Ethiopia's transformation from a largely agrarian toward a manufacturing-led economy. Second, rising wages have important implications on rural development and overall poverty reduction given the potential increase in the cost of food production and food prices resulting from rising wages. An important question is whether increases in wages in the future more than compensate for welfare losses due to subsequent food price increases. Third, changes in rural wages often trigger changes in farming systems by, for example, driving the introduction of labor-saving mechanization or bringing about changes in farm sizes, partly because the use of machinery reduces the advantage of small-scale farm operations in labor supervision (Otsuka, Liu, and Yamauchi 2014). In the case of Ethiopia, the latter depends on the extent to which land is consolidated and how this is facilitated in the future. To understand some of these issues, we use data on the daily wages of unskilled laborers collected by CSA on a monthly basis during the period July 2004 to September 2018. We return to each of these points below.

We discuss trends in wages expressed in US dollars (US$) and in US Consumer Price Index (CPI) deflated real US dollars. However, this deflation method might be imperfect because of changes in the overvaluation of the birr–dollar exchange rate over time (World Bank 2014). We therefore also rely on other deflation methods. First, we use the regional General CPI (GCPI) from CSA (Ethiopia, CSA 2018b). Second, we use the Poor Persons' General CPI (PP-GCPI). The PP-GCPI is constructed using expenditure shares on 26 food and nonfood categories of households in the lowest two income quintiles, which we define as poor households, obtained from CSA (Ethiopia, CSA 2007) and retail prices for those items (Ethiopia, CSA 2018a; see Headey et al. 2012).

Figure 11.6 depicts the daily wages of casual laborers expressed in nominal US dollars and real US dollars. These figures are important as they show what international investors would pay to employ unskilled labor in the country. Therefore, they are indicators of the competitiveness of Ethiopia in labor-intensive industries. Figure 11.6 shows that wages of unskilled laborers increased dramatically during July 2004 through September 2018, with daily wages nearly quadrupling in the period considered. Specifically, wages expressed in US dollars and real US dollars per person per day, which averaged 0.86 and 0.92 in the third quarter of 2004, grew to 3.28 and 2.63 in the third quarter of 2018, respectively. Growth in the respective wages averaged 10 percent and 7.8 percent per year, or 0.8 percent and 0.6 percent per month. However, the growth was not consistent over this period, and there was a slight decline in wages between January 2009, September 2010, and

FIGURE 11.6 Wages of unskilled laborers per day in Ethiopia in nominal US$ and real (December 2006) US$, July 2004–September 2018

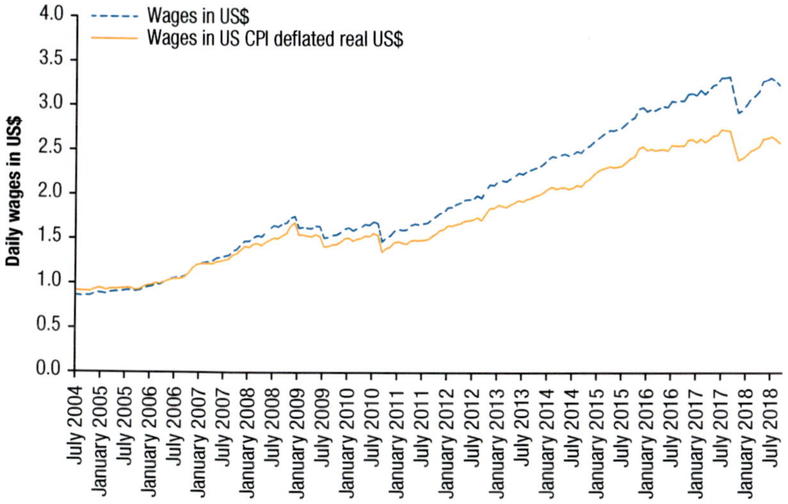

Source: Authors' computation using CSA nominal wages data (Ethiopia, CSA 2018a), National Bank of Ethiopia (2018), and US Bureau of Labor Statistics (2018).

October 2017. This might have been linked to the rapid devaluation of the birr in those periods (World Bank 2014).

To improve our assessment of local wage and price situations and to evaluate changes in the purchasing power of agricultural wage labor, we study real wages of unskilled laborers obtained by deflating nominal wages using both the regional average General CPI (GCPI) and the Poor Person's General CPI.

In Figure 11.7 we present urban and rural wages deflated by the General CPI. We note significant increases in real wages during the period: GCPI deflated real wages of unskilled labor in rural areas increased from 27.3 birr in the third quarter of 2004 to 45.3 birr in the third quarter of 2018 (total growth of 69 percent), while in urban areas it increased from 28.4 birr to 48.4 birr (total growth of 70 percent). Generally, the urban-rural wage gap has been low during most of the period. Wages in urban areas were on average 6 percent higher. This wage gap is significantly lower than those noted in other countries—for example, in Bangladesh real urban wages were double those of rural wages at the end of the 1990s but the gap gradually declined to 25 percent in 2012 (Zhang et al. 2014; Yang et al. 2013). However, the wage gap has been widening in recent years.

FIGURE 11.7 Regional General CPI (GCPI) deflated daily wages of unskilled laborers in rural and urban areas in Ethiopia, in December 2011 birr per day, July 2004–September 2018

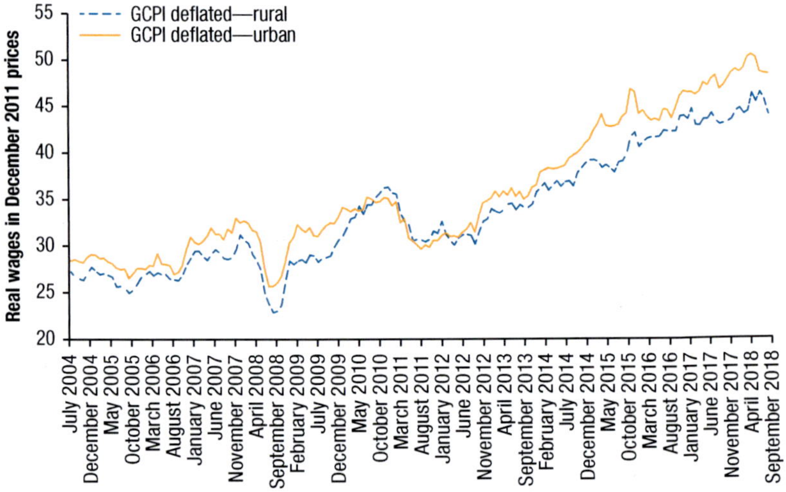

Source: Authors' computation using CSA nominal wages data (Ethiopia, CSA 2018a) and CSA regional GCPI data (Ethiopia, CSA 2018b).

In Figure 11.8 we provide the Poor Persons' General CPI (PP-GCPI) deflated daily real wages of unskilled labor.[8] PP-GCPI deflated average real wages of unskilled labor were 29.3 birr in rural areas and 31.3 birr in urban areas in the third quarter of 2004. This increased by an average of nearly 18.4 birr and 17.2 birr in the second quarter (that is, April–June) of 2017, or by about 3.9 percent and 3.5 percent per year in rural and urban areas, respectively. Similar to GCPI deflated wages, the PP-GCPI deflated urban-rural real wage gap was higher during 2008 and 2009, during which it averaged 11 percent, whereas the wage gap has been 3.4 percent during the rest of the period. Growth in PP-GCPI deflated real wages were mostly similar to the growth pattern in GCPI deflated real wages, indicating significant welfare improvements for such labor over time. However, the difference in PP-GCPI deflated real wage levels between urban and rural areas have in recent periods become smaller, indicating higher prices of goods for consumption in urban areas. We note a significant decline in real wages during 2008 and between

8 PP-GCPI could not be computed for the period after June 2017 because expenditure shares of the 26 food and nonfood categories have changed and data on the new shares are currently unavailable.

FIGURE 11.8 Poor Persons' General CPI (PP-GCPI) deflated daily wages of unskilled laborers in rural and urban areas in Ethiopia, in December 2011 birr per day, July 2004–June 2017

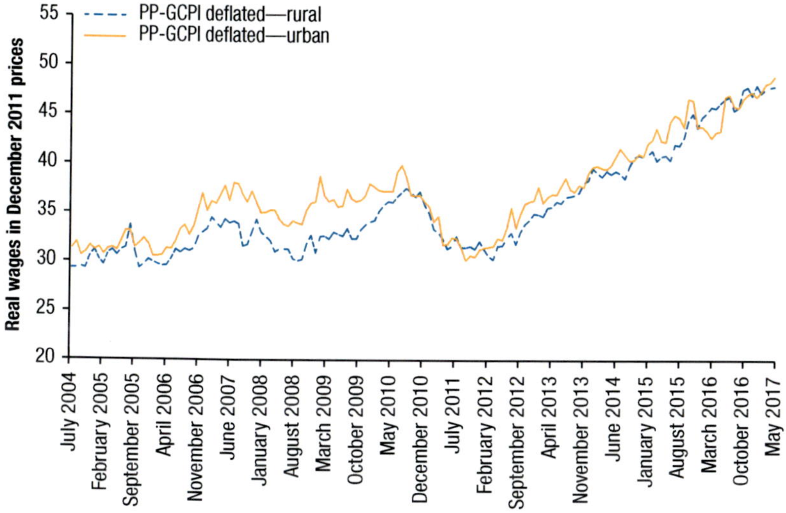

Source: Authors' computation using nominal wages data from CSA (Ethiopia, CSA 2018a).

the middle of 2010 and 2011, when inflation rates in the country were high and wage adjustments were significantly lower (Headey et al. 2012).

Drivers of Change in Wage Levels for Unskilled Labor

The Ethiopian economy has shown significant growth over the past decade—growth in real gross domestic product (GDP) averaged 11 percent during the 2004–2013 period (World Bank 2015a). We investigate whether real wages of unskilled labor in rural areas were impacted by this growth in the overall and sectoral GDP. For this purpose we use the World Development Indicators data (World Bank 2018) on real (aggregate) GDP and real value-added in agriculture, manufacturing and industry, and services, together with PP-GCPI deflated real wages. In these analyses we use national averaged real wage as dependent variable, and we interchangeably use nationally aggregated real GDP and real value-added in agriculture, manufacturing and industry, and services as explanatory variables.

Results of the econometric analyses are shown in Table 11.10. Each of the entries in the table are obtained by regressing the explanatory variables listed in the first column on average annual PP-GCPI deflated real wages of rural areas. We categorize zones as *rural* using the four criteria defined in the first

TABLE 11.10 Estimates of growth—unskilled real wage elasticity for rural populations in Ethiopia between 1999 and 2014 using different definitions of rural areas, by economic sector

Dependent variable (real wages in rural areas where rural areas are defined as)				
Explanatory variable	Average woreda population in zone less than 30,000	Average woreda population in zone less than 50,000	Rural areas[a]	Nonurban regions[b]
Real GDP	0.22***	0.18***	0.15***	0.12***
Real agricultural GDP	0.24***	0.20***	0.17***	0.15***
Real manufacturing GDP	0.19***	0.16***	0.14***	0.12***
Real industry GDP	0.17***	0.14***	0.12***	0.10***
Real services GDP	0.15***	0.12***	0.11***	0.09***

Source: Authors' analysis using Ethiopia, CSA (2018a), Ethiopia, CSA (2007), and World Bank (2015b).

Note: Coefficients with superscripts ***, **, and * are significant at 1 percent, 5 percent, and 10 percent levels, respectively. GDP = gross domestic product.

a. Woredas other than those that comprise the urban centers of Addis Ababa, Harari, or Dire Dawa regions or the cities of Mekele, Gondar, Bahir Dar, Dessie, Adama, Bishoftu, Kersa, Jigjiga, and Hawassa.

b. Regions other than Addis Ababa, Harari, or Dire Dawa.

row of the table. The data cover the 1999–2014 period, and each of the analyses comprise 16 observations. The results in Table 11.10 indicate that real wages in rural areas increase at 22 percent of the increase in aggregate GDP in rural administrative zones where zones are defined as rural if an average woreda in the zone has a population of fewer than 30,000. For the remaining three criteria used to define rural zones, the elasticity of real wages with respect to GDP is generally lower, ranging from 12 percent to 18 percent. The results indicate that the elasticity of rural real wages with respect to real agriculture sector value-added (GDP) was 24 percent in predominantly rural areas with woredas averaging fewer than 30,000. The elasticity ranged between 15 percent and 20 percent when this relationship was computed including more zones with larger populations. The elasticity of wages with respect to manufacturing, industry, and services were 19 percent, 17 percent, and 15 percent, respectively, in predominantly rural zones, with the elasticity being lower in zones with larger populations.

Two observations are notable about these results. First, the elasticity of real wages is highest with respect to agricultural value-added. This is likely because of the large importance of agriculture in most of the zones considered. Second, despite the value-added in agriculture accounting for less than 50 percent of the GDP during the period analyzed, the elasticity of real rural wages with respect to aggregate GDP is close but less than the elasticity of agricultural GDP, particularly in predominantly rural areas. The latter means that wages in rural areas respond most to changes in output that is locally produced, seemingly as most of the local population is engaged in these activities.

Implications of Wage Changes

Here we point to some of the likely implications of trends in rural wages and patterns in agricultural labor use observed in the sections above. In particular, we discuss the link between agricultural wages and rural poverty on the one hand and adjustments in agricultural production practices, such as labor-saving modern inputs, agricultural mechanization, and increase in the likelihood of using of modern inputs due to alleviated seasonal liquidity constraints on the other hand.

First, an important implication of the transformation of rural and agricultural labor markets is its role in poverty reduction (Ravallion 2000). We investigate this effect using Figure 11.9, which plots a quadratic function of the predicted Poverty Head Count index against real wages, both of which are zonally aggregated and pertain to the years 1996, 2000, 2005, and 2011. The figure demonstrates that increases in real wages are strongly correlated with a decline in poverty. This negative relationship is particularly strong at lower real wages. Similar patterns have been observed in other settings (for example, Ravallion 2000).

We use the same data to econometrically test whether zonal-level poverty HCI is related with real wages levels following Lanjouw and Murgai (2009). The general equation we estimate is given as:

$$\text{In } Poverty\ HCI_{j,t} = \beta_0 + \beta_1 \text{ In } Real\ wage_{j,t} + \eta_j + \tau + e_{j,t} \qquad (1)$$

where $Poverty\ HCI_{j,t}$ stands for poverty HCI of zone j, such that $j \in [1,2,...,56]$ in year t, with $\in[1996,2000,2005,2011]$, η_j represents fixed effects of zone j; and τ stands for a time trend variable. We estimate OLS and fixed-effects specifications of equation (1) whereby in the OLS specification the zone fixed-effect variables (η_j) are replaced by region dummies. We provide results of the analyses in Table 11.11.

The results indicate that the elasticity of poverty HCI with respect to real wages is negative and significant in both specifications. Moreover, the elasticities from the OLS (−0.57) and fixed-effects (−0.44) specifications are close to each other. Elasticities obtained from specifications without time trends (not shown) are unit elastic in both models.

Second, we note increasing substitution of labor with labor-saving modern inputs such as herbicides over time in Ethiopia (Bachewe et al. 2015; Minten et al. 2013). This is partly driven by an increase of real wages, as documented

FIGURE 11.9 Correlation of prevailing real wages and Poverty Head Count Index by administrative zone in Ethiopia

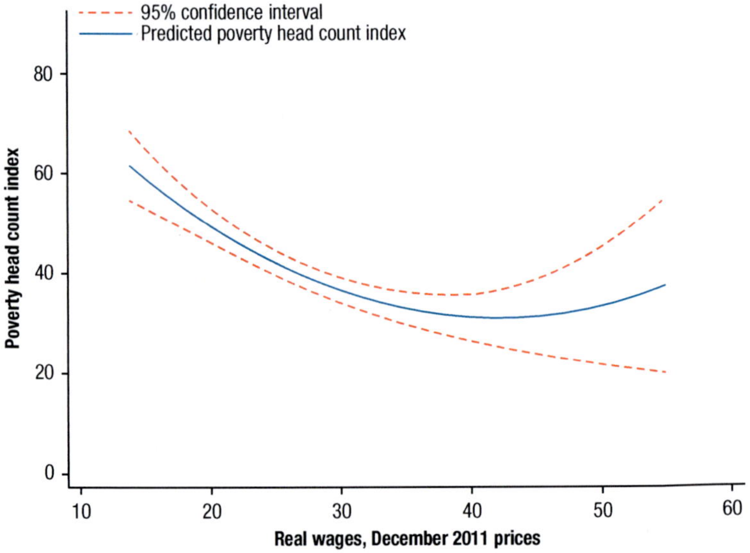

Source: Authors' computations using Ethiopia, CSA (2014a; 2014b); Hill and Tsehaye (2018).

TABLE 11.11 Strength of the association between zonal-level poverty head count index and real wages in Ethiopia

Dependent variable: Zonal poverty head count index in 1996, 2000, 2005, and 2011

Variables	OLS	Fixed effects
Log of real wages	−0.574***	−0.444*
	(0.179)	(0.229)
Time trend	−0.141***	−0.149***
	(0.031)	(0.033)
Constant	6.027***	5.288***
	(0.615)	(0.719)
F statistics	8.4	28.5
Probability of F	0.00	0.00

Source: Authors' analysis using Ethiopia, CSA (2014a; 2014b) and Hill and Tsehaye (2018). Coefficients with superscripts ***, **, and * are significant at 1 percent, 5 percent, and 10 percent levels, respectively.

above, as well as by a decrease of real prices for herbicides (Tamru et al. 2017). Figure 11.10 shows that when herbicides are used, the weeding efforts are significantly lower, indicating substitution between these inputs. The tightening of the labor markets might contribute to the increasing use of herbicides in the country. For example, Figure 11.10 shows the increasing adoption of herbicides in major commercial teff areas in the country: 31 percent of the teff farmers used herbicides 10 years prior to the survey, but this figure increased to 63 percent in 2012. Commercially oriented and well-connected farming areas show a particularly strong uptake of such herbicides, possibly linked to the higher wages that are paid in such areas.

Third, higher rural wages often provide incentives for mechanization in agriculture (Binswanger 1986; Diao et al. 2014; Yang et al. 2013). Most surveys in Ethiopia do not collect data on mechanization, so we cannot assess the extent to which mechanization is used in different parts of the agricultural production process. In a rare exception, Berhane, Hirvonen, and Minten (2015) analyze the use of mechanized services in major production zones in the country. They find that about 9 percent of smallholder farmers used mechanized services at some period during the agricultural season, mostly for plowing (5 percent) and relatively less for harvesting (3 percent) and threshing (2 percent). While the adoption of these mechanized services is still relatively low, Berhane, Hirvonen, and Minten (2015) show a strong threshold effect in the use of mechanized services and the reported daily wages in the community (Figure 11.11). While a slight upward trend is seen in the use of mechanization by smallholders for wage levels up to 100 birr per day (just over US$5), use of mechanized services significantly expand after that wage threshold. Yang et al. (2003) have found similar strong effects of wage increases on the adoption of mechanization in agriculture in China. Unfortunately we do not have data on the cost of mechanization services over time in Ethiopia, and therefore we cannot compare to what extent possible price changes for these services might have been driving increasing adoption.

Fourth, as shown in other studies, off-farm income might lead to a relaxation of credit constraints in the period of the cropping season when agricultural inputs are required and therefore might lead to improved agronomic practices. Reardon, Crawford, and Kelly (1994) and Lamb (2003) show that such off-farm income sources might lead to significant farm investments and higher use of modern inputs, leading to higher agricultural productivity. We use the AGP baseline dataset to test the extent to which households that have off-farm income in Ethiopia are more likely to use modern inputs, such as fertilizer, improved seeds, and agrochemicals. We find that only in the case of

FIGURE 11.10 Association between use of weeding labor and value of herbicides used (left); prevalence of herbicide use by teff producers in 2002 and 2012 as a function of transport costs to Addis Ababa (right)

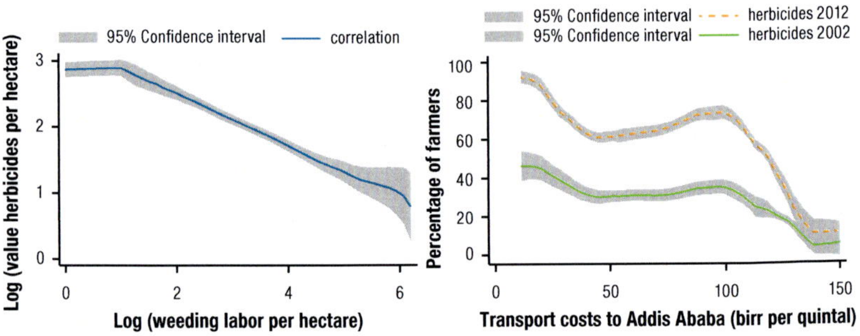

Source: Authors' computations using IFPRI and EDRI (2020).

FIGURE 11.11 Prevalence of use of agricultural mechanization and prevailing daily agricultural wage rates in Ethiopia

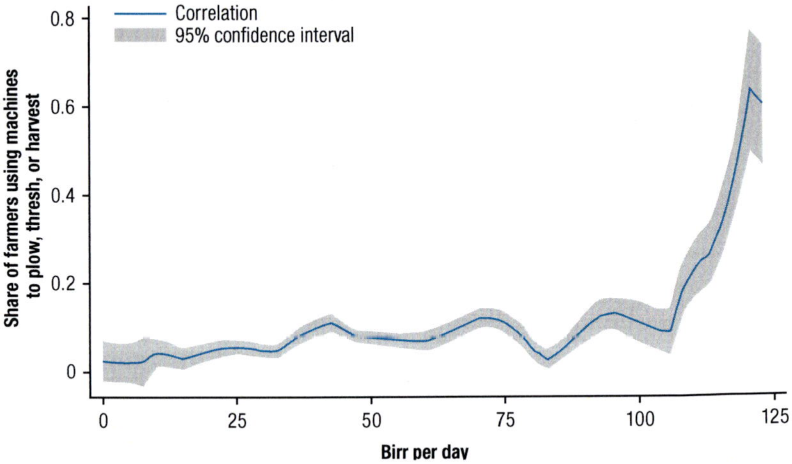

Source: Berhane, Hirvonen, and Minten (2015), citing data from FtF (2015).

Note: Local polynomial regression.

chemical fertilizer use is off-farm income positively associated with a higher likelihood of using chemical fertilizer. However, the association is weak, as the coefficient is only significant at the 10 percent level.

Conclusion

As economies develop and shift away from agrarian subsistence to commercial agriculture and increasingly to nonagriculture-based economic activities, labor markets emerge and increasingly become a critical part of the economy in providing livelihoods for many people (Fox et al. 2013). In the initial stages of this transformation, and especially so in rural areas, these labor markets are important for poverty reduction as well as for assuring employment for the rural youth, an increasing concern in Africa, given its rapidly growing population and its "youth bulge." In the case of Ethiopia, the economy is undergoing rapid transformation, mostly driven by rapid agricultural growth. However, priorities are shifting in the planning process, and there is increasing emphasis on stimulating the manufacturing sector as an anticipated source of growth in future decades. This is illustrated in the second phase of the Ethiopian government's Growth and Transformation Plan. Given that such shifts will have important implications on labor markets, it is important to understand rural and agricultural labor markets. This chapter aims to fill this knowledge gap and contributes to greater insight into understanding likely future trends in these markets.

We find that off-farm income in the high potential agricultural areas in Ethiopia where the AGP is being implemented makes up almost 18 percent of household income and that wage labor income accounts for 10 percent. Wage income is estimated to be as important as livestock income in these areas. We estimate that in these high-potential agricultural zones of the country, only 7 percent of the agricultural production is carried out by hired labor. We find that there is considerable heterogeneity in rural unskilled labor markets over space and time. Agricultural labor markets are relatively more important in better connected areas and in urban areas, which is reflected in higher wages in these areas. This suggests that connectedness and urbanization are among the driving forces toward improved labor markets.

Moreover, we note strong seasonality with peaks in hired labor use and wages in the harvest period. Income from off-farm sources are shown to be relatively more important for the poor than for the rich, suggesting that push factors might still be relatively more important in rural areas for engagement by households in the rural off-farm income sector than are pull factors. While wages of unskilled labor in rural areas of Ethiopia are still low relative to the international context, real wages are on the rise, measured both in US dollars and birr. This is seemingly driven by an improvement in agricultural performance, which leads to a reduction in poverty. This increase in real wages is gradually leading to an increasing push toward the adoption of labor-saving technologies in agriculture, such as mechanized services—though starting from a low base—as well as to a widespread use of herbicides that save on weeding labor.

These findings have important implications on policy. First, low wages have been an asset for attracting investors to Ethiopia, and they have contributed to the success of investments in labor-intensive industries, such as in floriculture. However, our results indicate that Ethiopia may gradually lose that edge before it further matures as the wage gap between rural and urban areas narrows and future cheap labor sources disappear. Ethiopia therefore needs to make sure that the skills of its young population are upgraded so that industries can develop in those areas where labor productivity is higher. Second, the higher costs of labor as well as the increasing commercialization of agriculture will require increasing adoption of labor-saving technologies in the sector. It is therefore important that Ethiopia proactively implements policies that allow for the provision of such appropriate technologies at low cost. Third, flexible and responsive labor markets require easier migration to those areas with employment opportunities. Therefore migration needs to be encouraged through more flexible land tenure and more secure land rental rules. Such policies could also facilitate the consolidation of small farms, given the introduction of labor-saving technologies. Furthermore, this may subsequently reduce the incentives to operate small farms, which have the traditional advantages over larger farms of decreased costs of labor supervision.

References

Bachewe, F., G. Berhane, B. Minten, and A. S. Taffesse. 2015. *Agricultural Growth in Ethiopia (2004–2014): Evidence and Drivers.* IFPRI-ESSP Working Paper 81. Addis Ababa: International Food Policy Research Institute/Ethiopia Strategy Support Program (IFPRI/ESSP).

Berhane, G., M. Dereje, J. Hoddinott, B. Koru, F. Nisrane, F. Tadesse, A. S. Taffesse, I. Worku, and Y. Yohannes. 2013. *Agricultural Growth Program (AGP) of Ethiopia-Baseline Report 2011.* ESSP-EDRI Report. Addis Ababa: IFPRI.

Berhane, G., K. Hirvonen, and B. Minten. 2015. *Mechanization of Agriculture in Ethiopia: Evidence from the 2015 Feed the Future Survey.* IFPRI-ESSP II Research Note 41. Addis Ababa: IFPRI/ESSP II.

Binswanger, H. 1986. "Agricultural Mechanization: A Comparative Historical Perspective." *World Bank Research Observer* 1 (1): 27–56.

Block, S., and P. Webb. 2001. "The Dynamics of Livelihood Diversification in Post-Famine Ethiopia." *Food Policy* 26 (4): 333–350.

Deichmann, U., F. Shilpi, and R. Vakis. 2009. "Urban Proximity, Agricultural Potential and Rural Non-Farm Employment: Evidence from Bangladesh." *World Development* 37 (3): 645–660.

Dercon, S., and P. Krishnan. 1996. "Income Portfolios in Rural Ethiopia and Tanzania: Choices and Constraints." *Journal of Development Studies* 32 (6): 850–875.

Diao, X., F. Cossar, N. Houssou, and S. Kolavalli. 2014. "Mechanization in Ghana: Emerging Demand and the Search for Alternative Supply Models." *Food Policy* 48: 168–181.

Escobal, J. 2001. "The Determinants of Nonfarm Income Diversification in Rural Peru." *World Development* 29 (3): 497–508.

Ethiopia, CSA (Central Statistical Agency). 2007. *Household Income, Consumption and Expenditure (HICE) Survey 2004/05.* Volumes 1 and 2. Statistical Bulletin 394. Addis Ababa.

———. 2011. Unpublished Baseline Survey data. Addis Ababa, Ethiopia: The Agricultural Growth Program of Ethiopia.

———. 2013. Unpublished Ethiopia Rural Socioeconomic Survey (ERSS) data. Addis Ababa, Ethiopia: Central Statistical Agency.

———. 2014. *Population Projection of Ethiopia for All Regions at Wereda Level from 2014–2017.* Addis Ababa.

———. 2018a. *Consumer Price Survey.* Addis Ababa.

———. 2018b. *Country and Regional Level Consumer Price Indices.* Addis Ababa.

Fafchamps, M., and F. Shilpi. 2003. "The Spatial Division of Labour in Nepal." *Journal of Development Studies* 39 (6): 23–66.

———. 2005. "Cities and Specialisation: Evidence from South Asia." *Economic Journal* 115 (503): 477–504.

Fox, L., C. Haines, J. H. Munoz, and A. Thomas. 2013. *Africa's Got Work to Do: Employment Prospects in the New Century.* IMF Working Paper WP/13/201. Washington, DC: International Monetary Fund.

FtF (Feed the Future). 2013. Unpublished Ethiopia Feed the Future Survey data.

———. 2015. Unpublished Impact Evaluation Survey data. Addis Ababa, Ethiopia.

Haggblade, S., P. B. Hazell, and T. Reardon. 2007. *Transforming the Rural Nonfarm Economy: Opportunities and Threats in the Developing World*. Baltimore: Johns Hopkins University Press.

Headey, D., F. N. Bachewe, I. Worku, M. Dereje, and A. S. Taffesse. 2012. *Urban Wage Behavior and Food Price Inflation: The Case of Ethiopia*. IFPRI-ESSP Working Paper 41. Addis Ababa: IFPRI/ESSP.

Hill, R., and E. Tsehaye. 2018. *Growth, Safety Nets and Poverty: Assessing Progress in Ethiopia from 1996 to 2011*. Policy Research working paper; no. WPS 8380. Washington, DC: World Bank Group.

Holden, S., B. Shiferaw, and J. Pender. 2004. "Non-farm Income, Household Welfare, and Sustainable Land Management in a Less-Favoured Area in the Ethiopian Highlands." *Food Policy* 29 (4): 369–392.

IFPRI (International Food Policy Research Institute) and EDRI (Ethiopian Development Research Institute). 2019. Coffee Value Chain Survey in Ethiopia: Producer/Household Survey. Washington, DC: IFPRI [dataset]. https://doi.org/10.7910/DVN/GRIAV8. Harvard Dataverse, VERSION 1.

———. 2020. Teff Value Chain in Ethiopia: Producer/Household Survey. Washington, DC: IFPRI [dataset]. https://doi.org/10.7910/DVN/NLT2MO, Harvard Dataverse, V1.

Jacoby, H., and B. Minten. 2009. "On Measuring the Benefits of Lower Transport Costs." *Journal of Development Economics* 89 (1): 28–38.

Lamb, R. L. 2003. "Fertilizer Use, Risk, and Off-Farm Labor Markets in the Semi-Arid Tropics of India." *American Journal of Agricultural Economics* 85 (2): 359–371.

Lanjouw, J. O., and P. Lanjouw. 2001. "The Rural Non-Farm Sector: Issues and Evidence from Developing Countries." *Agricultural Economics* 26 (1): 1–23.

Lanjouw, P., and R. Murgai. 2009. "Poverty Decline, Agricultural Wages, and Nonfarm Employment in Rural India: 1983–2004." *Agricultural Economics* 40: 243–263.

Minten, B., S. Tamru, E. Engida, and T. Kuma. 2013. *Ethiopia's Value Chains on the Move: The Case of Teff*. IFPRI-ESSP Working Paper 52. Addis Ababa: IFPRI/ESSP II.

National Bank of Ethiopia. 2018. Daily Exchange Rate. Accessed December 1, 2018. https://nbebank.com/inter-bank-daily-foreign-exchange-rate-in-usd/.

Otsuka, K., Y. Liu, and F. Yamauchi. 2014. *The Future of Small Farms in Asia*. Accessed February 19, 2016. www3.grips.ac.jp/~esp/wp-content/uploads/2014/11/112014.pdf.

Ravallion, M. 2000. "Prices, Wages and Poverty in Rural India: What Lessons Do the Time Series Data Hold for Policy?" *Food Policy* 25 (3): 351–364.

Reardon, T., J. Berdegué, C. Barrett, and K. Stamoulis. 2006. "Household Income Diversification into Rural Nonfarm Activities." In *Transforming the Rural Nonfarm Economy*, edited by S. Haggblade, P. Hazell, and T. Reardon, 115–140. Baltimore: Johns Hopkins University Press.

Reardon, T., E. Crawford, and V. Kelly. 1994. "Links between Nonfarm Income and Farm Investment in African Households: Adding the Capital Market Perspective." *American Journal of Agricultural Economics* 76 (5): 1172–1176.

Tamru, S., B. Minten, F. Bachewe, and D. Alemu. 2017. "The Rapid Expansion of Herbicide Use in Smallholder Agriculture in Ethiopia: Patterns, Drivers, and Implications." *European Journal of Development Research* 29 (3): 628–647.

United States, Bureau of Labor Statistics. 2018. Historical Consumer Price Index Data. Accessed December 1, 2018. www.bls.gov/cpi/home.htm.

USAID (United States Agency for International Development). 2011. "Feed-the-Future Ethiopia: Multi-year Strategy (2011–2015)." Accessed February 9, 2016. www.usaid.gov/sites/default/files/documents/1860/USAID%20FtF%20MYS%20Final%20Version.pdf.

Wiggins, S., and S. Keats. 2014. *Rural Wages in Asia*. London: Overseas Development Institute.

World Bank. 2014. *Ethiopia: Poverty Assessment*. Report No. AUS6744. Poverty Global Practice. Africa Region. World Bank Group. Washington, DC.

———. 2015a. *Ethiopia's Great Run: The Growth Acceleration and How to Pace It*. Report no. 99399-ET. Washington, DC.

———. 2015b. World Development Indicators. Accessed October 15, 2015. http://data.worldbank.org/data-catalog/world-development-indicators.

Yang, J., Z. Huang, X. Zhang, and T. Reardon. 2013. "The Rapid Rise of Cross-Regional Agricultural Mechanization Services in China." *American Journal of Agricultural Economics* 95 (5). 1245–1251.

Zhang, X., S. Rashid, K. Ahmad, and A. Ahmed. 2014. "Escalation of Real Wages in Bangladesh: Is It the Beginning of Structural Transformation?" *World Development* 64: 273–285.

Chapter 12

URBANIZATION AND STRUCTURAL TRANSFORMATION

Emily Schmidt, Paul Dorosh, Mekamu Kedir Jemal, and Jennifer Smart

Introduction

Ethiopia is urbanizing rapidly but from a low base. While the absolute numbers of the rural population in Ethiopia are projected to rise for the next several decades, the urban population growth rate (4.2 percent between 1994 and 2015) exceeds the overall population growth rate (2.5 percent). As a result, the share of the population residing in urban areas (using the official Central Statistical Agency of Ethiopia definition of urban areas) rose from 13.7 percent to 19.4 percent between 1994 and 2015.[1] This share is expected to increase to 27.8 percent by 2025 (assuming a 5.4 percent urban population growth rate, per World Bank [2015]) but is still far below the average for Africa south of the Sahara (SSA) (45.3 percent in 2020), for which urban populations are growing by 4.4 percent per year (UNDESA 2015).

Urbanization has major implications for overall growth. Long-term economic development is generally associated with a movement of workers from low-productivity agricultural labor (in rural areas) to high-productivity manufacturing labor (in more urbanized areas).[2] This structural transformation

1 Official estimates from the government of Ethiopia of urban population are based on administrative boundaries that exclude peri-urban areas but include small cities. In this chapter we present data on two other measures of urbanization as well—one based on agglomerations (that is, large, high-population density areas) and another focusing on cities with populations greater than 50,000.

2 Whereas within-sector productivity growth and structural change go hand-in-hand, there are two primary views as to where the process of growth originates (Rodrik, McMillan, and Sepúlveda 2013). One argues that agricultural productivity is a precondition for growth (Gollin 2018; Johnston and Mellor 1961; Timmer 1988, 2002; Mellor 1995, 2017); the other argues that growth is industrialization-led, and agricultural productivity improves when nonfarm jobs absorb rural population growth, maintaining the same level of agricultural output with a smaller agricultural labor force (Lewis 1954).

not only boosts incomes of individuals earning a higher wage, but it also provides capital to invest in inputs and mechanization for greater agricultural productivity growth in rural areas. Similarly, assuming a greater share of individuals move toward off-farm labor, this frees up agricultural land for more productive farmers (that choose to stay in agriculture), resulting in greater productivity effects among agricultural laborers (Mellor 1995). Moreover, urbanization, if supported by appropriate infrastructure, has the potential to boost total factor productivity and economic output through positive agglomeration effects. For these reasons urbanization has been an important aspect of Ethiopia's development strategy for the past two decades, including the agricultural development–led industrialization (ADLI) strategy of the early 2000s and the more recent Growth and Transformation Plans I and II (Ethiopia, MoFED 2007, 2010; National Planning Commission 2016).

This chapter examines the urbanization process in Ethiopia and places it in the context of structural transformation and recent public investments to promote industrialization. The chapter is organized as follows: After a brief overview of the data and methodology used in the chapter, we describe the historical growth of total and urban populations in Ethiopia and project future urban population growth. The chapter builds on this analysis, presenting data on both population growth and the structural transformation of Ethiopia's economy, in comparison with other countries in Africa south of the Sahara, as well as with the experience of India and China. Next we focus on key aspects of Ethiopia's development strategy related to structural transformation and urbanization. In doing so, we estimate the direct and multiplier effects of industrial parks and agro-industrial centers, which represent the major public investments designed to implement Ethiopia's current development strategy. The chapter concludes with a summary and highlights areas for future research and analysis.

Urbanization in Ethiopia

Over the past three decades, the size and number of cities dramatically increased across Ethiopia. In 1994 the country had a population of more than 53.6 million people but housed only 13 cities with populations greater than 50,000. By 2007 the total population increased by 20 million, and the number of cities doubled to 26. By 2015 the total population increased another 16 million from 2007, and the number of cities almost doubled again with 45 cities of at least 50,000. Urbanization rates differ across regions, with Oromia and Southern Nations, Nationalities, and Peoples (SNNP) regions

FIGURE 12.1 Number of cities in Ethiopia of at least 50,000 people

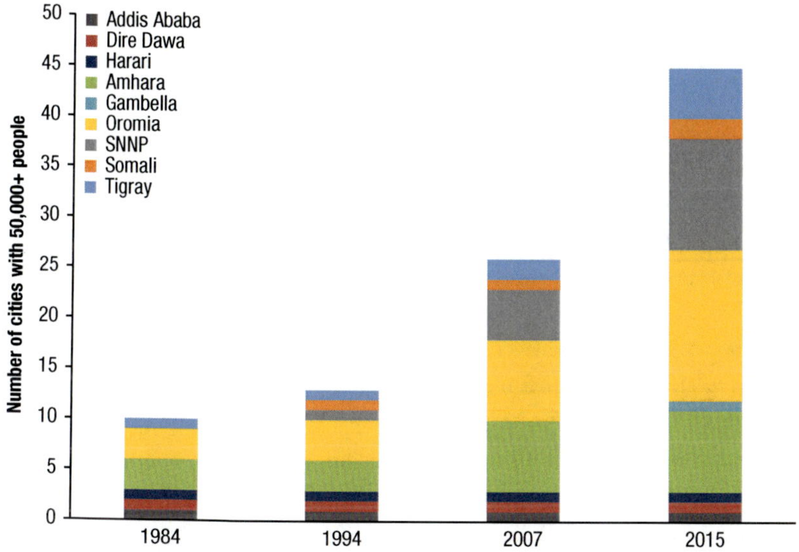

Source: Authors, based on data from Central Statistical Agency of Ethiopia (Ethiopia, CSA 1991, 1996, 2010, 2013b). See Schmidt et al. (2018), Annex Table A2.1.

Note: SNNP = Southern Nations, Nationalities, and Peoples.

experiencing the largest growth in city numbers from 8 to 15 and 5 to 11 cities, respectively, between 2007 and 2015 (Figure 12.1).[3]

Correlated with this increase in the number of cities over time is an overall increase in urbanization throughout Ethiopia. We evaluate urban population and urban growth rates using two methodologies. First, we review the official rural and urban population figures based on official census data collected in 1984, 1994, and 2007 (Ethiopia, CSA 1991, 1996, 2010, 2013b). In addition, we simulate urbanization trends into the future using CSA's published population projection figures and the urban growth rate of 5.4 percent per year estimated in the World Bank's Ethiopian urbanization review (World Bank 2015).[4] Second, we evaluate urbanization over time using the agglomeration index methodology developed by Uchida and Nelson (2008). Compared to administrative definitions of urban and rural locations (defined by census

3 Addis Ababa, Dire Dawa, and Harari are regions defined by an urban administrative unit and thus remain constant over time.

4 The average urban growth rate estimated by the World Bank's Ethiopian urbanization review (World Bank 2015) incorporates natural population growth, rural-urban migration, migration to mega project sites, and urban area expansion.

designations and used for official CSA documents) that change over time due to redistricting and redefinition, the agglomeration index defines "urban" based on travel time to cities of a defined population size and surrounding areas with a defined population density.[5] For this analysis urban areas are locations that have a population density of at least 150 people per square kilometer and are located within one-hour travel time to a city of at least 50,000. This analysis builds on earlier work by Schmidt and Kedir (2009) that modeled travel times to major cities (see Appendix 12A for further detail on the estimation of travel time and the agglomeration index).[6] We compare previous estimates reported by Schmidt and Kedir (2009) to updated 2015 agglomeration values. A comparison of results generated from these two methodologies reveals important differences in overall urban growth rates and, in particular, urban growth trends in and around the capital city of Addis Ababa.

Comparing Ethiopia's urban share of population as measured by the agglomeration index with official urbanization rates based on administrative boundaries suggests that urban growth is occurring faster than officially projected. Until 2007, official CSA estimates of the share of Ethiopia's population living in urban areas was greater than the agglomeration index estimation. In 1984 and 1994 few urban centers and limited road networks resulted in lengthy travel times to urban areas, hence a lower urbanization rate (as per the agglomeration index estimation). However, by 2015 the compounded effect of city and population growth, along with ongoing investments in road infrastructure and improved accessibility led to a higher urbanization rate as measured by the agglomeration index than the CSA official estimate (Figure 12.2). As urban centers proliferate, and transportation infrastructure facilitates the movement of goods and people, it is expected that agglomerations extending past official urban boundaries will continue to expand.

Focusing specifically on the official population figures published from the census, we evaluate urbanization over time by city size, separating cities into three categories: (1) cities with between 50,000 and 100,000 people; (2) cities with between 100,000 and 500,000 people; and (3) the city of Addis Ababa. To understand growth rates by city size, we identify cities that were projected to have at least 50,000 people by 2015 based on the 5.4 percent growth rate published by the World Bank's Ethiopian urbanization review (World Bank

5 A detailed explanation of administrative definitions and urbanization calculations in Ethiopia can be found in Schmidt and Kedir (2009).

6 The methodology to summarize travel time and calculate agglomeration is slightly modified compared to Schmidt and Kedir (2009). See Appendix 12A for an in-depth description of the methodology and modifications.

FIGURE 12.2 Comparison of the agglomeration index for Ethiopia and official estimates of the share of the total population residing in urban areas, 1984–2015

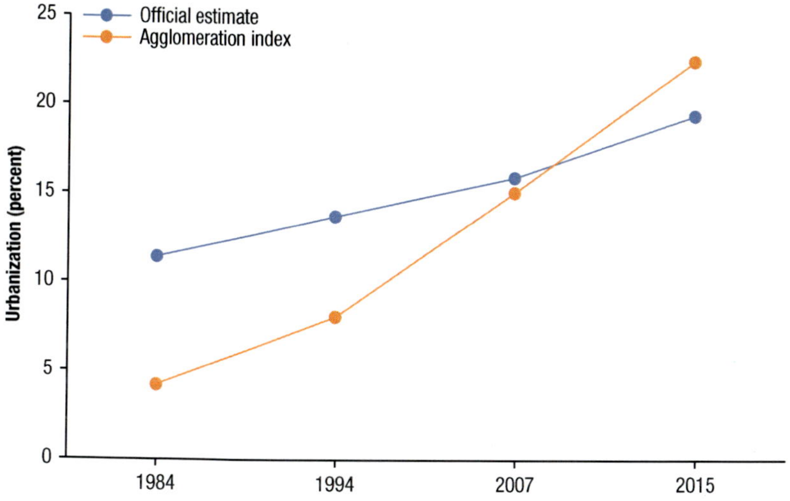

Source: Ethiopia, CSA (2013b) and agglomeration index estimations.

2015) using the most recently published census data from 2007. We then create a database of those specific cities dating back to 1984 (the first census year of the analysis) to evaluate urban growth over time, holding the number of cities constant throughout the period from 1984 to 2015. Similarly, we use the same consistent set of cities to project urban growth to 2035.

Census data, reported by administrative unit, suggest that the population of Addis Ababa, Ethiopia's capital and largest city, grew at 2.02 percent per year from 1994 to 2007 and at 2.25 percent per year between 2007 and 2015. The population of cities between 100,000 and 500,000 persons grew at twice the rate of Addis Ababa (Table 12.1; see also Appendix Tables 12B.1 and 12B.2). Similarly, the growth rates of cities between 50,000 to 100,000 people grew at more than twice the rate of Addis Ababa between 1994 and 2015. Projecting urbanization rates into the future using the official census data suggests that the population growth rate of medium-sized and large cities will continue to outpace the growth rate of Addis Ababa. These projections suggest that the total urban population living in cities greater than 50,000 people will increase by more than 10 million by 2025 compared to 2007 levels and increase by an additional 11 million by 2035 (Table 12.1 and Figure 12.3).

However, these data can be misleading given the urban area expansion that has occurred in and around Addis Ababa. Addis Ababa official population

TABLE 12.1 Urban center census populations, population projections, and growth rates in Ethiopia, 1984–2035

	Census			Population projection[a]		
City size	May 1984	October 1994	May 2007	July 2015	2025	2035
Population size (thousands)						
50,000–100,000[a]	337	650	1,139	1,782	3,284	5,593
100,000–500,000[a]	847	1,340	2,276	3,488	9,225	20,242
Addis Ababa	1,413	2,113	2,740	3,273	4,561	5,810
Subtotal	2,597	4,102	6,154	8,543	17,070	31,646
Total urban (CSA)	4,869	7,385	11,863	17,459	31,072	52,574
Total population	42,617	53,764	73,751	90,076	111,644	132,701
Annual population growth rate since previous measurement (%)						
50,000–100,000[a]		6.77	4.41	5.76	6.31[c]	5.47[c]
100,000–500,000[a]		4.69	4.16	5.48	10.21[c]	8.18[c]
Addis Ababa		4.11	2.02	2.25	3.37[c]	2.45[c]
Subtotal		5.33	3.13	4.37	6.94[c]	6.20[c]
Total urban (CSA)		4.25	3.71	4.95	5.40[b]	5.40[b]
Total population		2.35	2.46	2.53	2.17[d]	1.74[d]

Source: CSA population and housing census data for 1984, 1994, 2007, and 2015 (Ethiopia, CSA 1991, 1996, 2010, 2013b); EGIS International (2016); and authors' calculations.

Note: a. City size categories are defined in terms of 2015 population using Ethiopia, CSA urban population projections regardless of population size in years prior to or following 2015. This permits examination of changes in population for a consistent set of geographic localities over time.

b. 5.4 percent is the overall urban growth rate used for projections from 2012 forward by World Bank (2015). This projection rate is also used by the Ethiopian Development Research Institute (EDRI).

c. Growth rates derived from the EGIS national urban system study report projections (EGIS International 2016).

d. These total population growth rate projections are the log estimate annual growth rate of the respective period years of the full 2008–2037 series of World Bank projected populations, medium variant.

statistics are confined to the Addis Ababa census administrative boundary designation and do not consider the urban expansion occurring along major transportation corridors leading from and adjacent to the Addis Ababa administrative boundaries. A paper by Zeleke et al. (2018) shows that the expansion of Addis Ababa in terms of land use is exponential. In absolute terms the expansion in Addis Ababa is larger than the combined land use of Adama, Mekele, Hawassa, and Bahir Dar with 2015 populations of 324,000 (Adama), 324,000 (Mekele), 300,000 (Hawassa), and 282,000 (Bahir Dar) (Ethiopia, CSA 2013b; City Population 2018). From 1986 to 2010 the average rate of growth in area for Addis Ababa was calculated at 9.1 square kilometers per year, while the combined average rate of growth in area for Adama, Mekele, Hawassa, and Bahir Dar was 3.2 square kilometers per year (Zeleke et al. 2018).

FIGURE 12.3 Breakdown of urban population in Ethiopia by city size categories for urban centers with population of more than 50,000, 1984–2035

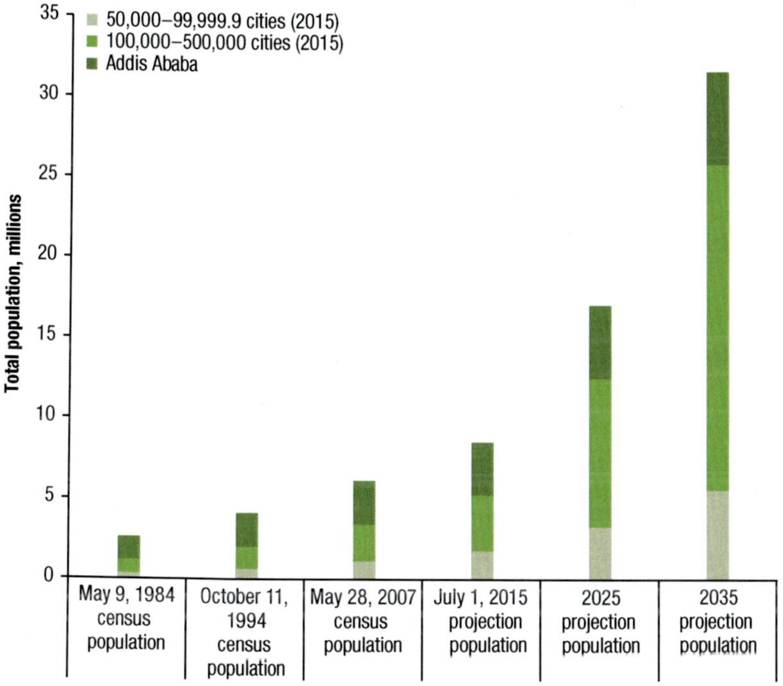

Source: CSA population and housing census data for 1984, 1994, 2007, and 2015 (Ethiopia, CSA 1991, 1996, 2010, 2013b); EGIS International (2016); and authors' calculations.

Note: City size categories are defined in terms of 2015 population (using Ethiopia, CSA urban population projections) regardless of population sizes in years prior to or following 2015.

Comparing urbanization over the past four decades using the agglomeration index methodology highlights the immense investments in road infrastructure as well as the increasing growth of secondary cities across Ethiopia. While in 1984 only 4.2 percent of the total population was considered urban according to the agglomeration index, 22.5 percent of the population was urban in 2015 (Table 12.2). Of the population increase in Ethiopia of 36 million between 1994 and 2015, the urban share of annually added population increased from 34 percent between 1994 and 2007 to 56 percent between 2007 and 2015 (Figure 12.4).

Urbanized networks are forming between important secondary cities throughout the country. For example, between 1994 and 2015 the emergence of upgraded or newly constructed roads along the main corridors connecting

TABLE 12.2 Percentage of people residing in urban areas in Ethiopia, by region (1984–2015)

	Urban share of total population (%)			
Regions	1984[a]	1994	2007	2015
Addis Ababa	57.5	91.1	91.9	99.9
Afar	0.0	0.0	0.0	0.6
Amhara	3.2	4.9	7.9	14.0
Benishangul-Gumuz	0.0	0.0	0.0	0.0
Dire Dawa	16.9	55.8	59.5	67.9
Gambella	0.0	0.0	0.0	11.9
Harari	44.7	75.8	81.0	94.3
Oromia	2.0	5.2	10.3	15.9
SNNP	—	2.6	23.7	39.5
Somali	0.3	2.4	2.7	2.4
Tigray	2.7	4.3	8.1	19.1
Ethiopia	**4.2**	**8.0**	**15.1**	**22.5**

Source: CSA population and housing census of 1984, 1994, 2007, and 2015 (Ethiopia, CSA 1991, 1996, 2010, and 2013b); and EGIS International (2016).

Note: See Appendix 12A for details on urban population calculations using the agglomeration index. SNNP = Southern Nations, Nationalities, and Peoples.

a. Population figures for 1984 were approximated due to changes in administrative boundaries after 1984. To maintain consistency across all years, we geographically allocated population to the current administrative boundaries.

FIGURE 12.4 Population added in Ethiopia since 1994 and annual population growth (millions)

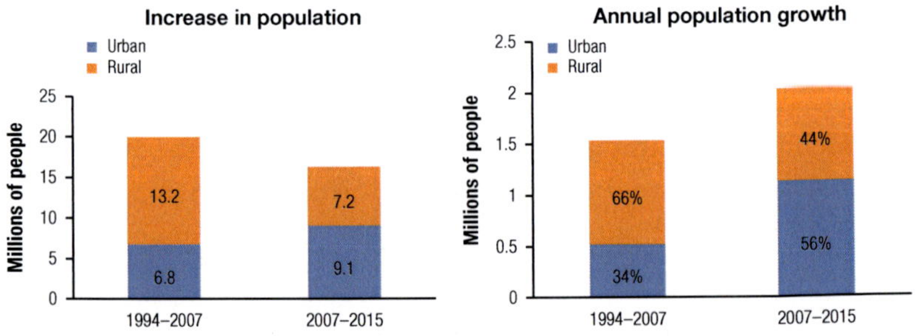

Source: CSA population and housing census data for 1984, 1994, 2007, and 2015 (Ethiopia, CSA 1991, 1996, 2010, 2013b); EGIS International (2016); and authors' calculations.

Addis Ababa to Dukem, Bishoftu, Mojo, Adama, and Assela to the southeast of Oromia region all significantly improved the travel time between cities and accelerated the rate of urbanization. Particularly striking is the growth in urban corridors in and around Addis Ababa over time (Figure 12.5). In 1984 urban population (as classified by the agglomeration index) was limited to the western part of the city with limited agglomeration around the five major transportation corridors, while in 2015 the greater urban area and corridors of Addis Ababa fully comprise the Addis Ababa administrative unit and reach to other neighboring smaller cities in Oromia and Amhara regions (Figure 12.5). According to the agglomeration index analysis, approximately 20 percent of Addis Ababa's urban population lives outside of the Addis Ababa administrative region boundary along major transportation corridors and in urban feeder towns (Figures 12.5 and 12.6).

Investments and improvements in major transportation arteries as well as the development of regional secondary cities within Ethiopia have improved the accessibility of remote rural areas to urban centers. Between 2007 and 2015 an additional 15 percent of the population were within three hours' travel time of a city of at least 50,000 people—the share of the population with this level of access to urban centers rising from 44.5 percent in 2007 to 58.6 percent in 2015 (Table 12.3; Appendix Tables 12A.2, 12A.3, and 12A.4 and Figure 12A.3). Even the most remote populations that were more than 10 hours away from a city decreased from 11 percent of the population in 2007 to 5.4 percent in 2015. One-fifth of the population of Ethiopia (nearing 20 million people now) are estimated to be farther than five hours from an urban center of 50,000 or greater, suggesting that further investments are needed to link these individuals to important public infrastructure and services (see discussions in Chapters 3, 5, and 11).

While urban centers continue to grow, total population and labor force growth in Ethiopia are slowing. Annual population growth is expected to decelerate from 2.5 percent per year (between 2011 and 2015) to 1.7 percent per year between 2025 and 2035 (Table 12.4).[7] Likewise, annual labor force growth, which was 4.9 percent between 2011 and 2015, is expected to drop to 2.1 percent in the 2025–2035 period.[8] Given the recent surge in population and subsequent deceleration, Ethiopia is currently experiencing a "youth

7 Population growth averaged 2.5 percent from 2010 to 2020 and is expected to decline to 1.8 percent between 2020 and 2030, and further to 1.4 percent by 2040.

8 Annual labor force growth was an average 3.7 percent from 2010 to 2020, down from 5.8 percent during the 1990–2000 period.

FIGURE 12.5 Estimates of the urban extents for 1984, 1994, 2007, and 2015: Addis Ababa

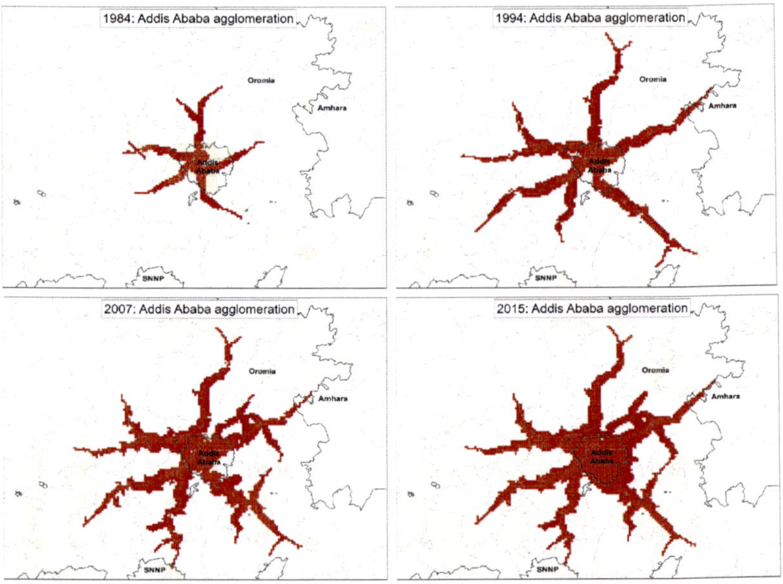

Source: Authors' calculations.

FIGURE 12.6 Urban population within the Addis Ababa regional administrative unit and additional urban population based on agglomeration index calculations, 1984–2015

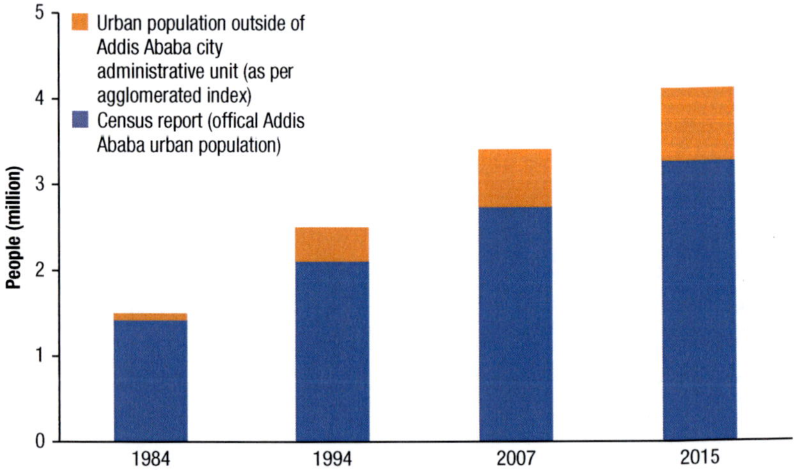

Source: Authors' calculations.

TABLE 12.3 Travel time to nearest city in Ethiopia of at least 50,000 people in 2015, share of total population, by region (%)

Region	Less than 1 hour	1–3 hours	3–5 hours	5–10 hours	More than 10 hours
Tigray	24.7	34.9	20.0	18.1	2.4
Afar	1.2	18.3	26.3	41.7	12.6
Amhara	16.2	39.3	24.9	17.6	2.1
Oromia	18.3	39.7	24.5	15.4	2.2
Somali	4.6	11.0	12.8	29.7	42.0
Benishangul-Gumuz	0.1	1.3	18.3	49.0	31.4
SNNP	40.4	36.2	11.2	9.6	2.5
Gambella	15.7	18.0	16.3	27.6	22.4
Harari	94.3	5.7	0.0	0.0	0.0
Addis Ababa	99.9	0.1	0.0	0.0	0.0
Dire Dawa	69.0	16.8	8.8	5.5	0.0
Ethiopia	**24.5**	**34.1**	**19.8**	**16.2**	**5.4**

Source: Authors' calculations using the agglomeration index.
Note: SNNP = Southern Nations, Nationalities, and Peoples.

bulge" in the labor force as people born between 1998 and 2002—those ages 15 to 19 years old in 2015—enter the labor market. Until this cohort retires, Ethiopia has an opportunity to benefit from a "demographic dividend" as the ratio of workers to dependents rises (Figure 12.7). Although a greater share of laborers (compared to dependents) will characterize the Ethiopian demographic for coming decades, this "dividend" will only pay off if workers find work. Filmer and Fox (2014) outline the constraints for rural youth to enter into paid labors, listing a variety of factors, including insufficient capital or credit or both, information asymmetries of markets and input supplies, and lack of education and skills. A concerted effort will need to be placed on improving job opportunities in terms of demand for labor and supply of skilled labor to reap the potential benefits of Ethiopia's "youth bulge."

Ethiopia's labor force, defined as the sum of females and males between the ages of 15 and 64 years, is increasing faster than the overall population. It will continue to do so over the next two decades, rising from a 51.8 percent participation rate in 2007 to an expected 72.1 percent in 2035 (Table 12.5). Although the population of females of any age is slightly lower than the population of males, female participation in the labor force is slightly greater than that of males. Moreover, population growth of females is taking place at a

TABLE 12.4 Urban and rural populations in Ethiopia, 1994–2035

	Population					Labor Force				
	Addis Ababa	Other urban[a]	Total urban[b]	Rural	Total	Addis Ababa	Other urban[a]	Total urban[b]	Rural	Total
Population and labor force size (thousands)										
1994	2,113	1,990	4,102	49,375	53,764					
2007	2,740	3,415	6,154	67,597	73,751	1,420	1,770	3,190	35,039	38,229
2011[c]	2,994	4,242	7,237	74,269	81,506	1,724	2,442	4,166	42,753	46,919
2015	3,273	5,270	8,543	81,533	90,076	2,064	3,323	5,387	51,416	56,804
2025	4,561	12,148	16,709	94,935	111,644	3,169	8,441	11,611	65,969	77,580
2035	5,810	24,677	30,487	102,214	132,701	4,187	17,781	21,967	73,649	95,616
Annual growth rate (%)										
2011–2015	2.25	5.57	4.24	2.36	2.53	4.61	8.01	6.64	4.72	4.90
2015–2025	3.37[d]	8.71	6.94[d]	1.53	2.17	4.38	9.77	7.98	2.52	3.17
2025–2035	2.45[d]	7.34	6.20[d]	0.74	1.74	2.82	7.74	6.59	1.11	2.11

Source: CSA population and housing census data for 1994, 2007, and 2015 (Ethiopia, CSA 1996, 2010, 2013b); EGIS International (2016); and authors' calculations.

Note: Urban population in each year is calculated as the sum of inhabitants living in cities that had a population of at least 50,000 in 2015.

a. City size categories are defined in terms of 2015 population using Ethiopia, CSA urban population projections regardless of population size in years prior to or following 2015. This permits examination of changes in population for a consistent set of geographic localities over time.

b. Urban is defined as settlements with a population of 50,000 and greater. All other areas are rural for the purposes of this table.

c. 2011 figures are imputed using a steady growth rate between 2007 and 2015.

d. Growth rates derived from the EGIS national urban system study report projections (EGIS International 2016).

FIGURE 12.7 Ethiopia's census population and population projections by age bracket, 1984–2035

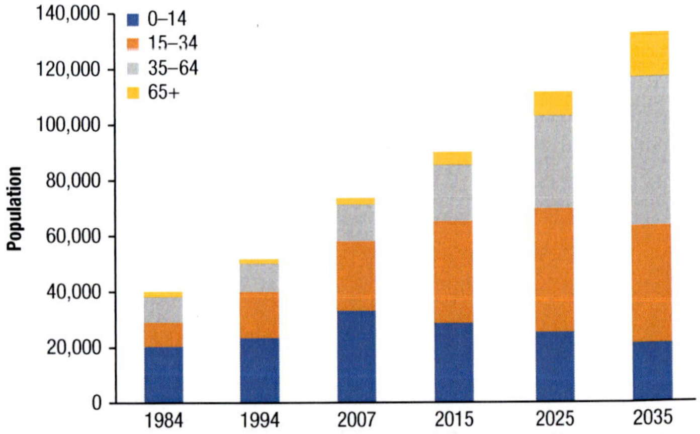

Source: Authors' calculations using historical CSA data and projections (Ethiopia, CSA 2013b) and EGIS growth rates (EGIS International 2016).

TABLE 12.5 Female and male labor force participation and annual growth in Ethiopia, 2007–2035

	Population			Labor force			
	Female	Male	Total	Female	Male	Total	Participation rate (%)
Population and labor force size (thousands)							
2007	36,534	37,217	73,751	19,312	18,917	38,229	51.8
2015	44,825	45,250	90,075	28,486	28,318	56,804	63.1
2025	55,705	55,939	111,644	38,856	38,724	77,580	69.5
2035	66,311	66,389	132,700	47,833	47,783	95,616	72.1
Annual growth rate (%)							
2007–2015	2.6	2.5	2.5	5.0	5.2	5.1	n.a.
2015–2025	2.2	2.1	2.2	3.2	3.2	3.2	n.a.
2025–2035	1.8	1.7	1.7	2.1	2.1	2.1	n.a.

Source: Authors' estimates based on CSA data and projections (Ethiopia, CSA 2013b).
Note: n.a. = not applicable.

slightly higher rate of 0.1 percentage points above the annual male population growth, while the male annual labor force growth is about equal to or slightly higher than growth of the female labor force.

Migration and Structural Transformation

Migration in Ethiopia

According to the last three censuses in Ethiopia—those of 1984, 1994, and 2007—the share of migrants moving to another district is increasing. From 1984 to 1994 migrant population grew from 4.5 million people, or 11 percent of the population in 1984, to 6.9 million, or 13 percent of the population in 1994. Total migrants increased to 12.2 million (17 percent) in 2008 (Ethiopia, CSA 1991, 1996, 2010). Almost half of the migrants in 2008 migrated from a rural area to another rural area, while 27 percent of migrants moved from a rural to an urban area (Dorosh et al. 2011). The remaining migrated within urban areas (20 percent) or from an urban to a rural area (7 percent). More recent data collected by the National Labor Force Survey (NLFS) in 2013 (Ethiopia, CSA 2013a) suggests that rural-rural and rural-urban migration shares were almost equivalent at 35 and 33 percent, respectively, of total migrants (Table 12.6).

Given that the most recent census data are from 2007, we use the National Labor Force Survey of Ethiopia to evaluate migration rates within the country.[9] Although measures of the contribution of rural-urban migration toward total urban growth are reported in various datasets, including the 2007 census and the National Labor Force Survey, the sampling design used to generate these calculations are constrained to administrative boundaries. Therefore the rural-urban migration rate of 32.5 percent as reported by the NLFS in Table 12.6 masks a substantial share of rural-to-peri-urban and rural-to-urban migration. Thus rural-urban migration as reported by the NLFS can be considered as a lower bound because it is defined using administrative boundaries and does not capture the agglomerated urban areas as presented in this chapter.

The NLFS data provide a national picture of migration and suggest that migration destinations vary by regions. Greater rural-rural migration occurs in Amhara, Oromia, and SNNP regions; less than a third of migrants moved from rural to urban areas in these regions (Table 12.6).[10] Youth between the ages of 15 and 35 make up a large share of the migrant population. Overall, 53 percent of rural-urban migrants and 44 percent of rural-rural migrants were between the ages of 15 and 35 years (Figure 12.8).

According to the NLFS data, 32 percent of migrants at least 15 years old move in search of work or due to job transfers. While 45 percent of males move for work, a majority of female migration is attributed to marriage, followed by migration in search of work at 32 and 22 percent, respectively. When disaggregating migrants by origin and destination, of all migrants that move from rural to urban areas, fewer than half—58 percent of men and 34 percent of women—are motivated to migrate in search of work.

More detailed studies of migrants in Ethiopia suggest that migrants are predominantly pushed from their homes rather than attracted by an urban pull of higher returns on human capital investments. For example, the Ethiopian Urban Migration Study (Moller 2012) reports that more than 42 percent of migrants stated that they would not have migrated if they had been able to make a living in their original home. In comparison to nonmigrants, migrants who moved to Addis Ababa came from families with much

9 The NLFS is a nationally representative dataset that provides information on labor trends in the country and is collected every five years.

10 Data from the Ethiopia Socioeconomic Survey (ESS) for 2012–2013 suggest that the rural-urban migration rate in Ethiopia was approximately 2 percent of rural households per year.

TABLE 12.6 Forms of migration by region in Ethiopia, percentage of region migrants (%)

Current region	Rural-rural	Rural-urban	Urban-urban	Urban-rural	Total (thousands)
Amhara	40.0	32.9	18.0	9.2	2,628.0
Oromia	39.8	27.4	17.4	15.4	4,368.4
SNNP	39.2	27.5	18.1	15.3	2,067.4
Tigray	31.1	35.5	24.4	9.1	768.4
Benishangul-Gumuz	55.4	22.5	15.1	6.9	227.0
Gambella	40.6	25.6	22.9	11.0	86.9
Afar	25.5	37.3	25.6	11.6	190.2
Somali	20.5	23.8	35.7	20.0	101.9
Addis Ababa	0.0	58.7	41.3	0.0	1,217.1
Dire Dawa	3.9	37.7	52.8	5.6	112.8
Harari	6.5	38.5	52.5	2.4	50.2
Ethiopia	**34.5**	**32.5**	**21.3**	**11.6**	**11,818.3**

Source: Authors' calculations using NLFS 2013 (Ethiopia, CSA 2013a).
Note: SNNP = Southern Nations, Nationalities, and Peoples.

FIGURE 12.8 Forms of migration by age group and by migration type in Ethiopia, total and percentage of migrants

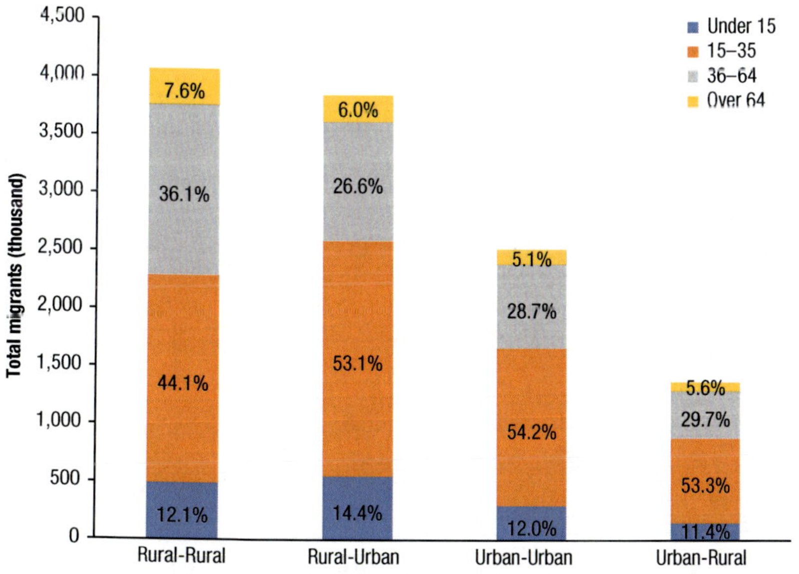

Source: Authors' calculations using NLFS 2013 (Ethiopia, CSA 2013a).

lower educational levels than nonmigrants. Female migrants had three times the illiteracy rate of nonmigrant females.

Other analyses also suggest that migrants in Ethiopia are pushed from rural areas as a means to diversify rural incomes. Asfaw, Degaga, and Gete (2010) report that young men are the most common migrants in Amhara region, with respondents citing a lack of sufficient means of subsistence, shortage of land, and shortage of employment opportunities in the rural areas as their primary reasons for migrating. Dorosh et al. (2011) reported that households with more agricultural land were less likely to send out migrants, while poorer households and households afflicted by a communitywide drought were more likely to have migrant household members. Similarly, Kosec et al. (2017) found that larger land inheritances led male youths in Ethiopia to migrate less to urban centers and increased the likelihood of them being employed in the agricultural sector. These findings are echoed by Gray and Mueller (2012), who reported men's labor migration in Ethiopia increases with drought.

In theory, migration can potentially improve the sending household's well-being via remittances. However, studies of Ethiopian migrants suggest remittances are uncommon among both rural and urban migrants. Results from the Ethiopia Urban Migration Survey suggest that only 13 percent of respondents send remittances home (Martins 2014). A majority of migrants report that it is too costly to remit. Analysis by De Brauw, Mueller, and Woldehanna (2013), using a panel survey of 1,800 households, reported that only 33 percent of rural-urban migrants in Ethiopia send remittances. Given the relatively low remittance rate of internal migrants in Ethiopia and the reported push factors for migration (such as climate shock, land scarcity, and the like), it may be important for households to diversify income sources via migration (seasonal or permanent) to take pressure off of the sending households during times of agricultural production shortfalls.

Migrating can be costly, risky, and difficult for any individual regardless of ex ante human or capital endowments. To the extent that these workers have skills, find employment, and increase market demand, they can contribute to agglomeration economies that raise overall productivity (Lucas 1988; Moretti 2004; Ciccone and Peri 2006). Nonetheless, given that rural-rural migration is more prevalent than rural-urban migration in Ethiopia and that rural-urban migration is characterized predominantly by push factors, an increased focus on the rural and small-town, nonfarm economy may be a more viable avenue to absorb excess rural labor, diversify rural income sources, and reduce seasonal consumption shortfalls as a medium-term rural development strategy. Christiaensen et al. (2018) find that in Tanzania, migration to towns contributed more to poverty reduction compared to migration to cities.

Benefits of Secondary City Development and the Rural Nonfarm Economy

Much of the future growth in Ethiopia's urban population will likely be in secondary cities that will have the potential to serve as regional markets for agricultural products and to provide seasonal and full-time employment for nearby rural areas. However, current labor statistics in Ethiopia suggest a low rural nonfarm participation rate. Recent work by Schmidt and Bekele (2016) evaluated labor activities in Ethiopia to understand labor diversification in rural areas. They used the Ethiopia Socioeconomic Survey (ESS) to compose work portfolios for each individual based on hours reported working in each sector. The analysis suggested that approximately 76 percent of the total working-age population (15–64 years old) reported working on their own farm, for a wage, or in a nonfarm enterprise (Table 12.7). Of those, more than three-quarters of the working population reported working solely on their own farm. About 13 percent of the population reported working in a mix of own-farm and off-farm activities, while 12 percent reported working exclusively off-farm in wage labor or in nonfarm enterprises.

To understand work portfolios by location, we disaggregate the labor data in the ESS (2012/2013) into workers who live in rural areas, small towns (defined as those living in centers with fewer than 10,000 people), or urban centers of greater than 10,000 people. Given the growth in secondary cities and urban agglomerations along major transportation corridors, it is surprising that only 11 percent of the rural working-age population reports working in a nonfarm activity in addition to working on their own farm. Even in small towns, only 12 percent of the population works in a mixture of off- and on-farm labor, whereas one-third of the population of small towns reports working solely in off-farm work.

Although the government of Ethiopia has made significant investments in education during the past decade, a large share of nonagricultural work is comprised of low-skill jobs. For example, according to the National Labor Force Survey (Ethiopia, CSA 2013a), 30 percent of nonagricultural work is comprised of sales workers, of which street and local market vendors comprise 44 percent (Schmidt and Bekele 2016). The second largest employment type is construction and mining activities that account for only 10 percent of total

TABLE 12.7 Distribution of labor type in Ethiopia by spatial domain, 2013

	People ages 15–64 years, thousands (% share)			
Labor type	Rural	Small town[c]	Urban[c]	Total
Own farm only	27,304.3	55.4	216.4	27,576.1
	(67.8)	(13.8)	(3.1)	(57.8)
Own farm and off-farm	4,404.5	47.8	166.1	4,618.3
	(10.9)	(11.9)	(2.4)	(9.7)
Off-farm[a]	922.7	130.6	3,189.7	4,243.0
	(2.3)	(32.6)	(45.5)	(8.9)
Not working	6,620.3	121.9	2,077.5	8,819.8
	(16.4)	(30.4)	(29.6)	(18.5)
Student[b]	1,010.5	45.3	1,357.6	2,413.4
	(2.5)	(11.3)	(19.4)	(5.1)
Working population	32,631.4	233.8	3,572.2	36,437.5
	(81.0)	(58.3)	(51.0)	(76.4)
Total population (ages 15–64 years)	**40,262.2**	**401.1**	**7,007.4**	**47,670.7**

Source: Adapted from Schmidt and Bekele (2016).

Note: a. Off-farm work comprises individuals that work in off-farm enterprise or wage work or both.

b. Students are defined as those that do not report time working in own farm, wage, or off-farm enterprise activities and report activity as student.

c. Small towns were defined as those under 10,000 inhabitants, while urban areas were defined as having a population of 10,000 inhabitants or more at the time of the survey year (2013/2014).

nonagricultural employment in Ethiopia. The specific service activities that individuals are engaged in across Ethiopian small towns and cities reflect a low level of development with limited labor demand.

Contextualizing Ethiopia's Story: Patterns of Population Growth in Asia and Africa

Migration and employment in Africa appear consistent with recent patterns observed elsewhere. We take China and India as interesting comparisons to structural transformation processes in Ethiopia. In both countries initial investment policy focused on industrialization at the expense of rural areas. These strategies were paired with restrictions on labor mobility, particularly in China. Eventually, both countries enacted agricultural policy reform—China in the late 1970s and India in the early 1990s—and relaxed rural-urban migration restrictions. Although China has experienced a more classic economic transformation (via rapid industrial and agricultural growth leading to significant rural-urban migration), India's transformation has been constrained by insufficient labor demand from the industrial sector and is more characterized by rural-rural migration (Fan, Chan-Kang, and Mukherjee 2005; Binswanger-Mkhize 2012; Cai 2013; Cai and Wang 2010).

China began experiencing negative rural population growth starting in the 1990s, resulting in greater urbanization. It is projected that in India, even by 2020, rural population growth is still expected to be positive. In 1990 the urban share of the population in India was 27.7 percent, while that in China was 36 percent; the agriculture share of GDP was 30.1 percent for India and 26.6 percent for China.[11] In Africa south of the Sahara, while urbanization is still taking place at a rapid pace, rural population growth is projected to be positive in 2020 at 1.7 percent. While India's labor mobility has been driven by manufacturing, Africa south of the Sahara has largely urbanized without industrialization (Gollin et al. 2016). Ethiopia has lower urbanization compared with other countries in SSA, with a rural population growth of 1.6 percent, close to the average level for Africa south of the Sahara.

Over the past several decades China experienced a massive exodus of people from rural areas driven by the overall slowing of population growth due to restrictive population control policies early on, a resultant drop in the youth population, and significant urban economic growth. In contrast, trends in Africa south of the Sahara, including Ethiopia, are more similar to India's experience. The rural population is continuing to increase, and any absolute population declines are projected to take place decades from now (Figure 12.9). Whereas India's labor force is expanding at close to the total population's growth rate of about 1.2 percent, Ethiopia's labor force population exceeds its total population growth rate at 3.5 percent and 2.5 percent, respectively. The proportion of population 15 to 34 years old to the entire population is currently at its greatest in Ethiopia (having grown from a share of 22 percent in 1984 to 41 percent in 2015 and projected to be 40 percent in 2025 and 32 percent in 2035, derived from the same data presented in Figure 12.7), and many of these young adults are just entering the labor force now. In comparison, China's young working population is decreasing in size (Thurlow et al. 2018). In Ethiopia the number of young adults ages 15–35 is still peaking and is only expected to decrease after 2025.

11 A brief summary of the growth pathways that China and India experienced during their economic structural transformations is included in Schmidt et al. (2018).

FIGURE 12.9 Population growth and share of population that is urban by global region, 1950–2050

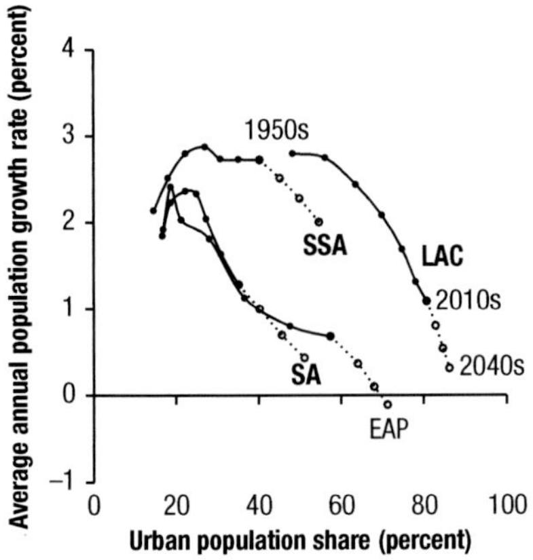

Source: Thurlow, Davis, and Dorosh (2018).

Note: Figure includes historical data (solid lines) and projections (dotted lines). Future populations are based on the median variant fertility and life expectancy projections. EAP = East Asia and the Pacific; LAC = Latin America and the Caribbean; SA = South Asia; SSA = Africa south of the Sahara.

Figure 12.9 shows total population growth and urban population shares from 1950 to 2050 for the regions of Latin America and the Caribbean (LAC), Africa south of the Sahara (SSA), South Asia (SA), and East Asia and the Pacific (EAP). Every region has passed its peak population growth rate, and population growth continues to slow in each region going forward to 2050 (Thurlow, Davis, and Dorosh 2018). SSA in the 2040s will have a population growth rate similar to South Asia in the 1990s and to East Asia and the Pacific in the 1970s, indicating a comparable trend, although about six decades later and at a larger scale. Figure 12.9 also shows that South Asia in 2010 is only as urbanized as East Asia and the Pacific was in 1990, and that the gap between these two regions is expected to stay roughly the same over the next several decades.

Compared to the majority of other countries in SSA, where population growth peaked at an average rate of 2.9 percent in 1980, Ethiopia's peak population growth took place around 1990 and at a higher growth rate of 3.3 percent (Figure 12.10). Uganda and Kenya, two of Ethiopia's East African neighbors, similarly peaked at population growth rates higher than the average SSA rate. However, both of these countries reached their peaks closer

FIGURE 12.10 Population growth and share of population that is urban in Africa south of the Sahara, 1950–2050

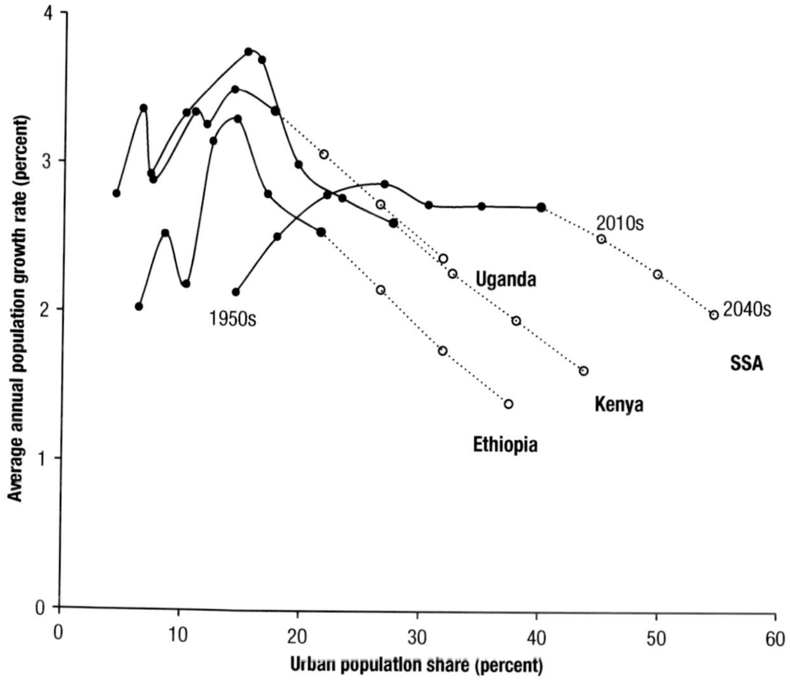

Source: Authors' calculations using population estimates for developing countries from UNDESA (2015, 2017).
Note: SSA = Africa south of the Sahara. Figure includes historical data (dark lines) and projections (dotted lines). Future populations are based on the median variant fertility and life expectancy projections.

to when the rest of SSA did—Kenya's peak took place closer to 1970, and Uganda's took place around 1980.

Today, Ethiopia is only as urbanized as SSA was in the 1980s—Ethiopia's urban share in 2020 will be 26.8 percent according to these data, while the SSA urban share reached 27.1 percent by 1980, up from a 22.3 share in 1970. However, Ethiopia's urbanization growth rate is similar to Kenya and Uganda—Ethiopia's urban population share lags consistently between 0.9 to 6.3 percentage points behind Kenya and leads consistently between 1.5 and 5.5 percentage points ahead of Uganda in urban population share in every decade from 1950 to 2040. By 2040 projections suggest that Ethiopia will be about 40 percent urban compared with SSA's average of 55 percent urban, Kenya at 45 percent, and Uganda at only 35 percent.

Industrial and Agro-industrial Parks

Government Policy Related to Industrial Development: The Investment Administration of Ethiopia and the Growth and Transformation Plan II

In an effort to incentivize investments in the face of low levels of private sector development and industrialization, Ethiopia established its first investment office in 1992, with the objective "to promote and facilitate domestic and foreign investment" (Oqubay 2015). A high priority was placed on the manufacturing sector and the development of the industrial parks as well as on addressing the inefficiencies of the nascent policy environment.[12]

Taking into account the performance and lessons learned from the first Growth and Transformation Plan (GTP I), the GTP II reaffirmed its commitments and recognized its shortcomings in the area of promoting industrial development and encouraging foreign and local investment. Many of the GTP I targets related to the development of the industrial sector were not met, both in the promotion of small and medium-sized enterprises as well as in the development of industrial parks. The importance of continued investments in the industrial sector was recognized in the context of the high performance of the sector compared with other sectors during the GTP I implementation period.

The expansion of the country's industrial parks comes with three primary expectations. The parks are expected to (1) create a conducive environment for increasing investments in the manufacturing sector, (2) promote higher export processing, and (3) enhance linkages between domestic and foreign firms to facilitate transfer of technology, skills, and other externalities. Furthermore, to achieve these expectations for competitive industrial development, the GTP II outlines several provisions the government commits to offer the parks to allow for adequate delivery:

---

12 In 1996 the name for the office, chaired by the prime minister, changed to the Ethiopian Investment Authority (EIA), which later became the Federal Investment Agency (FIA) in 2006 (then chaired by the minister of trade and industry). In 2010 the cabinet was restructured, and the FIA was made accountable to the Ministry of the Interior. From 2008 to 2012, 84 percent of the board's 446 decisions related to duty-free privileges or duty exemptions, while only 17 decisions focused on amendments to various directives. The manufacturing sector and the development of the industrial parks have become a priority of investment law, benefiting from increased and generous incentives, while other sectors' incentives have been reduced. Oqubay (2015, 91) cites that "because of lack of effective control, there were many instances where investors abused their duty-free privileges, selling duty-free vehicles and goods for hotel and touring operations at market prices and transferring and reselling land given under concessional terms."

- All required infrastructure;
- Streamlined public procedures;
- Fiscal and trade policy incentives to attract export-oriented foreign direct investment as well as domestic industrial investment;
- Sector and subsector policies to ensure sustainable and competitive industrial development;
- Follow-up and support to ensure the effectiveness of existing policies; and
- To the extent possible, creation of the required capacity needed in the park development and management by using international best practices to inform them.

As part of its support to the industrial parks, and the manufacturing industry as a whole, the GTP II envisions that the industrial parks are to be developed based on feasibility studies and investment demand in coming years. Implicitly, these new investments build on the major prior investments in road infrastructure and electricity described earlier in this chapter. The plan aims to strengthen the institutional arrangement and ownership of the parks as well as the capacity in these institutions to develop, administer, and regulate the parks, to enable them to perform their role more effectively. The ongoing land administration reforms would also be accelerated during GTP II, prioritizing land supplied to industrial parks with the goal of encouraging investments in industrialization and export development. Finally, there is mention throughout the GTP II strategy that any emphasis on developing, expanding, and supporting the parks is certainly not to be perceived as a detriment for medium-sized, small, or micro enterprises, but rather that these changes should also strengthen and increase the market share of these businesses.

The Current Status of Ethiopia's Industrial Parks

Five industrial parks are currently operational in the country (Figure 12.11), including the following:

- Addis Industry Village Park, which has existed since the 1980s, although it was renovated in 2015;
- Bole Lemi I, which was completed during or shortly after the conclusion of the GTP I;

- Hawassa, inaugurated in July 2016;
- Kombolcha, inaugurated in July 2017;
- Mekelle, inaugurated in July 2017.

The lead subsector in all of these parks is apparel and textiles. Secondary subsectors include food processing. Addis Industry Village has a land area of 8.7 hectares, and the other four parks currently have a combined area of 427 hectares. Approximately 30,000 people are currently employed by Hawassa park, and 11,000 are employed by Bole Lemi I park. The four together are expected to attain an area of 2,147 hectares, with 189,000 people employed over the next 10 years.[13] The current developed area of these four parks is about 20 percent of future targets, while employment currently sits at about 22 percent of targeted future employment.

Seven industrial parks are currently under construction but not yet operational. These include Jimma, Bole Lemi II, Kilinto, Dire Dawa, Adama, Bahir Dar, and Arerti. They are anticipated to have a pilot phase total combined area of 1,018 hectares, with an eventual combined area of 9,979 hectares. A wide spectrum of sectors will be operating in these industrial parks, including pharmaceuticals, agro-processing, electric and electronics products, wood and furniture, textiles, apparel, vehicle assembly, food processing, chemicals, paper and allied products, and heavy industries, machinery, and parts.

Jimma, Dire Dawa, Adama, and Bahir Dar industrial parks are included in the GTP II, as are the industrial parks in Hawassa, Kombolcha, and Mekelle that are already in operation. The construction of Kilinto and Bole Lemi II has been commissioned by the Ethiopian Industrial Parks Development Corporation (IPDC) with the support of a loan by the World Bank. Adama Industrial Park is being constructed as a joint venture between the governments of Ethiopia and the Hunan Province of China, with the help of a $250 million Chinese loan.[14] Every park is being constructed by a Chinese firm. The seven parks currently being constructed are expected to be operational by June 2018.

Up to seven more parks are in planning for development over the next five years with construction on some now under way. These include industrial parks in Debre Birhan, Aysha, Awsh Arba, Andido, Bishoftu, and Asayta,

---

13 More details concerning the current status of Ethiopia's industrial parks is included in Schmidt et al. (2018).

14 All references to dollars in the chapter refer to US dollars.

FIGURE 12.11 Geographic locations of selected industrial parks in Ethiopia, 2017

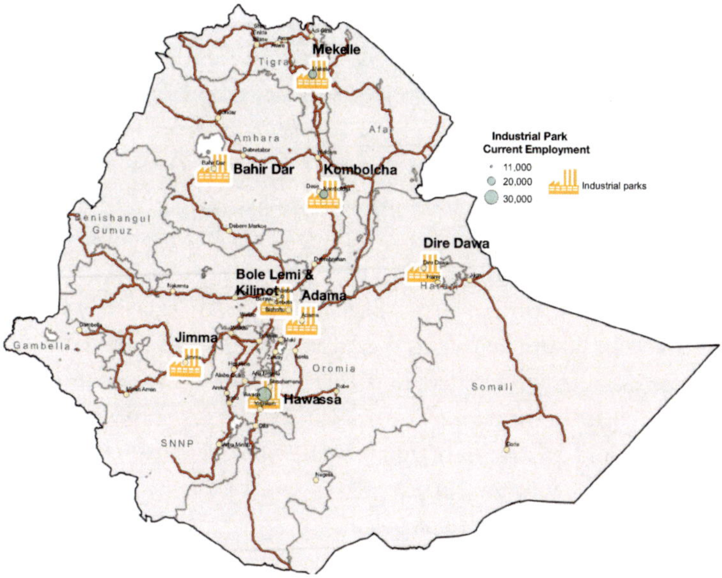

Source: Compiled by authors.

and the Airlines Logistics park. The sectors to be active in these new parks are yet to be entirely determined, but the combined area is expected to be approximately 1,500 hectares. Some estimates have projected that two million jobs will ultimately come from the 17 to 19 industrial parks that are in current plans. These jobs would constitute between 3 percent and 5 percent of Ethiopia's total labor force. Expected employment per urban labor force is highest for Hawassa industrial park in SNNP region at 42.9 percent of Sidama zone's urban labor force and 3.8 percent of the zone's total labor force (rural and urban combined). Bole Lemi I, Bole Lemi II, and Kilinto—the three industrial parks in Addis Ababa—are expected to employ 90,000 workers, which amounts to a combined 28.5 percent of East Shewa zone's urban labor force and 8.2 percent of its total labor force. The total labor force is estimated at 63 percent of total population. Excluding Addis Industrial Village, 11 of these parks are expected to employ a combined minimum total of 352,000 persons. Total investment on the 11 parks is estimated at US$1.8 billion, with the value of public investment being US$1.1 billion. This yields about one job per US$5,000 of total investment, or about one job per

US$3,000 in public investment. Annual wages across the 11 parks is estimated at US$280 million.[15]

There are 17 agro-industrial parks planned for construction across the country, focused on livestock, sesame, cereals, coffee, sugar plantations, pulses, fruits, and vegetables. Activities revolve around agrifood processing as well as the production of agri-inputs (for example, bioenergy, greenhouse cultivation, agrichemicals, renewable energy), agri-infrastructure (for example, energy management, mechanization and transport, storage facilities), capacity building and others, such as agritourism, rural financing, and crop insurance. Only one park is currently operational in its pilot phase, Yirgalem in eastern SNNP. Along with Yirgalem, three others—Baeker in western Tigray, Burie in southwest Amhara, and Bulbula in central-eastern Oromia—are part of the first four-year investment development horizon timeline. Beginning with a combined 997 hectares in the pilot phase, these four parks are expected to cover 4,000 hectares and employ more than 2.3 million individuals in 120 firms. By 2025, 13 additional agro-industrial parks are expected to be constructed over a combined area of 100,000 hectares.

Industrial parks also have been developed by foreign companies, primarily Chinese. These include Huajian Industrial Park on the outskirts of Addis Ababa, specializing in shoes; Modjo Industrial Park, specializing in "George Shoes"; and the Eastern Industrial Zone, located 37 kilometers south of Addis Ababa, which hosts miscellaneous manufacturing industries and was operational in 2015. It currently spans 40 hectares but will expand to an expected 207 hectares. Also, a fertilizer plant has been constructed at the Dire Dawa industrial park that was cofinanced by the Moroccan and Ethiopian governments, with 60 percent of financing from private investment.

In summary, Ethiopia is developing and supporting industrial zones to foster higher-value manufacturing and service activities, according to the GTP II. These industrial zones will benefit from the large investments in transportation (primarily road) infrastructure and maintenance that Ethiopia has invested in during the past several decades (shown above by the agglomeration index analysis). However, a focus on increasing human capacity and labor mobility will be necessary to ensure that rural farmers are able to take advantage of the new labor opportunities outside of the agriculture sector that the industrial parks offer. The following section models a variety of scenarios to

15 For a list of the 12 parks that are either operational or are currently under construction and the 2015 CSA projected population of both the city and zone(s) in which they are located, as well as the total expected employment each park is anticipated to generate, expected public and total investments and wage estimates, see Schmidt et al. (2018).

understand how Ethiopia's focus on industrialization may impact GDP and employment in the future.

Estimates of Multiplier Effects on GDP and Employment

Table 12.8 presents data on investment and production for Ethiopia's industrial, agro-industrial parks, and sugar factories, along with calculations of the multiplier effects of these investments. For these estimates we use value-added coefficients consistent with Ethiopia's national accounts—0.3 for Bole Lemi I and Hawassa, and 0.25 for the remaining industrial parks. For the agro-industrial parks we assumed a coefficient of value-added of 0.4 for Baeker and Bulbula, where a combination of agricultural products is processed (livestock, sesame, coffee, cereals, sorghum, sugar, fruits and vegetables, and pulses) and a coefficient of value-added of 0.22 for Burie and Yirgalem. A 0.925 coefficient of value-added is used for the sugar factories.

Two estimates of the multiplier effects are presented. The first estimate, *total value-added I*, is based on a multiplier of 1.75, an average of national economywide multiplier estimates using endogenous prices from computable general equilibrium models (Haggblade, Hazell, and Reardon 2007). The second estimate, *total value-added II*, uses a multiplier of 2.10, based on national semi-input-output fixed-price models. These larger multipliers may be appropriate for local economy effects where there is substantial underemployment.[16]

For the industrial parks the value-added per investment ranged from 1.75 to 2.10 in the case of the endogenous price multipliers, excluding Bole Lemi I, whose production appears to be an outlier. Using the larger fixed-price model multiplier, the value-added per investment ranged from 2.10 to 2.52, again excluding Bole Lemi I. The value-added per investment for the agro-industrial parks ranged from 0.21 to 0.77 using the lower endogenous price multiplier, and from 0.25 to 0.93 using the larger fixed-price multiplier. For the sugar factories the value-added per investment ranged from 0.56 to 4.72 in the case of the endogenous price multipliers and from 0.68 to 5.66 using the larger fixed-price model multiplier.

Given that the total investment in the industrial parks is $1.74 billion, estimated production and value-added are $6.57 and $1.69 billion, respectively. Including estimated multiplier effects, total value-added is $3.0 billion to $3.6 billion (equal to 4 percent to 5 percent of GDP in 2016, $72.4 billion).

---

16 We use two alternative multipliers taken from a literature review of economies with widely varying infrastructure and level of economic development. In this way we capture the likely range of multiplier effects as Ethiopia's economy develops.

TABLE 12.8 Estimates of multiplier effects of investments in industrial parks and sugar factories in Ethiopia (US$ millions)

	Invest-ment	Produc-tion	Value-added[a]	Total value-added I[b]	Total value-added I / Investment	Total value-added II[c]	Total value-added II / Investment
Industrial parks							
Bole Lemi I	102	20	6	11	0.10	13	0.12
Hawassa	250	1,000	300	525	2.10	630	2.52
Kombolcha	125	500[d]	125	219	1.75	263	2.10
Mekelle	125	500[d]	125	219	1.75	263	2.10
Jimma	64	256[d]	64	112	1.75	134	2.10
Bole Lemi II	149	596[d]	149	261	1.75	313	2.10
Kilinto	234	937[d]	234	410	1.75	492	2.10
Dire Dawa	190	760[d]	190	333	1.75	399	2.10
Adama	500	2,000[d]	500	875	1.75	1,050	2.10
Total	**1,739**	**6,569**	**1,693**	**2,963**	**1.70**	**3,556**	**2.04**
Agro-industrial parks							
Baeker (western Tigray)	634[e]	700	280	490	0.77	588	0.93
Burie (southwestern Amhara)	620[e]	558	123	216	0.35	259	0.42
Bulbula (eastern Oromia)	635[e]	591	236	414	0.65	496	0.78
Yirgalem (eastern SNNP[f])	480[e]	260	57	101	0.21	121	0.25
Total	2,370	2,109	697	1,220	0.51	1,464	0.62

For agro-industrial parks the total investment is $2.3 billion with total production of $2.1 billion and value-added of $0.70 billion. Including multiplier effects, the total value-added ranges from $1.2 billion to $1.5 billion, approximately 2 percent of GDP in 2016.

Public and private investment in industrial and agro-industrial parks may provide a catalyst for future growth by facilitating the transfer of technology and contributing a significant share of export earnings. However, this analysis shows that even if the plans for these industrial parks are successful, they are likely to provide only a small share of total output and employment.

Investments in the sugar factories are anticipated to total $5.2 billion, with estimated production of $3.6 billion and value-added of $3.3 billion, which is 9.4 times 2014/2015 sugar production and value-added. If these targets were achieved, the sugar investments would have a larger effect on GDP than the

	Investment	Production	Value-added[a]	Total value-added I[b]	Total value-added I / Investment	Total value-added II[c]	Total value-added II / Investment
Sugar factories							
Wonji Shoa Modern Factory	528	206	191	334	0.63	400	0.76
Finchaa sugar crushing mill	132	206	191	334	2.53	400	3.03
Metehara (expanded)		85	78	137		165	
Tendaho (second phase not started)	769	472	437	765	0.99	918	1.19
Arjo Dediessa	152	444[g]	411	719	4.72	863	5.66
Kesem	297	202[g]	187	327	1.10	393	1.32
Wolkayit	200	376[g]	348	609	3.05	731	3.65
Kuraz Sugar Development Project	1,475	1,041[g]	963	1,685	1.14	2,022	1.37
Tana Belles Sugar Project	1,620	564[g]	522	914	0.56	1,096	0.68
Total	5,173	3,597	3,327	5,823	1.13	6,988	1.35

Source: Authors' calculations.

Note: a. We assume a coefficient of value-added of 0.3 for Bole Lemi I and Hawassa, and a coefficient of value-added of 0.25 for the remaining industrial parks listed. For the agro-industrial parks, we assumed a coefficient of value-added of 0.40 for Baeker and Bulbula, and a coefficient of value-added of 0.22 for Burie and Yirgalem. For the sugar factories, we assume a coefficient of value-added of 0.925, which accounts for 0.075 fertilizer share of sugar cane value, the value found in the Pakistan study of Anderson, Cockburn, and Martin (2010) in 2004/2005 (assuming US$1 cane = US$1.6 processed sugar).

b. The total value-added I is based on the multiplier 1.75, which is the national endogenous price estimation method given by Haggblade, Hazell, and Reardon (2007) in Table 7.3 of their study.

c. The total value-added II is based on the multiplier 2.10, which is based on national semi-input-output fixed price models.

d. For all parks but Bole Lemi I, the production figure is scaled to the total investment made in the park, using the share of expected production to investment for Hawassa industrial park.

e. For the agro-industrial parks, we assume private investment is allocated according to shares of public investments.

f. SNNP = Southern Nations, Nationalities, and Peoples.

g. For all the sugar factories but Wonji Shoa, Finchaa, Metehara, and Tendaho, the expected production is calculated using an average of the expected production in dollars per kilogram of sugar output per year of Wonji Shoa, Finchaa, Metehara, and Tendaho, multiplied by the tons of sugar per year capacity of each respective factory.

industrial and agro-industrial parks combined. Including estimated multiplier effects, the total value-added of the planned sugar factories in Ethiopia comes to between $5.8 billion and $7.0 billion. However, an increase in sugar output of this magnitude would imply massive sugar exports that would not necessarily be profitable financially. As summarized by Arkebe Oqubay (2015, 284): "Industrial clustering and industrial parks have played an insignificant role till now but could play a much bigger future role in the overall industrial development strategy. Again, some of these issues point to dilemmas that the government will need to address. For instance, there is the tension between industrial clustering and agglomeration and the political commitment to spreading resources and opportunities across federal regions."[17]

Conclusion

Ethiopia's urbanization is driven by various demographic, economic, and political factors. Higher net birth rates, rural-urban migration, and expanding urban centers contribute to overall urban population growth (Farrell 2017). This rapid urbanization has major implications for structural transformation.

Ethiopia is urbanizing rapidly. However, the country remains behind many countries in Africa south of the Sahara. Urbanization is expected to reach 40 percent by 2035. According to official CSA projections, the population of Addis Ababa will increase from 3.27 million in 2015 to 7.17 million in 2035; however, much of the expected 56 percent increase in urban population nationally over the next 20 years will likely be in secondary cities. Between 1984 and 2015 the number of secondary cities expanded rapidly. By 2015, 35 cities had populations over 50,000, and 9 cities had populations over 100,000. These secondary cities have the potential to play a key role in Ethiopia's future development, as they serve as major regional markets for local agricultural products, help spread the benefits of growth across more of Ethiopia, and contribute to rapid poverty reduction, in part by providing seasonal and full-time employment opportunities for new migrants and for residents in surrounding areas.

Government policy will play a key role in these developments in terms of infrastructure investments. In addition, macroeconomic and trade policy will have major implications for private-sector incentives. Public and private investment in industrial and agro-industrial parks may provide a catalyst for future growth by facilitating the transfer of technology and contributing a significant share of export earnings, even though they are likely to provide only a small share of total output and employment in the Ethiopian economy.

17 Oqubay is a former minister and special adviser to the prime minister of Ethiopia and a high-level official in policy implementation in the country.

Appendix 12A: Estimations of Travel Time and the Agglomeration Index

Travel Time Estimation

In order to calculate travel time to the nearest city of 50,000 people, the model built here combines transport infrastructure (road network) and landscape features (land use, rivers, lakes, and slope). The traverse modality from point of origin to the nearest city assumes that all movement along the road is by vehicle, whereas off-road people walk through different land cover types before reaching the nearest road. This analysis uses the 2007 road network as the base data (Table 12A.1), with updated road surface types for the 2015 year from Open Street Maps and Google Earth. Access to the nearest city is mainly determined by road infrastructure, and a specific speed limit was assigned to the three road types in our GIS data (asphalt/concrete, gravel, and earth), ranging from 90 kilometers per hour to 10 kilometers per hour. For off-road travel, however, an average walking speed of 5 kilometers per hour is used through various land cover types.

Prior to combining the GIS layers into a friction (or impedance) grid, each layer was assigned the designated speed limit and vector layers, including roads, rivers, and lakes, and were converted into 1-kilometer grid cell raster layers. Slope is also taken into account to model uphill and downhill movement. All input layers are combined into a friction grid using a Python (arcpy) script that runs a "Mosaic to New Raster" function in ArcGIS 10.5.

The travel time raster grid is generated by calculating the travel time from any point within Ethiopia to the nearest city of at least 50,000 people. The Python (arcpy) script used the COSTDISTANCE function, where the friction grid and cities of at least 50,000 people locations are inputs to calculate travel time in minutes from each grid cell to the nearest city. Given the broad assumptions and the quality of data used, the travel time model provides an estimated measure of varying market access across the country.

The travel time computation was summarized differently for the research here from how it was done in Schmidt and Kedir (2009). Schmidt and Kedir (2009) used woreda population and average woreda-level travel time to calculate population by travel time. In contrast, the analysis presented in this chapter classifies each grid cell's travel time values (without averaging cells at woreda level) and accordingly summarizes population values. Examples in Figure 12A.1 and Figure 12A.2 show this difference.

TABLE 12A.1 Estimating travel time to nearest city with population of at least 50,000 in Ethiopia (data sources summary)

Layers	Description
Road network in 2015	The 2007 road network surface updated using Open Street Map and verified on Google Earth.
Cities	Cities/towns of at least 50,000 as reported by CSA population projection report (Ethiopia CSA 2013b).
Landscape features	Slope factor: raster value generated for each cell based on slope rise or run according to van Wagtendonk and Benedict (1980) formula
	$v = v0e^{-ks}$
	where:
	v = off-road foot-based velocity over sloping terrain $v0$ = the base speed of travel over flat terrain, 5 kilometers per hour in this case s = slope in gradient (meters per meter) k = a factor that defines the effect of slope on travel speed (k is 3.0 in this case)
	Rivers and lakes: a 1 kilometer per hour traverse speed assigned for both features
	Land cover: an average of 5 kilometers per hour is used for entire land cover area other than waterbodies.
Population density	GRUMP 2000: The Global Rural-Urban Mapping Project, Version 1 (GRUMPv1) gridded population estimated based on country's census unit and derived density at resolution of 30 arc-second (1 kilometer) grid for the year 2000.
	GRUMP 2015: Gridded Population of the World, Version 4 (GPWv4) density grid estimated for 2015 using national census and adjusted to UN population projection for 2015.
	LandScan 2000 and 2012: LandScan spatially models population density by allocating population with respect to land use/land cover, transportation infrastructure, landscape (elevation and slope), and so on.[a]
	For the research years 1984,1994, and 2007, GRUMP and LandScan 2000 was projected to each year using a 3 percent average annual growth rate.
	For 2015, LandScan 2012 projected to 2015 using woreda-level annual population growth rate calculated from 2007 census and 2015 projected population.

Source: Authors.

Note: a. The LandScan 2012™ High Resolution Global Population Data Set is copyrighted by UT-Battelle, LLC, operator of Oak Ridge National Laboratory under Contract No. DE-AC05-00OR22725 with the US Department of Energy. See Bright, Rose, and Urban (2012).

FIGURE 12A.1 Travel time to nearest city in Ethiopia calculation methods, Example 1

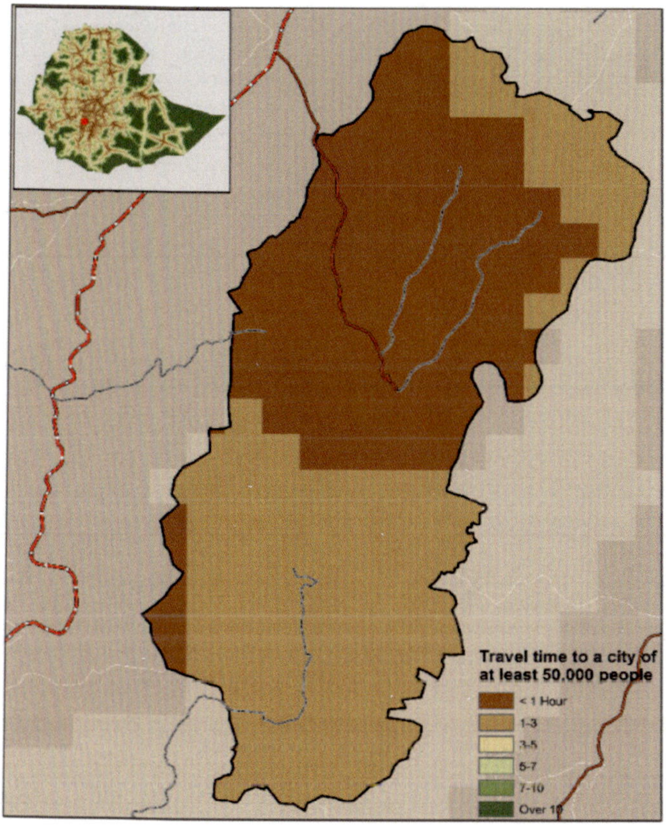

Previous method: Average travel time grid-cell values in the woreda used to set the entire woreda population into one travel time category. This either underestimates (see example 1) or overestimates (example 2) population within a travel time band.

New method: Each travel time grid-cell value is used to summarize the corresponding population grid values.

Example 1: The woreda average travel time to the nearest city is 59 minutes, and therefore the whole woreda population falls under the 1 hour travel time category. The summary from grid population, however, shows 49 percent of the population is under 1 hour and the rest between 1 and 3 hours.

	Percentage of population calculated (%)	
Anigacha woreda	Woreda population	Grid population
Under 1 hour	100	49.5
1–3 hours	0	50.5

Source: Authors' calculations.

FIGURE 12A.2 Travel time to nearest city in Ethiopia calculation methods, Example 2

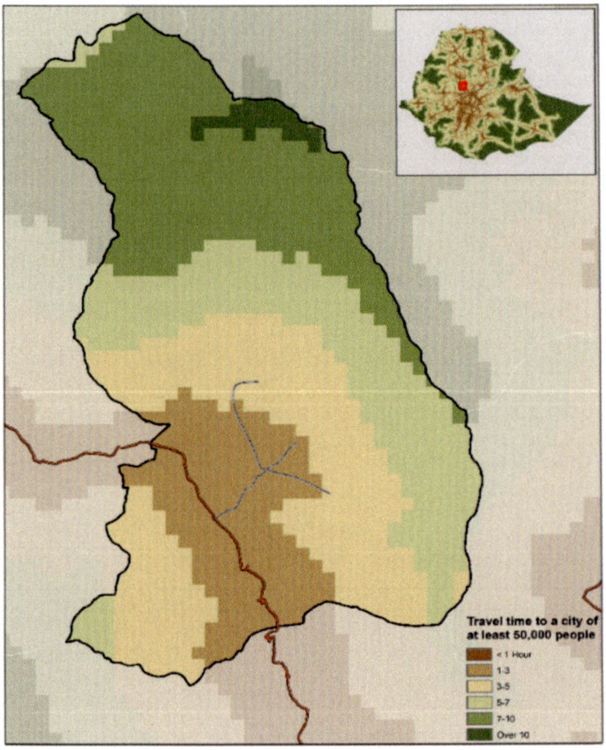

The average woreda travel time to the nearest city is 5.5 hours. Using the average woreda travel time for the entire woreda population results in 100 percent of the population being within the 5 to 7 hours travel time category.

However, if a gridded population surface is used, only 20 percent of the woreda population is within the 5 to 7 hours category, 53 percent are closer than 5 hours, and 27 percent face a travel time of more than 7 hours to the nearest city.

	Percentage of population calculated (%)	
Abuna Gindebere	Woreda population	Grid population
Under 1 hour	0	0
1–3 hours	0	25
3–5 hours	0	28
5–7 hours	100	20
7–10 hours	0	26
Over 10 hours	0	1

Source: Authors' calculations.

TABLE 12A.2 Travel time to nearest city of at least 50,000 in Ethiopia in 1984, share of population (%)

Region	Less than 1 hour	1–3 hours	3–5 hours	5–10 hours	More than 10 hours
Tigray	2.7	4.1	9.5	45.0	38.6
Afar	0.0	0.3	4.0	22.2	73.6
Amhara	3.7	8.8	13.9	37.1	36.5
Oromia	2.3	10.0	13.6	40.1	34.0
Somali	0.3	1.5	5.3	11.4	81.5
Benishangul-Gumuz	0.0	0.0	0.0	4.8	95.2
SNNP	0.0	0.4	4.2	50.4	45.0
Gambella	0.0	0.0	0.0	35.0	65.0
Harari	45.3	44.9	8.6	1.3	0.0
Addis Ababa	57.5	41.3	1.2	0.0	0.0
Dire Dawa	17.1	52.7	14.6	11.3	4.3
Ethiopia	**4.4**	**8.2**	**10.2**	**37.4**	**39.7**

Source: Authors' calculations.

Note: SNNP = Southern Nations, Nationalities, and Peoples.

TABLE 12A.3 Travel time to nearest city of at least 50,000 in Ethiopia in 1994, share of population (%)

Region	Less than 1 hour	1–3 hours	3–5 hours	5–10 hours	More than 10 hours
Tigray	5.3	10.9	17.6	37.6	28.6
Afar	0.0	1.8	4.3	31.0	63.0
Amhara	5.8	13.8	17.2	36.2	26.9
Oromia	6.2	16.0	15.7	33.9	28.2
Somali	2.9	4.7	5.1	13.7	73.6
Benishangul-Gumuz	0.0	0.0	1.6	16.0	82.4
SNNP	2.6	18.1	24.1	34.4	20.8
Gambella	0.0	0.0	0.0	16.4	83.6
Harari	84.0	14.1	2.0	0.0	0.0
Addis Ababa	91.1	8.9	0.0	0.0	0.0
Dire Dawa	57.1	11.7	12.5	15.7	3.1
Ethiopia	**8.7**	**14.0**	**16.1**	**31.9**	**29.3**

Source: Authors' calculations.

Note: SNNP = Southern Nations, Nationalities, and Peoples.

TABLE 12A.4 Travel time to nearest city of at least 50,000 in Ethiopia in 2007, share of population (%)

Region	Less than 1 hour	1–3 hours	3–5 hours	5–10 hours	More than 10 hours
Tigray	10.0	15.9	28.0	33.8	12.3
Afar	0.0	3.5	14.3	38.2	44.1
Amhara	9.3	31.4	26.8	27.0	5.6
Oromia	11.8	28.6	24.5	28.3	6.8
Somali	3.3	5.5	5.9	19.1	66.3
Benishangul-Gumuz	0.0	1.9	8.1	28.1	62.0
SNNP	24.1	42.9	14.5	14.3	4.2
Gambella	0.0	0.0	0.0	14.4	85.6
Harari	85.3	11.5	3.2	0.0	0.0
Addis Ababa	91.9	8.1	0.0	0.0	0.0
Dire Dawa	62.3	20.3	11.1	6.3	0.0
Ethiopia	**16.2**	**28.3**	**20.8**	**23.9**	**10.8**

Source: Authors' calculations.

Note: SNNP = Southern Nations, Nationalities, and Peoples.

FIGURE 12A.3 Maps of travel time to nearest city in Ethiopia of at least 50,000 in 1984, 1994, 2007, and 2015

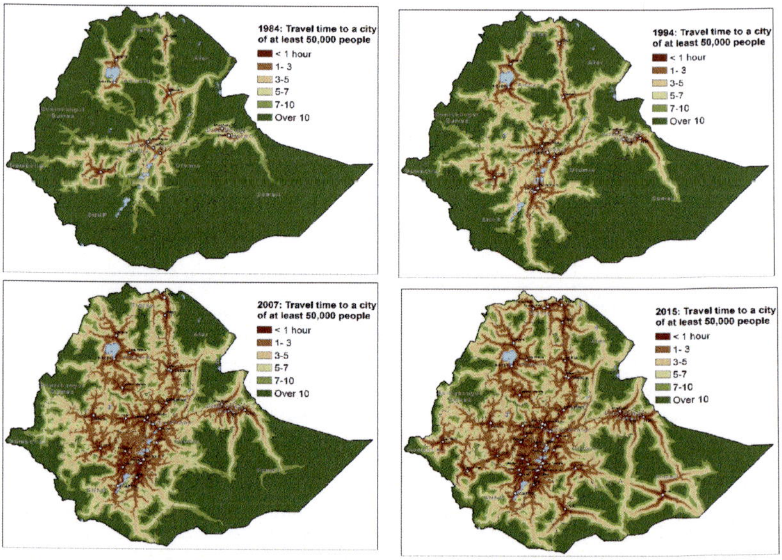

Source: Agglomeration index estimations.

Agglomeration Index

Agglomerated areas are places characterized by relatively large population concentrations with economic interactions (Lemelin, Rubiera-Morollón, and Gómez-Loscos 2016). For countries like Ethiopia where urbanization is relatively low, we classify agglomerated areas as those with a population density above 150 persons per square kilometer and within one-hour travel time from an urban center of at least 50,000 people (Schmidt and Kedir 2009). Accordingly, all areas that fall above the minimum population density (150 people per square kilometer) and under the maximum travel time to an urban center (one hour) are considered an agglomerated area, even if the area transcends the defined urban administrative boundary. The shape of city growth follows the pattern of infrastructure, which may also be hindered by natural physical barriers. The agglomeration index measurement requires travel time and demographic datasets of population density.

The agglomeration index generated for this analysis was built by Schmidt and Kedir (2009) for three consecutive census years (1984, 1994, 2007) using projected average gridded population density of 1-kilometer resolution from the Global Rural and Urban Mapping Project (GRUMP) as well as from LandScan (Figure 12A.4). Given that the underlying population distribution models for GRUMP and LandScan differ and each has pros and cons in how the population is spatially allocated, we average the two datasets for this analysis to better capture the true pattern of population distribution in Ethiopia (Figure 12A.5).[18]

For the three index (census) years 1984, 1994, and 2007, Schmidt and Kedir (2009) adjust the GRUMP and LandScan grids from 2000 with a 3 percent per year growth rate. For the 2015 projection year, unlike the previous analysis, we use LandScan 2012 gridded data and project the population to 2015 by a growth rate calculated from the 2007 census (Ethiopia, CSA 2007) and the 2015 projected population figure (Ethiopia, CSA 2013b). The woreda-level growth rate was then applied to each grid that was located within each woreda boundary (Figures 12A.4 and 12A.5). Areas that met the criteria set to define agglomerated area with a population density of at least 150 persons per square kilometer and less than 1 hour to the nearest city with 50,000 or more people were identified using the Python (arcpy) script using the CON function of ArcGIS 10.5.

18 GRUMP data spatially allocates population based on the census population at the lowest available administrative unit, whereas LandScan data uses roads, land cover, slopes, and other terrain features to model population density across geographic space.

FIGURE 12A.4 GRUMP and LandScan population density grids for Ethiopia, 2015

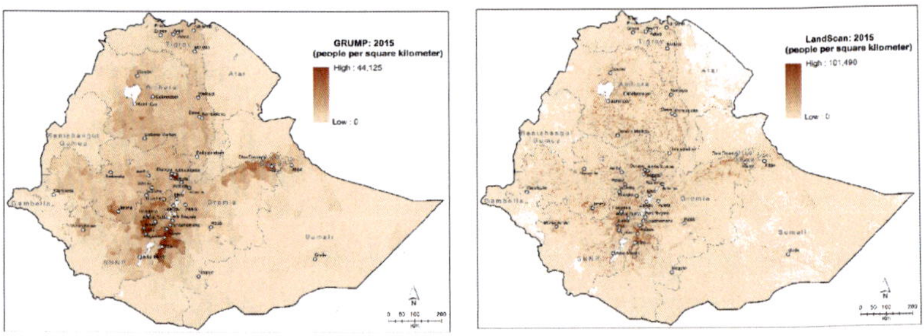

Source: Authors' calculations.

FIGURE 12A.5 Average of GRUMP and LandScan population density grids for Ethiopia, 2015

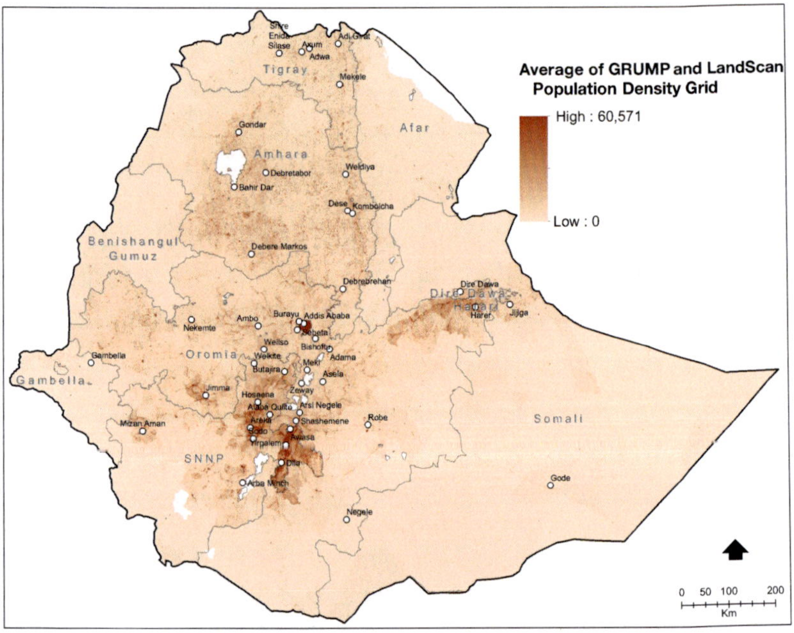

Source: Authors' calculations.

Appendix 12B: Population Projections for Ethiopia

TABLE 12B.1 Population figures and projections for selected urban centers in Ethiopia, 1984–2035

	Census			Projection		
Urban center	May 1984	October 1994	May 2007	July 2015	2025	2035
Addis Ababa	1,412,600	2,112,700	2,739,600	3,273,000	4,561,149	5,810,263
Adama (Nazret)	76,300	127,800	220,200	324,000	856,789	1,880,020
Gondar	80,900	112,200	207,000	323,900	856,525	1,879,440
Mek'ele	61,600	96,900	215,900	323,700	855,996	1,878,280
Hawassa	36,200	69,200	157,100	300,100	793,588	1,741,340
Dire Dawa	98,100	164,900	233,200	277,000	732,502	1,607,301
Bahir Dar	54,800	96,100	155,400	243,300	643,385	1,411,756
Dessie	68,800	97,300	120,100	187,900	496,885	1,090,296
Jimma	61,000	88,900	121,000	177,900	470,441	1,032,270
Jijiga	23,200	56,800	125,900	159,300	421,255	924,343
Shashemene	31,500	52,100	100,500	147,800	390,844	857,614
Bishoftu (Debre Zeit)	51,100	73,400	99,900	147,100	388,993	853,552
Sodo	24,600	36,300	76,100	145,100	383,704	841,947
Arba Minch	23,000	40,000	74,900	142,900	377,886	829,182
Hosaena	15,200	31,700	70,000	133,800	353,822	776,379
Harari	62,200	76,400	99,400	129,000	341,129	748,527
Dila	23,900	33,700	59,200	112,900	298,554	655,106
Nekemte	28,800	47,300	75,200	110,600	292,578	641,992
Debre Birhan	25,800	38,700	65,200	102,100	269,994	592,438
Total population for centers with population 50,000 to 100,000 (thousands)[a]	337	650	1,139	1,782	3,284	5,593
Total population for centers with population 100,000 to 500,000 (thousands)[a]	847	1,340	2,276	3,488	9,225	20,242
Addis Ababa (thousands)	1,413	2,113	2,740	3,273	4,561	5,810
Total urban (thousands)	**4,869**	**7,385**	**11,863**	**17,459**	**31,072**	**52,574**

(continued)

TABLE 12B.1 Continued

Urban center	May 1984	Census: October 1994	May 2007	Projection: July 2015	2025	2035
Population growth from last census (%)						
For centers with population 50,000 to 100,000 (thousands)[a]	n.a.	6.77	4.41	5.76	6.31	5.47
For centers with population 100,000 to 500,000 (thousands)[a]	n.a.	4.69	4.16	5.48	10.21	8.18
Addis Ababa	n.a.	4.11	2.02	2.25	3.37	2.45
Total urban	n.a.	**4.25**	**3.71**	**4.95**	**5.40**	**5.40**

Source: CSA population and housing census of 1984, 1994, 2007, and 2015 (Ethiopia, CSA 1991, 1996, 2010, and 2013b); EGIS International (2016); World Bank (2015); and authors' calculations using growth rates from EGIS national urban system study report projections, and a growth rate of 5.4 percent for the total urban population projection from the World Bank (2015) for years 2012 and forward.

Note: n.a. = not applicable.

a. City size categories are defined in terms of 2015 population using CSA urban population projections regardless of population size in years prior to or following 2015.

TABLE 12B.2 Urban population projections for Ethiopia, 2025 and 2035

	City size categories, by 2015 population: 20,000–50,000[a]	50,000–100,000	100,000–500,000	Addis Ababa	Greater than 50,000
Cities (number)	95	26	18	1	45
1994 total urban population	1,168	650	1,340	2,113	4,102
2007 total urban population	1,901	1,139	2,276	2,740	6,154
2015 total urban population	**2,982**	**1,782**	**3,488**	**3,273**	**8,543**
Annual change, 1994–2007 (%)	3.82	4.41	4.16	2.02	3.17
Annual change, 2007–2015 (%)	5.79	5.76	5.48	2.25	4.18
Projected 2025 total urban population	5,531	3,284	9,225	4,561	16,709
Projected 2035 total urban population	9,792	5,593	20,242	5,810	30,487
Annual change 2015–2025 (%)[b]	6.37	6.31	10.21	3.37	6.94
Annual change 2025–2035 (%)[b]	5.88	5.47	8.18	2.45	6.20

Source: CSA population and housing census of 1994, 2007, and 2015 (Ethiopia, CSA 1996, 2010, and 2013b); EGIS International (2016); and authors' calculations.

Note: a. City size categories are defined in terms of 2015 population (using CSA urban population projections) regardless of population sizes in years prior to or following 2015.

b. Growth rates derived from the EGIS national urban system study report projections (EGIS International 2016).

References

Anderson, K., J. Cockburn, and W. Martin. 2010. *Agricultural Price Distortions, Inequality, and Poverty*. Washington, DC: World Bank.

Asfaw, W., D. Degaga, and Z. Gete. 2010. "Causes and Impacts of Seasonal Migration on Rural Livelihoods: Case Studies from Amhara Region in Ethiopia." *Norsk Geografisk Tidsskrift (Norwegian Journal of Geography)* 64: 58–70.

Binswanger-Mkhize, H. P. 2012. "India 1960–2010: Structural Change, the Rural Nonfarm Sector, and the Prospects for Agriculture." Paper presented at the Center on Food Security and the Environment Stanford Symposium Series on Global Food Policy and Food Security in the 21st Century, Stanford University, Berkeley, CA, US, May 11.

Bright, E. A., A. N. Rose, and M. L. Urban. 2012. LandScan database. Oak Ridge, TN, US: Oak Ridge National Laboratory. Accessed September 1, 2017. https://landscan.ornl.gov/downloads/2012.

Cai, F. 2013. "Approaching a Neoclassical Scenario: The Labor Market in China after the Lewis Turning Point." *China Finance and Economic Review* 1 (1): 1-15. http://www.chinafinanceandeconomicreview.com/content/1/1/1.

Cai, F., and M. Wang. 2010. "Growth and Structural Changes in Employment in Transition China." *Journal of Comparative Economics* 38 (1): 71–81.

Christiaensen, L., J. De Weerdt, B. Ingelaere, and R. Kanbur. 2018. *Migrants, Towns, Poverty and Jobs: Insights from Tanzania*. World Bank Policy Research Working Paper 8340. Washington, DC: World Bank.

Ciccone, A., and G. Peri. 2006. "Identifying Human-Capital Externalities: Theory with Applications." *Review of Economic Studies* 73 (2): 381–412.

CIESIN (Center for International Earth Science Information Network, Columbia University). 2016. *Gridded Population of the World, Version 4 (GPWv4): Population Density Adjusted to Match 2015 Revision UN WPP Country Totals, Dataset*. Palisades, NY, US: NASA Socioeconomic Data and Applications Center SEDAC. Accessed June 3, 2016. http://dx.doi.org/10.7927/H4HX19NJ.

City Population. 2018. "Cities & Towns." Accessed February 12, 2018. www.citypopulation.de/Ethiopia.html.

De Brauw, A., V. Mueller, and T. Woldehanna. 2013. "Motives to Remit: Evidence from Tracked Internal Migrants in Ethiopia." *World Development* 50: 13–23.

Dorosh, P., G. Alemu, A. De Brauw, M. Malek, V. Mueller, E. Schmidt, T. Kibrom, and J. Thurlow. 2011. *The Rural-Urban Transformation in Ethiopia*. IFPRI-ESSP and EDRI Report. Addis Ababa: IFPRI/Ethiopia Strategy Support Program (IFPRI/ESSP II).

EGIS International. 2016. *National Urban Development Spatial Plan: National Urban Systems Study Final Report.* Addis Ababa: Ministry of Urban Development and Housing.

Ethiopia, CSA (Central Statistical Agency). 1991. *The 1984 Population and Housing Census of Ethiopia: Analytical Report at National Level.* Addis Ababa: Office of the Population and Housing Census Commission.

———. 1996. *The 1994 Population and Housing Census of Ethiopia: Results at Country Level.* Volume 1, *Statistical Report.* Addis Ababa: Office of Population and Housing Commission.

———. 2010. *The 2007 Population and Housing Census of Ethiopia: Results for Country Level.* Statistical Report. Addis Ababa: Population Census Commission.

———. 2013a. *Ethiopia National Labour Force Survey 2013 (2005 E.C.) Dataset.* Addis Ababa.

———. 2013b. *Population Projection of Ethiopia for All Regions at Woreda Level from 2014–2017.* Addis Ababa.

Ethiopia, CSA and LSMS (Living Standards Measurement Study), World Bank. 2015. *Ethiopia Socioeconomic Survey Report.* Addis Ababa.

Ethiopia, MoFED (Ministry of Finance and Economic Development). 2007. *Ethiopia: Building on Progress: A Plan for Accelerated and Sustained Development to End Poverty (PASDEP).* Addis Ababa.

———. 2010. *Growth and Transformation Plan: 2010/11–2014/15.* Volume 1. Addis Ababa. http://www.ethiopians.com/Ethiopia_GTP_2015.pdf.

Fan, S., C. Chan-Kang, and A. Mukherjee. 2005. Rural and Urban Dynamics and Poverty: Evidence from China and India. FCND Discussion Paper 196 and DSG Discussion Paper 23. Washington, DC: IFPRI.

Farrell, K. 2017. *The Rapid Urban Growth Triad: A New Conceptual Framework for Examining the Urban Transition in Developing Countries.* Stockholm, Sweden: Urban and Regional Studies, Department of Urban Planning and Environment, KTH Royal Institute of Technology.

Filmer, D., and L. Fox. 2014. *Youth Employment in Sub-Saharan Africa.* Washington, DC: World Bank.

Gollin, D. 2018. *Structural Transformation without Industrialization.* Pathways for Prosperity Commission Background Paper 2. Oxford, UK: Oxford University Press.

Gray, C., and V. Mueller. 2012. "Drought and Population Mobility in Rural Ethiopia." *World Development* 40 (1): 134–145.

Haggblade, S., P. B. Hazell, and T. Reardon, eds. 2007. *Transforming the Rural Non-farm Economy: Opportunities and Threats in the Developing World.* Washington, DC: IFPRI; Baltimore: Johns Hopkins University Press. http://citeseerx.ist.psu.edu/viewdoc/download?doi=10.1.1.644.299&rep=rep1&type=pdf.

Johnston, B. F., and J. W. Mellor. 1961. "The Role of Agriculture in Economic Development." *American Economic Review* (51): 566–593.

Kosec, K., H. Ghebru, B. Holtemeyer, V. Mueller, and E. Schmidt. 2017. "The Effect of Land Access on Youth Employment and Migration Decisions: Evidence from Rural Ethiopia." *American Journal of Agricultural Economics* 100 (3): 931–954. https://doi.org/10.1093/ajae/aax087.

Lemelin, A., F. Rubiera-Morollón, and A. Gómez-Loscos. 2016. "Measuring Urban Agglomeration: A Refoundation of the Mean City–Population Size Index." *Social Indicators Research: An International and Interdisciplinary Journal for Quality-of-Life Measurement* 125 (2): 589–612 (January).

Lewis, W. A. 1954. "Economic Development with Unlimited Supplies of Labour." *Manchester School* 22 (2): 139–191.

Lucas, R. E. 1988. "On the Mechanics of Economic Development." *Journal of Monetary Economics* 22 (1): 3–42.

Martins, P. 2014. *Structural Change in Ethiopia: An Employment Perspective.* World Bank Policy Research Working Paper 6749. Washington, DC: World Bank.

Mellor, J. W. 1976. *New Economics of Growth: A Strategy for India and the Developing World.* Ithaca, NY, US: Cornell University Press.

———. 1995. *Agriculture on the Road to Industrialization.* Baltimore: IFPRI and Johns Hopkins University Press.

———. 2017. *Agricultural Development and Economic Transformation: Promoting Growth with Poverty Reduction.* New York: Springer.

Moller, L. C. 2012. *The Ethiopian Urban Migration Study 2008: The Characteristics, Motives and Outcomes to Immigrants to Addis Ababa.* Washington, DC: World Bank.

Moretti, E. 2004. "Estimating the Social Return to Higher Education: Evidence from Longitudinal and Repeated CrossSectional Data." *Journal of Econometrics* 121 (1): 175–212.

National Planning Commission. 2016. *Growth and Transformation Plan II (GTP II)—(2015/16–2019/20).* Addis Ababa: National Planning Commission.

Oqubay, A. 2015. *Made in Africa: Industrial Policy in Ethiopia.* New York: Oxford University Press.

Rodrik, D., M. McMillan, and C. Sepúlveda. 2013. "Structural Change, Fundamentals, and Growth: An Overview." In *Structural Change, Fundamentals and Growth: A Framework and Case Studies,* edited by M. McMillan, D. Rodrik, and C. Sepúlveda, 1–38. Washington, DC: IFPRI.

Schmidt, E., and F. Bekele. 2016. *Rural Youth and Employment in Ethiopia*. IFPRI-ESSP Working Paper 98. Addis Ababa: IFPRI/ESSP.

Schmidt, E., P. Dorosh, M. Kedir, and J. Smart. 2018. *Ethiopia's Spatial and Structural Transformation: Public Policy and Drivers of Change*. IFPRI-ESSP Working Paper 119. Washington, DC: IFPRI; Addis Ababa: Ethiopian Development Research Institute (EDRI).

Schmidt, E., and M. Kedir. 2009. *Urbanization and Spatial Connectivity in Ethiopia: Urban Growth Analysis Using GIS*. IFPRI-ESSP Working Paper 3. Addis Ababa: IFPRI/ESSP.

Thurlow, J., B. Davis, and P. Dorosh. 2018. "Demographic Change, Agriculture and Rural Poverty." Unpublished, IFPRI, Washington, DC.

Timmer, C. P. 1988. "The Agricultural Transformation." *Handbook of Development Economics* (1): 275–331.

———. 2002. "Agriculture and Economic Development." *Handbook of Agricultural Economics* (2): 1487–1546.

Uchida, H., and A. Nelson. 2008. "Agglomeration Index: Towards a New Measure of Urban Concentration." Background paper for the WDR 2009 (unpublished manuscript). Accessed March 12, 2020. http://siteresources.worldbank.org/INTWDR2009/Resources/4231006-1204741572978/Hiro1.pdf.

UNDESA (United Nations, Department of Economic and Social Affairs), Population Division. 2015. *World Urbanization Prospects: The 2014 Revision (ST/ESA/SER.A/366)*. New York.

———. 2017. *World Population Prospects: The 2017 Revision, DVD Edition.*

van Wagtendonk, J., and B. James. 1980. "Travel Time Variation on Backcountry Trails." *Journal of Leisure Research* 12: 99–106.

World Bank. 2009. *The World Development Report: Reshaping Economic Geography*. Washington, DC: World Bank.

———. 2015. *Ethiopia Urbanization Review: Urban Institutions for a Middle-Income Ethiopia*. Washington, DC: World Bank.

Zeleke, G., T. Kassawmar, W. Asfaw, V. Gete, A. Heinimann, K. Hurni, H. Hurni, and B. Debele. 2018. "Land Use and Land Cover Change Trends in Ethiopia and Policy Imperatives." Unpublished, IFPRI, Washington, DC.

Chapter 13

PUBLIC INVESTMENTS AND POVERTY REDUCTION

Paul Dorosh, James Thurlow, Frehiwot Worku Kebede,
Tadele Ferede, and Alemayehu Seyoum Taffesse

Introduction

Ethiopia's agricultural sector has grown rapidly in the past decade due to a doubling in the use of modern inputs (chemical fertilizers and improved seeds), significant land expansion, increased labor use, and a 2.3 percent per year growth in total factor productivity (TFP) (see Chapter 2; Bachewe et al. 2018). At the same time, there has been a substantial spatial and structural transformation of the economy. Ethiopia's urban population has more than doubled in the past 20 years from 7.3 million in 1994 to 18.3 million in 2016 (see Chapter 12), nonagricultural employment and output have grown rapidly, and the shares of nonagricultural sectors (especially services) in both overall employment and GDP have risen (see Chapter 11). Moreover, household welfare indicators have improved dramatically: rural poverty fell from 45.0 percent to 23.5 percent from 1999/2000 to 2015/2016 (NPC 2016), and the prevalence of stunting in young children, an indicator of chronic malnutrition, fell from 58 percent in 2000 to 40 percent in 2014.

Agricultural growth played a crucial role in Ethiopia's economic development over the past two decades, particularly in the early 2000s under the agricultural development–led industrialization (ADLI) strategy that formed the basis for subsequent medium-term development plans, such as the Plan for Accelerated and Sustained Development to End Poverty (PASDEP) (2005/2006–2009/2010), the Growth and Transformation Plans (GTP I, 2010/2011–2014/2015 and GTP II, 2015/2016–2019/2020), and Ethiopia's participation in the Comprehensive Africa Agriculture Development Programme (CAADP) of the African Union. Substantial public investment in agriculture and rural development has continued under GTP I and GTP II; agricultural expenditures grew by an average of 10.1 percent per year from 2009/2010 to 2014/2015 (Figure 13.1 and Table 13.1). Nonetheless, the focus of the development strategy has increasingly shifted toward commercial

FIGURE 13.1 Real public expenditures on agriculture and rural development in Ethiopia, 2007/2008–2015/2016 (billion 2015/2016 birr)

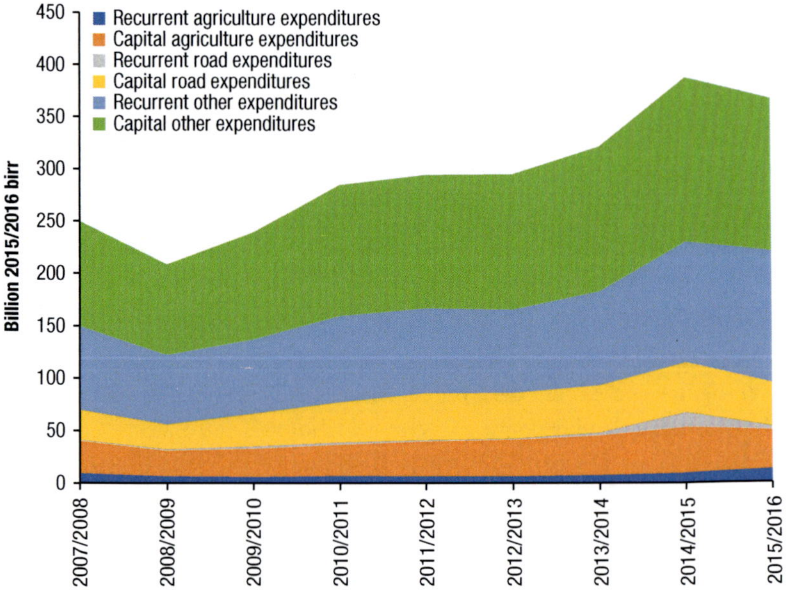

Source: Ethiopia, MoFED (various years) and authors' calculations.

TABLE 13.1 Real public expenditures on agriculture and rural development in Ethiopia, 2009/2010 and 2014/2015 (billion 2015/2016 birr)

Sector	2009/2010	2014/2015	Annual growth (%)
Agriculture	33.3	53.7	10.1
Recurrent	6.4	10.0	9.3
Capital	26.8	43.7	10.2
Roads	33.2	61.2	13.0
Other	172.6	271.9	9.5
Total	**239.1**	**386.8**	**10.1**
Agriculture share (%)	13.9	13.9	0.0
Roads share (%)	13.9	15.8	0.4

Source: Ethiopia, MoFED (various years) and authors' calculations.

agriculture and nonagricultural growth and employment as the key engines of economic development.

Spurring economic growth was not the only rationale for heavy investments in agriculture in the past two decades, however. Increasing agricultural productivity and incomes was also seen as a key means of reducing poverty. Economic analysis of the impacts of public spending on agriculture suggested that investing 10 percent of public expenditures on agriculture and rural development would lead to major reductions in poverty, not only in Ethiopia but in most of Africa south of the Sahara (Diao et al. 2013; Dorosh and Thurlow 2013b). Other analyses showed that because most of Ethiopia's poor lived in rural areas and derived much of their incomes from agriculture, investments that raised productivity in agriculture (and in small towns) reduced poverty faster than did investments in urban areas (Dorosh and Thurlow 2013a, 2014).

However, the effectiveness of agricultural growth in reducing poverty at the national level depends on several factors, including the productivity of the investments themselves as well as the structure of the economy—the level of urbanization, the share of agricultural sector in overall output and employment, and the main sources of incomes of the poor, among other factors. As these factors change over time, so will the impacts of agricultural investments on growth and poverty. Thus, while agricultural investments may have been highly effective in reducing poverty in the early 2000s, there remains a question as to whether this is still the case.

This chapter explores these issues using an economywide computable general equilibrium (CGE) model. First, the chapter describes the structure of Ethiopia's economy as reflected in the database used for the model—a 2010/2011 social accounting matrix (SAM). The basic structure of the model is outlined, and the design of the base run and simulations are described. We present the results of four alternative investment scenarios with faster investment in (1) economic sectors in large cities; (2) crop agriculture; (3) rural nonfarm sectors; and (4) livestock. The final section includes a summary and conclusion, along with important caveats and areas for future research.

Policy Context and the Structure of Ethiopia's Economy

Priorities for public investments in Ethiopia have gradually changed over time along with the changing structure of Ethiopia's economy and the success of past development plans in raising incomes and lowering poverty. In light of these achievements, Ethiopia's overarching growth target is now to become a lower-middle income country by 2025 with improved agricultural and manufacturing productivity. Part of this overall economic vision for the country, the GTP II, aims for Ethiopia to be a hub for light manufacturing in Africa.

Nonetheless, the ADLI strategy adopted in the early 1990s remains a core part of Ethiopia's development strategy. ADLI's main objective has been to enhance productivity growth in agriculture through better access to modern agricultural inputs, extension services, and road connectivity, and to use the resulting growth in agriculture as the basis for industrial expansion. The underlying premise is exploiting the country's comparative advantage in agro-processing industries as productivity growth in agriculture will facilitate the reliable and low-cost supply of agricultural inputs for manufacturers (NPC 2016). To facilitate these cross-sectoral linkages, the GTP II outlines investments and policies related to agriculture, (agro-)industry, regional development, and urbanization.

Sector Specific Investment Initiatives

Within agriculture, GTP II focuses on several key aspects, including the following:

- Promoting the productivity and quality of staple food crops production;
- Supporting development of high-value crops, industrial inputs, and export commodities;
- Improving investment in irrigated agriculture, horticulture, fruits and vegetables, and livestock and in fisheries development; and
- Facilitating participation of educated young farmers and private investors in the sector.

To improve the competitiveness, quality, and productivity of the agricultural sector, the GTP II continues policies of development and dissemination of available technologies, investments in irrigation, provision of extension services, and scaling up of best practices of model smallholder farmers (NPC 2016). In particular, the plan includes development of more than 4 million hectares of smallholder irrigated land and to complement investments in dams and medium- and large-scale irrigation. These small-, medium-, and large-scale irrigation interventions are expected to enhance agricultural productivity and ensure stability of agricultural production. Investments in livestock include improving the genetics of livestock, expanding livestock health coverage, enhancing service quality and control, and improving supply of livestock feed.

Complementary to these agricultural investments, the Ministry of Industry's industrial development roadmap (IDR) for 2013–2025 and the Industrial Development Strategic Plan (IDSP) together define strategies, programs, and projects to promote industrialization. To complement the public investments, the strategic plan also aims to encourage domestic private-sector investment as well as to attract Foreign Direct Investment (FDI) through the following:

- Expansion of special economic zones or specialized industrial parks with the required infrastructure and streamlined public procedures as well as fiscal and trade policy incentives;
- Provision of more efficient trade logistics and transport services and energy supply; and
- Supply of land, credit, and foreign exchange mainly through an industrial parks approach.

In particular, the establishment of integrated industrial development parks across the country is designed to ease logistics hurdles and other constraints though investments in physical infrastructure and the establishment of support services, including one-stop-shop services for investors located in the designated industrial parks.[1] These parks are dedicated for specific sectors, such as textile and apparel, leather and leather products, pharmaceuticals, and agro-processing. The Ethiopian government also plans to establish additional industrial parks to boost manufacturing industries at Bahir Dar, Jimma, and Adama. The government of Ethiopia with support from the United Nations Industrial Development Organization (UNIDO) has developed a master plan for integrated agro-industrial parks (IAIPs) in four regions—namely Bure (Amhara region), Baeker (Tigray region), Bulbula (Oromia region), and Yirgalem (SNNP region)—to advance inclusive and sustainable industrialization and rural transformation (UNIDO 2018).

1 Recently, new industrial parks have become operational, including Bole Lemi I, Hawassa, Kombolcha, and Mekelle. There are also a number of upcoming industrial parks in Ethiopia, including Bole Lemi II, Kilinto, Adama, Dire Dawa, Arerti, and Debre Berhan industrial parks. See Schmidt et al. (2018).

The government of Ethiopia has also developed a master plan for rural transformation centers (RTCs), which serve as raw material aggregation points in the catchment areas (100 kilometer radius) of each IAIP. Rural transformation centers are expected to provide several functions, including warehouses, input supply, sorting, grading, extension services, pre-processing activities, and microfinance. Development of industrial parks enjoys strong government support. These parks are located along key economic corridors, are connected to ports, and are supported by various types of infrastructure, including airports, railway lines, dry ports, and universities.

These industrial sectoral policies and investments implicitly focus on urban centers (including secondary cities) and relate closely to the national urban development policy of 2005. Although Ethiopia's urban population share is low, this is set to change dramatically. The urban population is projected to nearly triple from 15.2 million in 2012 to 42.3 million in 2037, growing at 3.8 percent a year. This growth rate places Ethiopia among the fastest urbanizing countries in Africa south of the Sahara (Ethiopia, CSA 2013). According to the Ethiopia Urbanization Review (EUR), Ethiopia will experience rapid urbanization at about 5.4 percent a year, higher than CSA's estimate (World Bank 2015b). Key factors driving urbanization in Ethiopia include migration to new towns and urban centers, reclassification of rural villages as urban centers, and formal expansion of existing urban boundaries to incorporate nearby settlements. These all accelerate urbanization, which will change the structure and location of economic activity from rural agriculture to more diversified and larger urban industrial and service sectors.

To support sustainable urban development, the GTP II outlines an integrated urban housing development program, an integrated urban infrastructure program (including urban green infrastructure), urban land development and management, solid waste management, development of small towns, and strengthening of rural-urban linkages. Among the major programs is the Urban Local Government Development Program (ULGDP), financed by the World Bank, that supports improved institutional performance in the planning, delivery, and sustained provision of urban services and infrastructure by local governments.[2] The second phase of ULGDP (2014–2019) is aligned with new government urban policies and strategies, such as the Ethiopian Cities Prosperity Initiative, 2013/2014–2025, and is designed to sustain urban

2 The first phase of the program included 18 urban local governments and Addis Ababa with a budget envelope of US$416 million. The first phase of the program begun in 2008 was supported by a US$300 million Specific Investment Loan (World Bank 2015a).

infrastructure and services. The program intends to address urban infrastructure gaps, improve urban productivity, and strengthen the institutions of both the urban local governments involved in service delivery and the regions and the Ministry of Urban Development and Housing.

The Structure of the Ethiopian Economy

The main database for the economywide analysis conducted for this study is a 2010–2011 social accounting matrix (SAM) for Ethiopia, a consistent set of accounts representing economic flows for production activities; commodity flows; incomes and expenditures of households, government, and other institutions; transactions with the rest of the world; savings; and investment.[3] This 2010–2011 SAM was constructed using data from the national accounts, sectoral production and prices, the balance of payments, and the 2010–2011 national Household Consumption Expenditures Survey (HCES) (Ethiopia, CSA 2012).[4]

This 2010–2011 Ethiopia SAM includes 69 production sectors split across six geographical areas: five rural areas (including towns), based on agroecologies (elevation and rainfall) and cropping patterns, and one urban category (cities with populations of 50,000 or more). In each of these six areas, rural and urban labor are disaggregated by education/skill levels (no education, primary, secondary, and tertiary education) and households are disaggregated by per capita expenditure quintiles, with poor households defined as those in the bottom 40 percent of the per capita expenditure distribution (see Appendix 13A).

Agricultural land in the SAM is disaggregated across five agroecological zones: (1) drought prone, highland; (2) moisture reliable, highland-cereal; (3) moisture reliable, highland-enset (mostly in the SNNP region); (4) humid moisture reliable, lowland; and (5) drought prone, lowland and pastoralist (Figure 13.2). The highland moisture reliable zones (2 and 3) account for more than 90 percent of the national cereal area and production. We define highlands as land 1,500 or more meters above sea level. Highland areas with a ratio of mean to standard deviation of rainfall greater than or equal to 7.5 are deemed moisture reliable. For the lowlands, areas with average rainfall less than 500 millimeters per year are classified as drought prone/pastoralist.

3 To a large extent, the regional structure of this SAM follows that of the EDRI/IFPRI 2005–2006 Ethiopia SAM (EDRI 2009).

4 Household-level data from the 2015–2016 HCES was not available at the time when this SAM was constructed.

FIGURE 13.2 Agroecological zones in Ethiopia in the 2010–2011 SAM

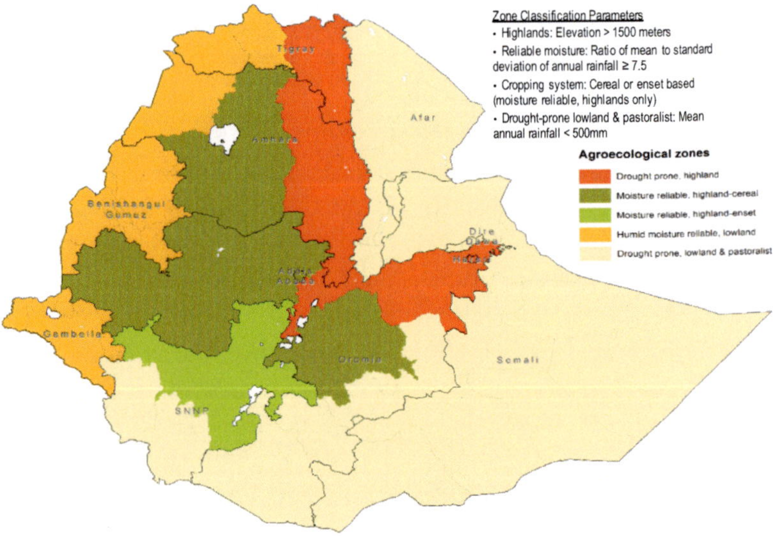

Source: EPAU and IFPRI (2017).
Note: SNNP = Southern Nations, Nationalities, and Peoples.

Table 13.2 presents estimates of incomes and expenditures by household group by geographical area for 2015/2016.[5] As shown, rural households accounted for 81.8 percent of the population (70.9 million of 86.7 million people) and 85.6 percent of the country's poor (29.7 million of 34.7 million poor people). Overall, total expenditures by the poor (the poorest 40 percent of the population) are estimated to have accounted for less than 10 percent of total household expenditures in Ethiopia (566.6 billion birr of 6,534 billion birr). Per capita expenditures of the rural poor (3,332 birr per person) were only 51 percent of the national average (6,534 birr per person) and 25 percent of urban nonpoor (13,553 birr per person) (Figure 13.3).

As shown in Table 13.3, both rural poor and nonpoor households obtain most of their incomes from labor. For rural poor households, almost all of this is from unskilled labor (which is 50.2 percent of their total incomes).[6] Returns to land and livestock capital account for another 24.1 percent and 9.7 percent of their incomes, respectively. They get very little of their income from nonagricultural capital (enterprise incomes) or other sources. Urban

5 These estimates are the results of the baseline simulation described below in this chapter.
6 In the SAM, "unskilled labor" is defined as workers who have not completed primary school.

TABLE 13.2 Population and expenditures per capita of household groups in Ethiopia, 2015/2016 estimates, by region

	All households				Poor households			
	Total expenditures (billion birr)	Population (millions)	(%)	Expenditures per capita (thousand birr per person)	Total expenditures (billion birr)	Population (millions)	(%)	Expenditures per capita (thousand birr per person)
Rural	402.3	70.9	8[illegible].8	5,674.9	98.9	29.7	34.2	3,331.8
R1: Dry highlands	91.6	16.8	19.4	5,455.2	21.8	6.4	7.4	3,407.3
R2: Dry lowlands	51.8	8.3	9.5	6,276.3	10.8	3.2	3.7	3,397.1
R3: Moist lowlands	31.8	4.9	5.7	6,479.7	6.7	1.6	1.9	4,087.2
R4: Moist high-cereals	156.7	28.3	32.6	5,547.1	42.7	12.4	14.3	3,434.2
R5: Moist high-enset	70.4	12.7	14.6	5,547.6	16.9	6.0	6.9	2,801.2
Urban	164.3	15.8	18.2	10,385.7	17.9	5.0	5.8	3,563.8
National	**566.6**	**86.7**	**100.0**	**6,534.1**	**116.8**	**34.7**	**40.0**	**3,365.3**

Source: Ethiopia model simulations EPAU and IFPRI (2017).

FIGURE 13.3 Estimated per capita expenditures of poor and nonpoor households in Ethiopia, 2015/2016

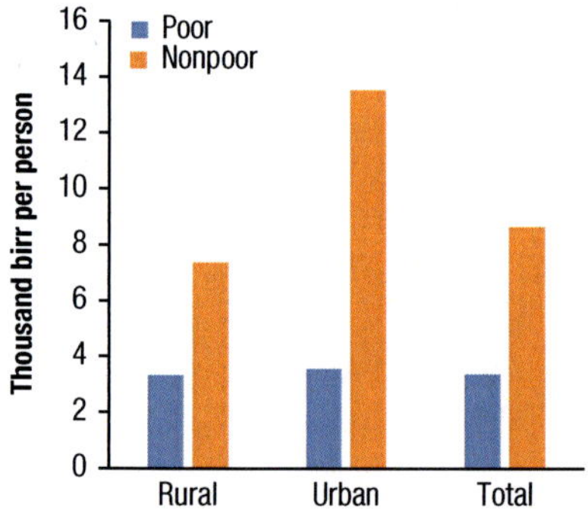

Source: Model simulations based on EPAU and IFPRI (2017).

TABLE 13.3 Factor income sources by household groups in Ethiopia, shares of total income, 2010/2011

	Unskilled labor	Skilled labor	Land	Agri-cultural capital	Livestock capital	Enter-prises	Govern-ment	Rest of world	Total
Rural poor	0.502	0.050	0.241	0.021	0.097	0.082	0.000	0.006	1.000
Rural nonpoor	0.296	0.081	0.123	0.010	0.048	0.416	0.001	0.025	1.000
Urban poor	0.359	0.238	0.000	0.000	0.000	0.399	0.000	0.004	1.000
Urban nonpoor	0.071	0.217	0.000	0.000	0.000	0.691	0.001	0.019	1.000
All households	0.291	0.105	0.118	0.010	0.046	0.408	0.001	0.020	1.000
Total	0.291	0.068	0.081	0.006	0.031	0.264	0.047	0.013	1.000

Source: Dorosh et al. (2018).

Note: Poor households are defined as those in the bottom 40 percent of the national per capita expenditure distribution. Agricultural capital includes tractors, plows, and so on. Livestock capital includes both livestock and poultry. Enterprises include informal household enterprises.

poor households likewise derive a large share of their incomes from labor, 35.9 percent from unskilled labor and 23.8 percent from skilled labor. Urban poor households also differ from the rural poor in that they obtain a large share of their incomes (39.9 percent) from enterprises (almost exclusively informal enterprises).

Rural nonpoor households differ from rural poor households, not only in the overall amount of land cultivated, livestock capital, and the magnitude of their incomes but also in the relatively large share of income deriving from enterprises (41.6 percent), including implicit returns on capital from incomes earned in marketing. Finally, the urban nonpoor get only 28.8 percent of their incomes from labor (only 7.1 percent of their incomes from unskilled labor); 69.1 percent of their incomes derive from returns to enterprise capital. (Note that remittances and foreign transfers account for only 2.0 percent of household incomes overall.)

Methodology: Model Structure and Equations

In this analysis we use a neoclassical recursive dynamic computable general equilibrium (CGE) model (Dervis, de Melo, and Robinson 1982; Diao and Thurlow 2013; Dorosh and Thurlow 2013b) with modifications for effects of agglomeration and congestion on total factor productivity in urban areas (Dorosh and Thurlow 2012, 2014). The model consists of both behavioral equations that describe the economic decisions related to production, marketing, consumption, and so on of economic agents (firms, households, and institutions) and structural equations that specify accounting relationships between the incomes and expenditures of individual agents and within the macroeconomy.[7]

In the model, producers choose levels of inputs and outputs, given the technology available, as determined by constant elasticity of substitution functions that allow substitution between factors based on relative factor price changes. Demand by producers for intermediate inputs is determined by fixed input-output coefficients. Consumer demand is determined by (linear expenditure system) demand equations that implicitly maximize utility given budget constraints. Income (expenditure) elasticities are estimated using data from the 2010/2011 HCES. National market prices adjust to clear overall supply and demand for each product.

Domestically produced goods and services are modeled as imperfect substitutes with goods and services that are exported or imported.[8] World prices are fixed (exogenous) under the assumption that changes in Ethiopia's demand for imports or supply of exports do not affect world prices (that is, a small country assumption). We model six regional labor markets and four types of labor. Total supply of labor is fixed with each year's simulation, with wage rates

7 See Diao and Thurlow (2013) and Dorosh and Thurlow (2013b and 2014) for more details on the basic model equations and parameters.

8 We use Armington (Constant Elasticity of Substitution) functions to model imperfect substitution between imported and domestic goods and services (Armington 1969) and constant elasticity of transformation (CET) functions to model imperfect substitution between exported and domestic goods. Parameters are based on Dimaranan (2006).

adjusting in each period to equate supply and demand in each region and each type of labor.

Investment is determined by the total supply of savings in the economy. Given fixed marginal propensities to save for each household type, increases in incomes result in higher levels of savings and investment. Tax rates and level of government recurrent spending are also fixed, with net government savings (the difference between government revenues and recurrent spending) adding to the total pool of savings in the economy. Foreign capital inflows (foreign savings) are assumed to be fixed. The nominal exchange rate adjusts to clear the foreign exchange market. The domestic price index is the model's numeraire.

Between periods we update parameters based on long-term trends to reflect changes in factor supplies and productivity, household population growth, government spending, and foreign capital inflows. Capital stocks within each sector and region are updated each year to reflect depreciation and investments from the previous period. Given that investment in capital responds to relative returns across sectors, sectors with above-average profits receive a larger share of new capital stocks than their share of installed capital in the previous period.

We model internal migration across regions between periods (years) in response to wage differentials. The migration rates are estimated by assuming the same natural population growth (fecundity minus morbidity) in urban and rural areas and then scaling migration to match projected population growth rates (natural population growth plus net migration). We model urban agglomeration effects of increased urban population density using an agglomeration elasticity of 0.08 (the percentage change in urban TFP resulting from a 1 percent increase in population density) (Henderson and Wang 2005; Dorosh and Thurlow 2012). Urban TFP also depends on congestion, with a congestion elasticity of 0.10 (the percentage increase in urban TFP gain from a 1 percent increase in public capital per capita).

We model the return on agricultural and rural investments using an agricultural TFP to public agricultural spending elasticity of 0.3 from Benin, Fan, and Johnson (2012). A similar rate of return is used for the rural nonfarm TFP (note that nonfarm GDP is 42 percent of rural GDP). Population and labor (by zone) grow at different (exogenous) rates. Total cropland (by zone) grows at fixed rates, but individual crops' areas are endogenously allocated according to relative profitability.[9]

9 Note that in the base run and most simulations, the growth rate of area cultivated in all three highland zones is set to zero for 2027–2040, reflecting severe land constraints.

Model Simulations

To analyze the implications of alternative public investment strategies, we first construct a baseline simulation and then model four alternative scenarios. The baseline, as well as each alternative scenario, assumes the same overall level of foreign savings (net capital inflows) and public investment; only the allocation of investments to sectors changes across simulations.[10] The five simulations are the following:

- *Baseline.* Business-as-usual but with reduced foreign savings since growth over the period 2011–2016 was driven to a large extent by large-scale foreign borrowing and other capital inflows and likely is not sustainable.
- *Cities.* Faster urbanization in cities with populations of more than 50,000 in 2011; increased foreign savings relative to baseline.
- *Agriculture.* Increased investment in agriculture with baseline foreign savings.
- *Rural nonfarm (RNF).* Faster growth in the rural nonfarm sector (concentrated in small towns with populations of fewer than 50,000) and increased foreign savings.
- *Livestock and crop shift.* Increased concentration of livestock production in the highlands with a corresponding shift in crop area from the rainfall-sufficient highlands to the rainfall-sufficient lowlands.[11]

Design of Simulations

We run the model over the period 2011/2012–2039/2040 (2012–2040).[12] The first subperiod, 2012–2016, is designed to replicate historical trends and thus is the same for the base run and all policy simulations. Two other subperiods are modeled: period 2 (2017–2026) and period 3 (2027–2040). The baseline simulation models a slowdown in growth of agricultural production,

10 In terms of the modeling of investments, the difference between the cities and the rural nonfarm scenario productivity shocks is the physical location of the capital. Implicitly, however, the actual public investments that raise productivity of industry, transport, trade, and other sectors will differ by location (for example, building feeder roads in rural areas versus flyovers in urban areas).

11 No changes in yields due to changes in rainfall and temperature are assumed in the simulations. As noted in Chapter 4 (Table 4.2), estimated effects of changes in rainfall and temperature on cereal yields are small (1.2 percent or less through 2035).

12 For ease of presentation we adopt the convention that year $t/t + 1$ is denoted year $t + 1$ (that is, 2010/2011 = 2011).

consistent with recent data. Increases in area cultivated, labor use, use of fertilizer and improved seeds, and TFP accounted for much of the 8.3 percent annual average grain output growth from 2004/2005 to 2015/2016, as area cultivated and yields increased by 2.6 percent and 4.8 percent per year, respectively. However, growth in cereal production decelerated somewhat to only 6.7 percent per year in the second half of this period, as TFP growth slowed to only 0.85 percent per year, a marked deceleration from average TFP growth of 2.02 percent per year in the 2004/2005–2015/2016 period.[13] Annual increases in cereal area cultivated have also declined, from more than 3 percent in the early 2000s to less than 1 percent in recent years (Figure 13.4).

Moreover, opportunities for future land expansion are limited, however, particularly in the moisture reliable highland zones (including both cereal and enset-based cropping systems) that accounted for 92 percent of cereal area cultivated in 2013/2014 (Schmidt and Thomas 2018). Thus the model simulations assume slow annual growth in area cultivated over the 2017–2040 period: an average of 0.73 percent in the moisture-sufficient highlands and 1.76 percent in the relatively low population density moisture-sufficient lowlands (Tables 13.4 and 13.5; Figure 13.5).

The rate of growth in land area cultivated of most individual crops is a function of the exogenous overall growth in area cultivated in each region and the relative profitability of the crop, although for some crops (sorghum, barley, rice, and industrial crops) the area cultivated is set exogenously. Yield growth rates by crop and region are largely determined by exogenous growth in total factor productivity that approximate recent actual yield growth rates.[14]

The moist highland cereal-producing areas hosted a large proportion (about 42 percent) of total livestock (measured in tropical livestock units) in 2015/2016 due to the concentration of rural population in these areas where the interaction of farming-livestock appears to be strong. Dry and moist lowland areas together accounted for 24 percent of total livestock (Table 13.6 and Figure 13.6). The model assumes a declining average growth rate in livestock numbers from 1.8 percent between 2011 and 2016 to 0.40 percent between

13 These average growth rates do not include 2014/2015, a year for which the data show an 11.0 percent increase in land cultivated. Yield growth is calculated from production and area growth. These figures are adapted from Bachewe et al. (2017). For the 2011–2016 period (without 2014/2015), labor (number of holders), cattle, and fertilizer use grew at 2.89 percent, 1.84 percent, and 12.7 percent per year, respectively.

14 Growth in area cultivated in the base simulation is set to zero for sorghum in all agroecological zones in all three periods. Area growth in barley is set to −1.1 percent per year. Area growth rates for rice and industrial crops (by region) are set to the overall exogenous growth rate of total area in each region (see Appendix Table 13B.1).

FIGURE 13.4 Annual growth in grain area cultivated and yields in Ethiopia, 2005/2006–2016/2017 (%)

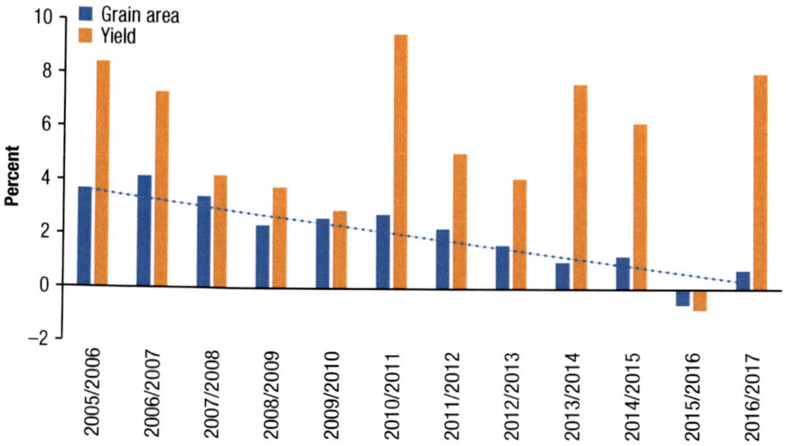

Source: Authors' calculations from agricultural sample survey data, Ethiopia, CSA (various years).

TABLE 13.4 Total area cultivated estimates from selected model simulations in Ethiopia by region, 2039/2040 (million hectares)

	2015/2016	S0-Baseline	S2-Agriculture	S4-Livestock
R1: Dry highlands	4.16	3.85	3.85	3.70
R2: Dry lowlands	1.14	1.11	1.25	1.29
R3: Moist lowlands	1.28	1.66	2.34	4.83
R4: Moist high cereals	10.66	10.32	10.32	8.08
R5: Moist high enset	3.22	2.86	2.86	2.72
Total	**20.46**	**19.81**	**20.62**	**20.63**

Source: Dorosh et al. (2018).
Note: S4-Livestock denotes the livestock and crop shift scenario.

2026 and 2040. This assumption of a declining herd size is consistent with the government's focus on improving the productivity of the livestock sector and reducing the number of herds as articulated in the Climate Resilient Green Economy (CRGE) strategy (Ethiopia 2011).

Labor supply and population growth are set exogenously to an average of 2.04 percent in all simulations (Table 13.7).[15] Labor supply in rural areas

15 Total labor supply growth is set to 2.58 percent, 2.33 percent, and 1.83 percent per year in periods 1, 2 and 3, respectively.

TABLE 13.5 Annual growth rates and share of total area cultivated in Ethiopia, estimates from selected model simulations, by region and subperiod

	S0-Baseline		S2-Agriculture		S4-Livestock[a]	
Growth rate (%)	2015/2016–2025/2026	2025/2026–2039/2040	2015/2016–2025/2026	2025/2026–2039/2040	2015/2016–2025/2026	2025/2026–2039/2040
R1: Dry highlands	0.40	0.00	0.40	0.00	0.00	0.00
R2: Dry lowlands	1.00	0.80	2.00	1.00	3.00	0.40
R3: Moist lowlands	2.00	1.50	4.00	3.00	9.90	4.80
R4: Moist high cereals	1.00	0.50	1.00	0.50	−0.20	−0.75
R5: Moist high enset	0.50	0.00	0.50	0.00	0.00	0.00
Total: All Ethiopia	**0.86**	**0.41**	**0.99**	**0.57**	**1.13**	**0.28**
Share of total area (%)	**2025/2026**	**2039/2040**	**2025/2026**	**2039/2040**	**2025/2026**	**2039/2040**
R1: Dry highlands	20.3	19.5	19.9	18.7	19.1	17.9
R2: Dry lowlands	5.4	5.6	5.8	6.0	6.4	6.3
R3: Moist lowlands	7.5	8.4	9.0	11.3	15.6	23.4
R4: Moist high cereals	51.7	52.1	50.6	50.0	44.9	39.2
R5: Moist high enset	15.1	14.5	14.8	13.9	14.0	13.2
Total: All Ethiopia	**100.0**	**100.0**	**100.0**	**100.0**	**100.0**	**100.0**
Total area (million hectares)	**19.0**	**19.8**	**19.4**	**20.6**	**19.4**	**20.6**

	Total area (million hectares, 2039/2040)			Growth rate: 2015/2016–2039/2040 (%)		
	S0-Baseline	S2-Agriculture	S4-Livestock[a]	S0-Baseline	S2-Agriculture	S4-Livestock[a]
R1: Dry highlands	3.85	3.85	3.70	−0.32	−0.32	−0.48
R2: Dry lowlands	1.11	1.25	1.29	−0.13	0.37	0.52
R3: Moist lowlands	1.66	2.34	4.83	1.09	2.54	5.68
R4: Moist high cereals	10.32	10.32	8.08	−0.14	−0.14	−1.15
R5: Moist high enset	2.86	2.86	2.72	−0.49	−0.49	−0.69
Total: All Ethiopia	**19.81**	**20.62**	**20.63**	**−0.14**	**0.03**	**0.03**

Source: Dorosh et al. (2018).

Note: Growth rates for area cultivated for simulations S1 (urban) and S3 (rural nonfarm) are the same as the base simulation.

a. S4 Livestock denotes the livestock and crop shift simulation.

FIGURE 13.5 **Estimated area cultivated by region in Ethiopia, from selected model simulations, 2039/2040 (million hectares)**

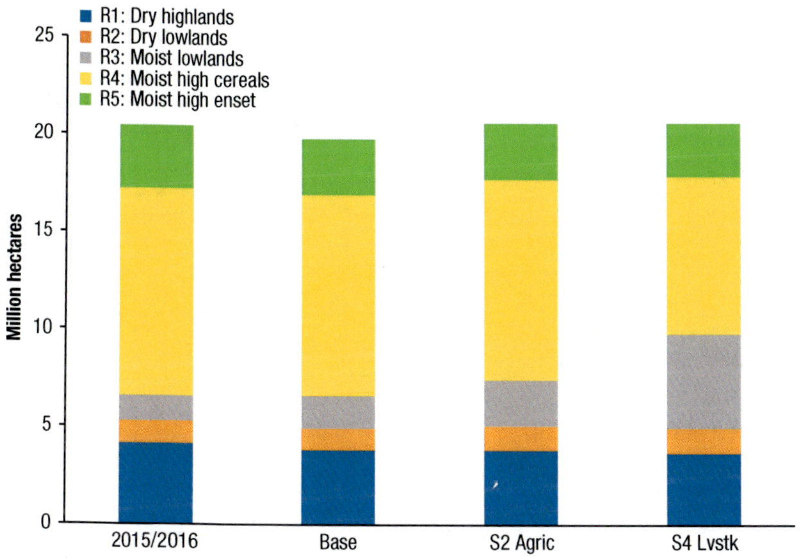

Source: Model simulations.

Note: Growth rates for area cultivated for Simulations S1 (cities) and S3 (rural nonfarm) are the same as the base simulation. S2 Agric = Increased crop agriculture productivity. S4 Lvstk = Increased livestock productivity with expanded crop area in lowlands.

grows at an average annual rate of 1.70 percent per year and 3.30 percent per year in urban areas. Growth rates by rural versus urban regions in the policy simulations vary somewhat, however, because of endogenous changes in migration.[16]

Growth in nonagricultural sectors is a function of growth in factors of production (land, labor, and capital) along with exogenous TFP growth (technical change) specified for each sector and region. In the base scenario, average TFP growth rates for agriculture are 2.50 percent per year from 2016/2017 to 2026/2027, but then slow to only 1.00 percent per year from 2026/2027

16 Population growth rates in the base simulation are derived from the CSA population projections. Overall, population growth slows from an average of 2.41 percent per year in period 1 (2010/2011–2016/2017) to 2.13 percent in period 2 (2017/2018–2026/2027) and 1.81 percent in period 3 (2027/2028–2036/2037). Annual average urban population growth rates are much higher, reflecting substantial rural-urban migration (4.94 percent in period 1 and 3.86 percent in period 3) as compared with rural population growth rates (1.98 percent in period 1 and 1.16 percent in period 3). See Schmidt et al. (2018) for a discussion of population growth and urbanization projections.

TABLE 13.6 Livestock by region in Ethiopia, 2015/2016 (thousands of tropical livestock units)

Region	Cattle	Sheep and goats	Other large animals	Poultry	Total
R1: Dry highlands	6,950	1,371	1,656	130	10,107
R2: Dry lowlands	4,809	1,524	1,358	38	7,728
R3: Moist lowlands	4,221	582	455	97	5,354
R4: Moist high cereals	17,629	1,764	2,913	247	22,554
R5: Moist high enset	6,346	547	614	91	7,598
Total: All Ethiopia	**39,955**	**5,788**	**6,997**	**601**	**53,341**
Shares within zone					
R1: Dry highlands	0.688	0.136	0.164	0.013	1.000
R2: Dry lowlands	0.622	0.197	0.176	0.005	1.000
R3: Moist lowlands	0.788	0.109	0.085	0.018	1.000
R4: Moist high cereals	0.782	0.078	0.129	0.011	1.000
R5: Moist high enset	0.835	0.072	0.081	0.012	1.000
Total: All Ethiopia	**0.749**	**0.109**	**0.131**	**0.011**	**1.000**

Source: Ethiopia, CSA data and authors' calculations.

FIGURE 13.6 Number of livestock in Ethiopia, 2015/2016 (thousands of tropical livestock units)

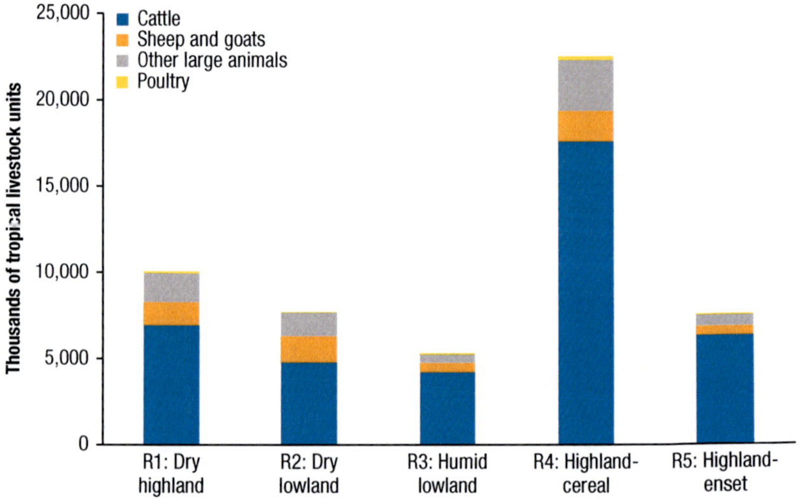

Source: Ethiopia, CSA data and authors' calculations.

TABLE 13.7 Economywide model results: Growth drivers

	Average annual growth rates, 2016/2017–2039/2040 (%)				
Variable	S0-Baseline	S1-Cities	S2-Agriculture	S3-Rural nonfarm	S4-Livestock[a]
National population	2.04	2.04	2.04	2.04	2.04
Rural	1.70	1.70	1.71	1.71	1.71
Urban	3.30	3.32	3.29	3.27	3.29
Total GDP growth	6.15	6.47	5.54	6.26	5.46
Labor supply	2.04	2.04	2.04	2.04	2.04
Land supply	0.67	0.67	0.78	0.67	0.76
Capital accumulation	7.53	7.85	6.87	7.82	6.79
Total factor productivity growth	6.15	6.47	5.54	6.26	5.46
Shares in 2039/2040					
Foreign capital/GDP (%)	24.18	24.26	22.39	24.53	21.82
Investment/GDP (%)	25.90	27.37	20.94	26.53	20.25

Source: Model simulations.
Note: a. Livestock and crop shift scenario.

to 2039/2040. TFP growth rates for nonagriculture are 0.50 percent for both periods (Appendix Table 13B.2).[17]

Growth in capital is determined by overall (domestic and foreign) savings and investment. Much of the growth in the Ethiopian economy in recent years was due to capital growth financed by unsustainable foreign capital inflows that grew at 36.2 percent per year from 2010/2011 through 2015/2016. Moreover, as a result of foreign and domestic borrowing in this period, Ethiopia's total foreign and domestic debt reached 60.3 percent of GDP in 2016/2017.

For this forward-looking analysis, we assume that the balance of payments and debt problems are resolved over time through reduction in foreign borrowing and domestic spending. Thus in the model simulations (2016/2017–2039/2040), foreign savings and government foreign income are assumed to decrease by 4.0 percent per year to model moderate levels of investment and overall GDP growth.

17 These TFP growth rates are adjusted to meet target growth rates in overall agricultural GDP and total GDP.

Baseline Simulation Results

In the baseline scenario, investment growth slows to 3.7 percent per year and real GDP increases by 6.2 percent per year (Table 13.9). Given the high growth rate of GDP, foreign savings as a share of GDP falls steadily, from 20.9 percent in 2016 to only 0.1 percent in 2039/2040, and the ratio of investment to GDP declines from 49.6 percent to 25.9 percent between 2016 and 2040 (Table 13.8). Agricultural GDP increases by 2.9 percent per year—much

TABLE 13.8 Model simulations: Macroeconomic outcomes

Model outcome	Base 2016	Base 2040	S1-Cities 2040	S2-Agriculture 2040	S3-Rural nonfarm 2040	S4-Livestock[a] 2040
Level (billion 2016/2017 birr)						
Absorption	1,015.5	3,292.7	3,463.7	2,913.1	3,396.7	2,842.6
Households	568.1	1,858.6	1,928.2	1,732.0	1,912.4	1,694.1
Investment	384.6	910.3	1,011.6	657.2	960.5	624.7
Government	62.8	523.8	523.8	523.8	523.8	523.8
Exports	128.2	1,179.1	1,278.5	961.5	1,226.0	928.3
Imports	336.8	1,193.8	1,293.2	976.3	1,240.7	943.1
Trade deficit/GDP (%)	44.5	24.2	24.3	22.4	24.5	21.8
Real exchange rate	48.1	31.5	28.7	36.6	30.5	36.9
Nominal exchange rate	99.4	62.6	56.9	73.2	60.5	73.8
Investment/GDP (%)	49.6	25.9	27.4	20.9	26.5	20.2
Private savings/GDP (%)	22.3	18.6	18.3	19.4	18.3	19.4
Foreign savings/GDP (%)	20.9	0.1	0.1	0.2	0.1	0.2
Government savings/GDP (%)	6.3	7.2	9.0	1.4	8.1	0.7
Growth rate (%)	2011–2016	2016–2040				
Absorption	11.9	5.0	5.2	4.5	5.2	4.4
Households	6.6	5.1	5.2	4.8	5.2	4.7
Investment	25.1	3.7	4.1	2.3	3.9	2.0
Government	9.2	9.2	9.2	9.2	9.2	9.2
Exports	8.9	9.7	10.1	8.8	9.9	8.6
Imports	15.5	5.4	5.8	4.5	5.6	4.4
Real exchange rate	0.1	−1.7	−2.1	−1.1	−1.9	−1.1
Nominal exchange rate	−0.1	−1.9	−2.3	−1.3	−2.0	−1.2
Domestic price index	−0.2	−0.2	−0.2	−0.1	−0.2	−0.1

Source: Model simulations.

Note: a. Livestock and crop shift scenario.

slower than for industry (6.4 percent per year) or services (7.5 percent per year) (Table 13.9).

Within agriculture, differences in assumed TFP growth by crop (and region) largely determine the growth outlooks for individual crops. In the base simulation, average annual TFP growth rates for maize and wheat (1.68 percent and 1.46 percent, respectively) are significantly higher than those for teff (1.00 percent). Annual growth rates of production for wheat

TABLE 13.9 Model simulations: Sectoral output for Ethiopia

				Annual average growth rate (2016–2040)				
Sector	Base 2016 (billion birr)[b]	Base 2040 (billion birr)[b]	Base 2040 (%)[c]	S0-Baseline (%)	S1-Cities (%)	S2-Agriculture (%)	S3-Rural nonfarm (%)	S4-Livestock[a] (%)
All sectors (GDP)	**727.4**	**3,045.4**	**318.7**	**6.15**	**6.47**	**5.54**	**6.26**	**5.46**
Agriculture	253.1	499.0	97.1	2.87	2.56	3.28	2.85	3.15
Crops	177.3	306.8	73.1	2.31	1.96	2.80	2.26	2.75
Cereal crops	81.9	148.0	80.8	2.50	2.19	2.92	2.46	2.82
Maize	13.7	27.1	98.2	2.89	2.67	3.16	2.91	3.01
Sorghum	18.8	38.8	106.2	3.06	2.81	3.34	3.07	3.18
Teff	25.8	40.6	57.2	1.90	1.58	2.41	1.83	2.38
Wheat	15.5	25.6	65.3	2.11	1.54	2.89	1.96	2.88
Export crops	39.7	78.0	96.4	2.85	2.40	3.47	2.78	3.49
Coffee	27.7	60.9	119.3	3.33	2.91	3.90	3.27	3.94
Other crops	55.7	80.8	45.0	1.56	1.24	2.04	1.52	2.03
Livestock	53.2	121.9	129.2	3.52	3.26	3.88	3.54	3.42
Other agriculture	22.7	70.3	210.2	4.83	4.64	4.97	4.86	4.98
Industry	123.1	550.3	347.2	6.44	6.89	5.32	6.62	5.18
Mining	19.4	77.6	299.1	5.94	6.63	4.47	5.80	4.43
Manufacturing	28.7	239.1	731.9	9.23	9.62	8.23	9.48	8.06
Agro-processing	13.7	46.7	241.5	5.25	5.30	4.95	5.24	4.78
Other manufacturing	15.1	192.4	1,176.3	11.19	11.67	9.99	11.51	9.82
Other industry and electricity	84.2	284.4	237.7	5.20	5.63	4.13	5.39	3.99
Construction	65.5	182.9	179.2	4.37	4.82	3.14	4.61	2.97
Services	351.2	1,996.1	468.3	7.51	7.94	6.74	7.64	6.68

Source: Model simulations.

Note: a. Livestock and crop shift scenario.

b. 2016/2017 birr.

c. Percentage change, 2016/2017–2039/2040.

TABLE 13.10 Model simulations: Growth of cereal production, prices, and imports, 2016–2040 (%)

Cereal	S0-Baseline	S1-Cities	S2-Agriculture	S3-Rural nonfarm	S4-Livestock[a]
Maize					
Total factor productivity	1.68	1.36	2.11	2.11	2.11
Production	2.89	2.67	3.16	2.91	3.01
Price	−0.01	0.05	−0.13	0.02	−0.15
Teff					
Total factor productivity	1.00	0.67	1.42	1.42	1.42
Production	1.90	1.58	2.41	1.83	2.38
Price	1.40	1.52	1.12	1.42	1.14
Wheat					
Total factor productivity	1.46	1.13	1.88	1.88	1.88
Production	2.11	1.54	2.89	1.96	2.88
Price	−1.65	−1.94	−1.20	−1.75	−1.25
Imports	6.40	6.66	5.80	6.38	5.62
Imports/availability	20.8	24.2	16.1	21.4	15.5

Source: Model simulations.
Note: a. Livestock and crop shift scenario.

(2.11 percent) and maize (2.89 percent) are thus significantly higher than for teff (1.90 percent) (Table 13.10). For teff, steady income growth raises demand faster than the increases in supply, so prices rise steadily by 1.40 percent per year. Demand for maize increases at the same rate as supply, so maize prices are essentially unchanged over time. For wheat, however, increases in demand lead to an increase in imports, which rise by 6.4 percent per year. By 2040 wheat imports account for 20.8 percent of supply. In the Cities simulation (S1), with higher public investment in cities but lower agricultural investment, growth in TFP and production of all three major cereal crops is lower and prices are higher, except for wheat, for which imports increase. Agricultural production growth rates are highest and wheat imports are lowest in the scenarios involving more agricultural investment (S2 and S4).

In the base run, as well as in the other scenarios, the share of agriculture in GDP falls over time (from more than 40 percent in 2010 to under 10 percent in 2040 in the base run). The share of the broader agriculture and food system (AFS) also declines over time. However, the share of downstream activities (processing, trade, and transport costs) in total AFS GDP increases steadily, from about 20 percent in 2010 to more than 30 percent in 2040, as a result of

FIGURE 13.7 Baseline agrifood system dynamics in Ethiopia, 2015/2016–2039/2040

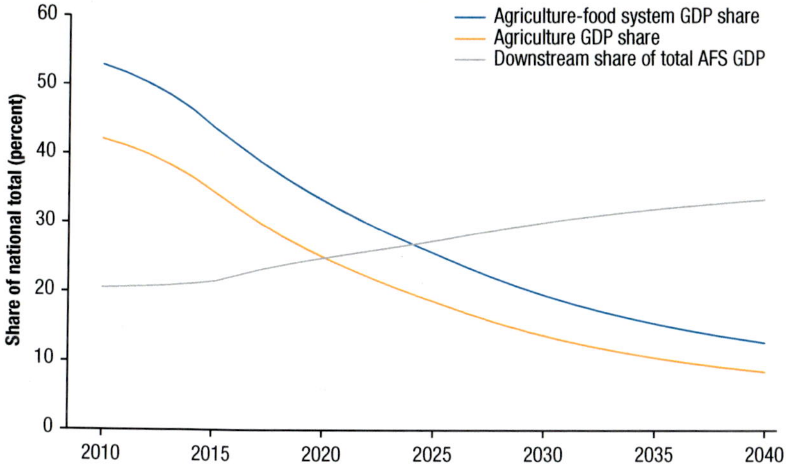

Source: Model simulations.

TABLE 13.11 Agrifood system outcomes in Ethiopia, annual average growth rates, 2015/2016–2039/2040 (%)

Sector	S0-Baseline	S1-Cities	S2-Agriculture	S3-Rural nonfarm	S4-Livestock[a]
National economy	6.1	6.5	5.5	6.3	5.5
Agriculture-food system	3.3	3.1	3.6	3.3	3.4
Agriculture	2.9	2.6	3.3	2.8	3.1
Agro-processing	5.3	5.3	4.9	5.2	4.8
Input production	4.8	4.7	4.8	4.9	4.5
Trade and transport services	4.3	4.2	4.3	4.2	4.2

Source: Model simulations.

Note: a Livestock and crop shift scenario.

sharp increases in marketed sales, urban demand, and overall demand for processed products (Figure 13.7 and Table 13.11).

Table 13.12 presents simulation results for per capita expenditure growth by household group and population growth. Overall, per capita household incomes increase by 2.35 percent per year overall (2.60 percent per year in rural areas and 1.30 percent per year in urban areas). Slower per capita income growth in urban areas is due in part to rural-urban migration that contributes to a faster population growth rate in urban areas (3.30 percent per year) than in rural areas (1.70 percent per year). Within rural areas, faster agricultural

TABLE 13.12 Model simulations: Per capita expenditure growth in Ethiopia, 2016–2040 (%)

Indicator	All Households					Poorest 40%				
	S0-Baseline	S1-Cities	S2-Agriculture	S3-Rural non-farm	S4-Livestock[a]	S0-Baseline	S1-Cities	S2-Agriculture	S3-Rural non-farm	S4-Livestock[a]
Rural	2.60	2.54	2.62	2.80	2.51	3.79	3.79	3.64	3.98	3.56
R1: Dry highlands	2.54	2.48	2.52	2.73	2.42	3.51	3.53	3.30	3.71	3.25
R2: Dry lowlands	1.89	1.67	2.22	2.14	2.14	2.99	2.82	3.17	3.19	3.18
R3: Moist lowlands	2.13	1.98	2.46	2.43	2.55	3.42	3.36	3.63	3.68	3.95
R4: Moist high cereals	2.97	2.98	2.85	3.14	2.70	4.24	4.30	3.98	4.42	3.81
R5: Moist high-enset	2.52	2.46	2.54	2.76	2.46	3.56	3.51	3.48	3.74	3.41
Urban	1.30	1.37	1.11	0.98	0.97	0.73	0.78	0.52	0.46	0.38
National	2.35	2.33	2.30	2.40	2.19	3.28	3.28	3.12	3.41	3.03
Rural population growth	1.70	1.70	1.71	1.71	1.71	2.19	2.19	2.20	2.20	2.20
Urban population growth	3.30	3.32	3.29	3.27	3.29	4.47	4.48	4.44	4.43	4.44
Total population growth	2.04	2.04	2.04	2.04	2.04	2.58	2.58	2.58	2.58	2.58

Source: Model simulations.

Note: a. Livestock and crop shift scenario.

productivity growth in the highlands contributes to more rapid per capita income growth in the highland regions than in the lowlands. Incomes of poor households (the bottom 40 percent of the income distribution) increase slightly faster at 3.28 percent per year overall, but at 3.79 percent per year for the rural poor. Incomes of poor urban households increase at only 0.73 percent per year.

Policy Scenarios: The Implications of Alternative Public Investment Strategies

Simulations of alternative public investment strategies using an economy-wide model of the Ethiopian economy highlight the trade-offs between various policy options. Investments in urban infrastructure result in the highest growth rate of an average of 6.47 percent per year from 2017 through 2040. Agricultural sector investments, whether in crops or livestock, result in lower growth rates of 5.54 percent and 5.46 percent per year, respectively. In part, this is due to lower productivity of agricultural investments as well as to

FIGURE 13.8 Model simulations: Growth outcomes in Ethiopia by total, rural, urban, agricultural, and nonagricultural GDP, 2015/2016–2039/2040

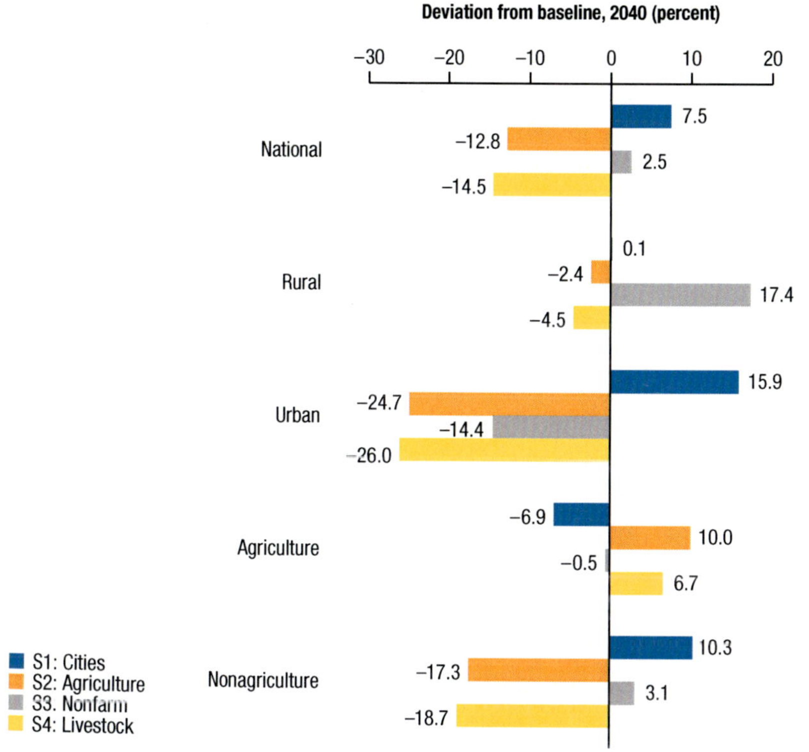

Source: Model simulations.

Note: S4 = the livestock and crop shift scenario.

increases in overall investment, as the simulations assume that total investment as a share of GDP is constant across simulations. Thus greater increases in GDP in the urban investment scenario not only generates more consumption, they also result in increased investment and more future growth. By 2040 real GDP in the urban investment scenario is 7.5 percent higher than in the baseline simulation, and real GDP in the agricultural investment scenarios is 12.8 percent lower.

Not surprisingly, the largest gains in agricultural GDP arise from investments in crop agriculture (S2) or the livestock and crop shift (S4) scenarios, with GDP levels in these scenarios that are 10.0 percent and 6.7 percent greater than in the baseline, respectively, by 2040 (Figure 13.8). However, because these investments draw resources away from rural nonagriculture activities, overall rural GDP is actually somewhat lower than in the baseline at −2.4 percent and −4.5 percent in crop and livestock investment simulations, respectively. Real nonagricultural GDP is also 17.3 percent and 18.7 percent lower in 2040 than in the baseline in the increased crop and livestock investment scenarios. The biggest gains in overall rural GDP, however, arise from investments in the rural nonfarm sector—rural GDP is 17.4 percent higher under this scenario relative to the baseline. By contrast, increased public investment in urban sectors, because it crowds out public investment in rural sectors, results in a decline in agricultural GDP by 6.9 percent in 2040.

National welfare improves in all scenarios, including the baseline and, in general, rural welfare increases faster than urban welfare, although from a lower initial level. Overall, the urban economy grows faster in all scenarios, but increased rural-urban migration in the urban investment scenario dampens urban per capita income growth. Faster urban growth in the urban investment scenario widens the rural-urban income divide; greater rural investment in agriculture and small-town scenarios narrows the rural-urban divide.

Household welfare outcomes differ considerably across scenarios, however. Investments in agriculture lead to similar changes in national household consumption in 2040 as do investments in cities. The largest gains in national welfare come from investments in the rural nonfarm sector. Although a shift toward urban investments benefits urban households by 1.5 percentage points relative to the baseline, national average welfare falls by 0.7 percentage points relative to the baseline (Figure 13.9). In part, this is because gains in incomes of urban households lead to greater demand for imported products, including imported agricultural products. In contrast, rural households who gain from higher rural nonfarm incomes generally spend a higher share of their incomes on local nonagricultural goods and services, spurring rural incomes.[18]

Given the concentration of poverty in rural areas and linkages between agriculture and small towns, investments in agriculture and rural nonfarm are more pro-poor than investments in cities. In fact, rural nonfarm investment is more pro-poor than both agricultural investments and investments

18 Analysis by Diao et al. (2007) showed that rural households spend about 0.60 birr for each additional birr of income on nonagricultural commodities and that low-income rural households have a higher marginal budget share (MBS) of nonagricultural spending compared to high-income groups. In rural areas the MBS of nonagricultural products and services for lowest income quintiles is about 0.74 compared with 0.47 for the top income quintile, indicating that additional income for the lowest income groups will stimulate demand for nonfarm products and other urban-based products and services and thereby spur economywide growth. Given a large market size in rural areas, investment in nonfarm activities would generate large welfare gains due to strong income and demand linkages.

FIGURE 13.9 Model simulations: Annual per capita consumption growth in Ethiopia, 2015/2016–2039/2040

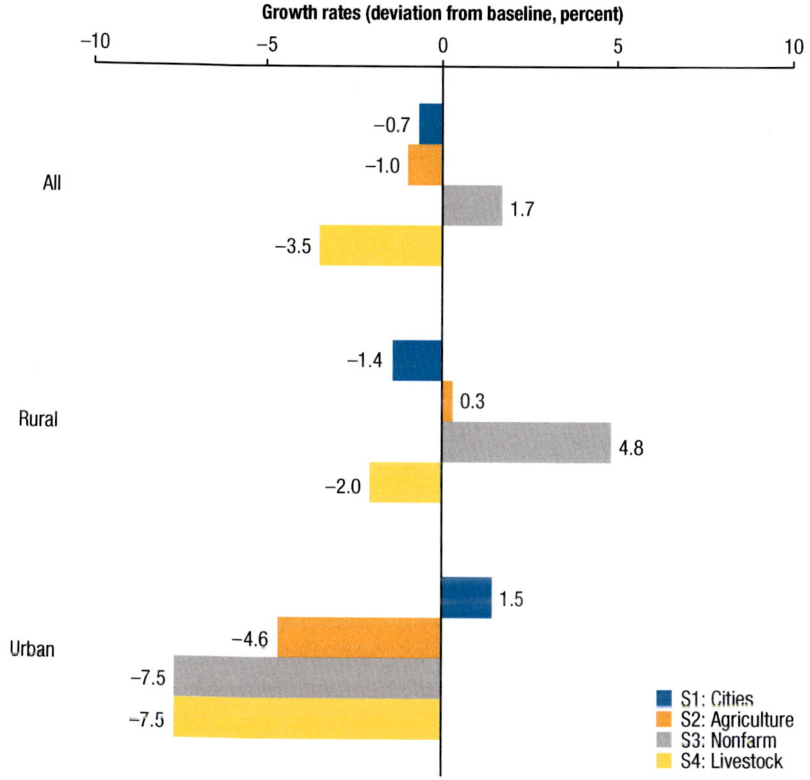

Source: Model simulations.
Note: S4 = the livestock and crop shift scenario.

in cities. This is because rural nonfarm investments result in large gains in incomes in 2040 to poor households (2.5 percent) as well as nonpoor households (1.4 percent) (Figure 13.10). Investments in cities draw resources away from the agrifood system, which hurts both poor and nonpoor consumers. Note that households in small towns mainly consume agricultural commodities, products of agro-processing, and services. In particular, poor households in both small and large urban centers spend a higher share of their income on agricultural commodities as compared to nonpoor households (Ferede and File 2018). This strong demand stimulus is expected to spur increased output and productivity of rural nonfarm activities, which would benefit mainly poor households in rural and small towns.

FIGURE 13.10 Model simulations: Annual per capita consumption growth for poor and nonpoor households in Ethiopia, 2015/2016–2039/2040

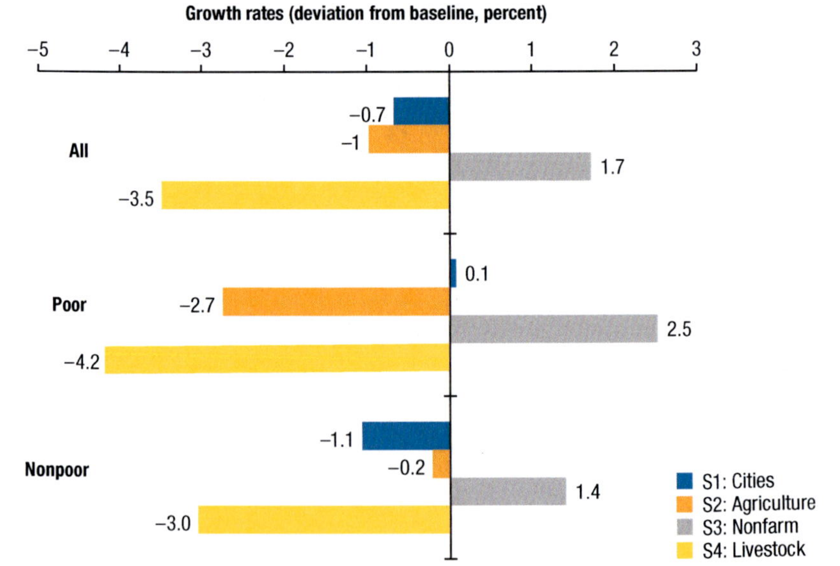

Source: Model simulations.

Note: S4 = the livestock and crop shift scenario. Poor households = those in the lowest 40 percent of the income distribution.

The above estimates of poverty impacts, however, measure the overall impacts of a consistent investment policy over more than 20 years, a period in which the economy becomes increasingly urban and the share of agriculture in GDP and employment declines steadily. Considering only the marginal impact of each year's public investments on poverty in the four policy scenarios shows that at some point urban investments dominate in terms of both growth and poverty outcomes (Figure 13.11).

The model simulations suggest that the poverty reducing effects of investments in agriculture may diminish as the structure of the economy changes. Thus, while investments in agriculture lead to a 0.8 percent greater reduction in poverty than urban investments in 2017, by 2025 investments in cities will have a greater effect on poverty. Likewise, investments in the rural nonfarm sector have a larger impact on poverty than investments in cities only through 2026; afterward, investments in cities dominate in terms of both growth and equity. Moreover, as indicated in Figure 13.10, which reflects the cumulative impact of these alternative investments as reflected in the average annual

FIGURE 13.11 Annual per capita consumption growth for poor households in Ethiopia

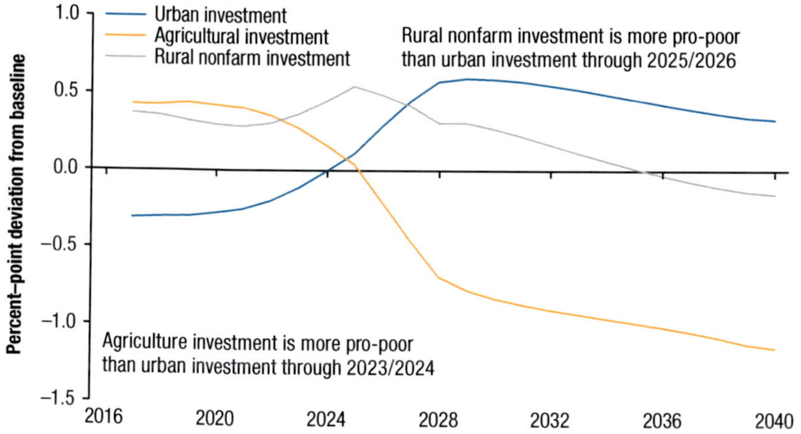

Source: Model simulations.

Note: Poor households = those in the lowest 40 percent in the income distribution.

growth rates from 2016/2017 to 2039/2040, investments in cities modeled in Simulation 1 ultimately results in higher income growth for both the overall economy and the poor. This suggests that an optimal strategy for growth and equity may involve continued high rates of public investment in agriculture and the rural nonfarm sectors through the mid-2020s with a shift toward investments in cities thereafter.

Sensitivity Analysis: Poverty Impacts of Alternative Investments

This analysis of the changes in the impacts of public investments on poverty over time suggests that these impacts depend significantly on the structure of the economy and the relative pace of agricultural and nonagricultural growth. To quantify these effects, we conduct sensitivity analysis assuming higher foreign savings and three alternative sets of agricultural and nonagricultural TFP growth rates (see Appendix 13C for a description of the alternative baseline simulations).

As shown in Figure 13.12, speeding up nonag TFP growth from 0.50 in period 1 (2016–2027) and in both periods 1 and 2 (2016–2040) has little effect on the relative impacts of agricultural, rural nonfarm, and investments in cities in the initial years of the simulation, but the effects increase over time. In all cases, agricultural growth is more pro-poor than urban growth from 2017 through 2023. Increased foreign savings and faster growth in nonagricultural TFP result in a gradual increase in the impact of agricultural growth

FIGURE 13.12 Relative effects of sectoral investments on incomes of the rural poor in Ethiopia, 2017–2040: Sensitivity analysis

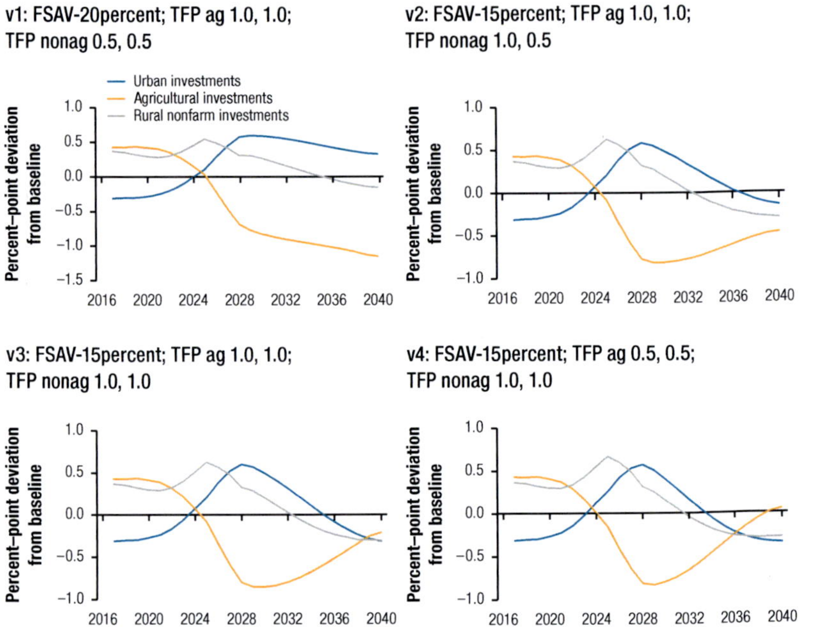

Source: Model simulations.

on incomes of the rural poor relative to the baseline, as a larger nonagricultural sector provides a bigger market and higher sales prices for agricultural products.[19] Whereas in version 1 (v1) of the simulations, the gap between the income effect of agricultural investments relative to the baseline gap widens over time, in v2 and v3, the gap narrows after 2028. With fast TFP growth in nonagricultural sectors for the entire 2017–2040 period (v3), incomes of the rural poor are actually greater after 2039 with agricultural investments than with alternative investments. Under a slow trend agricultural TFP growth scenario (v4), marginal increases in agricultural investments lead to higher incomes of the rural poor from 2037.

These results suggest that the dominance of agricultural investments relative to nonagricultural investments for raising incomes of the poor hold through about 2024 even under alternative assumptions regarding growth

19 Note that foreign savings, though higher than in the main simulations, still declines sharply over time.

rates of total factor productivity in the nonagricultural and agricultural sectors. Structural change in the Ethiopian economy is likely to diminish the relative dominance of agriculture for reducing poverty over the next ten years, but in the shorter run of, say, five years, agricultural sector investments appear to be the most pro-poor.

Conclusion

Several broad conclusions emerge from this analysis. First, agricultural growth is likely to decelerate as land constraints become increasingly binding, especially in the highlands. Given that the rural population is projected to increase despite continued urbanization, average farm size will likely increase. Continued investments in research and technology—for example, continued improvement and dissemination of high-yielding varieties and increased supply of fertilizer—to increase agricultural productivity will be crucial to avoid a large increase in imports or declining food availability. Increased supplies of fertilizer, both in quantity and quality (composition), targeted to specific crops and localities will also be needed, along with significant levels of public investment in irrigation and rural infrastructure—for example, rural road networks through the universal rural road access program. Second, rapid growth in the nonfarm economy will likely result in a sharp increase in market demand for agricultural products (see Chapter 7). Increases in agricultural production (see Chapter 3) will be needed to avoid major increases in imports or a sharp increase in real food prices that would harm the poor. Facilitating the expansion of this broader agrifood system through well-designed policies and public and private investments will be crucial to meet the growing market demand.

Third, economywide analysis suggests that investments in cities generate faster economic growth and structural transformation but that, given the large share of the population with incomes linked to agriculture and the rural economy, investments in the rural economy are likely to continue to be more pro-poor than public investments in cities through the mid-2020s (see Chapter 12). Further work is needed on understanding the dynamics of the agrifood system, including the effects of changing household demand behavior, especially the shift from own-consumption to marketed output. Other priority areas for research include the implications for productivity growth and income distribution of changes in technology—for example, expansion of solar and other non-fossil-fuel energy, increases in use and further advances in information/digital technology, and so on.

In summary, even with rapid urbanization most of Ethiopia's population, as well as the poor in the bottom two income quintiles, will likely still live in rural areas and small towns in 2040, where agricultural production, processing, and trade remain the main livelihoods, especially for the poor. Though rapid economic growth and structural transformation have diminished the relative importance of the agricultural sector in Ethiopia's economy, continued public investments in agriculture and the broader agrifood system will remain crucial for equity and poverty alleviation in Ethiopia.

Appendix 13A: Accounts in the 2015/2016 Ethiopia NEXUS Social Accounting Matrix

Activities/Commodities (45 Sectors)

AGRICULTURE (23)

- Crops (15): maize, sorghum, teff, barley, wheat, pulses, oils, roots, vegetables, sugar cane, tobacco, fruit;
- Livestock and other primary sectors (8): cattle, milk, poultry, sheep and goats, other livestock, forestry, fish, mining.

INDUSTRY (16)

- Meat processing, dairy, grain milling, sugar refining, other food processing, beverages, textiles, leather, wood, chemicals, metals, machinery, other manufacturing, electricity, water, construction.

SERVICES (6)

- Trade, transport, financial services, real estate, government services, other services.

FACTORS OF PRODUCTION (47)

- Labor (18): three skill levels (no education, primary, secondary) in six regions;
- Land (5): in five rural regions;
- Capital (24): four types of capital (crops, livestock, mining, nonagriculture) in six regions.

INSTITUTIONS (38)

- Households (30): five quintiles in six regions;
- Other institutions (8): enterprises (in six regions); government, rest of world.

REGIONS (6)

- R1 Drought prone highlands
- R2 Moisture reliable highland cereal
- R3 Moisture reliable highland enset
- R4 Moisture reliable lowland
- R5 Drought prone lowland and pastoralist
- U1 Urban

Appendix 13B: Assumptions in Growth Projections

TABLE 13B.1 Assumed annual growth rates of land and livestock capital in Ethiopia, by rural region and scenario (%)

	Base	Base	S1, S2, and S3	S1, S2, and S3	S4	S4
Category	2017–2027	2027–2040	2017–2027	2027–2040	2017–2027	2027–2040
Land	0.40	0.00	0.40	0.00	0.30	−0.10
R1: Dry highlands	1.00	0.80	1.00	0.80	1.10	0.90
R2: Dry lowlands	2.00	1.50	2.00	1.50	2.10	1.60
R3: Moist lowlands	1.00	0.50	1.00	0.50	0.90	0.40
R4: Moist high cereals	0.50	0.00	0.50	0.00	0.40	−0.10
R5: Moist high enset	0.00	0.00	0.00	0.00	0.00	0.00
Livestock						
R1: Dry highlands	0.40	0.00	0.40	0.00	1.40	1.00
R2: Dry lowlands	1.00	0.80	1.00	0.80	0.00	−0.20
R3: Moist lowlands	2.00	1.50	2.00	1.50	1.00	0.50
R4: Moist high cereals	1.00	0.50	1.00	0.50	2.00	1.50
R5: Moist high enset	0.50	0.00	0.50	0.00	1.50	1.00

Source: Model simulations.

TABLE 13B.2 Assumed annual growth rates in total factor productivity in Ethiopia, by scenario, subsector, and analytical period (%)

Sector	Base	S1: Cities	S2: Agriculture	S3: Rural nonfarm	S4: Livestock
Agriculture					
2017–2027	2.50	1.79	3.43	2.50	3.43
2027–2040	1.00	1.00	1.00	1.00	1.00
Crops (all regions)					
2017–2027	2.50	1.79	3.43	2.50	3.43
2027–2040	1.00	1.00	1.00	1.00	1.00
Livestock (regions R1, R4, and R5)					
2017–2027	1.00	0.29	1.93	1.00	0.93
2027–2040	1.00	1.00	1.00	1.00	0.00
Livestock (regions R2 and R3)					
2017–2027	1.00	0.29	1.93	1.00	2.93
2027–2040	1.00	1.00	1.00	1.00	2.00
Nonagriculture (all regions)*					
2017–2027	0.50	0.50	0.50	1.43	0.50
2027–2040	0.50	0.50	0.50	0.50	0.50

Source: Model simulations.

Note: S4 = the livestock and crop shift scenario. * Nonagriculture includes urban regions.

Appendix 13C: Alternative Base Simulations

Several alternative base simulations were constructed as part of the sensitivity analysis of poverty impacts (Table 13C.1). Base 1 is the main base simulation used in the analysis in which agricultural TFP grows by 1.0 percent per year and nonagricultural TFP grows by 0.5 percent for the entire 2015–2040 period. Given that foreign borrowing was unsustainably high in the 2011–2016 period, we model a sharp reduction in foreign savings over time, reducing foreign savings by 20 percent per year from 2016 through 2027 and by 10 percent per year from 2028 through 2040.

In this base simulation, nonagricultural GDP increases much faster (7.25 percent per year) than does agricultural GDP (2.87 percent per year), and total GDP increases by 6.15 percent per year. Given the still substantial inflow of foreign savings in the economy, the real exchange rate appreciates by an average of 1.75 percent per year, reducing incentives for exports. Prices of wheat, a major importable good, and coffee, a major exportable

TABLE 13C.1 Alternative base simulation assumptions and resultant key average annual growth rates in Ethiopia, 2014–2040 (%)

Model variable	Base 1	Base 1a	Base 2	Base 3	Base 4
Assumptions (annual growth rates)					
Foreign savings					
2016–2027	–20.00	–15.00	–15.00	–15.00	–15.00
2028–2040	–10.00	–10.00	–10.00	–10.00	–10.00
Agricultural total factor productivity					
2016–2027	1.00	1.00	1.00	1.00	0.50
2028–2040	1.00	1.00	1.00	1.00	0.50
Nonagricultural total factor productivity					
2016–2027	0.50	0.50	1.00	1.00	1.00
2028–2040	0.50	0.50	0.50	1.00	1.00
Scenario results (annual growth rates)					
Total GDP	6.15	6.51	7.07	7.47	7.25
Agricultural GDP	2.87	2.87	2.86	2.86	2.22
Nonagricultural GDP	7.25	7.69	8.35	8.82	8.65
Real exchange rate	–1.75	–1.97	–2.28	–2.48	–2.61
Wage (R2 unskilled)	1.87	1.95	2.06	2.13	1.66
Per capita income rural poor	2.99	3.19	3.48	3.67	3.22
Agricultural prices	–0.14	–0.16	–0.18	–0.20	–0.03
Maize	–0.01	0.03	0.09	0.13	0.42
Wheat	–1.65	–1.82	–2.07	–2.24	–2.33
Teff	1.40	1.47	1.56	1.61	1.63
Coffee	–0.99	–1.07	–1.16	–1.21	–0.78
Cattle	0.39	0.48	0.58	0.63	0.69

Source: Model simulations.

good, decline over time by 1.65 percent and 0.99 percent per year, respectively. However, given the rapid economic growth, the wage rate for rural unskilled labor increases by 1.87 percent per year and per capita incomes of the rural poor (the bottom 40 percent in the per capita expenditure distribution) rise by 2.99 percent per year.

In Base 1a, increasing the foreign savings inflow by reducing the rate of decline of foreign savings from –20 percent per year to –15 percent per year in the 2016–2027 period results in faster GDP growth (6.51 percent per year in Base 1a versus 6.15 percent per year in Base 1), as the increased foreign savings funds faster investment growth. The real exchange rate appreciates more rapidly, contributing to greater declines in wheat and coffee prices, but annual growth in wage rates for rural unskilled labor and per capita incomes of the rural poor accelerate slightly, by 0.08 and 0.20 percentage points, respectively.

Increasing the growth rate of nonagricultural TFP from 0.50 in period 1 (2016–2027) and in both periods 1 and 2 (2016–2040), results in further increases in GDP and nonagricultural GDP growth rates but has little effect on agricultural GDP growth rates (Base 2 and Base 3). Nonetheless, faster nonagricultural GDP productivity growth results in larger gains in rural wage rates for unskilled labor and per capita incomes of the rural poor relative to Base 1 and 1a.

Finally, a reduction in the growth rate of agricultural TFP from 1.0 percent per year to 0.5 percent per year (Base 4) leads to a sharp slowdown in agricultural GDP growth (from 2.86 percent to 2.22 percent per year in Base 3). Nonagricultural GDP growth also slows due to a reduction in demand for nonagricultural goods and services. The annual growth rates of the wage rate of unskilled rural labor and of per capita incomes of the rural poor fall steeply by 0.47 and 0.45 percentage points relative to Base 3, respectively. Lower productivity growth in agriculture also results in higher agricultural prices—an increase in the annual growth rate of average prices of five major agricultural commodities by 0.17 percentage points.

References

Armington, P. A. 1969. "A Theory of Demand for Products Distinguished by Place of Production." *IMF Staff Papers* 16 (1): 159–178.

Bachewe, F., G. Berhane, B. Minten, and A. S. Taffesse. 2018. "Agricultural Transformation in Africa? Assessing the Evidence in Ethiopia." *World Development* 105 (c): 286–298.

Benin, S., S. Fan, and M. Johnson. 2012. *Estimating Public Agricultural Expenditure Requirements.* Washington, DC: International Food Policy Research Institute (IFPRI).

Dervis K., J. de Melo, and S. Robinson. 1982. *General Equilibrium Models for Development Policy.* New York: Cambridge University Press.

Diao, X., B. Fekadu, S. Haggblade, A. S. Taffesse, K. Wamisho, and B. Yu. 2007. *Agricultural Growth Linkages in Ethiopia: Estimates Using Fixed and Flexible Price Models.* IFPRI Discussion Paper 695. Washington, DC: IFPRI.

Diao, X., A. S. Taffesse, P. Dorosh, J. Thurlow, A. N. Pratt, and B. Yu. 2013. "Ethiopia." In *Strategies and Priorities for African Agriculture: Economywide Perspectives from Country Studies,* edited by X. Diao, J. Thurlow, S. Benin, and S. Fan, 107–139. Washington, DC: IFPRI.

Diao, X., and J. Thurlow. 2013. "A Recursive Dynamic Computable General Equilibrium Model." In *Strategies and Priorities for African Agriculture: Economywide Perspectives from Country Studies,* edited by X. Diao, J. Thurlow, S. Benin, and S. Fan, 17–50. Washington, DC: IFPRI.

Dimaranan, B. 2006. *Global Trade, Assistance and Production: The GTAP 6 Data Base.* West Lafayette, IN, US: Center for Global Trade Analysis, Purdue University.

Dorosh, P., and J. Thurlow. 2012. "Agglomeration, Growth and Regional Equity: An Analysis of Agriculture- versus Urban-led Development in Uganda." *Journal of African Economies* 21 (1): 94–123.

———. 2013a. "Agriculture and Small Towns in Africa." *Agricultural Economics* 44 (3): 435–445.

———. 2013b. "Implications of Accelerated Agricultural Growth on Household Incomes and Poverty in Ethiopia: A General Equilibrium Analysis." In *Food and Agriculture in Ethiopia: Progress and Policy Challenges,* edited by P. Dorosh and S. Rashid, 219–255. Philadelphia: University of Pennsylvania Press.

———. 2014. "Can Cities or Towns Drive African Development? Economywide Analysis for Ethiopia and Uganda." *World Development* 63: 113–123.

Dorosh, P., J. Thurlow, F. Worku-Kebede, T. Ferede, and A.S. Taffesse. 2018. *Structural Change and Poverty Reduction in Ethiopia: Economy-wide Analysis of the Evolving Role of Agriculture.* Working Paper 123 (September). Addis Ababa: IFPRI Ethiopia Strategy Support Program.

EPAU (Economic and Policy Analysis Unit) and IFPRI (International Food Policy Research Institute). 2017. *2010/11 Social Accounting Matrix for Ethiopia.* Washington, DC: IFPRI.

Ethiopia. 2011. *Ethiopia's Climate-Resilient Green Economy: Green Economy Strategy.* Federal Democratic Republic of Ethiopia, Addis Ababa.

Ethiopia, CSA (Central Statistical Agency). 2012. *Household Income, Consumption and Expenditure Survey 2010/11.* Addis Ababa.

———. 2013. *Population Projections for Ethiopia 2007–2037.* Addis Ababa.

———. Various years. *Report on Area and Production of Major Crops.* Addis Ababa.

Ethiopia, MoFED (Ministry of Finance and Economic Development). Various years. Unpublished data, Addis Ababa.

Ferede, T., and B. File. 2019. "Rural–Urban Linkages in Ethiopia." In *Oxford Handbook of the Ethiopian Economy,* edited by Fantu C., C. Cramer, and A. Oqubay, 763–784. New York: Oxford University Press.

Henderson, J. V., and H. G. Wang. 2005. "Aspects of the Rural-Urban Transformation of Countries." *Journal of Economic Geography* 5: 23–42.

NPC (National Planning Commission). 2016. *Growth and Transformation Plan II (GTP II) (2015/16–2019/20), Volume 1: Main Text.* Addis Ababa.

Schmidt, E., P. Dorosh, M. K. Jemal, and J. Smart. 2018. *Ethiopia's Spatial and Structural Transformation: Public Policy and Drivers of Change.* IFPRI-ESSP Working Paper 119. Addis Ababa: IFPRI/Ethiopia Strategy Support Program (ESSP).

Schmidt, E., and T. Thomas. 2018. *Running out of Room? Discovering the Economic and Climatic Boundaries for Cropland Expansion in Ethiopia.* IFPRI Discussion Paper 1723. Washington, DC: IFPRI.

UNIDO (United Nations Industrial Development Organization). 2018. "Integrated Agro-Industrial Parks (IAIPs) in Ethiopia." Accessed March 12, 2020. www.unido.org/sites/default/files/files/2018-08/Integrated-Agro-Industrial-Parks-in-Ethiopia-booklet.pdf.

World Bank. 2015a. *Ethiopia— Urban Local Government Development Project* (English). Washington, DC. http://documents.worldbank.org/curated/en/843841467991961808/Ethiopia-Urban-Local-Government-Development-Project.

———. 2015b. *Ethiopia Urbanization Review: Urban Institutions for a Middle-Income Ethiopia.* Washington, DC. http://documents.worldbank.org/curated/en/543201468000586809/Ethiopia-Urbanization-review-urban-institutions-for-a-middle-income-Ethiopia.

Chapter 14

TOWARD A MEDIUM-TERM AGRICULTURAL AND RURAL DEVELOPMENT STRATEGY

Paul Dorosh and Bart Minten

Introduction

Ethiopia has invested heavily in agriculture and the rural economy since the early 1990s. These investments have largely paid off. The provision of inputs and new technologies have enabled rapid agricultural production growth. Investments in road infrastructure and increased demand for agricultural products have led to a large expansion in the value of sales. Public investments in urban infrastructure have also contributed to a large expansion in economic activity and the number of secondary cities. Effective safety nets have enhanced household food security and improved resilience to droughts and other shocks. Lastly, macroeconomic policy has been broadly stable as adjustments to exchange rates, foreign borrowing, and public expenditures have kept macroimbalances from growing large enough to substantially hinder growth.

Looking forward, however, Ethiopia faces several major challenges related to rural (and overall) poverty reduction. Increasing population pressure will limit land availability and increase food demand, necessitating continued gains in both land and labor productivity. In the longer term, climate change is not expected to affect average yields, but the incidences of shocks are expected to increase (Chapter 4). At the same time, the Ethiopian food system is rapidly transforming from relatively low productivity cereal and small-scale livestock farms with little marketed output toward high productivity, diversified, market-oriented agriculture closely linked with major urban centers and international markets. As the food economy expands, the differences in resource constraints and market opportunities will become more pronounced. Addressing these challenges and managing structural changes to the rural economy will depend on key investments and policies and necessitate a regionally differentiated, food system approach.

Sustainable Growth in Agricultural Production

Although Ethiopia has succeeded in rapidly increasing agricultural production over the past decade (Chapter 3), a number of actions are needed to feed a growing population. Increasingly important is also the composition of agricultural production. Most agricultural growth has been concentrated in the cereal sector. As the Ethiopian population becomes richer and more urbanized, the demand of high-value products such as animal-sourced foods, fruits, and vegetables is expected to increase. However, production of these high-value products has been slow in the past decade, as evidenced by their increasing prices (Chapter 7). To ensure that this increased demand can be met by local supply and not by imports, several key investments and policies will be needed, especially those that are gender-sensitive as empowering women in agriculture has been shown to have payoffs for agricultural outcomes (Aguilar et al. 2015).

Most crops have not reached their yield potential, indicating that there is still a significant opportunity to increase production (Chapter 3). Fertilizers adapted to local soil conditions are needed as are higher adoption rates of improved seeds. To stimulate adoption of both, better supply and marketing conditions are required. On the supply side, more resources should be channeled to the development of improved seeds by local public researcher centers given the high returns (Alston et al. 2000). A more active role for the private sector is required as well, particularly in the development of hybrids. Publicly available soils maps and extension agents knowledgeable of local conditions can help tailor fertilizer recommendations for farmers to increase efficiency.[1] In the medium-term, increased coordination between knowledge centers, agricultural extension workers, and farmers will also be needed (Chapter 3). Moreover, better marketing that involves the private sector can help lower costs, enhance choice and improve distribution. This might imply, among others, better regulations and conditions for firms involved in the import of agricultural inputs such as seeds, agrochemicals, and machinery, creating a level playing field for the private and public sector, and embracing enhanced choice and competition in the distribution of agricultural inputs.

Intensifying agricultural productivity on limited land will also require the development of new technologies as well as more widespread adoption of improved agronomy/management practices already "on the shelf." Mechanized production and postharvest activities are currently low. However,

---

1 Soil mapping and microlevel soil recommendations have been developed by the Agricultural Transformation Agency (ATA), as discussed in Chapter 3.

as shown in Chapter 11, the transformation of Ethiopia's rural economy is rapidly leading to higher rural wages incentivizing labor-reducing technologies (Ruttan and Hayami 1984). This trend can already be seen by the increasing adoption of herbicides in commercial agricultural areas as a substitute for weeding labor, but it will also drive mechanization during planting and harvesting periods when demand is highest.

Again, both the public and private sector will have important roles to play in making inputs available to farmers at affordable prices. Ethiopia is currently underinvesting in agricultural research (Chapter 3), and more efforts are needed to develop and adapt innovations making them suitable for local users. As part of this effort, the innovation system's link with the extension system as well as the private sector should be institutionalized and strengthened (Berhane et al. 2018). The private sector understands what types of technologies are needed and ultimately will be responsible for their manufacture, distribution, and repair and must be seen as an essential partner.

Agricultural land expansion has decelerated in recent years, putting more pressure on available land in the highlands (Chapters 2 and 6). Important land reforms and the implementation of a large-scale land certification program have helped to ensure more secure property rights, and Kumar and Quisumbing (2015) show that reforms in the law and land registration have been an important avenue to improved gender equality in Ethiopia. This land certification program has been one of the largest, cheapest, and fastest in Africa (Deininger et al. 2008).[2] However, as populations continue to increase, it will be necessary to monitor land tenure rules to ensure their suitability with transforming agrifood systems. At the same time, land in less accessible regions but with sufficient rainfall such as parts of the western regions (Gambella and Benishangul-Gumuz, and the Oromia region) and in the dryer lowlands remains underexploited because of poor access to water, road infrastructure, and services. Infrastructure and irrigation investments that tap into these areas' latent potential will stimulate production but also support pastoral economies that play an important role in supplying animal-sourced foods. To reach their full potential, livestock systems will also require broader adoption of improved animal husbandry and feeding practices, increased production of genetically superior breeds of livestock, and the provision and use of appropriate veterinary health practices. Given the high perishability of these

2 Although land stayed the property of the state, these certificates have ensured more secure property rights as they have been found to have led to higher investments, more land rental market activity, higher productivity, and improved food security (Deininger, Ayalew, and Alemu 2011; Ghebru and Holden 2013).

products, investments will be needed toward new off-farm technologies, such as cold storage and processing.

Modernization of Agricultural and Food Markets

Most of the attention in past agricultural strategies has gone toward increasing agricultural supply, and the off-farm segment in agrifood systems has received relatively less attention. This was justified given the high level of autoconsumption of agricultural products by the average Ethiopian, who was typically a farmer living in a rural area. Such a low share of purchases (and sales) and very high transport and marketing costs limited potential growth linkages, particularly for small farmers, as increases in their production did not lead to substantially higher incomes for processors, transporters, and traders. However, as shown in Chapters 7 and 10, this is changing rapidly; already in 2016, the average Ethiopian acquired more than half of her calories from purchased food. Livestock product markets are also expanding rapidly. These trends are expected to continue and more policy attention toward ensuring efficiently functioning value chains is therefore required. While the performance of agricultural markets has improved enormously in the past decade, it is expected that growing incomes, population growth, urbanization, and better infrastructure will lead to a massive expansion of agricultural and food markets, and the agrifood system more broadly, and might therefore strain the existing system. A number of actions could be taken to facilitate that expected transformation and upscaling process and to improve agricultural and food markets.

Ethiopia's development model involving heavy public sector involvement was well-suited for addressing the major constraints on agricultural growth prior to 2010. At that time, weak road infrastructure required massive public investments. Reliance on traditional varieties of major cereals could be addressed with large-scale distribution of improved, open-pollenated varieties (not hybrids). And increased use of basic nitrogen and phosphate fertilizer was sufficient to produce major increases in yields.

However, this heavy public sector involvement was accompanied by policies that crowded out activities by the private sector, as seen in preferential treatment for access to credit allocations, foreign exchange, and land (World Bank 2015). While the public sector has an important role to play in stimulating agricultural growth, it typically lacks the information and incentives to allocate goods and services efficiently in agricultural and food markets. The public sector is best suited to ensure a competitive environment that enables the private sector to deliver goods and services more efficiently and with greater choice. However, private-sector development has been problematic as shown by a global assessment of the ease of doing business in which Ethiopia was ranked 159th of 190 countries in 2018 (World Bank 2019).

With the rapid transformation of the agriculture-food system, a new development model is needed—one led by the private sector supported by pro-farmer and pro-business policies and public investments. Agricultural and broad economic policy will require greater transparency and consistency. Rapid, broad-based growth will also require a level playing field across subsectors (various crops and livestock) and regions that encourages participation of thousands if not millions of actors, so as to promote investments in a very wide array of high-value products throughout many parts of Ethiopia. The establishment of industrial and agro-industrial parks was a good first step (Chapter 12), but there is also scope for investment outside these centers. As part of this, an increasing role for more modern supply chains and the retail trade could deliver a number of these foods at significantly lower prices to consumers, with potential benefits for producers as well (Minten and Reardon 2008; Reardon et al. 2009).

Food quality and safety will also become more important requirements in the agricultural food system given the changes in products, in value chains with increasing long-distance transport and storage requirements, and in consumer demands (Chapter 7). Making sure that appropriate institutions, regulations, certification and traceability mechanisms, infrastructure, and awareness are present to allow for the delivery of safe products is an important challenge for the future in these rapidly transforming and emerging markets.

As Ethiopia's economy grows and urbanization takes off, more attention needs to be paid to the off-farm aspect of agricultural value chains. These off-farm activities have important impacts on employment (often of young people) as well as on prices and food security for both urban and rural populations. Rapidly emerging agrifood small and medium-sized enterprises in this off-farm segment add significant value, but they are often neglected in the African food debate (Reardon et al. 2015).

Reducing Poverty and Malnutrition

Although economic growth has lifted a significant number of people further out of poverty, there is still a need to have well-functioning and targeted safety nets in place that will address the needs of vulnerable and poor households, especially in light of increased climate variability. Ethiopia has had important successes in the implementation of the Productive Safety Net Program (PSNP) in the past decade, but further adjustments might be needed moving forward.[3] First, given rapid urbanization, a large number of poor people are now living in urban areas, and further ensuring coverage by urban safety nets—as has recently been started—is seemingly needed. Second, as agricultural markets are now functioning better than before, infrastructure (communication as well as road) has significantly improved, and there are cost savings in delivering cash payments (compared with food) to targeted households, a further move toward cash payments for safety-net recipients seems warranted. Third, an improved early warning system, disaster risk management, and coordination between safety net and emergency aid relief seems needed. Finally, maintaining incentives for private-sector trade (including wheat imports) in times of crop shortages and high prices can potentially mitigate adverse price effects on consumers following droughts (Chapter 9).[4]

Moreover, despite high growth rates in agricultural production and significant reductions in poverty, the slow change in nutritional indicators and the high level of stunting in the country remain a major concern, especially in rural areas (Chapter 10; Hirvonen et al. 2019). More attention should therefore be paid to how agricultural growth can enhance food diversity and nutritional indicators. While there are still a number of unknowns on how this nutritional transformation can be most efficiently achieved, it seems that behavioral change communication, gender considerations, sanitation, improved market access, and agricultural production diversity, especially in less connected areas, should have a major role to play.

Macroeconomic Policy and Development Strategy

Maintaining macroeconomic incentives and stability is a prerequisite for sustained economic growth, which in turn is crucial to promote market demand for agricultural products. Broad economic growth, including growth in the nonagricultural economy, is thus a necessary condition for successful agricultural growth and poverty reduction. With appropriate policies and investments, Ethiopia's agriculture-food system and the overall economy could

3 The decline in poverty rates and malnutrition has not been accompanied by a major increase in average calorie consumption (mean caloric intake rose by only 1 percent between 2011 and 2017), reflecting in part the increase in meat consumption as well as the impact of timely targeted transfers. It may also reflect improvements in sanitation and feeding practices for young children.

4 As discussed in Chapter 9, wheat imports in most years were used almost exclusively for targeted transfers. In 2014/2015, government wheat imports accounted for 20.1 percent of net wheat supply (and 4.8 percent of net cereal supply). The volume of additional grain supplied in 2015/2016 in response to the drought was substantial (2.5 million metric tons of imports compared with 950,000 metric tons in 2014/2015) and lowered prices by an estimated 26.0 percent (Simulation S2, Table 9.12).

prosper. Avoiding a major surge in inflation and potential debt and balance-of-payments problems will require a major reduction in public sector borrowing and net foreign capital inflows—policy shifts that will likely entail a slowdown in economic growth as well.[5] Maintaining a strongly positive investment climate will require that enterprises and new investors have ready access to foreign exchange with minimal delays or costs of import licenses (currently a major constraint for many commercial firms).

Finally, managing the rural-urban transformation and the overall structural transformation of the economy will require appropriate public investments in rural as well as urban infrastructure. Policies outside the agricultural sector will be essential for this transformation and for long-term growth (Chapters 12 and 13). These policies include education policies and investments, strengthening the legal framework for business and investment, and investments in ports. Depending on broad political considerations, road and rail connections to Eritrea's ports may provide new opportunities to reduce costs of international trade. Particularly important will be electricity generation and distribution policies. Solar energy may offer important opportunities to provide electrification to small towns and relatively remote rural areas.

Implications for Development Strategy in Other African Countries

Although Ethiopia's highland agroecology differs from that of much of the rest of Africa, the basic ingredients of Ethiopia's successful agricultural and rural development are applicable to many other African countries. Most important is Ethiopia's sustained commitment to the agricultural sector through public investments in agricultural research and extension, ensuring wide access to fertilizer and improved seeds, and investment in rural roads. Second, Ethiopia has largely avoided market distortions such as imposition of official market prices and large-scale imports that have taxed producers in many other African countries. As discussed in Chapter 9, price interventions in wheat have helped Ethiopia limit price spikes following droughts, but the scale of interventions in other major cereal markets has been minimal. Third, Ethiopia's Productive Safety Net Program has effectively targeted food-insecure households, enhancing their food security. Finally, in comparison to other African countries (such as Nigeria, Zambia, Zimbabwe, and the

5 Note that all simulations in Chapter 13 assumed some slowdown in capital inflows, and sensitivity analysis showed that major results were not affected by moderate changes in capital flows.

Democratic Republic of the Congo), Ethiopia has enjoyed both broad macroeconomic and political stability over most of the past two decades.[6]

Conclusion

Given that significant yield gaps remain for major crops and livestock and dairy productivity remains low, public investments in agriculture are likely to continue to have a high payoff. Moreover, even with continued changes in economic structure due to increased incomes, urbanization, and improved infrastructure, agricultural investments are likely to continue to be strongly pro-poor given the concentration of poverty in rural areas. Agricultural investments will likely continue to result in greater poverty reduction than investments in urban sectors, at least through the mid-2020s. If economic growth slows in the coming decade, the dominance of agriculture as a pro-poor investment will likely continue until 2030 and beyond. Moreover, investments in the broader agrifood sector (including investment in roads and electricity in small towns and periurban areas) are even more pro-poor than investments restricted to agricultural production and should continue.

Massive changes have taken place in Ethiopia's economy. Land constraints are becoming increasingly binding, especially in the highlands, implying a likely deceleration in agriculture growth. With rural population projected to increase despite continued urbanization, average farm size will likely decrease. Rapid growth in the nonagricultural economy will likely result in a sharp increase in market demand for agricultural products. Continued massive urban investments can result in further economic growth and structural transformation, particularly if macroeconomic stability and incentives for private investment are maintained. But without considerable investments in the rural farm and nonfarm economy, it will be very difficult to achieve rapid poverty reduction.

In short, though rapid economic growth and structural transformation have diminished the relative importance of the agricultural sector in Ethiopia's economy, continued public investments in agriculture and the broader agrifood system (for example, agricultural research and extension, particularly in high-value crops and livestock, and rural and secondary city transport infrastructure and electricity supply) remain crucial for equity and poverty alleviation in Ethiopia. A favorable climate for private investment is also needed, including stable macroeconomic and trade policy, access to credit

6 There has been a recent upsurge in political instability since mid-2019, however.

and foreign exchange, and avoidance of policy changes that increase riskiness of investments. If poverty reduction is still a top priority, it is too soon to reduce the level of agricultural and rural investment in Ethiopia.

References

Aguilar, A., E. Carranza, M. Goldstein, T. Kilic, and G. Oseni. 2015. "Decomposition of Gender Differentials in Agricultural Productivity in Ethiopia." *Agricultural Economics* 46 (3): 311–334.

Alston, J. M., C. Chan-Kang, M. C. Marra, P. G. Pardey, and T. J. Wyatt. 2000. *A Meta-Analysis of Rates of Return to Agricultural R&D: Ex Pede Herculem?* Volume 113. Washington, DC: International Food Policy Research Institute (IFPRI).

Berhane, G., C. Ragasa, G. T. Abate, and T. W. Assefa. 2018. *The State of Agricultural Extension Services in Ethiopia and Their Contribution to Agricultural Productivity.* ESSP Discussion Paper 118. Adds Ababa: IFPRI.

Deininger, K., D. Ayalew, and T. Alemu. 2011. "Impacts of Land Certification on Tenure Security, Investment, and Land Market Participation: Evidence from Ethiopia." *Land Economics* 87 (2): 312–334.

Deininger, K., D. Ayalew, S. Holden, and J. Zevenbergen. 2008. "Rural Land Certification in Ethiopia: Process, Initial Impact, and Implications for Other African Countries." *World Development* 36 (10): 1786–1812.

Ghebru, H., and S. Holden. 2013. *Links between Tenure Security and Food Security: Evidence from Ethiopia.* IFPRI Discussion Paper 1288. Washington, DC. IFPRI.

Hirvonen, K., D. Headey, J. Golan, and J. Hoddinott. 2019. "Changes in Child Undernutrition in Ethiopia (2000–2016)." In *The Oxford Handbook of the Ethiopian Economy*, edited by C. Cramer, F. Cheru, and A. Oqubay, 399–411. New York: Oxford University.

Kumar, N., and A. R. Quisumbing. 2015. "Policy Reform Toward Gender Equality in Ethiopia: Little by Little the Egg Begins to Walk." *World Development* 67: 406–423.

Minten, B., and T. Reardon. 2008. "Food Prices, Quality, and Quality's Pricing in Supermarkets versus Traditional Markets in Developing Countries." *Review of Agricultural Economics* 30 (3): 480–490.

Reardon, T., C. B. Barrett, J. A. Berdegué, and J. F. Swinnen. 2009. "Agrifood Industry Transformation and Small Farmers in Developing Countries." *World Development* 37 (11): 1717–1727.

Reardon, T., D. Tschirley, B. Minten, S. Haggblade, S. Liverpool-Tasie, M. Dolislager, and C. Ijumba. 2015. "Transformation of African Agrifood Systems in the New Era of Rapid Urbanization and the Emergence of a Middle Class." In *Proceedings of the ReSAKSS Annual Conference, "Beyond a Middle Income Africa," Trends and Outlook Report,* 1–16. Conference in Addis Ababa, September 1–3.

Ruttan, V. W., and Y. Hayami. 1984. "Toward a Theory of Induced Institutional Innovation." *Journal of Development Studies* 20 (4): 203–223.

World Bank. 2015. *Ethiopia's Great Run: The Growth Acceleration and How to Pace It.* Report no. 99399-ET. Washington, DC.

———. 2019. *Doing Business 2019: Training for Reform,* 16th ed. Washington, DC.

GLOSSARY

belg: secondary rainy season (February–April).

chat: a stimulant product.

enset: "false banana" (*Ensete ventriculosum*), a perennial tuber grown as a staple food crop, especially in the southern Ethiopian highlands.

injera: Ethiopia's main national dish, a fermented pancake traditionally made with teff (although often made with cheaper grains).

kebele: the smallest administrative unit in the local government system, sometimes called a peasant association (or PA) in rural areas; a village.

kocho: a bread made out of processed enset.

meher: the primary rainy season and therefore the primary crop season.

teff: a cereal crop related to millet (*Eragrostis tef*).

woreda: administrative level under the zone level.

CONTRIBUTORS

Paul Dorosh is the director of the Development Strategy and Governance Division of the International Food Policy Research Institute (IFPRI), Washington, DC, and was the program leader of the Ethiopia Strategy Support Program in Addis Ababa, Ethiopia, from 2008 to 2010.

Bart Minten is a senior research fellow in the Development Strategy and Governance Division of IFPRI and was the program leader of the Ethiopia Strategy Support Program in Addis Ababa, Ethiopia, from 2011 to 2020.

Kibrewossen Abay was at the time of writing a research officer in the Ethiopia Strategy Support Program in the Development Strategy and Governance Division of IFPRI and the Ethiopian Development Research Institute (EDRI), Addis Ababa. He is now an MSc student at the Harvard Kennedy School, Cambridge, MA, US.

Fantu Bachewe is a research coordinator in the Ethiopia Strategy Support Program in the Development Strategy and Governance Division of IFPRI, Addis Ababa.

Guush Berhane is a research fellow in the Ethiopia Strategy Support Program in the Development Strategy and Governance Division of IFPRI, Addis Ababa.

Mekdim Dereje was at the time of writing a research officer in the Ethiopia Strategy Support Program in the Development Strategy and Governance Division of IFPRI and EDRI, Addis Ababa. He is now a PhD student in the Center for Development Research (ZEF), University of Bonn, Germany.

Tadele Ferede is an assistant professor in the School of Economics of Addis Ababa University, Ethiopia.

Ibrahim Worku Hassen was at the time of writing a research officer in the Ethiopia Strategy Support Program in the Development Strategy and Governance Division of IFPRI and EDRI, Addis Ababa. He is now a research fellow at the Growth Lab, Center for International Development at Harvard University, Cambridge, MA, US.

Kalle Hirvonen is a research fellow in the Ethiopia Strategy Support Program in the Development Strategy and Governance Division of IFPRI, Addis Ababa.

Mekamu Kedir Jemal is a research analyst in the Development Strategy and Governance Division of IFPRI, Washington, DC.

Frehiwot Worku Kebede was at the time of writing a researcher in the Economic Policy Analysis Unit of EDRI, Addis Ababa.

Bethelhem Koru is a research officer in the Ethiopia Strategy Support Program in the Development Strategy and Governance Division of IFPRI and EDRI, Addis Ababa.

Richard Robertson is a research fellow in the Environment and Production Technology Division of IFPRI, Washington, DC.

Emily Schmidt is a research fellow in the Development Strategy and Governance Division of IFPRI, Washington, DC.

Jennifer Smart is a senior research analyst in the Development Strategy and Governance Division of IFPRI, Washington, DC.

David Stifel is a professor and assistant head of the Department of Economics, Lafayette College, PA, US.

Fanaye Tadesse is a research officer in the Ethiopia Strategy Support Program in the Development Strategy and Governance Division of IFPRI and EDRI, Addis Ababa.

Alemayehu Seyoum Taffesse is a senior research fellow in the Ethiopia Strategy Support Program in the Development Strategy and Governance Division of IFPRI, Addis Ababa.

Seneshaw Tamru is a PhD candidate at LICOS—Center for Institutions and Economic Performance, University of Leuven, Belgium.

Timothy S. Thomas is a research fellow in the Environment and Production Technology Division of IFPRI, Washington, DC.

James Thurlow is a senior research fellow in the Development Strategy and Governance Division of IFPRI, Washington, DC.

Feiruz Yimer was at the time of writing a research officer in the Ethiopia Strategy Support Program in the Development Strategy and Governance Division of IFPRI and EDRI, Addis Ababa. She is now an MSc student at the Harvard Kennedy School, Cambridge, MA, US.

INDEX

Page numbers for entries occurring in figures are followed by an *f*; those for entries in notes, by an *n*; and those for entries in tables, by a *t*.

Made in United States
Orlando, FL
04 November 2026

90201391R00309